BASIC ELECTRICAL ENGINEERING

BASIC ELECTRICAL ENGINEERING

As per the New Syllabus of GBTU
(Common to all Branches of Engineering)

(SECOND EDITION)

Kuldeep Sahay
Professor
Department of Electrical Engineering
Institute of Engineering & Technology
Lucknow (UP)

An Imprint of

NEW AGE INTERNATIONAL (P) LIMITED, PUBLISHERS
New Delhi • Bangalore • Chennai • Cochin • Guwahati
Hyderabad • Kolkata • Lucknow • Mumbai
Visit us at **www.newagepublishers.com**

Published by New Age International (P) Ltd., Publishers
First Edition: 2006
Second Edition: 2014
Reprint: 2015

BRANCHES

- **Bangalore** 37/10, 8th Cross (Near Hanuman Temple), Azad Nagar, Chamarajpet, Bangalore-560 018 Tel.: (080) 26756823, Telefax: 26756820, **E-mail: bangalore@newagepublishers.com**
- **Chennai** 26, Damodaran Street, T. Nagar, Chennai-600 017 Tel.: (044) 24353401, Telefax: 24351463, **E-mail: chennai@newagepublishers.com**
- **Cochin** CC-39/1016, Carrier Station Road, Ernakulam South, Cochin-682 016 Tel.: (0484) 2377004, Telefax: 4051303, **E-mail: cochin@newagepublishers.com**
- **Guwahati** Hemsen Complex, Mohd. Shah Road, Paltan Bazar, Near Starline Hotel, Guwahati-781 008 Tel.: (0361) 2513881, Telefax: 2543669, **E-mail: guwahati@newagepublishers.com**
- **Hyderabad** 105, 1st Floor, Madhiray Kaveri Tower, 3-2-19, Azam Jahi Road, Near Kumar Theater Nimboliadda, Kachiguda, Hyderabad-500 027, Tel.: (040) 24652456, Telefax: 24652457 **E-mail: hyderabad@newagepublishers.com**
- **Kolkata** RDB Chambers (Formerly Lotus Cinema) 106A, 1st Floor, S.N. Banerjee Road, Kolkata-700 014 Tel.: (033) 22273773, Telefax: 22275247, **E-mail: kolkata@newagepublishers.com**
- **Lucknow** 16-A, Jopling Road, Lucknow-226 001 Tel.: (0522) 2209578, 4045297, Telefax: 2204098, **E-mail:lucknow@newagepublishers.com**
- **Mumbai** 142C, Victor House, Ground Floor, N.M. Joshi Marg, Lower Parel, Mumbai-400 013 Tel.: (022) 24927869, Telefax: 24915415, **E-mail: mumbai@newagepublishers.com**
- **New Delhi** 22, Golden House, Daryaganj, New Delhi-110 002 Tel.: (011) 23262368, 23262370, Telefax: 43551305, **E-mail: sales@newagepublishers.com**

ISBN: 978-81-224-3652-5

₹ 275.00

C-14-09-7918

Printed in India at Ajit Printing Press, Delhi.
Typeset at Subham Graphics, Delhi.

PUBLISHING GLOBALLY
NEW AGE INTERNATIONAL (P) LIMITED, PUBLISHERS
7/30 A, Daryaganj, New Delhi-110002
Visit us at **www.newagepublishers.com**

Dedicated to my
Father Sri Shripati Sahay and Mother Smt. Vijay Kumari,
&
Wife Smt. Anju Sahay, and Daughters Samridhi Sahay,
Prasidhi Sahay

—Kuldeep Sahay

Preface

This textbook is valuable for the course in Basic Electrical Engineering taught to nearly all the branches of engineering *i.e.*, Electrical, Electronics and Communication, Electronics and Instrumentation, Civil, Mechanical, Chemical, Information Technology and Computer Science and Engineering at undergraduate level. The purpose of this book is to provide firm theoretical and experimental knowledge in a lucid way. An effort has been made to provide a comprehensive coverage of the course as taught in various universities.

This book contains seventeen chapters, description as follows:

Chapter 1: Introduction & Chapter 2: DC Circuit Analysis and Network Theorem

Fundamental concepts of current, voltage, power energy and definitions are introduced. Ohm's law, Kirchhoffs voltage and current laws, resistance, inductance and capacitance have been discussed. Also the general methods provided are mesh current analysis and node voltage analysis, most of network theorems are covered.

Chapter 3: Basic Concepts of Alternating Current and Voltage, Chapter 4: Single Phase AC Series Circuits, Chapter 5: Single Phase AC Parallel Circuits & Chapter 6: Electric Resonance

Sinusoidal voltage and currents are studied and phasor representation of these variables is developed. Impedance forms and phasor diagrams for all the basic circuit elements are provided using complex numbers. Admittance is introduced and conversion between impedance and admittance forms is covered. Resonance in parallel as well as series cicuit has been discussed.

Chapter 7: Polyphase Circuits

The readers are introduced to the concepts of three phase power system. Relationships between line and phase voltages and currents in three-wire delta and three-and four-wire star (wye) system are shown. Measurement of power and power factor with single and two wattmeter method has been discussed. Also concepts of power factor and its improvement.

Chapter 8: Electromagnetism & Chapter 9 Magnetic Circuits

This chapter starts with introduction of magnetic and electromagnetic, including the concept of flux, flux density, and retentivity, permeability, and magnetic field intensity, including B-H curves. Fleming's left hand rule and Fleming's right hand rule, self inductance, mutual inductance has been discussed.

Chapter 10: Transformer

Introduction of the ideal transformer and the relation between turns ratio, Real Transformer and its Equivalent Circuit, Approximate Equivalent Circuit, Phasor Diagram, Name Plate Rating, Transformer Losses, Transformer Testing, Transformer Efficiency, Voltage Regulation and concept of Auto Transformer and been discussed.

Chapter 11: Three-Phase Induction Machine & Chapter 12: Single-Phase Induction Motor

An introduction of three phase induction motor is provided; including Construction details, Difference between Cage Rotor and Slip Ring IM, Equivalent Circuit of IM, Determination of Equivalent Circuit Parameter, Phasor Diagram of IM, Starting of Three Phase IM, and Speed control of Three Phase IM has been discussed.

Chapter 13: Direct Current Machine (Generators) & Chapter 14: D.C. Machine (Motor)

The relationship between a magnetic field and the velocity of a conductor in the field to produce generator action is introduced. The fundamental parts of a dc generator are shown and related characteristics of motor and generator are given. Speed controls of motor and efficiency test of dc machines are given.

Chapter 15: Synchronous Machines

An introduction of three phase alternator is provided; including Construction details, Equivalent Circuit Model and its phasor Diagram, Open Circuit Characteristic (OCC) Test, Short Circuit Characteristic (SCC) Test, Voltage Regulation, Synchronous Motors Working Principle, Power Developed by Synchronous motor and their applications.

Chapter 16: Electrical Measurements and Measuring Instruments

An introduction of measurement, classification, types of instruments development of torque and damping. Concepts of Moving Iron and Permanent Magnet Moving coil instruments, dynamometer type wattmeter extension of meters range have been briefly discussed.

Chapter 17: Introduction to Earthing and Electrical Safety

An introduction of earthing and its necessity effects on machines and personal, Essentials of Grounded System, Ground Electrode, Earthing Materials, Measurement Earth Resistance, Maintenance Of Earthing System, Bentonite & Marconite, Earth-Ground Grid Systems, Do's And Don'ts For Electrical Safety.

At this moments; I would like express my in-depth gratitude to all those who have directly or indirectly helped and encouraged to finish this task. Further I will welcome suggestions, comments and constructive criticism from readers for the betterment of the ' contents of this book.

A last, special thanks to staff of New Age International Pvt. Ltd. Publishers for their constant co-operation and support.

Syllabus

Basic Electrical Engineering

Unit-I

1. **DC Circuit Analysis and Network Theorems:** Circuit Concepts: Concepts of network, Active and passive elements, Voltage and current sources, Concept of linearity and linear network, Unilateral and bilateral elements, R, L and C as linear elements, Source transformation Kirchhoffs laws; Loop and nodal methods of analysis; Star-delta transformation.

   Network Theorems: Superposition theorem, Thevenin's theorem, Norton's theorem, Maximum Power Transfer theorem (Simple numerical problems). 9

Unit-II

2. **Steadystate Analysis of Single Phase AC Circuits:** AC fundamentals: Sinusoidal, square and triangular waveforms—Average and effective values, Form and peak factors, Concept of phasors, phasor representation of sinusoidally varying voltage and current, Analysis of series, parallel and series-parallel RLC Circuits, Resonance in series and parallel circuits, bandwidth and quality factor; Apparent, active and reactive powers, Power factor, Causes and problems of low power factor, Concept of power factor improvement (Simple numerical problems). 8

Unit-III

3. **Three Phase AC Circuits:** Three phase system—its necessity and advantages, Star and delta connections, Balanced supply and balanced load, Line and phase voltage/current relations, Three-phase power and its measurement (simple numerical problems). 3
4. **Measuring Instruments:** Types of instruments, Construction and working principles of PMMC and moving iron type voltmeters and ammeters, Single phase dynamometer wattmeter, Use of shunts and multipliers (Simple numerical problems on shunts and multipliers). 4

Unit-IV

5. **Introduction to Earthing and Electrical Safety:** Need of Earthing of equipment and devices, important electrical safety issues. 2
6. **Magnetic Circuit:** Magnetic circuit concepts, analogy between electric and magnetic circuits, B-H curve, Hysteresis and eddy current losses, Mutual coupling with dot convention, Magnetic circuit calculations. 3
7. **Single Phase Transformer:** Principle of operation, Construction, EMF equation, Equivalent circuit, Power losses, Efficiency (Simple numerical problems), Introduction to auto transformer. 3

Unit-V

8. **Electrical Machines:** Concept of electro mechanical energy conversion DC machines: Types, EMF equation of generator and torque equation of motor, Characteristics and applications of DC motors (simple numerical problems)

   Three Phase Induction Motor: Types, Principle of operation, Slip-torque characteristics, Applications (Numerical problems related to slip only)

   Single Phase Induction Motor: Principle of operation and introduction to methods of starting, applications.

   Three Phase Synchronous Machines: Principle of operation of alternator and synchronous motor and their applications. 8

Contents

Preface		vii
Syllabus		ix
CHAPTER 1.	**Introduction**	**1–15**
	1.1 Electric Charge and Electric Current	1
	1.2 Direction of Current	2
	1.3 Coulomb's Law	2
	1.4 Concept of Potential and Potential Difference	3
	1.5 Electromotive Force of a Source	4
	1.6 Voltage Drop	4
	1.7 Reference Polarity	4
	1.8 The Basic Circuit	4
	1.9 Ohm's Law	5
	1.10 Kirchhoff's Laws	5
	1.11 Power	6
	1.12 Energy	6
	1.13 Ampere's Law	6
	1.14 Resistance	7
	1.15 Inductance	8
	1.16 Capacitance	9
	1.17 Electrical Parameters	11
	1.18 Source Conversion	12
	Solved Numerical Problems	13
CHAPTER 2.	**DC Circuit Analysis and Network Theorem**	**16–52**
	2.1 Network Analysis	16
	2.2 Direct Method	16
	2.3 Network Reduction Method	16
	2.4 Kirchhoff's Laws	16
	2.5 Kirchhoff's Current Law (KCL)	16
	2.6 Kirchhoff's Voltage Law (KVL)	17
	2.7 Cramer's Rule	17
	2.8 Analysis of Network by Kirchhoff's Law	18
	2.9 Star Delta Transformation	20

2.10 Superposition Theorem 21
2.11 Thevenin's Theorem 22
2.12 Norton's Theorem 24
2.13 Maximum Power Transfer Theorem 25
2.14 Millman's Theorem 27
2.15 Compensation Theorem 29
2.16 Tellegen's Theorem 30
2.17 Voltage Divider Rule 31
2.18 Current Divider Rule 31
Solved Numerical Problems 32
Exercises 49

CHAPTER 3. Basic Concepts of Alternating Current and Voltage **53–73**

3.1 Introduction 53
3.2 Generation of Alternating Voltage 53
3.3 Cycle 55
3.4 Frequency 56
3.5 Angular Frequency 56
3.6 Instantaneous Value 56
3.7 Peak or Maximum Value 56
3.8 Determination of Average or Mean Value 57
3.9 MID-Ordinate Method 57
3.10 Method of Integration 58
3.11 Average Value of Sine Wave 59
3.12 Form Factor 62
3.13 Peak Factor 62
3.14 Phase and Phase Difference 62
3.15 Phasor Diagram of Alternating Quantity 64
3.16 Addition and Subtraction of Phasors 64
Solved Numerical Problems 65
Exercises 73

CHAPTER 4. Single Phase AC Series Circuits **74–94**

4.1 Introduction 74
4.2 Resistive Circuit Only 74
4.3 Power in Resistive Circuit 75
4.4 A.C. Circuits Containing Inductance 75
4.5 Power in Purely Inductive Circuit 77
4.6 A.C. Circuit Containing Capacitance 77
4.7 Power in Purely Capacitive Circuit 78

4.8	Power in Series R-L Circuit	80
	Solved Numerical Problems	84
	Exercises	93
CHAPTER 5.	**Single Phase AC Parallel Circuit**	**95–103**
5.1	Introduction	95
5.2	Methods to Solve Parallel Circuits	96
5.3	Admittance of a Series Circuit	100
5.4	Admittance of a Parallel Circuit	101
	Exercises	102
CHAPTER 6.	**Electric Resonance**	**104–120**
6.1	Introduction	104
6.2	Effect of Frequency Variation in R, L and C Series Circuit	104
6.3	Voltage Magnification	107
6.4	Bandwidth of Series Resonant Circuit	108
6.5	Resonance in Parallel Circuits	110
6.6	Current Magnification	112
	Solved Numerical Problems	114
	Exercises	120
CHAPTER 7.	**Polyphase Circuits**	**121–145**
7.1	Introduction	121
7.2	Generation of a Three-Phase Supply	122
7.3	Connection of Three-Fhase Supplies	123
7.4	Star (γ) Connection	123
7.5	Delta (Δ) Connection	124
7.6	Line Voltage, Phase Voltage and Current in Star Connection	124
7.7	Neutral Current	126
7.8	Line Voltage, Phase Voltages and Current in Delta Connection	127
7.9	Volt-Amperes, Power and Reactive Volt-Amperes in Three-Phase System	127
7.10	Measurement of Power	129
7.11	Power Factor	134
	Solved Numerical Problems	137
	Exercises	145
CHAPTER 8.	**Electromagnetism**	**146–166**
8.1	Introduction	146
8.2	Magnetic Field	146

8.3 Magnetic Flux 146
8.4 Magnetic Flux Density 147
8.5 Direction of Current in a Conductor 147
8.6 Magnetic Field Due to a Current Carrying Conductor 148
8.7 Direction of Magnetic Field/Flux 148
8.8 Representing of Uniform Magnetic Field 149
8.9 Flux Distribution of a Single Turn Coil 149
8.10 Permeability 151
8.11 Relation between Magnetic Flux Density and Field Intensity 151
8.12 Force on a Conductor in a Magnetic Field 151
8.13 Fleming's Left Hand (or Motor) Rule 152
8.14 Electromagnetic Induction 152
8.15 Faraday's Laws of Electromagnetic Induction 153
8.16 Lenz's Law 153
8.17 Fleming's Right Hand Rule 154
8.18 Magnitude of Induced E.M.F. in a Coil 154
8.19 Dynamically Induced E.M.F. 155
8.20 Statically Induced E.M.F. 155
8.21 Coefficient of Coupling 158
8.22 Inductances of Inductively Coupled Coils Connected in Series 159
Solved Numerical Problems 161
Exercises 166

Chapter 9. Magnetic Circuits **167–184**

9.1 Introduction 167
9.2 Magneto-Motive Force (M.M.F.) 167
9.3 Magnetic Field Strength or Intensity 167
9.4 Reluctance (S) 168
9.5 Laws of Magnetic Circuits 169
9.6 Ampere Turns for a Magnetic Circuit 170
9.7 Calculations of Ampere Turns 170
9.8 Leakage Flux 171
9.9 Fringing 171
9.10 Calculation of Ampere Turns for the Air Gap 172
9.11 Comparison of the Electric and Magnetic Circuit 172
9.12 Calculation of Ampere Turns of Series-Parallel Magnetic Circuit 173
9.13 Magnetic Curve or B-H Curve 174
9.14 Explanation of the Shape of B-H Curve 174
9.15 Hysteresis 175

9.16 Core Losses or Iron Losses 176
9.17 Application of Eddy Current 177
9.18 Stacking Factor 177
Solved Numerical Problems 177
Exercises 183

CHAPTER 10. Transformer **185–213**
10.1 Introduction 185
10.2 Ideal Transformer 187
10.3 Real Transformer and its Equivalent Circuit 190
10.4 Approximate Equivalent Circuit 193
10.5 Phasor Diagram 193
10.6 Name Plate Rating 194
10.7 Transformers Losses 195
10.8 Transformer Testing 195
10.9 Transformer Efficiency 199
10.10 Voltage Regulation 201
10.11 Autotransformers 202
Solved Numerical Problems 204

CHAPTER 11. Three-Phase Induction Machine **214–244**
11.1 Introduction 214
11.2 Construction of Induction Machine 214
11.3 Working Principle of a Three-phase Induction Motor 216
11.4 Production of Rotating Field 217
11.5 E.M.F. and Current in Induction Motor 219
11.6 Frequency of Rotor Voltage and Current 220
11.7 Equivalent Circuit of Induction Motor 222
11.8 Stator Circuit Diagram 223
11.9 Determination of Equivalent Circuit Parameters 224
11.10 Phasor Diagram of Induction Motor 225
11.11 Three-Phase Motor Characteristics 226
11.12 Starting of Three-Phase Induction Motor 229
11.13 Speed Control of Three-Phase Induction Motor 233
Solved Numerical Problems 235
Exercises 242

CHAPTER 12. Single-Phase Induction Motor **245–253**
12.1 Introduction 245
12.2 Single-Phase Induction Motor 245

12.3 Construction of Single-Phase Induction Motor 246
12.4 Working Principle of Single-Phase Induction Motor 246
12.5 Double Revolving Field Theory of Single-Phase Induction Motor 247
12.6 Starting of Single-Phase Induction Motor 248
Exercises 252

Chapter 13. Direct Current Machine (Generators) **254–272**
13.1 Introduction 254
13.2 Construction of D.C. Machine 254
13.3 Basic Magnetic Circuit of D.C. Generator 256
13.4 Excitation of D.C. Machine 257
13.5 Relationship between the Generated Voltage and Excitation Current of D.C. Generators 260
13.6 Application of Load 261
13.7 D.M.F. Equation of D.C. Machine 263
13.8 Armature Reaction 264
13.9 Commutation Process 265
13.10 Method for Decreasing the Effect of Commutation 266
Solved Numerical Problems 267
Exercises 271

Chapter 14. D.C. Machine (Motor) **273–296**
14.1 Introduction 273
14.2 Counter E.M.F. or Back E.M.F. in D.C. Motors 273
14.3 Speed of D.C. Machine 274
14.4 Torque Production in D.C. Machine 275
14.5 Classification of D.C. Motors 276
14.6 Starting of D.C. Motor 279
14.7 Generating Mode 280
14.8 Speed Control of D.C. Motor 282
14.9 Characteristics of D.C. Motors 284
14.10 Speed Armature Current Characteristics 285
14.11 Speed-Torque Characteristics (Shunt Motor) 285
14.12 D.C. Series Motor 286
14.13 D.C. Compound Motor 288
14.14 Losses in D.C. Machine 290
14.15 Efficiency of D.C. Machine 291
Solved Numerical Problems 291
Exercises 295

CHAPTER 15. **Synchronous Machines** **297–315**

15.1 Construction 297
15.2 Principle of Operation of Alternators 298
15.3 E.M.F. Equation 299
15.4 Alternator on Load 300
15.5 Synchronous Impedance 300
15.6 Equivalent Circuit Model and its Phasor Diagram 301
15.7 Lagging Power Factor 301
15.8 Unity Power Factor 302
15.9 Leading Power Factor 302
15.10 Voltage Regulation 304
15.11 Importance of Voltage Regulation is as follows 304
15.12 Power Delivered by Alternator 304
15.13 Synchronous Motors Working Principle 306
15.14 Power Developed by Synchronous Motor 308
Solved Numerical Problems 310
Exercises 315

CHAPTER 16. **Electrical Measurements and Measuring Instruments** **316–343**

16.1 Introduction 316
16.2 Classification of Measuring Instruments 316
16.3 Classification of Instruments 318
16.4 Torque in Instruments 318
16.5 Merits and Demerits of Moving Iron Instruments 323
16.6 Moving Coil Instruments 323
16.7 Shunts 328
16.8 Shunts for A.C. Instruments 330
16.9 Measurement of Electric Voltage, Current Power and Energy 331
Solved Numerical Problems 338
Exercises 342

CHAPTER 17. **Introduction to Earthing and Electrical Safety** **344–354**

17.1 Introduction 344
17.2 Shock Hazards 345
17.3 Essentials of Grounded System 346
17.4 Ground Electrodes 347
17.5 Types of Grounding Systems 348
17.6 Earthing Materials 348
17.7 Measurement of Earth Resistance 349

17.8 Maintenance of Earthing System 349
17.9 Bentonite 350
17.10 Marconite 350
17.11 Earth-Ground Grid Systems 350
17.12 Do's and Don'ts for Electrical Safety 352

Experiments **355–381**

Appendix **382–419**

1

Introduction

Primarily various laws of nature explained by theories put forth from time to time by scientists are based on observed facts. Similarly the electrical engineering is also science based on experimentally established fundamental laws. Necessity to study the fundamental laws, facilitates to understand the concepts of engineering where appropriate laws serve the purpose. The fundamental laws of electrical engineering are Coloumb's, Ohm, Faraday's, Kirchhoff's and Maxwell laws.

1.1 ELECTRIC CHARGE AND ELECTRIC CURRENT

Electric charge and energy form the two most fundamental electrical quantities. We know that an atom consists of electrons, protons and neutrons. Each electron has negative charge, each proton has a positive charge while a neutron has no charge, therefore called neutral. In an atom, the number of electrons are equal to the number of protons and the atom is electrically neutron. The electrons may be added to or taken away from an atom. This causes the atom to be charged negatively or positively. If the atom has a proton in excess of those needed to balance the effect of its electron; called a positive ion. If an electron is added to an atom, it attains a negative charge; called a negative ion. A body having ionised atom is said to be electrically charged.

The basic unit of charge is charge of an electron. The unit of charge in the international system of units is the Coulomb (C). An electron has a negative charge of 1.601×10^{-19} Coulomb. The charge may be at rest or be in motion. A charge in motion constitutes an electric current. Electric current is defined as the rate of flow of electrical charge passing through a specified area.

Electrons flow across the cross-section from left to right. This constitute an electron current from left to right within conductor. The conventional electric current, being the time rate of flow of positive charge will be considered to be flowing in the opposite direction *i.e.*, from right to left within the conductor.

Expressed mathematically

$$i = \frac{dq}{dt}$$

where q is the charge in motion and t is the time.

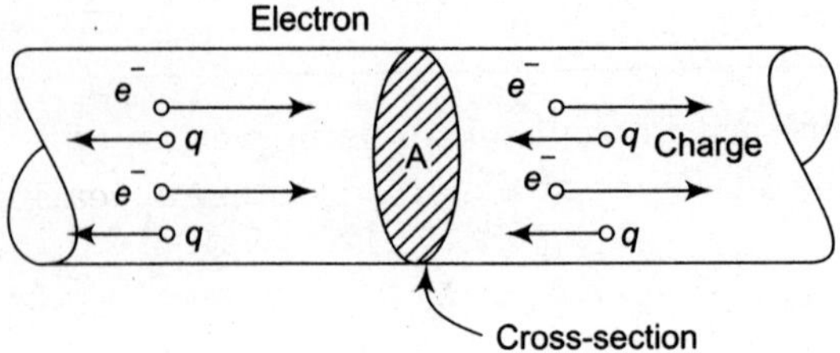

Fig. 1.1 *Motion of charge in a conductor.*

The current of one amperes signifies motion of

$$\frac{1}{1.601 \times 10^{-19}} = 6.25 \times 10^{18} \text{ electrons per second across the cross-section of the conductor.}$$

or $$I = \frac{Q}{t} \text{ amperes}$$

$$Q = It$$

Thus $$1 \text{ Coulomb} = 1 \text{ ampere second}$$

$$1 \text{ C} = 1 \text{ A sec}$$

$$1 \text{ A} = 1 \text{ C/sec.}$$

1.2 DIRECTION OF CURRENT

The current results from the motion of charges. The charge carriers may be electrons, holes or ions. The conventional direction of current flow is considered opposite to direction of electron flow. We assume positive direction of current to be the direction in which the positive charge is moving. Thus, when we say that there is a current of 1 A from a point a to point b in a conductor we actually mean that either positive charge is moving from a to b at the rate of 1 C/sec or negative charge is moving from b to a at the rate of 1 C/sec.

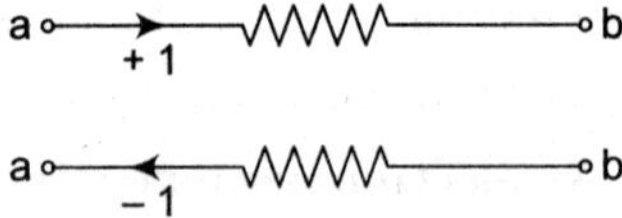

Fig. 1.2 *Two methods of representing the same current.*

1.3 COULOMB'S LAW

Based upon experiments of Coulomb the force of repulsion existing between two point charges Q_1 and Q_2 is

(*i*) directly proportional to the magnitude of the charges and

(*ii*) is inversely proportional to the distance between them.

Mathematically, the magnitude of the force is given by

$$F = \frac{Q_1 \cdot Q_2}{4\pi \varepsilon r^2} \text{ newton}$$

where Q_1 and Q_2 are expressed in Coulombs, r is in metres, ε is a constant related to the surrounding medium, while 4π is a proportionality constant appearing on account of our using rationalised IS units.

However, frequently written as

$$F = EQ_2 \text{ Newtons}$$

where $E = \dfrac{Q_1}{4\pi \varepsilon r^2}$ Newtons/Coulomb or volt/metre.

The quantity E is called the electric field intensity, while the quantity ε is called the permittivity.

$$\varepsilon = \varepsilon_o \, \varepsilon_r$$

where ε_r is called relative permittivity and ε_o is called permittivity of free space

$$\varepsilon_o = \frac{1}{36\pi \times 10^9} \text{ Farads/metre} \quad \text{or} \quad 8.85 \times 10^{-12} \text{ F/M}$$

1.4 CONCEPT OF POTENTIAL AND POTENTIAL DIFFERENCE

Energy is required for the movement of charge from one point to another. Work per unit charge is called voltage.

The potential difference V between two points is measured by the work required to transfer unit charge from one point to the other point or the energy required to move one Coulomb of charge from one point to the other point.

If W = Energy required to transfer the charge, Joule
Q = Charge transferred between the points, Coulomb
V = Potential difference, Volts

then $V = \dfrac{W}{Q}$ J/C

Now we say that a voltage exists between the two points. The voltage 'V' between two points may be defined in term of energy that would be required if a charge were transferred from one point to the other point. Thus, there can be a voltage between two points even if no charge is actually moving from one to the other. The voltage between a and b is given by, shown in Fig. 1.3.

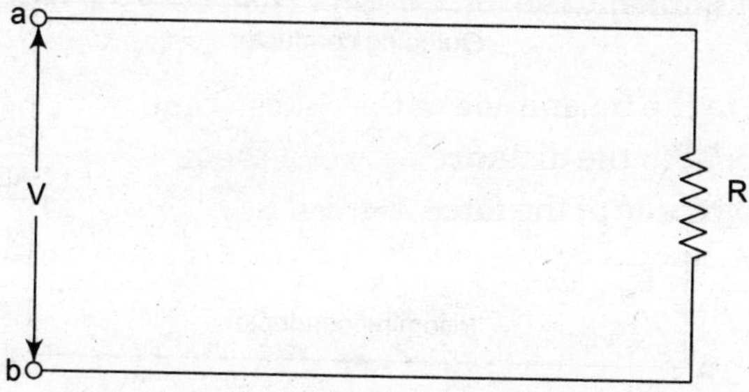

Fig. 1.3 *Concept of potential.*

1.5 ELECTROMOTIVE FORCE OF A SOURCE

The electromotive force of a source of voltage is the energy imparted by the source to each Coulomb of charge passing through it. The electromotive force is not the force, but it is the energy expended on each charge.

If, W = Energy imparted by the voltage source in Joules (J)

Q = Charge transferred through the source in Coulombs (C)

E = e.m.f. of the source

$$E = \frac{W}{Q} \text{ (J/C)}$$

1.6 VOLTAGE DROP

The voltage drop between two points of an elements is a decrease in the energy in transferring a charge of one Coulombs from one point to the other. The electromotive force, potential difference and voltage drop are all measured in units of voltage.

1.7 REFERENCE POLARITY

The reference polarities for voltage across an element can be provided by marking a + sign on one end and a – sign on the other. For example, in Fig. 1.4; point a is marked with $a+$ sign and point b is marked with – sign. Point a is at a higher potential and point b is at a lower potential. Thus, a positive amount of work is to be done to move a positive charge from b to a. Let a be more positive than b by volts, if we go from a to b so there will be voltage drop of V volts. When we go from b to a we experience a voltage rise of volts. Therefore V can either be positive or a negative number.

1.8 THE BASIC CIRCUIT

The path of current is called a circuit. The basic electric circuit is shown in Fig. 1.4. It consists of:

(a) a source of energy.

(b) a load which utilises the energy.

(c) two conductors connecting the source and load to transfer the energy.

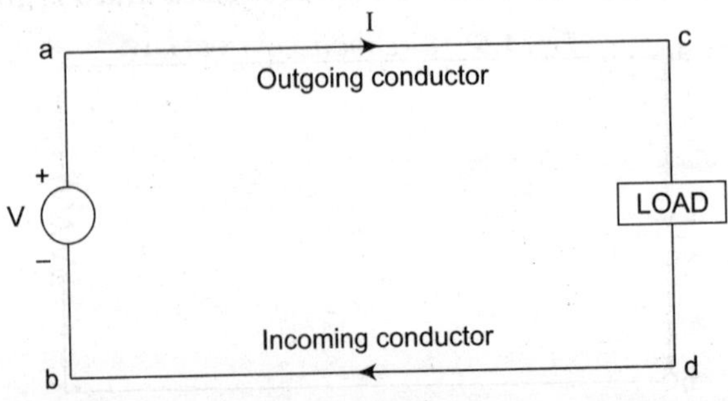

Fig. 1.4 *The basic circuit.*

1.9 OHM'S LAW

The law states that at constant temperature, potential difference V across, the ends of a conductor is proportional to the current I flowing in the conductor.

Ohm's law may be expressed as

$$V \propto I, \text{ or } V = IR$$

where I = Current in ampere
R = Resistance of conductor in Ohm
V = Potential difference in volts

Further

$R = \rho \dfrac{l}{A}$ = Electrical property of a material which opposes the current
ρ = Resistivity of conductor in Ohm-metre
l = Length of conductor, in metre
A = Cross-sectional area of conductor, in square metre

1.10 KIRCHHOFF'S LAWS

There are two laws due to Kirchhoff, Kirchhoff's current law (KCL) and Kirchhoff's voltage law (KVL).

1. *Kirchhoff's current law*: The sum of the currents entering a junction is equal to the sum of the currents leaving the junction. If the current towards a junction are considered positive and those away from the same junction negative, then this law states that the algebraic sum of all the currents meeting at a common junction is zero, shown in Fig. 1.5

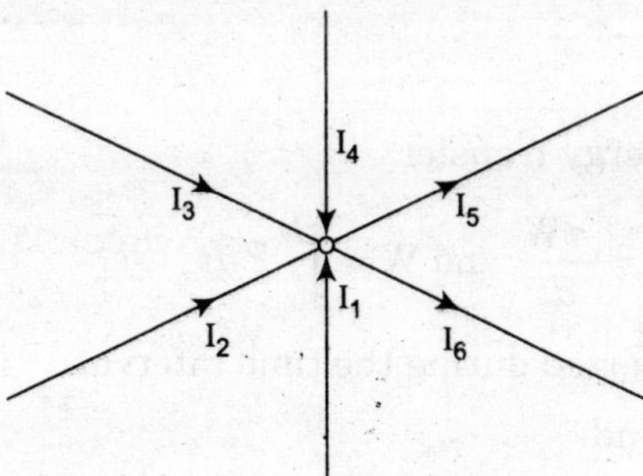

Fig. 1.5 *Kirchhoff's current law.*

$$I_1 + I_2 + I_3 + I_4 = I_6 + I_6$$

Generalised expression of KCL is

$$\sum_{j=1}^{k} I_j = 0$$

where k denotes the number of circuit elements connect to the node and Σ is greek symbol to indicate summation.

2. *Kirchhoff's voltage law*: The algebraic sum of all voltages around a closed path at any instant is zero.

$$V - IR_1 - IR_2 - IR_3 = 0$$

$$V - V_1 - V_2 - V_3 = 0$$

or

$$V = V_1 + V_2 + V_3$$

Generalized expression of KVL is

$$\sum_{J=1}^{k} V_j = 0$$

where V_j represents the voltage drop of the element in any given closed circuit which is assumed to have k elements, Fig. 1.6.

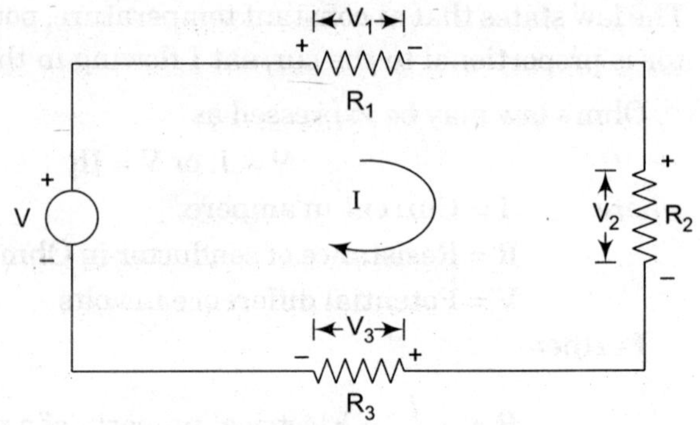

Fig. 1.6 *Kirchhoff's voltage law.*

1.11 POWER

The electrical power (P) is the product of impressed voltage (V) and resulting current (i).

$$P\,(\text{watts}) = V\,(\text{volts}) \times i\,(\text{amperes})$$

where P is a positive value of energy transferred to circuit.

If the power P is a periodic function time t with period T, then average power

$$P = \frac{1}{T}\int_0^T P\,dt$$

Unit of power is watt.

1.12 ENERGY

Since P is the time rate of energy transfer

$$P = \frac{dW}{dt} \text{ and } W = \int_{t_1}^{t_2} P\,dt$$

where W is the energy transferred during the time interval.

Unit of energy is watt-second.

1.13 AMPERE'S LAW

The force between two very long, parallel wires carrying current I_1 and I_2 is given by

$$F = \frac{\mu I_1 I_2}{2\pi r}\,l \text{ Newtons} \qquad ...(1)$$

where l = Length of conductors

r = Distance of separation between wires

μ = Property of the medium or permeability of medium.

Further shown by Ampere that the current I_1 could be considered as producing a magnetic field given by

$$B = \frac{\mu I_1}{2\pi r} \qquad ...(2)$$

Substituting equation (2) in equation (1); we have the well known relationship:

$$F = BlI_2$$

The above equation is valid for wires carrying current I_1 and I_2 lying in the same plane. If for some reasons, the conductor carrying I_2 is inclined by some angle θ with respect to the line perpendicular to the wire carrying I_1 then the force is given by

$$F = BlI_2 \sin\theta.$$

1.14 RESISTANCE

Resistance is one of the three basic parameters of electric circuit theory. It is a proportionality factor in Ohm's law relating current to potential difference.

Resistance, in general way described as that property of a circuit element which offers opposition to the flow of current and in so doing converts electrical energy into heat energy.

For a given conductor

$$V = Ri$$

or $$I = \frac{V}{R} = G.V$$

where $G = \frac{1}{R}$, is the conductance of the conductor.

Unit of resistance is Ohm (Ω)

Unit of conductance is Mho (℧)

The power absorbed by a resistor

$$P = Vi = (iR)\,i = i^2R$$

and the corresponding amount of energy converted to heat in time interval $t_2 - t_1$ is given as

$$W = \int_{t_1}^{t_2} i^2 R\, dt \text{ Joules}$$

when i is a constant quantity I

Time interval, $t = (t_2 - t_1)$ sec

Energy, $W = I^2 Rt$ Joules

Thus, resistance referred to as a 'sink' or energy dissipative element.

Although resistance is often considered a constant of proportionality, however it may vary due to temperature changes caused by the heat produced. Unless otherwise stated, resistance, is be assumed constant. It is therefore necessary to keep the temperature changes of the resistor within reasonable limits if the resistance is intended to remain constant.

1.15 INDUCTANCE

Inductance can be considered as that property of a circuit element by which energy is capable of being stored in a magnetic flux field. An important feature of this element is that it makes itself felt in a circuit only when there is a changing current in it. This aspect of inductance is particularly important from circuit view point because although a circuit may have inductance, its presence in the circuit is felt when there is a time rate of change of current.

$$V_L = L\frac{di}{dt}$$

In general both V_L and i are function of time.

An appropriate defining equation for inductance is

$$L = \frac{V_L}{\left(\frac{di}{dt}\right)} \text{ Volt-sec/amp or Henrys}$$

Any circuit element that exhibits the property of inductance is called an 'inductor'. In the ideal sense the inductor is considered to be resistance less, though practically it does contain the wire resistance out of which the inductor coil is formed.

A linear inductor is one for which the inductance parameter is independent of current. As current flows through an inductor it creates space flux. When this flux permeates air strict proportionality between current and flux prevails, so that the inductance parameter stays constant for all values of current. However, when the flux is made to penetrate iron the large current may upset the proportional relationship between the current and the flux produced. In such a case the inductor will be a non-linear one.

To express current in terms of potential difference across the inductor.

$$di = \frac{1}{L} V_L \, dt$$

Integrating, this becomes

$$\int_{i(0)}^{i(t)} di = \frac{1}{L}\int_0^t V_L \, dt$$

$$i(t) = \frac{1}{L}\int_0^t V_L \, dt + i(0)$$

The above equation shows that the current in an inductor is dependent upon the integral of the voltage across its terminals as well as the initial current in the coil at the start of integration.

Assuming zero initial current in an inductor, if a current i is made to flow through a coil across which appears the potential difference V_L the total energy received in the time interval from zero to t is

$$W = \int_0^t V_L \, i \, dt \text{ Joules}$$

putting, $$V_L = L\frac{di}{dt}$$

$$W = \int_0^t \left(L\frac{di}{dt}\right) i\,dt$$

$$= \int_0^t L\,i\,dt$$

$$\boxed{W = \frac{L}{2} i^2 \text{ Joules}}$$

Thus we find that this inductor stores an amount of energy proportional to inductance (L) as well as to the square of instantaneous value of current (i). The energy stored in the inductor is of finite value and retrievable.

Linear inductor with iron core inductance is given by

$$L = \frac{N^2 \mu a}{l} \text{ Henry}$$

where N = Number of turns,
μ = Permeability of the iron core,
a = Cross-sectional area of the core, and
L = The length of iron core.

1.16 CAPACITANCE

Capacitance can be considered as that property of the circuit element by which energy is capable of being stored in an electric field. It is felt in a circuit only when there is a changing voltage across the terminals of element.

Capacitance is a proportionally factor relating the charge between two metal surfaces or conductors to the corresponding potential difference existing between them.

$$q = C V_c$$

where q = Is the charge
V_c = Potential difference

as $$i = \frac{dq}{dt}$$

$$i = C\frac{dV_c}{dt}$$

So the capacitor is a circuit element which has the property of yielding a current directly proportional to the rate of change of voltage across its terminals.

Capacitor current in terms of capacitor voltage

$$dV_c = \frac{1}{C} i\,dt$$

Integrating both sides

$$\int_{V_c(0)}^{V_c(t)} V_c = \frac{1}{C}\int_0^t i\,dt$$

$$V_c(t) = \frac{1}{C}\int_0^t i\,dt + V_c(0)$$

The quantity V_c (0) denotes the initial voltage across the capacitor upon the start of integration process.

Assuming zero initial voltage across the capacitor if current i is allowed to flow in the circuit for a time interval i, the energy delivered to the capacitor during this time is

$$W = \int_0^t V_c\, i\,dt$$

Substituting the value of i

$$W = \int_0^t V_c\left(C\frac{dV_c}{dt}\right)dt$$

$$= \int_0^t C V_c\, dt$$

$$W = \frac{1}{2} C V_c^2 \text{ Joules}$$

In case of two parallel plates separated by distance of d meters the capacitance may be shown to be

$$C = \frac{\varepsilon A}{d} \text{ Farads}$$

where A is area of the plates in meters, d is distance of separation in meters and ε = permittivity of the material between the plates of the capacitor.

Circuit Response of Single Elements

Elements	*Voltage across the element*	*Current in element*
Resistance (R)	$V(t) = R\, i(t)$	$i(t) = \frac{V(t)}{R}$
Inductance (L)	$V(t) = L\frac{di}{dt}$	$i(t) = \frac{1}{L}\int V\,dt$
Capacitance (C)	$V(t) = C\int i\,dt$	$i(t) = C\frac{dV}{dt}$

During network system analysis some terms are frequently used. These several terms which are of prime importance during analysis of electrical circuit are defined under:

1. **Circuit:** A circuit is a closed conducting path through which an electric current either flows or is intended to flow.
2. **Linear Circuit:** A circuit which obeys superposition theorem (*i.e.*, additivity and homogeniety) and does not change with the change in either voltage or current.
3. **Non-linear Circuit:** A circuit which does not obeys superposition theorem and whose circuit condition changes with change in voltage or current.
4. **Electrical Network:** A combination of various electric elements like resistor, capacitor, inductor which are connected in any manner resulting a closed conducting path through which electric current flows is called as electrical network.
5. **Active Network:** These are the types of network which contains one or more than one e.m.f. source in it.
6. **Passive Network:** These are the types of network which does not contain, any e.m.f. source in it.
7. **Node:** It is a junction where various electrical elements are connected together.
8. **Branch:** It is that portion of network which lies between the two nodes.
9. **Loop:** It is a closed path in a circuit in which no elements or node is encountered twice.
10. **Mesh:** It is type of loop that contains no loop in it.

1.17 ELECTRICAL PARAMETERS

The various elements of an electric circuit are called its parameters. Basically two types of electrical elements are defined. The elements are active elements and passive elements.

1.17.1 Active Elements

The active elements are the basic source of energy to the electric circuit (like voltage source, current source) *e.g.* battery.

1.17.2 Passive Elements

The passive elements are those element which acts as sink for electrical energy. The example of passive elements are Resistor, Capacitor and Inductor.

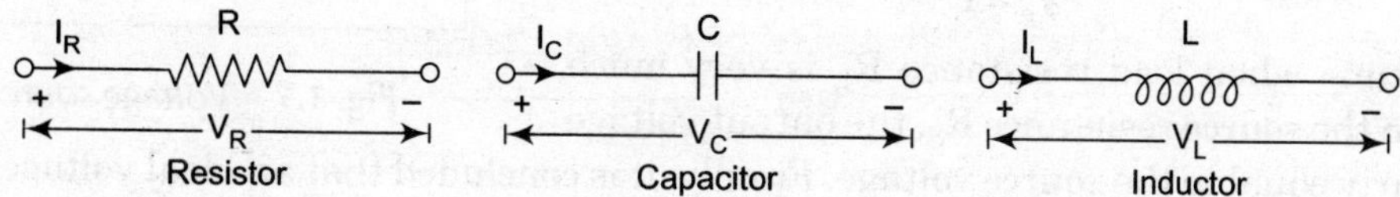

Resistor → It is the type of element which creates hinderance to the flow of current or restrict the flow of current.

Capacitor → It is the element which stores the electric charges for prescribed duration of time.

Inductor → In this type of element the magnetic firm of energy is stored for certain time limit.

1.18 SOURCE CONVERSION

For the voltage source of Fig. 1.7.

$$I_L = \frac{E}{R_S + R_L} = \frac{E}{R_S}\left(\frac{R_S}{R_S + R_L}\right)$$

or

$$= \frac{E}{R_S}\left(\frac{R_S}{R_S + R_L}\right) \qquad ...(3)$$

Equation (3) is similar to equation (2). Thus a voltage source having voltage E and source resistance R_S can be replaced by a current source with current $\frac{E}{R_S}$ and a source resistance R_S.

For current source of Fig. 1.8.

$$V_o = I_L R_L = I \frac{R_S}{R_L + R_S} \cdot R_L$$

$$= IR_S\left(\frac{R_L}{R_L + R_S}\right) \qquad ...(4)$$

Equation (4) is similar to equation (1). Thus a current source can also be converted into voltage source.

1.18.1 Voltage Source

The Fig. 1.7 shown below shows a battery having open circuit voltage E delivering or giving a current to a load resistance. The output voltage V_o across the load resistance R_L is given by

$$V_o = E \cdot \frac{R_L}{R_L + R_S} \qquad ...(1)$$

A battery source is converted into a constant voltage source by adding the internal resistance in series with the constant voltage source.

If $$R_S << R_L$$

$$V_o \simeq E$$

At the time when load resistance R_L is very much larger than the source resistance R_S, the output voltage will be nearly equal to the source voltage. Finally, it is concluded that an ideal voltage source is one which has zero source resistance.

Fig. 1.7 *Voltage source.*

1.18.2 Current Source

A current source is that type of source which delivers specified current through its terminals, irrespective of changes in the load resistance. The Fig. 1.8 shows a current source having source resistance R_S and feeding a load resistance R_L.

The load current I_L is given by

$$I_L = I \frac{R_S}{R_S + R_L} \quad ...(2)$$

If $\qquad R_L << R_S$

$$I_L \simeq I$$

When the load resistance is very much smaller than the source resistance, then the current in load resistance will be nearly equal to the source current I.

If source resistance R_S is infinite, the current source will be ideal current source.

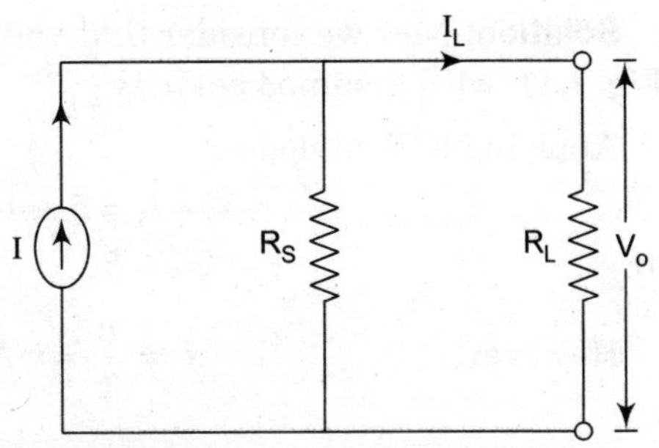

Fig. 1.8 *Current source.*

SOLVED NUMERICAL PROBLEMS

Example 1. *Obtain the value of i in the circuit of Fig. 1.9, if*

(a) $v = 1$ V $\qquad$ *(b)* $v = +2$ V

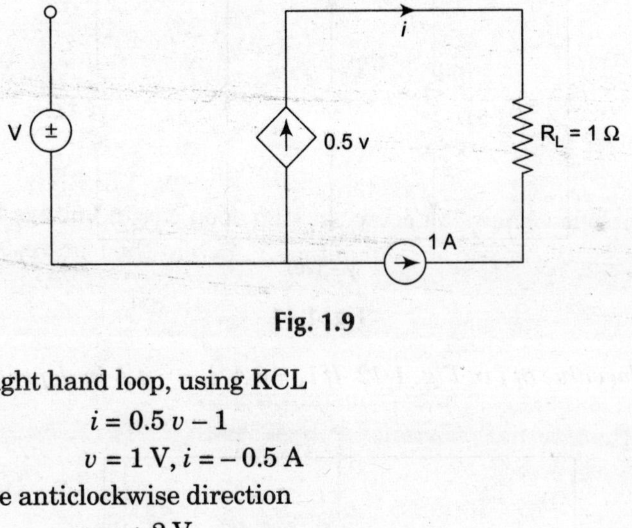

Fig. 1.9

Solution: In the right hand loop, using KCL

$$i = 0.5\,v - 1$$

for $\qquad v = 1\text{ V}, i = -0.5\text{ A}$

i.e., i is flowing in the anticlockwise direction

for $\qquad v = +2\text{ V}$

$i = 0$; *i.e.*, no current in the loop.

Example 2. *Find* i_1 *and* i_2 *in Fig. 1.10 below:*

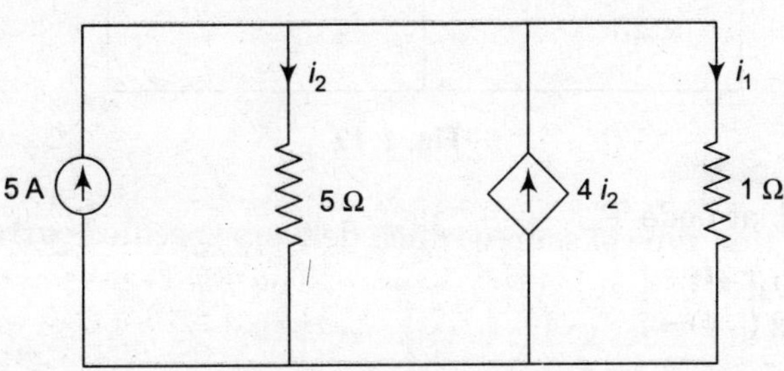

Fig. 1.10

Solution: Now we consider that voltage drop across 5 Ω and 1 Ω resistor be v volts shown in Fig. 1.11. with assumed polarity.

Applying KCL at node x

$$i_2 + i_1 = 5 + 4i_2$$

or $$i_1 - 3i_2 = 5 \quad ...(1.1)$$

However, $$i_1 = \frac{v}{1} = v \text{ A and } i_2 = \frac{v}{5} \text{ A}$$

Then from 1.11, $$v - \frac{3}{5}v = 5 \text{ or } v = 12.5 \text{ volt}$$

This gives $$i_1 = 12.5 \text{ A and } i_2 = 2.5 \text{ A}$$

$$5V - 3V = 25$$

$$2V = 25$$

$$V = 12.5$$

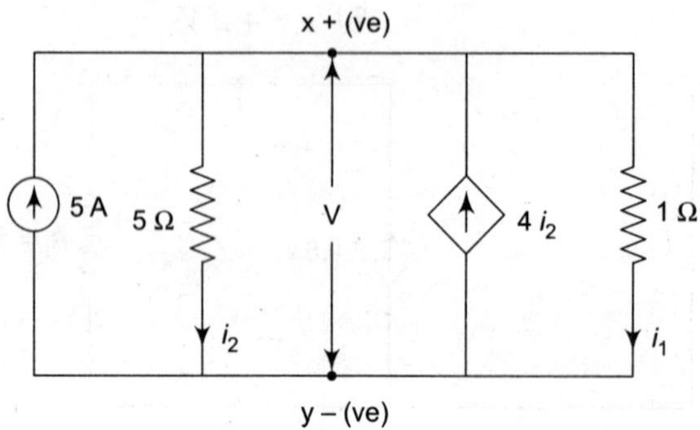

Fig. 1.11

Example 3. *Obtain the value of i in Fig. 1.12. If* $i_1 = 6$ *A,* $i_2 = -4$ *A in dependent current sources.*

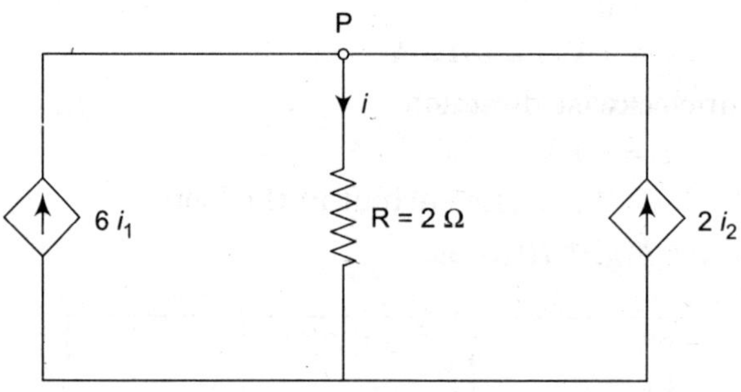

Fig. 1.12

Solution: Using KCL at node 'P'

$$6i_1 + 2i_2 = i$$

$$6 \times 6 + 2(-4) = i$$

$$i = 28 \text{ A.}$$

Example 4. *What is voltage V_s across the open switch in circuit of Fig. 1.13:*

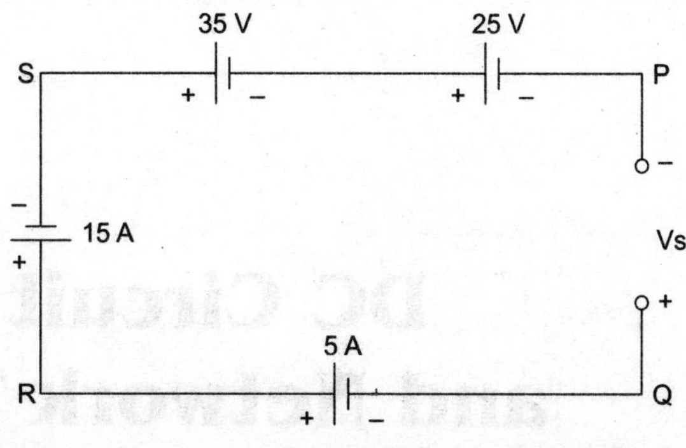

Fig. 1.13

Solution: Applying KVL to find V_s, starting from point P in the clockwise direction and using the sign convention.

We have $V_s + 5 - 15 - 35 - 25 = 0$

$\therefore \quad V_s = 70$ **V. Ans.**

Example 5. *By applying Kirchhoff's voltage law, find the value of current i and the voltage drop v_1 and v_2 in the below Fig. 1.14, which contains a current dependent voltage source. What is the voltage of the dependent source?*

Solution:

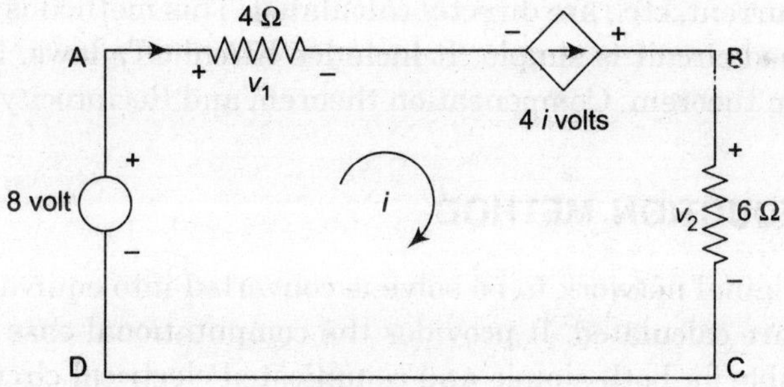

Fig. 1.14

Applying KVL to the circuit and starting from point A, we get

$$-v_1 + 4i - v_2 + 8 = 0$$

or $\quad v_1 - 4i + v_2 = 8$

Now $\quad v_1 = 4i$

and $\quad v_2 = 6i$

then $\quad 4i - 4i + 6i = 8$

$\therefore \quad i = \dfrac{8}{6} = 1.33$ A

$\therefore \quad v_1 = 4 \times 1.33 = 5.32$ V

and $\quad v_2 = 6 \times 1.33 = 7.98$ V

Voltage of dependent source $= 4i$

$= 4 \times 1.33 = 5.32$ V. **Ans.**

2

DC Circuit Analysis and Network Theorem

2.1 NETWORK ANALYSIS

There are two general approaches to network analysis:

2.2 DIRECT METHOD

In this method of analysis, the circuit is considered in its original form and the various unknowns like voltage, current, etc., are directly calculated. This method is applicable to limited area where the assigned circuit is simple. It includes Kirchhoff's laws, Nodal analysis, Loop analysis, Superposition theorem, Compensation theorem and Reciprocity theorem, etc.

2.3 NETWORK REDUCTION METHOD

In this method, the original network to be solve is converted into equivalent circuit and then different parameters are calculated. It provides the computational ease and save wastage of time. It can be applicable for both simple and complicated electrical circuits. This method includes Thevenen's theorem, Norton's theorem, Delta/Star and Star/Delta coversion etc.

2.4 KIRCHHOFF'S LAWS

A German Scientist Gustav R. Kirchhoff formulated two laws, in 1847, which are named after him. The electric circuit theory is based on these two laws.

2.5 KIRCHHOFF'S CURRENT LAW (KCL)

It states that at any instant the algebraic sum of currents flowing into a junction point (node) of an electrical circuit is zero.

The KCL can be expressed mathematically as

$$\sum_{J=1}^{h} I_J = 0 \qquad \text{...(1)}$$

where h denotes the number of circuit elements connected to the node and Σ denotes summation.

In given figure, when KCL is applied we get

$$i_1 - i_2 + i_3 - i_4 + i_5 + i_6 + i_7 - i_8 = 0$$

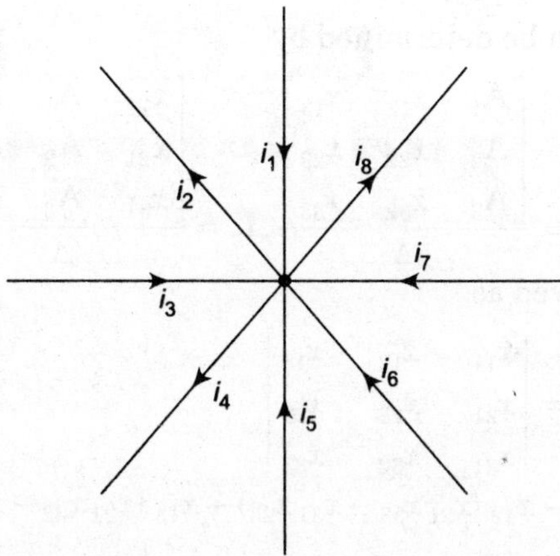

Fig. 2.1 *Kirchhoff's current law (KCL).*

The current going towards the node is considered to be positive and the outgoing current is considered to be negative. The opposite convention can also be used.

$$i_1 + i_3 + i_5 + i_6 + i_7 = i_2 + i_4 + i_8$$

Incoming current = Outgoing current.

2.6 KIRCHHOFF'S VOLTAGE LAW (KVL)

It states that the algebraic sum of all branch voltages around any closed loop of a network is zero at all instants of time. In other words, the algebraic sum of e.m.fs. around a closed loop equals the algebraic sum of IR drops around to loop. It is mathematically expressed as:

$$\sum_{K=1}^{h} V_J = 0 \qquad \text{...(1)}$$

where V_J is voltage in the Jth element of a closed loop having n elements.

Alternatively it can be also expressed as

$$\Sigma IR + \Sigma e.m.f. = 0$$

2.7 CRAMER'S RULE

Solution of linear equations by determinants :

Let us consider three linear equation in which three unknown current is to be determined.

$$x_{11}I_1 + x_{12}I_2 + x_{13}I_3 = A_1$$

$$x_{21}I_1 + x_{22}I_2 + x_{23}I_3 = A_2$$
$$x_{31}I_1 + x_{32}I_2 + x_{33}I_3 = A_3$$

In matrix firm above equation can be written as

$$\begin{bmatrix} x_{11} & x_{12} & x_{13} \\ x_{21} & x_{22} & x_{23} \\ x_{31} & x_{32} & x_{33} \end{bmatrix} \begin{bmatrix} I_1 \\ I_2 \\ I_3 \end{bmatrix} = \begin{bmatrix} A_1 \\ A_2 \\ A_3 \end{bmatrix}$$

The current I_1; I_2 and I_3 can be determined by

$$I_1 = \frac{\begin{bmatrix} A_1 & x_{12} & x_{13} \\ A_2 & x_{22} & x_{23} \\ A_3 & x_{32} & x_{33} \end{bmatrix}}{\Delta}, I_2 = \frac{\begin{bmatrix} x_{11} & A_1 & x_{13} \\ x_{21} & A_2 & x_{23} \\ x_{31} & A_3 & x_{33} \end{bmatrix}}{\Delta}, I_3 = \frac{\begin{bmatrix} x_{11} & x_{12} & A_1 \\ x_{21} & x_{22} & A_2 \\ x_{31} & x_{32} & A_3 \end{bmatrix}}{\Delta}$$

and determinant A can be solved as

$$\Delta = \begin{vmatrix} x_{11} & x_{12} & x_{13} \\ x_{21} & x_{22} & x_{23} \\ x_{31} & x_{32} & x_{33} \end{vmatrix}$$

$$x_{11}\,(x_{22}\,x_{33} - x_{32}\,x_{23}) - x_{12}\,(x_{21}\,x_{33} - x_{31}\,x_{23}) + x_{13}\,(x_{21}\,x_{32} - x_{31}\,x_{22}).$$

2.8 ANALYSIS OF NETWORK BY KIRCHHOFF'S LAW

Node Analysis

In node or junction method, the solution is analysed by the circuit as shown in Fig. 2.2. In nodal analysis it is essential to compute all branch currents. In writing the current expression the assumption is made that the node potential is always higher than the other voltages appearing in the equation. In node analysis the number of independent node pair equations needed in one less than the number of junction in the network. It is given by

$$n = J - 1$$

where n = Number of independent node equation

J = Number of junctions

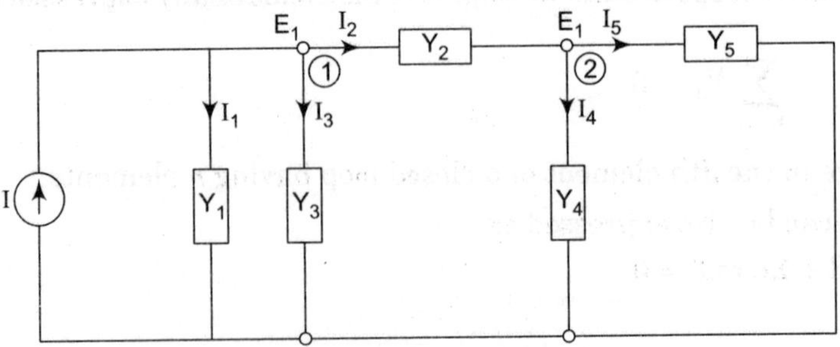

Fig. 2.2 *Node analysis.*

In above figure, the reference node is indicated by 0. Applying KCL at node (1) we get

$$I - I_1 - I_3 - I_2 = 0 \qquad \text{...}(i)$$

Equation (i) in terms of admittances and potentials can be written as

$$I - Y_1E_1 - Y_3E_1 - Y_2(E_1 - E_2) = 0$$

∴ E_1 is at higher potential

or $$(Y_1 + Y_2 + Y_3)\,E_1 + (-Y_2)\,E_2 = 0 \quad \ldots(ii)$$

Now considering node (2), we have

$$I_2 - I_5 - I_4 = 0 \quad \ldots(iii)$$

It can also be written as

$$Y_2(E_1 - E_2) - Y_5E_2 - Y_4E_2 = 0$$

or $$(-Y_2)\,E_1 + (Y_2 + Y_4 + Y_5)\,E_2 = 0 \quad \ldots(iv)$$

Equations (*ii*) and (*iv*) may be written in matrix form as

$$\begin{bmatrix} Y_1 + Y_2 + Y_3 & -Y_2 \\ -Y_2 & Y_2 + Y_4 + Y_5 \end{bmatrix}\begin{bmatrix} E_1 \\ E_2 \end{bmatrix} = \begin{bmatrix} 1 \\ 0 \end{bmatrix} \quad \ldots(v)$$

Using Cramer's rule, voltages at all the nodes can be obtained.

$$E_1 = \frac{1}{\Delta}\begin{bmatrix} I & -Y_2 \\ 0 & Y_2 + Y_4 + Y_5 \end{bmatrix}$$

$$E_2 = \frac{1}{\Delta}\begin{bmatrix} Y_1 + Y_2 + Y_3 & I \\ -Y_2 & 0 \end{bmatrix}$$

and determinant, $$\Delta = \begin{bmatrix} Y_1 + Y_2 + Y_3 & -Y_2 \\ -Y_2 & Y_2 + Y_4 + Y_5 \end{bmatrix}$$

As a general, node equation can be written as

$$[Y]\,[E] = [I]$$

Mesh Analysis

The Kirchhoff's voltage law is base in mesh analysis of the network. The name being derived from the similarities in appearance between the closed loops of a network and a physical 'fence' or mesh. In this method, a distinct current is assumed in the loop and the properties of drops in each element in the loop is determined by the assumed direction of loop current for that loop. In node analysis we have considered admittance but in mesh analysis replaced by impedance. For mesh analysis consider Fig. 2.3 as shown below:

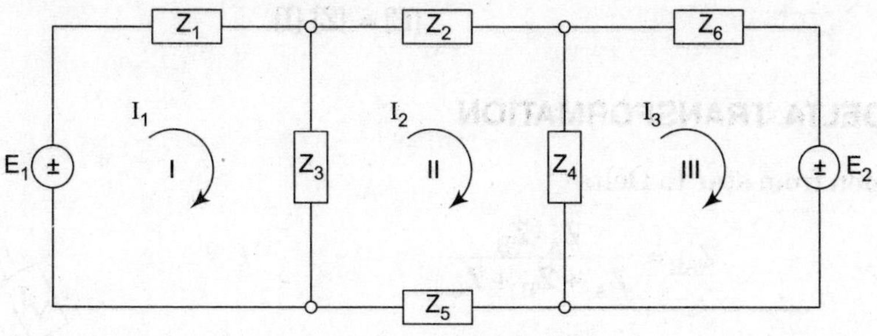

Fig. 2.3 *Mesh analysis.*

Apply KVL to above figure, we get

In mesh I, $$I_1Z_1 + (I_1 - I_2)\,Z_3 - E_1 = 0$$

or $$(Z_1 + Z_3)\,I_1 + (-Z_3)\,I_2 + (0)\,I_3 = E_1 \quad \ldots(i)$$

In mesh II, $(I_2 - I_1) Z_3 + I_2Z_2 + (I_2 - I_3) Z_4 + I_2Z_5 = 0$

$$I_2Z_3 - I_1Z_3 + I_2Z_2 - I_2Z_4 - I_3Z_4 + I_2Z_5 = 0$$

$$I_1 (- Z_8) + I_2 (Z_2 + Z_3 + Z_4 + Z_5) + I_3 (- Z_4) = 0 \quad \text{...(ii)}$$

and in mesh III, $(I_3 - I_2) Z_4 + I_3Z_6 + E_2 = 0$

or $(0) I_1 + (- Z_4) I_2 + (Z_4 + Z_6) I_3 = -E_2$ ...(*iii*)

Using equations (*i*), (*ii*) and (*iii*) as obtained in three meshes in matrix form as

$$\begin{bmatrix} Z_1 + Z_3 & -Z_3 & 0 \\ -Z_3 & Z_2 + Z_3 + Z_4 + Z_5 & -Z_4 \\ 0 & -Z_4 & Z_4 + Z_6 \end{bmatrix} \begin{bmatrix} I_1 \\ I_2 \\ I_3 \end{bmatrix} = \begin{bmatrix} E_1 \\ 0 \\ -E_2 \end{bmatrix}$$

Now, using Cramer's rule currents I_1, I_2 and I_3 can be determined *i.e.*,

$$I_1 = \frac{1}{\Delta} \begin{vmatrix} E_1 & -Z_3 & 0 \\ 0 & Z_2 + Z_3 + Z_4 + Z_5 & -Z_4 \\ -E_2 & -Z_4 & Z_4 + Z_6 \end{vmatrix}$$

$$I_2 = \frac{1}{\Delta} \begin{vmatrix} Z_1 + Z_3 & E_1 & 0 \\ -Z_3 & 0 & -Z_4 \\ 0 & -E_2 & Z_4 + Z_6 \end{vmatrix}$$

$$I_3 = \frac{1}{\Delta} \begin{vmatrix} Z_1 + Z_3 & -Z_3 & E_1 \\ -Z_3 & Z_2 + Z_3 + Z_4 + Z_5 & 0 \\ 0 & -Z_4 & -E_2 \end{vmatrix}$$

where,

$$\Delta = \begin{vmatrix} Z_1 + Z_3 & -Z_3 & 0 \\ -Z_3 & Z_2 + Z_3 + Z_4 + Z_5 & -Z_4 \\ 0 & -Z_4 & Z_4 + Z_6 \end{vmatrix}$$

So, in general the above can be concluded as

$$[E] = [Z]\,[I].$$

2.9 STAR DELTA TRANSFORMATION

(*i*) Conversion from star to Delta

$$Z_{AB} = \frac{Z_A \cdot Z_B}{Z_A + Z_B + Z_C}$$

$$Z_{BC} = \frac{Z_B \cdot Z_C}{Z_A + Z_B + Z_C}$$

$$Z_{CA} = \frac{Z_C \cdot Z_A}{Z_A + Z_B + Z_C}$$

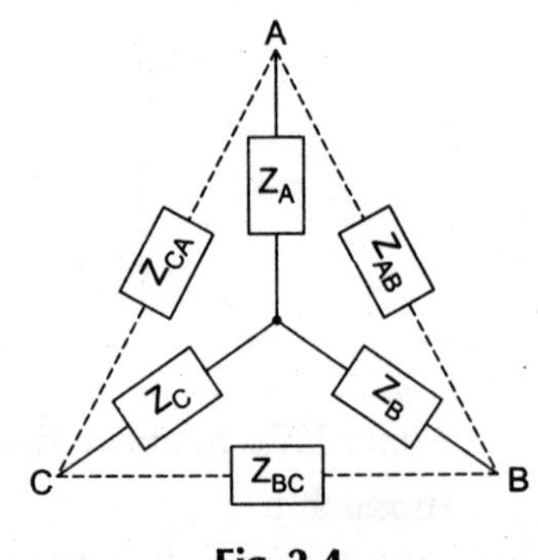

Fig. 2.4

(*ii*) Conversion from Delta to Star

$$Z_A = \frac{Z_{AB} \cdot Z_{CA}}{Z_{AB} + Z_{BC} + Z_{CA}}$$

$$Z_B = \frac{Z_{BC} \cdot Z_{AB}}{Z_{AB} + Z_{BC} + Z_{CA}}$$

$$Z_C = \frac{Z_{CA} \cdot Z_{BC}}{Z_{AB} + Z_{BC} + Z_{CA}}$$

2.10 SUPERPOSITION THEOREM

The superposition Theorem states that in any linear bilateral network containing several source (including the initial condition source), The overall response (loop current or node voltage) at any point in the network equals the sum of the responses of each individual source considered separately with all other sources made imperative i.e., replaced by impedances equal to their internal impedances.

Explanation: Consider figure shown in Fig. 2.5(a), applying the superposition theorem, let us first consider source V_1 alone and replace the V_2 by short-circuit in Step I.

(**Note:** Voltage source is short-circuited while current source is open-circuited).

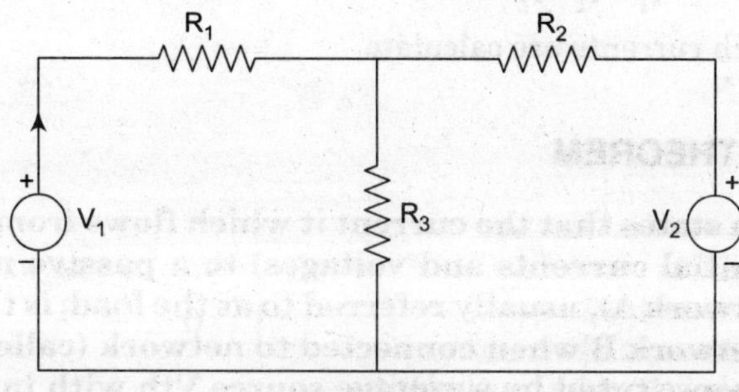

Fig. 2.5(*a*)

Step I: Consider only one source of e.m.f. or current and obtain the branch currents. Here in Fig. 2.5(*b*), we get

$$I'_1 = \frac{V_1}{\dfrac{R_2 \cdot R_3}{R_2 + R_3} + R_1}$$

and $$I'_2 = I'_1 \frac{R_3}{R_2 + R_3}$$

and finally $$I'_3 = I'_1 - I'_2 \qquad ...(i)$$

Step II: Likewise in Step I, repeat the same procedure for all the independent sources *i.e.*,

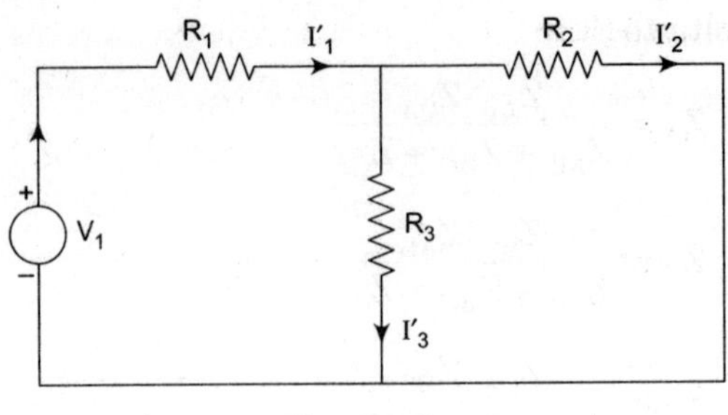

Fig. 2.5(*b*)

In Step II, V_1 source is short circuited and V_2 source is considered and then different branch currents are again calculated.

We get, $I_2'' = \dfrac{V_2}{\dfrac{R_1 \cdot R_3}{R_1 + R_3} + R_2}$ and $I_1'' = I_2'' = \dfrac{R_3}{R_1 + R_3}$

Also $I_3'' = I_2'' - I_1''$ ...(*ii*)

Now according to superposition theorem:

Using Steps I and II all branch currents are calculated.

Step III $I_3 = I_3' + I_1''$

$I_2 = I_2' - I_2''$

$I_1 = I_1' - I_1''$

Hence, all the branch currents are calculate.

2.11 THEVENIN'S THEOREM

Thevenin's Theorem states that the current it which flows from a given linear active network A (with initial currents and voltages) to a passive network B(not inductively coupled to network A), usually referred to as the load, is the same as one which flows to the same network B when connected to network (called Thevenin's equivalent of network A) constituted by a voltage source Vth with internal impedance Z_{th} where V_{th} is open circuited voltage of network A at load terminals and Z_{th} in the impedance of network 'A' seen looking back from the land terminals with all energy sources replaced by their internal impedances.

Explanation: Let us consider a circuit shown in Fig. 2.6(*a*):

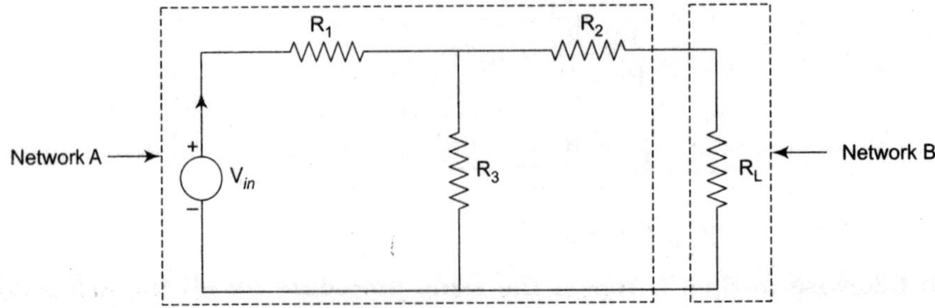

Fig. 2.6(a) *An active network A connected to passive network.*

Steps for solving and explaining Thevenin's theorem is given below:

Step I: Remove the load resistance R_L (resistance across which Thevenin's theorem circuit voltage is to be determined. So, R_L is open circuited as shown in Fig. 2.6(*b*).

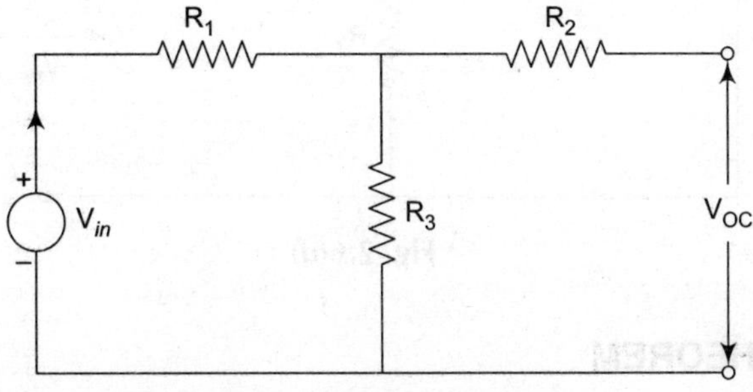

Fig. 2.6(*b*)

In order to find the V_{oc}, the load resistance R_L is removed and acts as open circuited.

$$V_{oc} = IR_3 \frac{V_{in}}{R_1 + R_3} \cdot R_3$$

Step II: In Step II all the sources either voltage source or current source is removed. The voltage source is replaced by its internal resistance and the current source is open circuited. Now, the internal resistance (Thevenin's resistance) of the source side looking through open circuit is obtained.

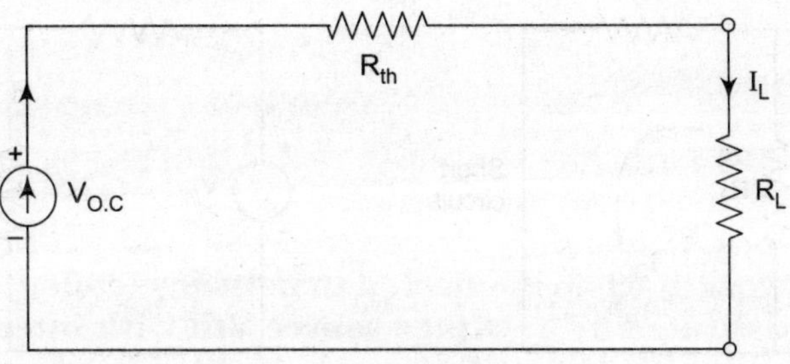

Fig. 2.6(c)

According to Fig. 2.6(*b*), we get

$$R_{th} = R_2 + \frac{R_1 \cdot R_3}{R_1 + R_3}$$

Step III: In this step we obtain Thevenin's equivalent circuit by using Step I and Step II. The load resistance across which Thevenin's equivalent is to be obtain is inserted in the Thevenin's equivalent circuit as shown in Fig. 2.6(*d*) and then I_L current across load resistance is obtained.

According to Thevenin's theorem,

$$I_L = \frac{V_{oc} \cdot V_{th}}{R_{th} + R_L} \text{ amp.}$$

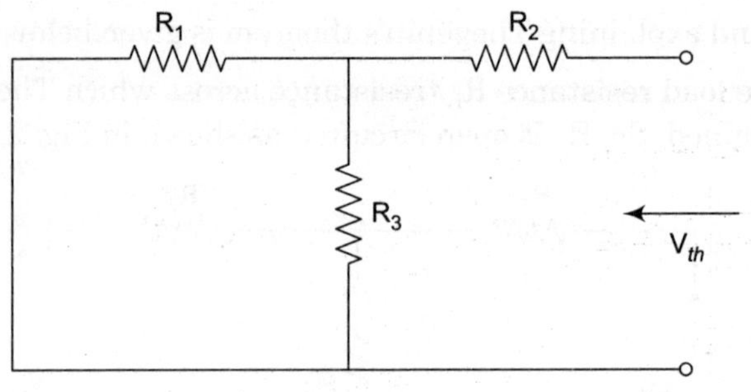

Fig. 2.6(*d*)

2.12 NORTON'S THEOREM

The Norton's theorem is called as dual of Thevenin's theorem because all the parameters of the network can be obtained either of the methods. It will be explained later on.

Norton's Theorem states that the current in any load impedance Z_l connected to the two terminals of a network is the same as it this load impedance Z_l were connected to a current source (called Norton's equivalent current source I_N) whose source current is the short-circuit current at the terminals and whose internal impedance of the network looking back into the terminals with all the sources replaced 6*j* impedances equal to their impedances (Z_n).

Explanation: Consider Fig. 2.7(*a*) drawn below (*b*)

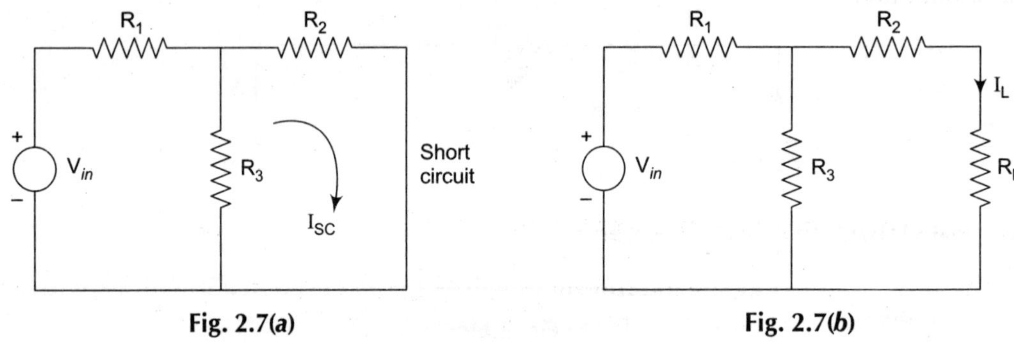

Fig. 2.7(*a*) Fig. 2.7(*b*)

(Simple network for obtaining I_{SC})

According to Norton's theorem remove load resistance R_L by short circuit, we get

$$I = \frac{V_{th}}{R_1 + \dfrac{R_2 \cdot R_3}{R_2 + R_3}} \text{ and } I_{SC} = I\frac{R_3}{R_3 + R_2}$$

Step I: The load resistance is removed and then find the internal resistance of the source by removing the source as short circuit.

As shown in Fig. 2.7(*b*), we calculate internal resistance as

$$R_{int} = R_2 + \frac{R_1 \cdot R_3}{R_1 + R_3}$$

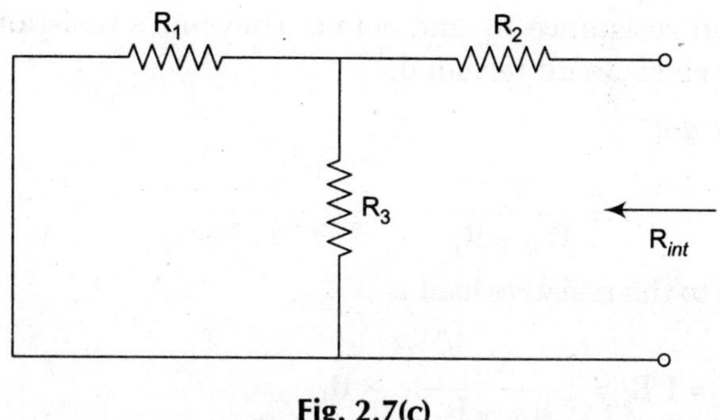

Fig. 2.7(c)

Step II: Now here, Norton's equivalent circuit model is drawn putting R_{int} in parallel to I_{SC} as shown in Fig. 2.7(*a*)

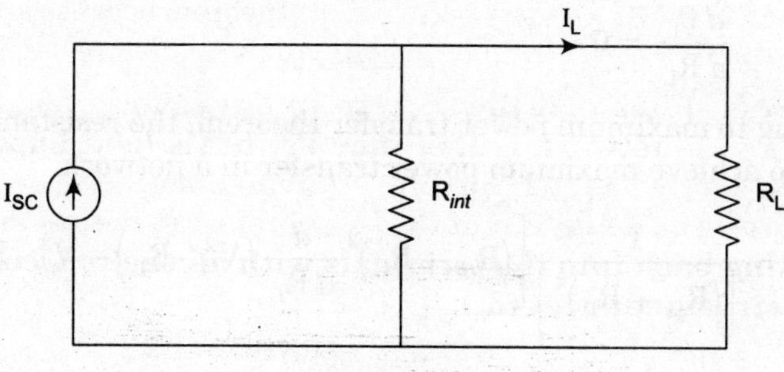

Fig. 2.7(*d*)

Step III: Connect the load resistance again in the circuit and calculate load current I_L.

$$I_L = I_{SC} \frac{R_{int}}{R_{int} + R_L}.$$

2.13 MAXIMUM POWER TRANSFER THEOREM

This theorem states that the maximum power is transferred in a linear network when the internal resistance of the source is equal to the load resistance.

Explanation: A resistance of varying nature R_L is connected to a d.c. source network as shown in Fig. 2.8(*a*) whereas Fig. 2.8(*b*) shows Thevenin's voltage V_{th} and theorem's resistance, R_{th}, of the source network.

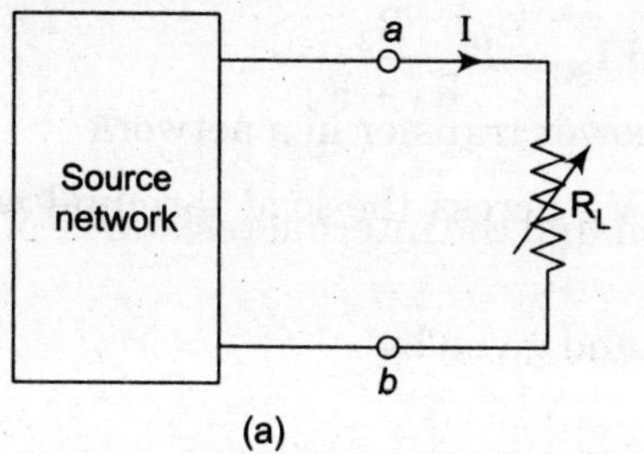

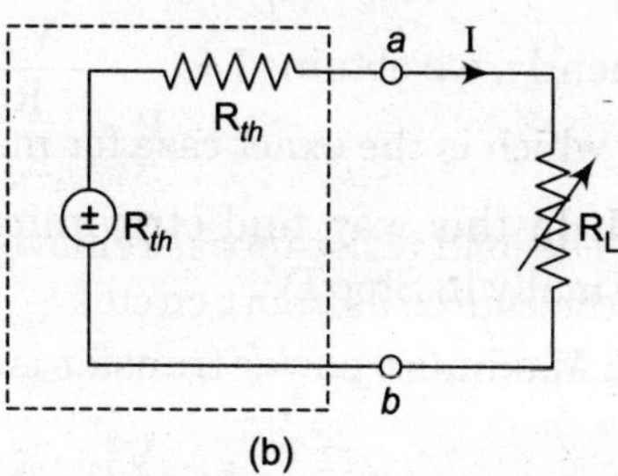

Fig. 2.8

Step I: Remove load resistance R_L and obtain Thevenin's resistance R_{th} of the source by viewing inword across open circuit terminal.

From Fig. 2.7(*b*), we get

$$I = \frac{V_{th}}{R_{th} + R_L}$$

while power delivered to the resistive load is

$$P_L = I^2 R_L = \left(\frac{V_{th}}{R_{th} + R_L}\right)^2 \times R_L$$

P_L can be changed by changing R_L and hence maximum power can be delivered at the condition when

$$\frac{dP_L}{dR_L} = 0$$

Step II: According to maximum power transfer theorem, the resistance R_L should be equal to the R_{th} in order to achieve maximum power transfer in a network.

However, $$\frac{dP_L}{dR_L} = \frac{1}{\left[(R_{th} + R_L)^2\right]}\left[(R_{th} + R_L)^2 \frac{d}{dR_L}\left(V_{th}^2 \cdot R_L\right) - V_{th}^2 \cdot R_L \frac{d}{dR_L}(R_{th} + R_L)^2\right]$$

or $$= \frac{1}{(R_{th} + R_L)^4}\left[(R_{th} + R_L)^2 V_{th}^2 - V_{th}^2 . R_L \times 2(R_{th} + R_L)\right]$$

$$= \frac{V_{th}^2 (R_{th} + R_L - 2R_L)}{(R_{th} + R_L)^3} = \frac{V_{th}^2 (R_{th} - R_L)}{(R_{th} + R_L)^3}$$

But for maximum transfer of power we know that

$$\frac{dP_L}{dR_L} = 0$$

Therefore, $$\frac{V_{th}^2 (R_{th} - R_L)}{(R_{th} + R_L)^3} = 0 \text{ which yields}$$

$$(R_{th} - R_L) = 0$$

Consequently, we obtain

$R_{th} = R_L$ which is the exact case for maximum power transfer in a network.

Step III: In this way find etrevinins voltage V_{th} across the load terminal which is open circuited. Finally in Step IV.

Step IV: Maximum power transfer is achieved and given by

$$\frac{V_{th}^2}{4R_{th}}.$$

2.14 MILLMAN'S THEOREM

Millman Theorem States that it n voltages sources $V_1, V_2, ..., V_n$ having internal impedances (or series impedances) $Z_1, Z_2, ..., Z_n$ respectively are connected in parallel as shown in Fig. 2.9(a), then these sources may be replaced by a single voltage source V_m having internal series impedance Z_m, where V_m and Z_m are given by the following equations:

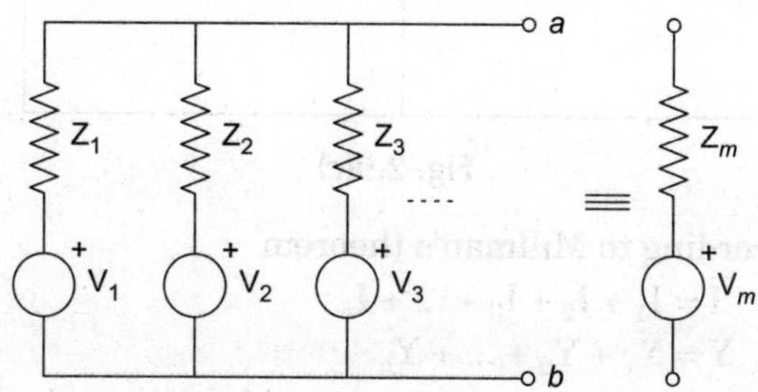

Fig. 2.9(*a*) *Diagram illustrating Millman Theorem.*

$$V_m = \frac{V_1Y_1 + V_2Y_2 + V_3Y_3 + ...}{Y_1 + Y_2 + Y_3}$$

Where $Y_1, Y_2, Y_3, ..., Y_n$ are the admittance corresponding to $Z_1, Z_2, ..., Z_n$.

According to Millman's theorem:

$$V = \frac{\pm V_1Y_1 \pm V_2Y_2 \pm ... \pm V_nY_n}{Y_1 + Y_2 + ... + Y_n}$$

where,
$$R = \frac{1}{Y} = \frac{1}{Y_1 + Y_2 + ... + Y_n}$$

For explaining Millman's theorem, consider a circuit diagram shown below:

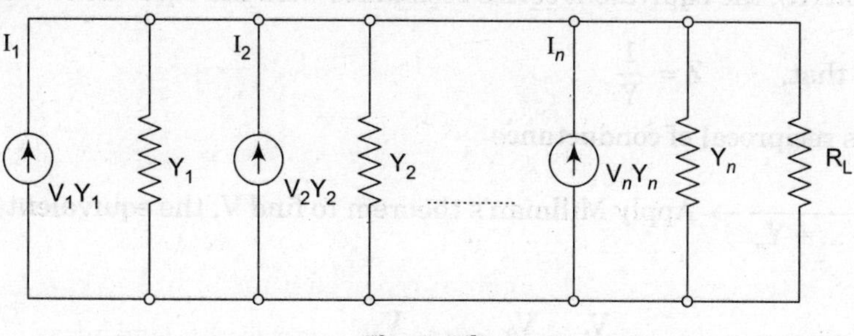

Fig. 2.9(*b*)

Conversion from parallel voltage to current source.

Step I: Calculate the conductance ($G_1, G_2, ...$) of each voltage source ($V_1, V_2, ...$) and obtain G while the load resistance is removed.

Let us consider Fig. 2.9(*c*) as shown.

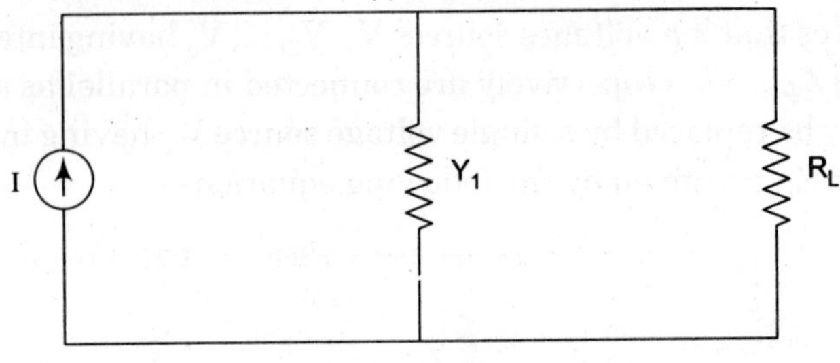

Fig. 2.9(c)

Here, we know that according to Millman's theorem

$$I = I_1 + I_2 + I_3 + ... + I_n$$

$$Y = Y_1 + Y_2 + ... + Y_n$$

At very next instant the current source is converted into its equivalent voltage source.

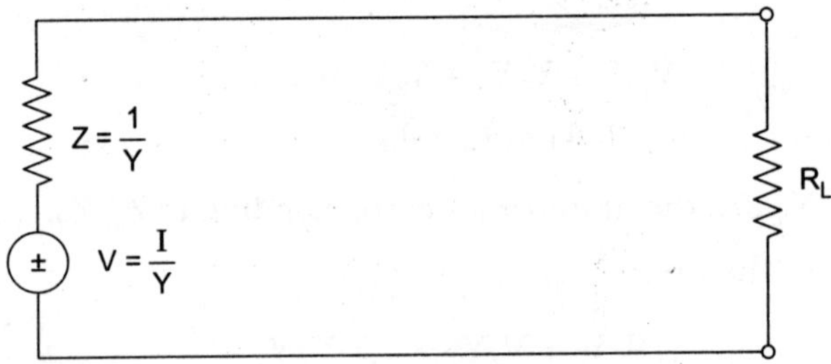

Fig. 2.9(*d*)

Therefore,

$$V = IZ = \frac{1}{Y} = \frac{\pm I_1 \pm I_2 \pm ... \pm I_n}{Y_1 + Y_2 + ... + Y_n}$$

Step II: Obtain (R), the equivalent series resistance with the equivalent voltage source (V).

Also we know that, $Z = \frac{1}{Y}$

i.e., Resistance is reciprocal of conductance

$Z = \frac{1}{Y_1 + Y_2 + ... + Y_n}$ → Apply Millman's theorem to find V, the equivalent voltage source given by

Step III :
$$V = \frac{\pm \frac{V_1}{Z_1} \pm \frac{V_2}{Z_2} \pm ... \pm \frac{V_n}{Z_n}}{\frac{1}{Z_1} + \frac{1}{Z_2} + ... + \frac{1}{Z_n}}$$

Therefore finally,
$$V = \frac{\pm V_1 Y_1 \pm V_2 Y_2 \pm ... \pm V_n Y_n}{Y_1 + Y_2 + ... + Y_n}$$

In summation form it can be given as

$$= \frac{\sum_{j=1}^{n} V_j \cdot Y_j}{\sum_{j=1}^{n} Y_j} \text{ and } G_j = \frac{1}{R_j}$$

Step IV: Finally the current through the load is expressed as follow *i.e.*,

$$I_L = \frac{V}{Z + R_L}, \text{ where } R_L \text{ is the load resistance.}$$

2.15 COMPENSATION THEOREM

In any network, linear or non-linear, any impedance may be replaced by a voltage source of zero internal impedance and voltage equal to the instantaneous potential difference produced across the replaced impedance by the current flowing through it.

Explanation: Consider a load resistance R_L is connected to a d.c. source network whose Thevenin's equivalent gives V_{th}, as Thevenin's resistance as shown in Fig. 2.10(a).

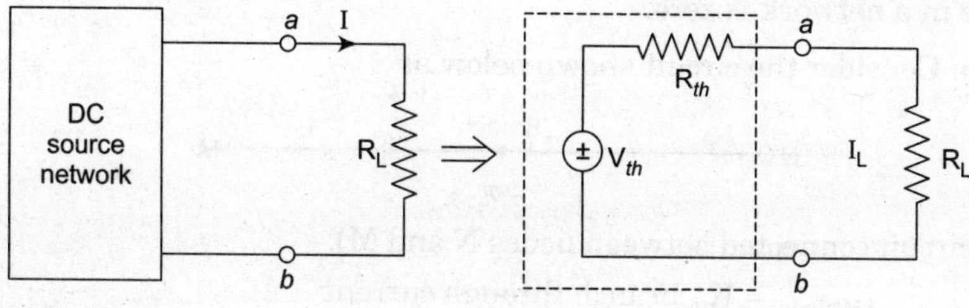

Fig. 2.10(*a*)

Thevenin's equivalent source.

Here, we have
$$I_L = \frac{V_{th}}{R_{th} + R_L}$$

Now, according to compensation theorem, the load resistance is changed to $(R_L + \Delta R_L)$ and the rest of the circuit remains unchanged, then the Thevenin's equivalent network remains the same as shown in Fig. 2.10(b).

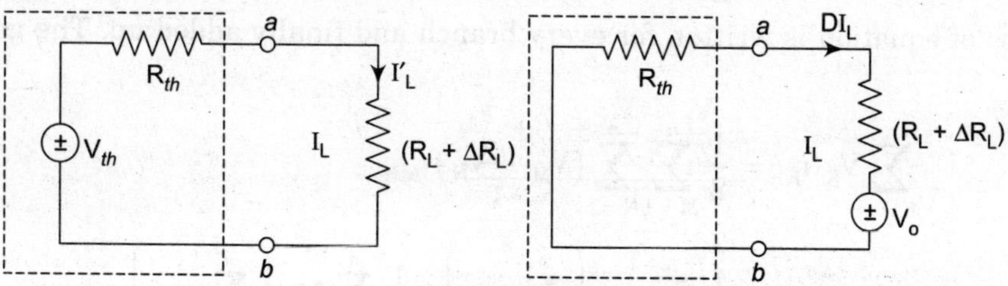

Fig. 2.10(*b*)

Therefore, the change in current is turned as ΔI_L.

So, $$\Delta I_L = I'_L - I_L$$

$$= \frac{V_{th}}{R_{th} + (R_L + \Delta R_L)} - \frac{V_{th}}{R_{th} + R_L}$$

$$= \frac{V_{th}\{R_{th} + R_L - (R_{th} + R_L + \Delta R_L)\}}{(R_{th} + R_L + \Delta R_L)(R_{th} + R_L)}$$

$$= -\left(\frac{V_{th}}{R_{th} + R_L}\right)\frac{\Delta R_L}{R_{th} + R_L + \Delta R_L} = -\frac{I_L\,\Delta R_L}{R_{th} + R_L + \Delta R_L}$$

$$= \frac{-V_0}{R_{th} + R_L + \Delta R_L}$$

The term V_0 is equal to the product of load current and changed resistance ΔR_L and it is known as compensation voltage. In this way the compensation theorem is verified.

2.16 TELLEGEN'S THEOREM

The Tellegen's theorem states that the sum of product of branch voltage drop and branch current at any time in a network is zero.

Explanation: Consider the circuit shown below as

N ——— Branch (i_{nm}) ———▶——— M

(Branch of a circuit connected between nodes N and M).

Consider $$i_{MN(=i_K)} = K_{th} \text{ branch through current}$$

$$V_K = \text{Voltage drop in } K_{th} \text{ branch}$$

$$= V_m - V_n \text{ while } V_M \text{ and } V_N \text{ are the node voltages.}$$

We have, $$V_K i_{MN} = (V_M - V_N)\, i_{MN} = V_K i_K$$

$$[\text{also } V_K i_K = (V_N - V_M)\, i_{NM}, \text{ So, } i_{MN} = -i_{NM}]$$

summing above two relation, we get

$$2V_K i_K = (V_M - V_N)\, i_{MN} + (V_N - V_N)\, i_{NM}$$

or $$V_K i_K = \frac{1}{2}\,[(V_M - V_N)\, i_{MN} + (V_N - V_M)\, i_{NM}]$$

Above type of equation is written for every branch and finally added up. The general form obtained as

$$\sum_{K=1}^{n} V_K\, i_K = \frac{1}{2}\sum_{M=1}^{n}\sum_{N=1}^{n}(V_M - V_N)\, i_{MN}$$

$$= \frac{1}{2}\sum_{M=1}^{n} V_M\left(\sum_{N=1}^{n} i_{MN}\right) - \frac{1}{2}\sum_{N=1}^{n} V_N\left(\sum_{M=1}^{n} i_{MN}\right)$$

i.e., $$\sum_{M=1}^{n} i_{MN} = 0 \; ; \; \sum_{N=1}^{n} i_{NM} = 0$$

With above result, finally we obtain

$$\sum_{K=1}^{n} V_K \, i_K = 0$$

Finally, we can conclude that, according to the Tellegen's theorem all the branch currents and branch voltages are calculated by conventional method and then sum of their product is calculated which will be equal to zero.

2.17 VOLTAGE DIVIDER RULE

It is applicable to a series circuit where the same current flows through each of the given resistor. Consider a series circuit with a constant voltage source shown in Fig. 2.11:

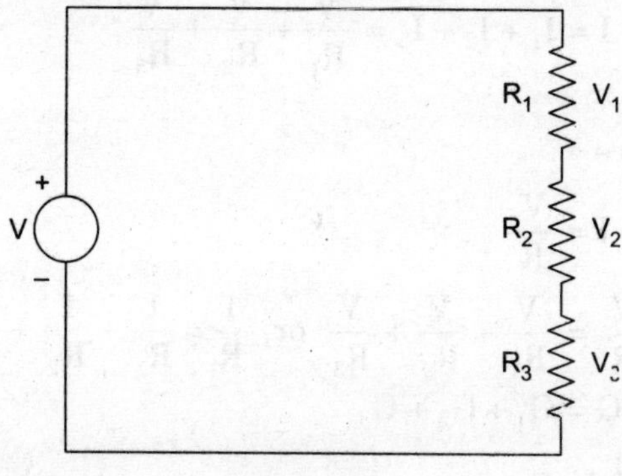

Fig. 2.11

Consider a source voltage denoted by V.

Total resistance $R = R_1 + R_2 + R_3$

According to Voltage Divider Rule, the voltage drop across all the resistors will be

$$V_1 = \frac{V}{R} R_1$$

$$V_2 = \frac{V}{R} R_2$$

and $$V_3 = \frac{V}{R} R_3$$

In this way, we obtain all the voltage drops across all the resistors in series.

2.18 CURRENT DIVIDER RULE

The current divider rule is applicable to the circuit where the several resistors are connected in parallel.

Consider the circuit diagram shown in Fig. 2.12, where a source voltage V is connected with several resistors which are parallel to each other.

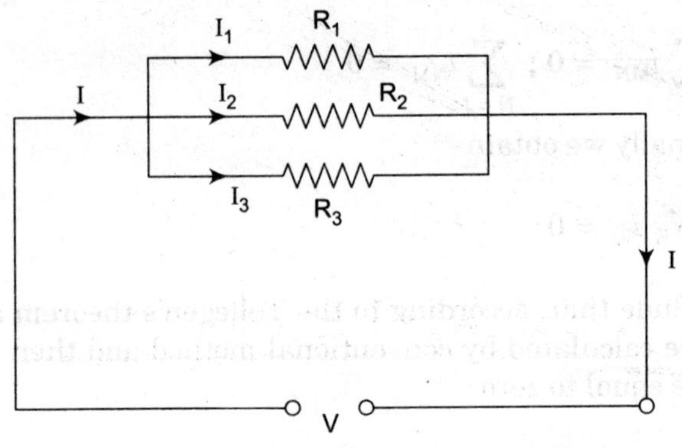

Fig. 2.12

The potential drop across all the resistor is same, current through each resistor is different and calculated by Ohm's law and total current is equal to the sum of all currents, *i.e.*,

$$I = I_1 + I_2 + I_3 = \frac{V}{R_1} + \frac{V}{R_2} + \frac{V}{R_3}$$

V → Applied voltage
R → Equivalent resistance

By Ohm's law $\qquad I = \frac{V}{R}$

$\therefore \qquad \frac{V}{R} = \frac{V}{R_1} + \frac{V}{R_2} + \frac{V}{R_3}$ or, $\frac{1}{R} = \frac{1}{R_1} + \frac{1}{R_2} + \frac{1}{R_3}$

Also, $\qquad G = G_1 + G_2 + G_3$

as $\qquad \frac{1}{R} = G.$

SOLVED NUMERICAL PROBLEMS

Example 1. *Determine the value of resistance R so that maximum power transfer takes place from the rest of the network to R in Fig. 2.13(a).*

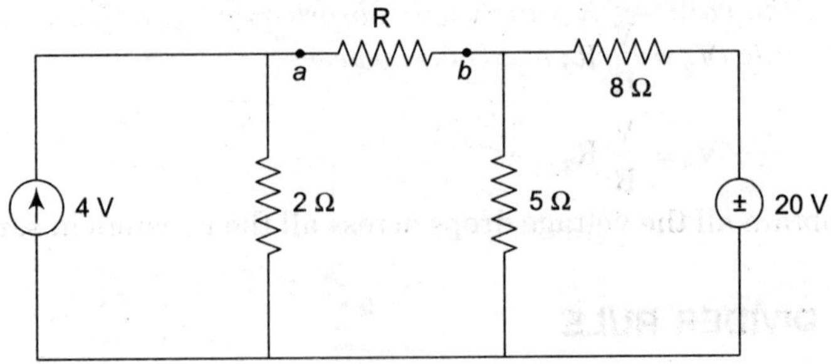

Fig. 2.13(*a*)

Solution: Let us convert the current source into a voltage source as shown in Fig. 2.13(*b*). Remove R from circuit as open circuit and obtain voltage V_{oc} as

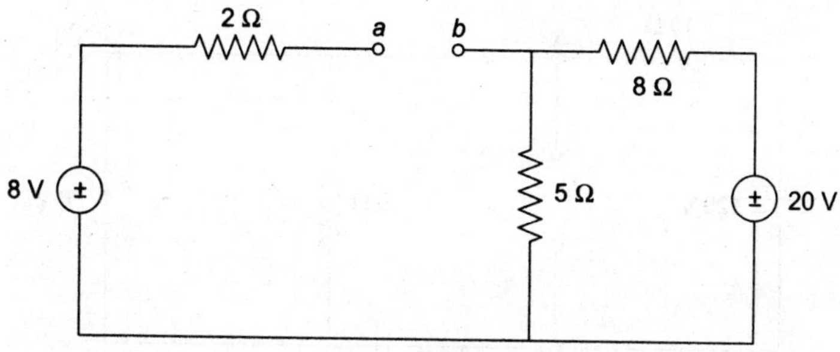

Fig. 2.13(*b*)

$$i = \frac{20}{13} = 1.53 \text{ A}$$

$$V_{th} = \text{Drop across } 5\ \Omega = 1.53 \times 5 = 7.69 \text{ V}$$

Thus in the left loop, $-8 + V_{oc} + 7.69 = 0$

or $$V_{oc} = 8 - 7.69 = 1.69 \text{ V}.$$

Now, again, circuit can be drawn removing all the sources from the circuit for obtaining R_{th} as

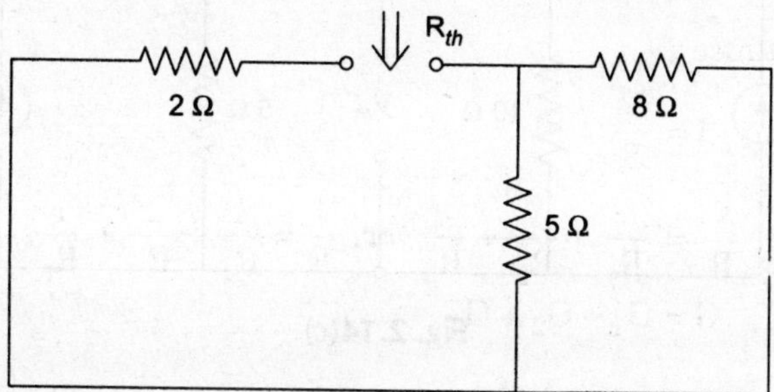

Fig. 2.13(c)

$$R_{th} = \frac{5 \times 8}{5 + 8} + 2 = \frac{40}{13} + 2 = 5.07\ \Omega$$

According to maximum power transfer theorem

$$R = R_{th} = 5.07\ \Omega$$

Example 2. *Calculate the value of R which will absorb maximum power from the circuit of Fig. 2.14(a). Also calculate the value of maximum power.*

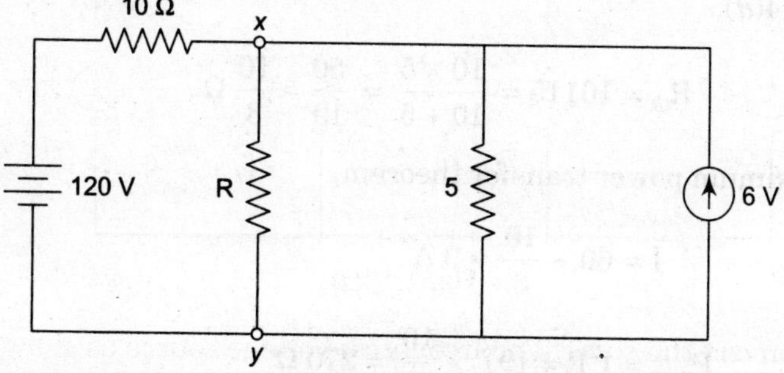

Fig. 2.14(a)

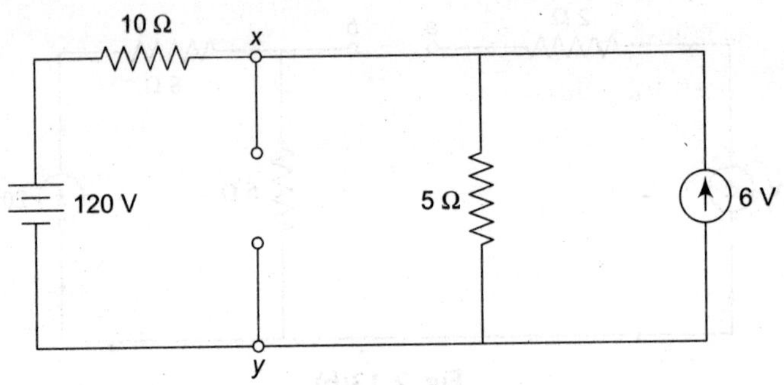

Fig. 2.14(b)

Solution: Let us remove R and calculate Thevenin's voltage V_{th} across terminal $x-y$ as shown in Fig. 2.14(*b*).

In very next step convert the voltage source to current source or vice-versa *i.e.*,

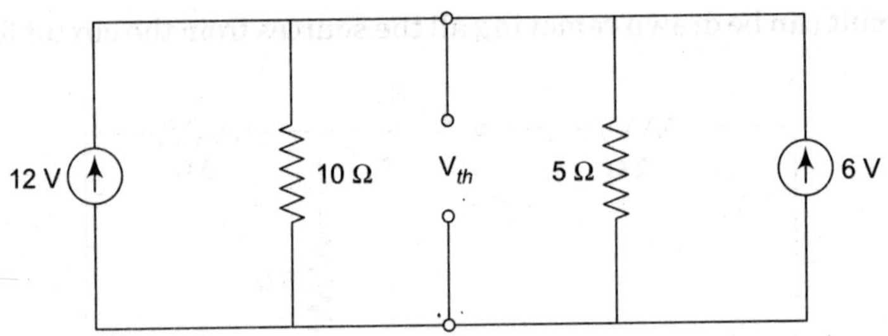

Fig. 2.14(c)

Applying KCL to Fig. 2.14(*c*), we get

$$\frac{V_{th}}{10} + \frac{V_{th}}{5} = 12 + 6$$

or

$$V_{th} = 60 \text{ V}$$

For finding R_{th}, all the sources are removed *i.e.*, current source is open circuited and voltage source if any is short circuited. The Fig. 2.14(*a*) takes the form as shown in Fig. 2.14(*d*).

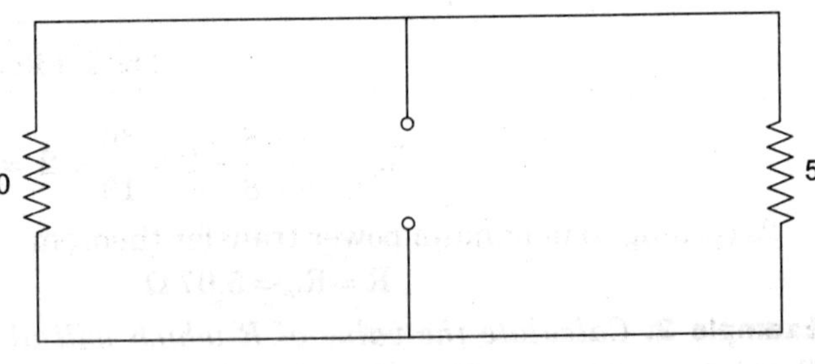

Fig. 2.14(d)

$$R_{th} = 10115 = \frac{10 \times 5}{10 + 5} = \frac{50}{10} = \frac{10}{3} \; \Omega.$$

According to maximum power transfer theorem,

$$I = 60 \div \frac{10}{3} = 9 \text{ A}$$

$$P_{max} = I^2R = (9)^2 \times \frac{10}{3} = 270 \; \Omega$$

Example 3. *Using Millman's theorem, calculate the current through R_L in the circuit of Fig. 2.15(a) and find voltage drop.*

[Given $r_1 = r_2 = r_3 = 2\ \Omega$, $R_L = 5\ \Omega$]

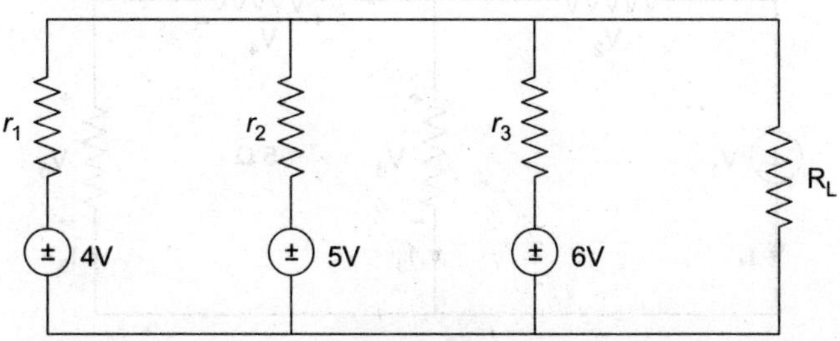

Fig. 2.15(a)

Solution: Assuming V be the equivalent voltage source and R be the equivalent resistance to be connected in series with the voltage source of Millman's equivalent network.

Here,
$$V = \frac{-V_1G_1 - V_2G_2 + V_3G_3}{G_1 + G_2 + G_3}$$

and
$$R = \frac{1}{G_1} = \frac{1}{G_1 + G_2 + G_3} = \frac{1}{\frac{1}{R_1} + \frac{1}{R_2} + \frac{1}{R_3}}$$

Putting the values, we get

$$V = \frac{-4 \times \frac{1}{2} - 5 \times \frac{1}{2} + 6 \times \frac{1}{2}}{\frac{1}{2} + \frac{1}{2} + \frac{1}{2}} = \frac{-2 - 2.5 + 3}{0.5 + 0.5 + 0.5}$$

$$= \frac{-4.5 + 3}{1.5} = -1\ \text{V}$$

and
$$R = \frac{1}{\frac{1}{2} + \frac{1}{2} + \frac{1}{2}} = \frac{1}{1.5}$$

$$= 6.67\ \Omega$$

Millman's equivalent circuit is drawn as below:

From Fig. 2.12(*b*)

6.67 Ω
R_L 5 Ω
− 1V

Fig. 2.15(*b*)

$$I = \frac{6.67}{6.67 + 5} = 5.7\ \text{A}$$

Example 4. *In the network of Fig. 2.15(a), established the validity of Tellegen's theorem provided $V_1 = 8$ V, $V_2 = 4$ V, $V_4 = 2$ V. Also $I_1 = 4$ A, $I_2 = 2$ A and $I_3 = 1$ A.*

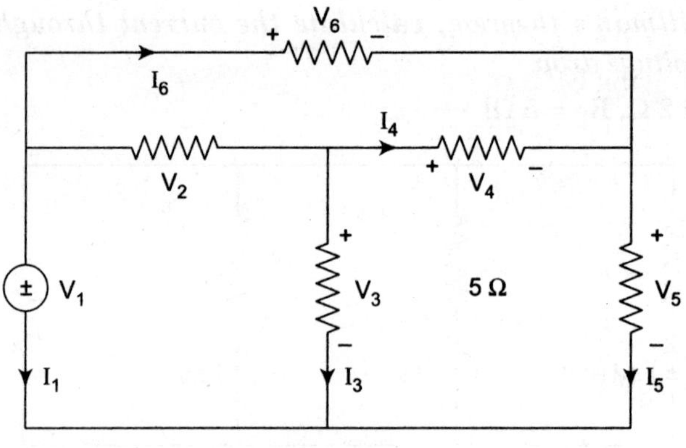

Fig. 2.16(a)

Solution: The primary step is to redraw Fig. 2.16(*a*) and clearly show the loop and nodes as drawn in Fig. 2.16(*b*).

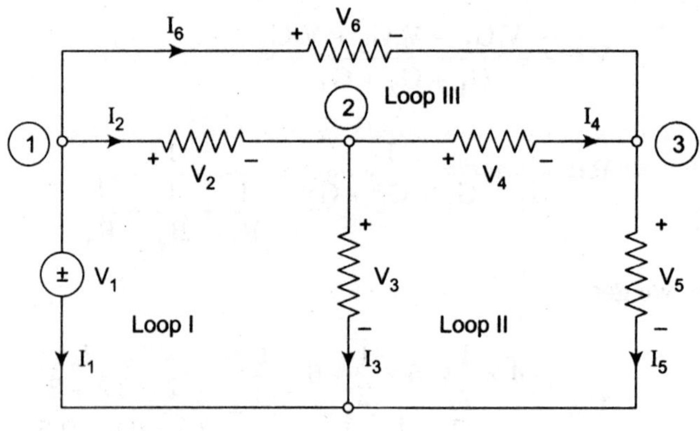

Fig. 2.16(b)

In loop I, applying KVL

$$-V_1 + V_2 + V_3 = 0$$

or $$V_3 = V_1 - V_2 = 8 - 4 = 4\text{ V}$$

Similarly

In loop II, applying KVL gives

$$-V_3 + V_4 + V_5 = 0$$

or $$V_5 = V_3 - V_4$$

$$= 4 - 2 = 2\text{ V}$$

In loop III, applying KVL yields

$$-V_2 + V_6 - V_4 = 0$$

$$V_6 = V_2 + V_4 = 4 + 2 = 6\text{ V}$$

Now, consider node 1 and applying KCL, we get

$$I_1 + I_2 + I_6 = 0$$

or $$I_6 = -I_1 - I_2$$

$$= -4 - 2 = 6\text{ A}$$

Again, at node 2, applying KCL

$$I_2 = I_3 + I_4,\ I_4 = I_2 - I_3$$
$$= 2 - 1 = 1\text{ A}$$

and at node 3,

$$I_4 + I_6 = I_5$$
$$I_5 = 1 + (-6) = -5\text{ A.}$$

According to Tellegen's theorem,

$$\sum_{n=1}^{n} V_n I_n = V_1I_1, + V_2I_2 + V_3I_3 + V_4I_4 + V_5I_5 + V_6I_6$$
$$= 8 \times 4 + 4 \times 2 + 4 \times 1 + 2 \times 1 + 2 \times (-5) + 6 \times (-6)$$
$$= 32 + 8 + 8 + 2 - 10 - 36$$
$$= 0$$

Hence verified.

Example 5. *Convert the delta connection shown in Fig. 2.17 into its equivalent star connection and also determine the current supplied by the battery. Given*

R_a = 15 ohm, R_b = 20 ohm, R_c = 45 ohm, R_d = 18 ohm and R_e = 5 ohm.

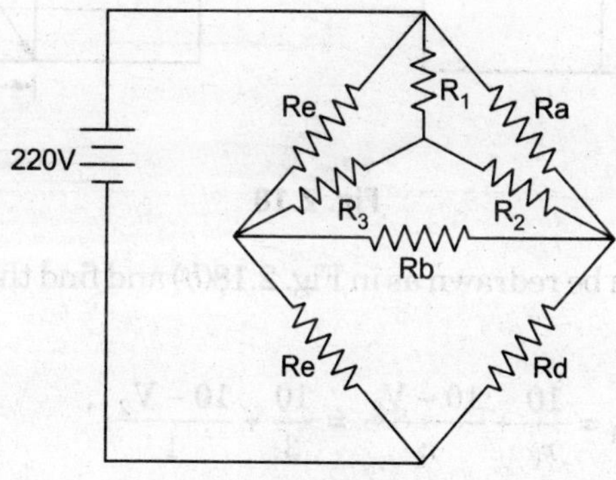

Fig. 2.17

Solution: From delta to star

$$R_1 = \frac{R_a \cdot R_c}{R_a + R_b + R_c} = \frac{15 \times 45}{15 + 20 + 45} = \frac{675}{80} = 8.43\text{ ohm}$$

$$R_2 = \frac{R_a \cdot R_b}{R_a + R_b + R_c} = \frac{15 \times 20}{15 + 20 + 45} = \frac{300}{80} = 3.75\text{ ohm}$$

$$R_3 = \frac{R_b \cdot R_c}{R_a + R_b + R_c} = \frac{20 \times 45}{15 + 20 + 45} = \frac{900}{80} = 11.25\text{ ohm}$$

The equivalent resistance of shown network is

$$R = R_1 + \frac{(R_2 + R_d)(R_3 + R_e)}{R_2 + R_d + R_3 + R_e}$$

$$= 8.43 + \frac{(3.75 + 18)(11.25 + 5)}{3.75 + 18 + 11.25 + 5}$$

$$= 8.45 + \frac{21.75 \times 16.25}{38}$$

$$= 8.45 + 9.30$$

$$R = 17.73 \text{ ohm.}$$

$$I = \frac{220}{17.75} = 12.39 \text{ amperes.}$$

Example 6. ***Node Analysis:*** *Find the current through the resistor r_2 by nodal method in given Fig. 2.18(a).*

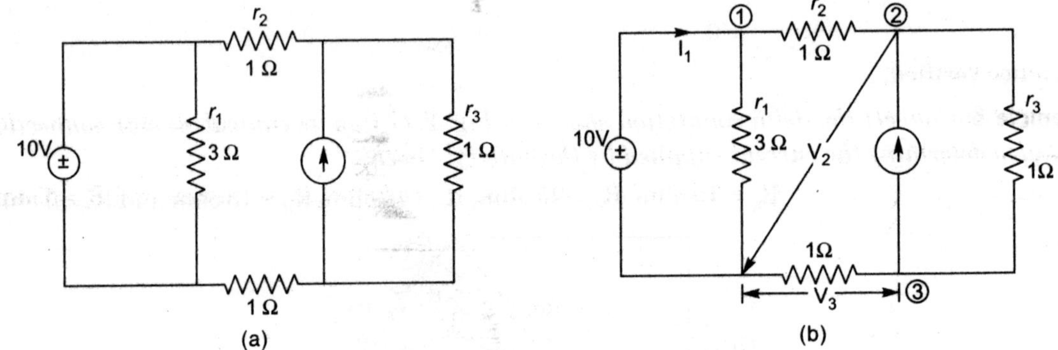

Fig. 2.18

Solution: Figure 2.18(a) can be redrawn as in Fig. 2.18(b) and find the current using node voltage.

At node (1),

$$I_1 = \frac{10}{r_1} + \frac{10 - V_2}{r_2} = \frac{10}{3} + \frac{10 - V_2}{1}$$

$$I_1 = 3.33 + \frac{10 - V_2}{1}$$

$$= 3.33 + 10 - V_2$$

$$V_2 + I_1 = 13.33 \qquad \text{...}(i)$$

At node (2),

$$\frac{V_2 - 10}{r_2} + \frac{V_2 - V_3}{r_3} = 10$$

$$\frac{V_2 - 10}{1} + \frac{V_2 - V_3}{1} = 10$$

$$= V_2 - 10 + V_2 - V_3 = 10$$

$$2V_2 - V_3 = 20 \qquad \text{...}(ii)$$

At node (3),

$$\frac{V_3 - V_2}{r_3} + \frac{V_3}{r_4} = + 10 = 0$$

or
$$V_3 - V_2 + V_3 + 10 = 0$$
$$2\,V_3 - V_2 = -\,10 \qquad ...(iii)$$

From equation (*iii*)

$$V_3 = \frac{V_2 - 10}{2} \qquad ...(iv)$$

Now from equation (*iv*), putting value of V_3 in (*ii*), we get

$$2\,V_2 - \frac{(V_2 - 10)}{2} = 20$$
$$4\,V_2 - V_2 + 10 = 40$$
$$3\,V_2 = 30$$
$$V_2 = 10\,V$$

So, the current through resistor r_2 will be

$$\frac{V_2 - 10}{r_2} = 0$$

Current through r_2 = 0 A.

Example 7. *Calculate the voltage V by mesh method such that the current through 5 V source is zero Fig. 2.19.*

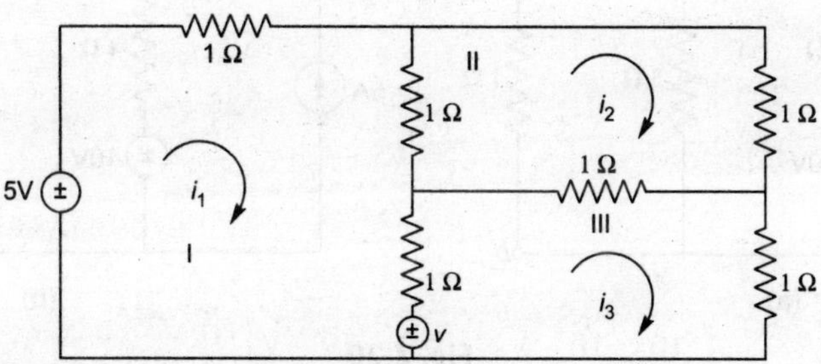

Fig. 2.19

Solution: In loop I, $-\,5 + i_1 \times 1 + (i_1 - i_2)\,1 + (i_1 - i_3)\,1 + V = 0$

or
$$3i_1 - i_2 - i_3 = -\,V + 5 \qquad ...(i)$$

In loop II,

$$i_2 \times 1 + (i_2 - i_3)\,1 + (i_2 - i_1)\,1 = 0$$

or
$$3i_2 - i_1 - i_3 = 0$$
$$-\,i_1 + 3i_2 - i_3 = 0 \qquad ...(ii)$$

In loop III,

$$(i_3 - i_2)\,i + i_3 \times 1 - V + (i_3 - i_1)\,1 = 0$$
$$-\,i_1 - i_2 + 3i_3 = V \qquad ...(iii)$$

According to the equation, $i_1 = 0$

(i_1 being the current through 5 V source)

Therefore, equations (*i*), (*ii*) and (*iii*) becomes

$$-i_2 - i_3 + V = 5 \quad ...(iv)$$

$$3i_2 - i_3 = 0 \quad ...(v)$$

$$-i_2 + 3i_3 = V \quad ...(vi)$$

From equations (*iv*) and (*vi*), we get

$$-4i_3 = 5 - 2V \quad ...(vii)$$

and from equations (*v*) and (*vi*),

$$i_3 = \frac{3}{8} V \quad ...(viii)$$

using equation (*viii*) in (*vii*)

$$-4 \times \left(\frac{3}{8} V\right) = 5 - 2V$$

or

$$-\frac{3}{2} V = 5 - 2V$$

or

$$V = 10 \text{ volts.}$$

Example 8. *In the circuit of Fig. 2.20(a), find the power loss in 1 Ω resistor by Thevenin's theorem.*

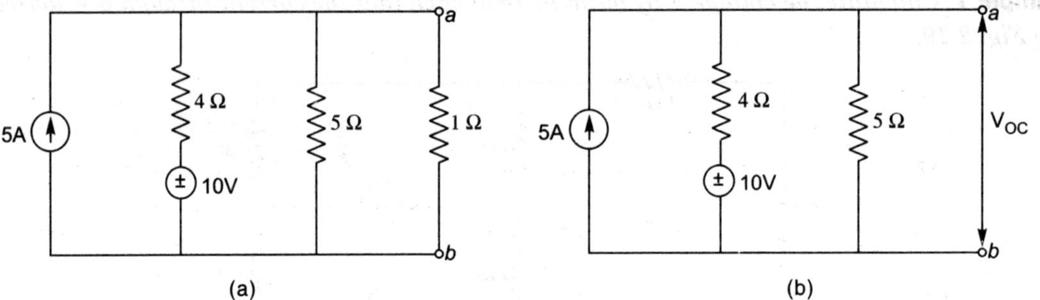

Fig. 2.20

Solution: Let us remove the load resistance (1 Ω) from $a - b$ terminal as shown in Fig. 2.20(*b*). Applying KCL at node a, we get

$$\frac{V_{oc}}{5} + \frac{V_{oc} - 10}{4} = 5$$

$$4 V_{oc} + 5(V_{oc} - 10) = 20 \times 5$$

$$4V_{oc} + 5V_{oc} - 50 = 100$$

$$9V_{oc} = 150$$

$$V_{oc} = \frac{150}{9} = 1.67$$

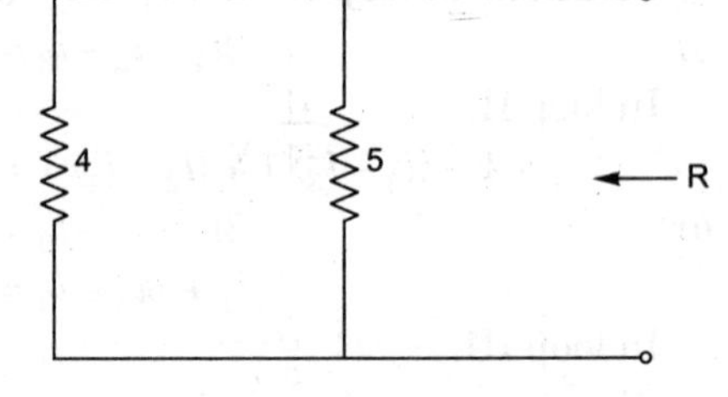

Fig. 2.20(c)

Figure 2.20(*d*) shows the circuit with independent sources deactivated

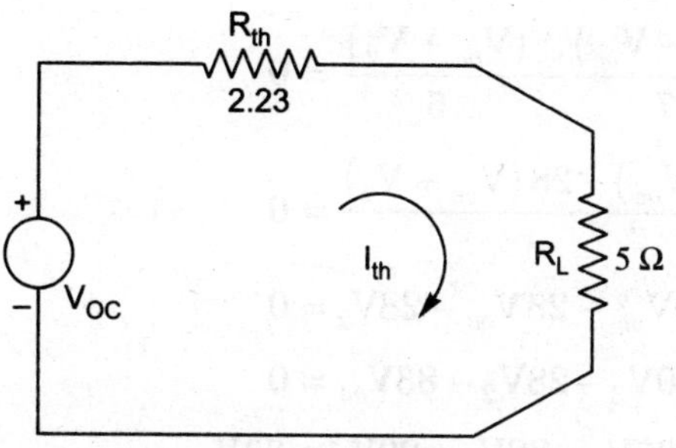

Fig. 2.20(*d*)

$$R_{th} = \frac{4 \times 5}{4 + 5} = \frac{20}{9} = 2.23\ \Omega$$

$$I_{th} = \frac{V_0C}{R_{th} + R_L} = \frac{1.67}{2.23 + 1} = \frac{1.67}{3.23} = 5.17 \text{ amperes.}$$

Power loss in 1 Ω resistor $= (5.17)^2 \times 1$

$= 2.67\ \Omega.$

Example 9. *In the given network as shown below, find the current through the 6 Ω resistor using Thevenin's theorem.*

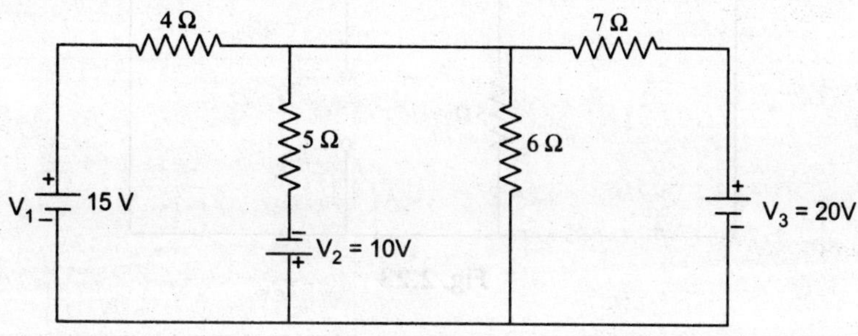

Fig. 2.21

Solution: Redrawing above circuit and removing the resistance of 6 Ω as open circuit, we get

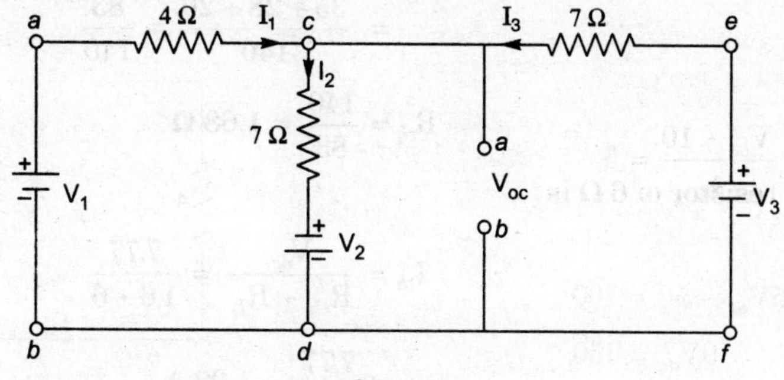

Fig. 2.22

Applying KCL at node C, we get

$$I_1 + I_3 - I_2 = 0$$

$$\frac{(V_1 - V_{oc})}{4} + \frac{(V_3 - V_{oc})}{7} - \frac{(V_{oc} + V_2)}{5} = 0$$

$$\frac{35(V_1 - V_{oc}) + 20(V_3 - V_{oc}) - 28(V_{oc} + V_2)}{140} = 0$$

$$35V_1 - 35V_{oc} + 20V_3 - 20V_{oc} - 28V_{oc} - 28V_2 = 0$$

$$\Rightarrow \quad 35V_1 + 20V_3 - 28V_2 - 83V_{oc} = 0$$

$$\Rightarrow \quad 35V_1 - 28V_2 + 20V_3 = 83V_{oc} \qquad \text{... (1)}$$

Putting value of V_1, V_2 and V_3 in (1)

$$35 \times 15 - 28 \times 10 + 20 \times 20 = 83V_{oc}$$

$$525 - 280 + 400 = 83V_{oc}$$

$$83V_{oc} = 645$$

$$V_{oc} = \frac{645}{83} = 7.77 \text{ volts.}$$

Now, for calculating R_{th} we open voltage source as shown in Fig. 2.23.

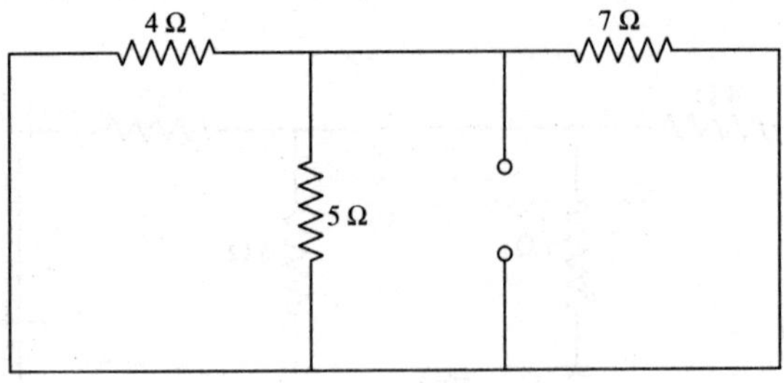

Fig. 2.23

Here,

$$\frac{1}{R_{th}} = \frac{1}{4} + \frac{1}{5} + \frac{1}{7}$$

$$= \frac{35 + 28 + 20}{140} = \frac{83}{140}$$

$$R_{th} = \frac{140}{83} = 1.68\ \Omega$$

Current through resistor of 6 Ω is

$$I_{th} = \frac{V_{oc}}{R_{th} + R_L} = \frac{7.77}{1.6 + 6}$$

$$= \frac{7.77}{7.6} = 1.02 \text{ A.}$$

Example 10. *Using superposition theorem, find the currents in the different branches of network shown in Fig. 2.24:*

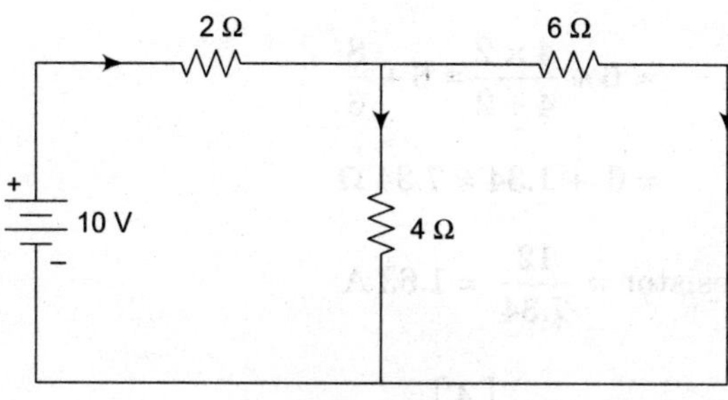

Fig. 2.24

Solution: In superposition theorem, the circuit above is drawn with one voltage source while other is open circuited.

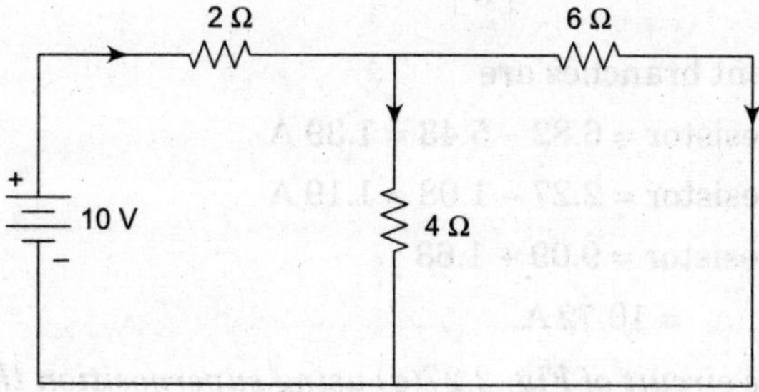

Fig. 2.25

Considering V_1 source only, the total resistance is circuit (2)

$$= 2 + \frac{4 \times 6}{4 + 6} = 2 + \frac{24}{10} = 4.4\ \Omega$$

Current through 2 Ω resistor $= \frac{10}{4.4} = 2.27$ A

Current through 6 Ω resistor $= 2.27 \times \frac{4}{4+6}$

$$= 2.27 \times 0.4 = 9.09 \text{ A}$$

Current through 4 Ω resistor

$$= 9.09 - 2.27 = 6.82 \text{ A.}$$

Now, considering voltage source V_2 and open circuiting V_1 the circuit becomes

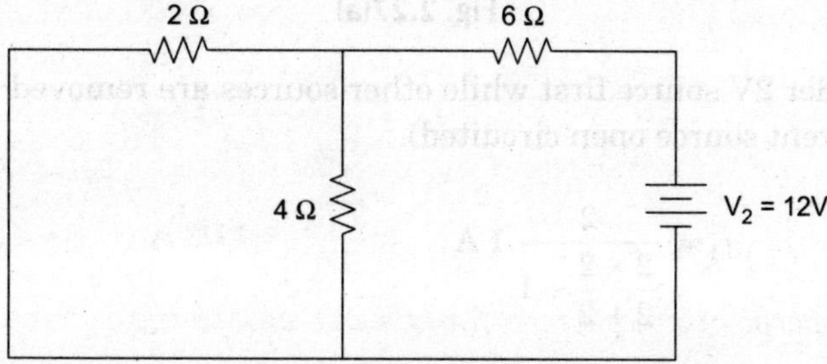

Fig. 2.26

Total resistance $= 6 + \frac{4 \times 2}{4 + 2} = 6 + \frac{8}{6}$

$= 6 + 1.34 = 7.34\ \Omega$

Current through 6 Ω resistor $= \frac{12}{7.34} = 1.63$ A

Current through 2 Ω resistor $= 1.63 \left[\frac{4}{6}\right] = 1.08$ A

Current through 4 Ω resistor $= 1.63 \left[\frac{2}{6}\right] = 5.43$ A

Total current in different branches are

Current through 4 Ω resistor = 6.82 – 5.43 = 1.39 A

Current through 2 Ω resistor = 2.27 – 1.08 = 1.19 A

Current through 6 Ω resistor = 9.09 + 1.63

= 10.72 A.

Example 11. *Find V in the circuit of Fig. 2.27(a) using superposition theorem.*

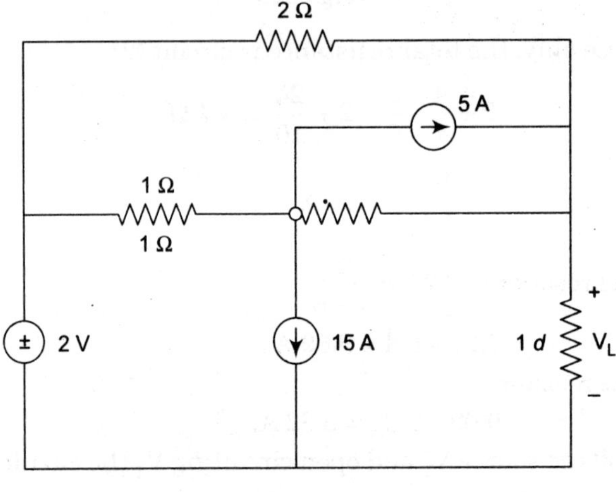

Fig. 2.27(a)

Solution: We consider 2V source first while other sources are removed (voltage source short circuited and the current source open circuited).

$$i_1 = \frac{2}{\frac{2 \times 2}{2 + 2} + 1} 1 \text{ A}$$

$\therefore$ V_1 (drop across r_L due to 2V source) $= 1 \times 1 = 1$ V.

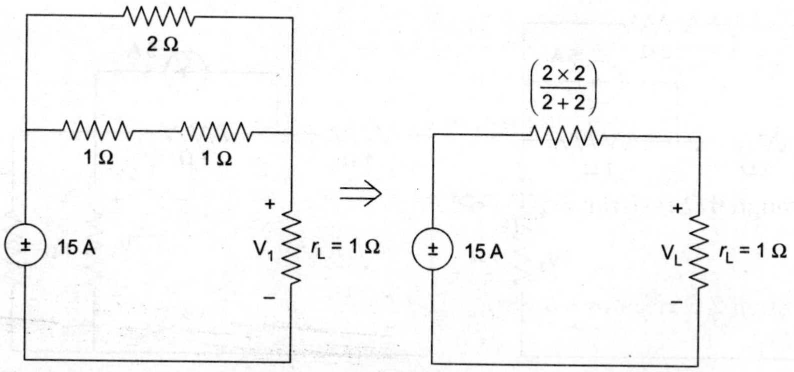

Fig. 2.27(*b*)

Taking the lower current source only

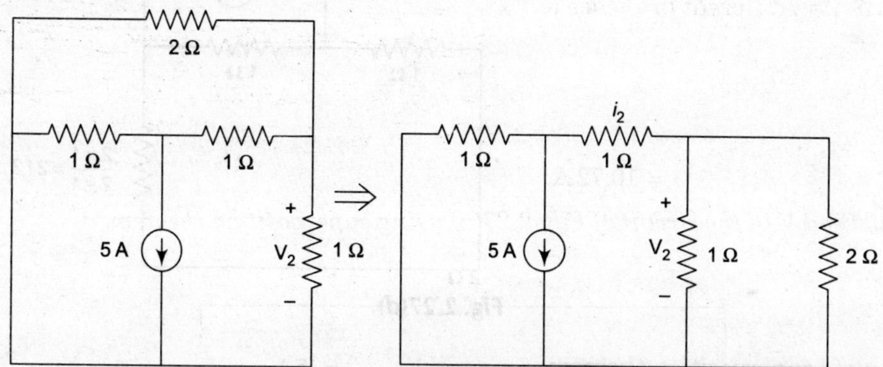

Fig. 2.27(c)

$$i_2 = (-5)\frac{1}{1+1+\dfrac{2}{3}} = (-5)\,\frac{3}{8} = -\frac{15}{8}\text{ A}$$

$$\therefore \qquad i_3 = -\left(\frac{15}{8}\right)\frac{2}{1+2} = -\left(\frac{15}{8}\right)\frac{2}{3} = -\frac{5}{4}\text{ A}$$

The results V_2 as $\qquad V_2 = -\frac{5}{4}\times 1 = -\frac{5}{4}\text{ V}$

In drawn Fig. 2.27(*d*).

$$\frac{2\times 1}{2+1} = \frac{2}{3}$$

$$i_4 = 5\times\frac{1}{\dfrac{3}{2}+1} = \frac{15}{5} = 3\text{ A}$$

This gives $\qquad i_{rL} = 3\times\frac{2}{2+1} = 2\text{ A}$

$$\therefore \qquad V_3 = 2\times 1 = 2\text{ V}$$

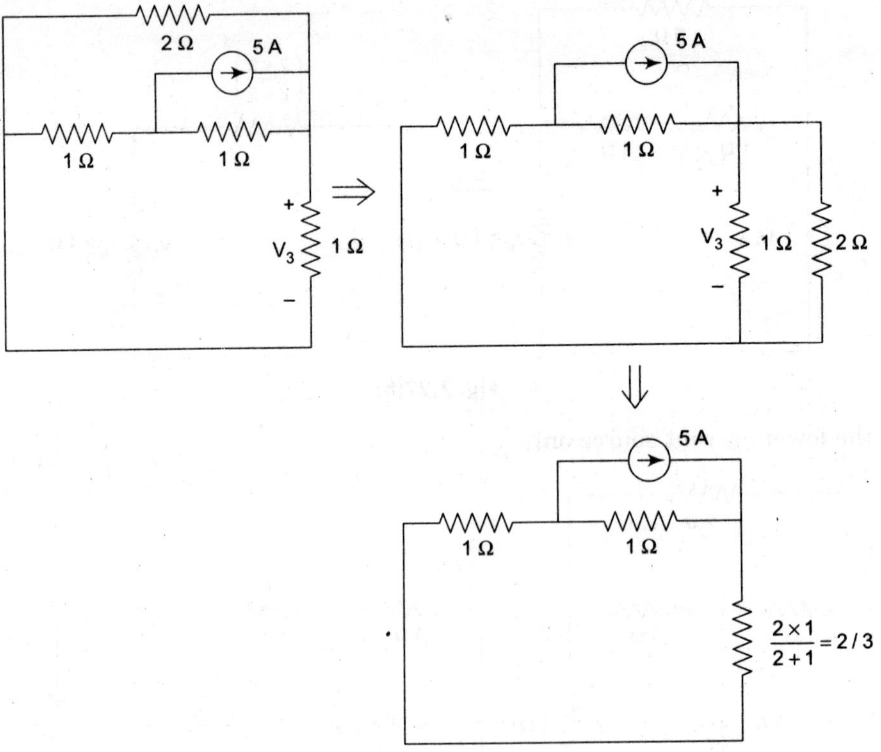

Fig. 2.27(*d*)

According to superposition theorem

$$V = V_1 + V_2 + V_3$$

$$= 1 + \left(-\frac{5}{4}\right) + 2 = \frac{7}{4}\ \text{V} = 1.75\ \text{V}.$$

Example 12. *Find Norton's equivalent circuit to the left of terminals a – b in the network shown in Fig. 2.28(a).*

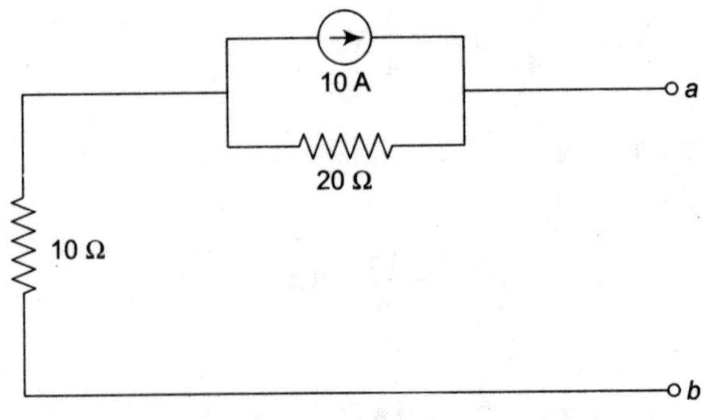

Fig. 2.28(a)

Solution: Let us first short the terminate $a - b$ of Fig. 2.28(a)

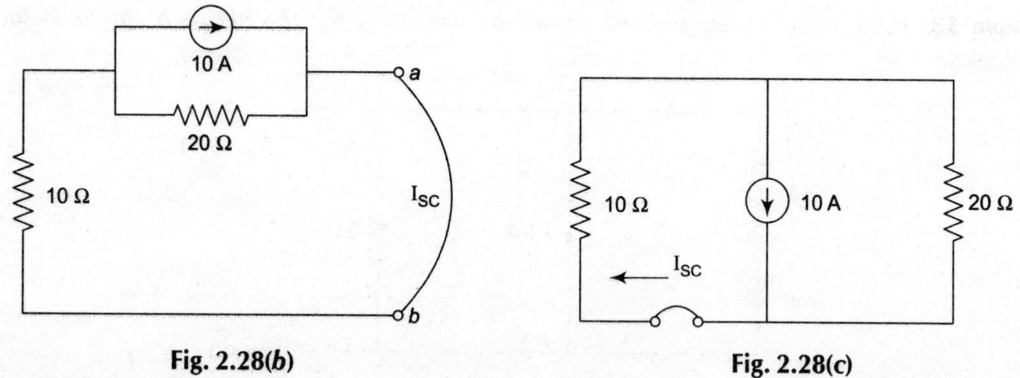

Fig. 2.28(b) **Fig. 2.28(c)**

Here I_{SC} is the short circuit current through 10 Ω resistor

$$\therefore \qquad I_{SC} = 10 \times \frac{20}{10 + 20} = 10 \times \frac{20}{30} = 6.67 \text{ A}$$

The equivalent resistance of circuit in Fig. 2.28(a) will be obtained by looking through $a - b$ and the constant source is removed as shown below:

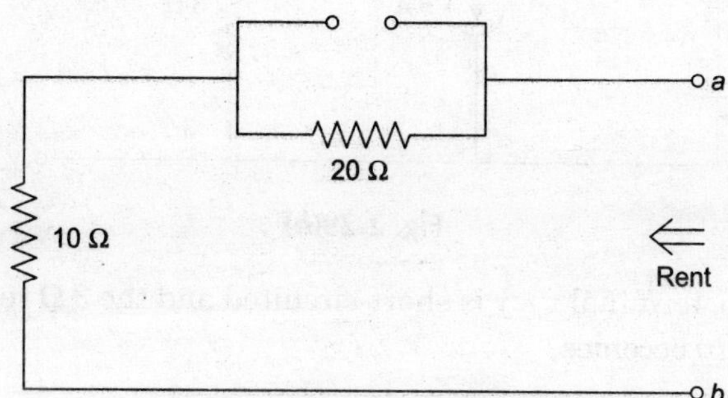

Fig. 2.28(d)

here,
$$R_{int} = 10 + 20$$
$$= 30\ \Omega$$

Norton's equivalent circuit is drawn as below:

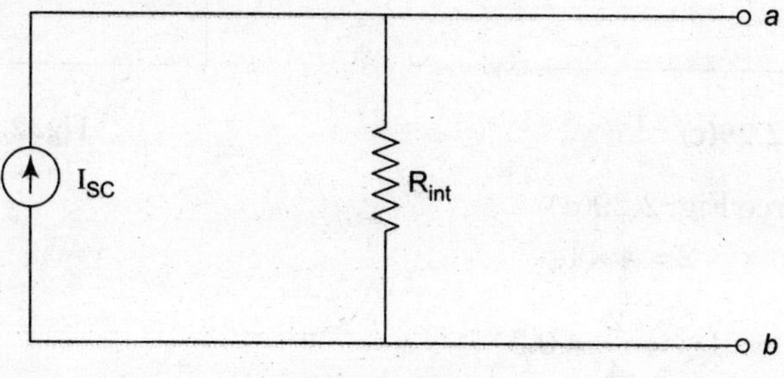

Fig. 2.28(e)

Therefore, $I_N = I_{SC} = 6.67$ A

Example 13. *Find Norton's equivalent circuit across x – y for the network shown below in Fig. 2.29(a).*

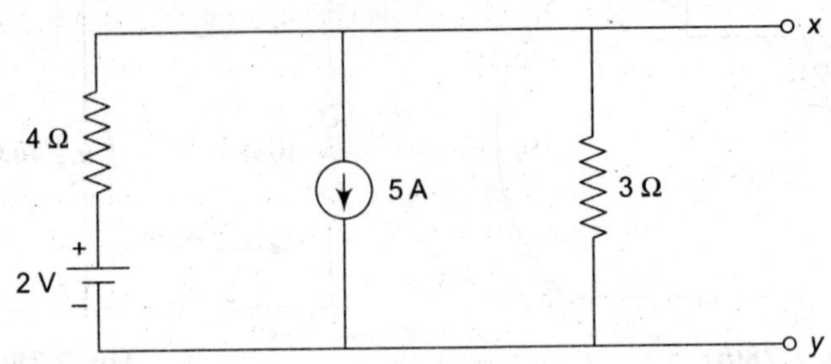

Fig. 2.29(a)

Simplifying Fig. 2.29(*a*)

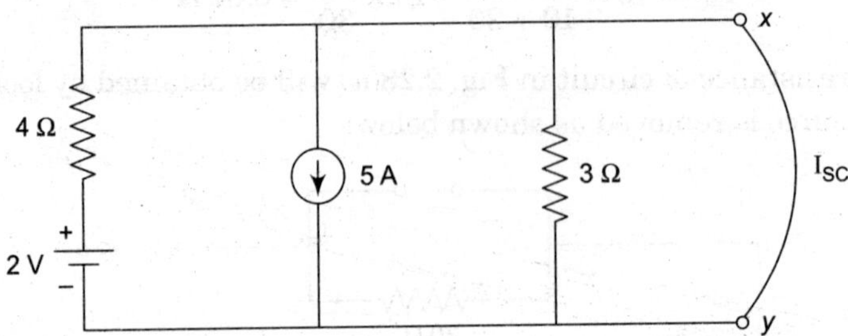

Fig. 2.29(*b*)

Solution: At first step, terminal $x - y$ is short-circuited and the 3 Ω resistor is then by passed and circuit in Fig. 2.29(*b*) becomes,

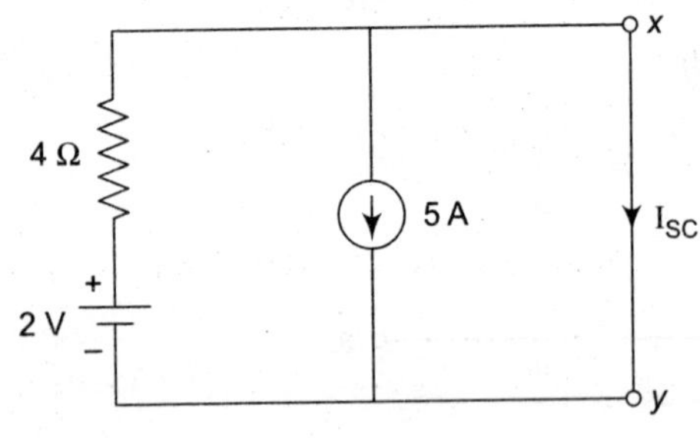

Fig. 2.29(c)

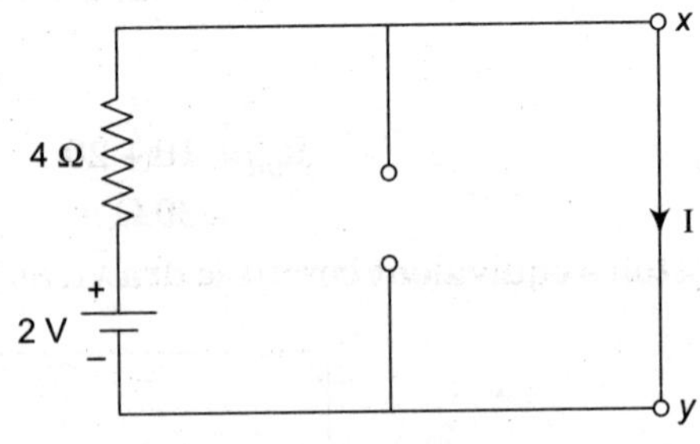

Fig. 2.29(*d*)

Considering 2V source Fig. 2.29(*d*)

$$2 = 4 \times I_{SC}$$

$$\therefore \qquad I_{SC} = \frac{2}{4} = 0.5$$

Now, considering 5A source only as shown in Fig. 2.29(*e*)

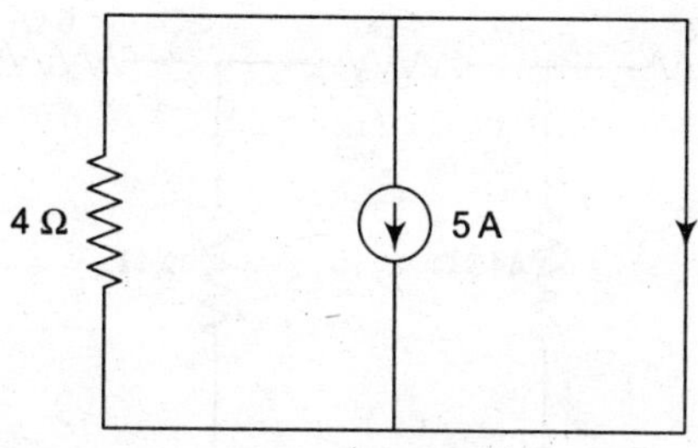

Fig. 2.29(e)

$I''_{SC} = -5A$

Using superposition theorem,

$$I_{SC} = I'_{SC} + I''_{SC} = 0.5 - 5 = -4.5 \text{ A}$$

For obtaining equivalent resistance R_{int}, all the sources in the given network are removed shown in Fig. 2.29(*f*),

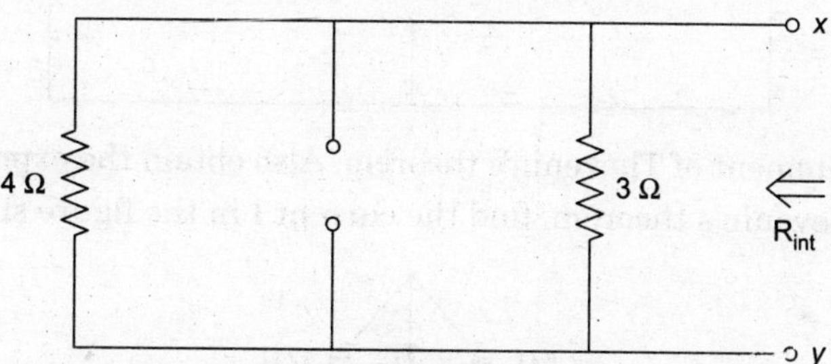

Fig. 2.29(f)

So, $$R_{int} = \frac{4 \times 3}{4 + 3} = \frac{12}{7} 1.72\ \Omega$$

Norton's equivalent circuit can be drawn as Fig. 2.29(*g*).

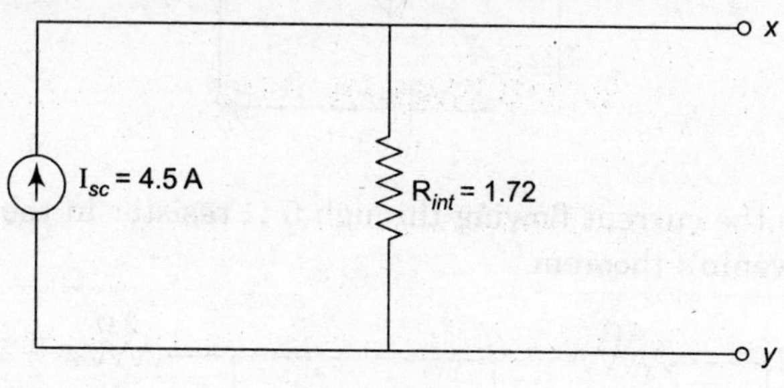

Fig. 2.29(g)

EXERCISES

1. State and prove superposition theorem.

   (*a*) In the circuit shown below. Calculate current I using superposition theorem.

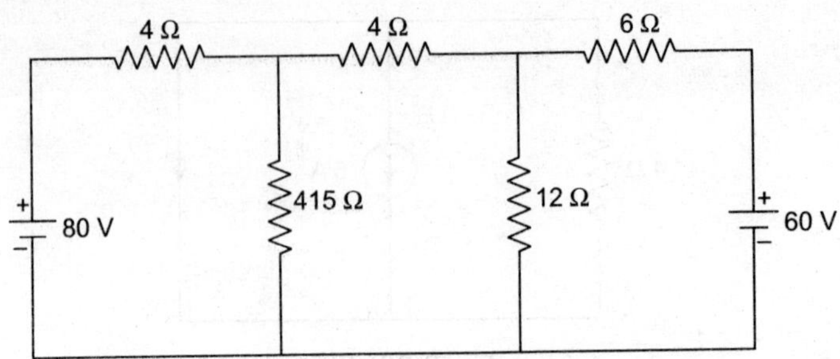

(*b*) Calculate current I in the given circuit using superposition theorem.

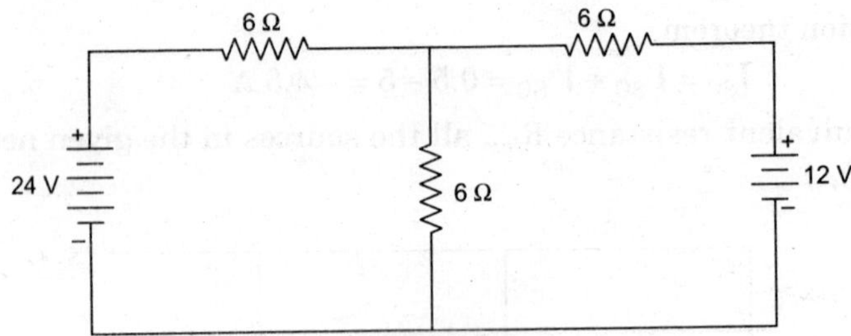

2. Explain the statement of Thevenin's theorem. Also obtain the expression for I_{th}.

(*a*) Using Thevenin's theorem, find the current I in the figure shown below :

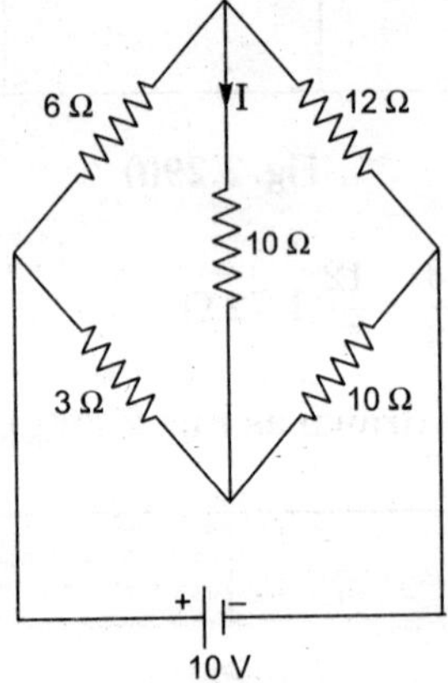

(*b*) Determine the current flowing through 5 Ω resistor in the network shown below using Thevenin's theorem.

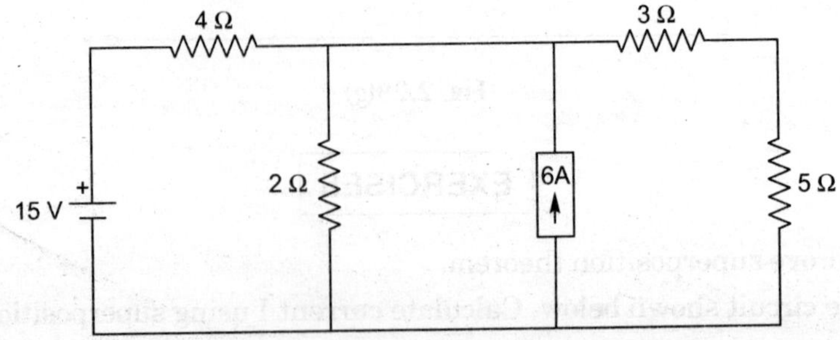

3. (*i*) **Consider figure shown below, which shows me node of an electric circuit. Find V_4 using KCL, the given parameters are $V_2 = 5e^{-2t}$, $V_3 = 2e^{-2t}$, $i_1 = 2e^{-2t}$.**

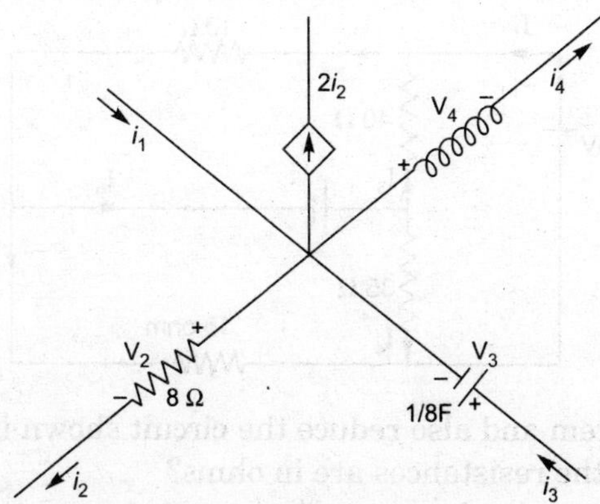

(*ii*) **The voltage and current through a circuit element are:**

$$V = 100 \sin (314\,t + 45°) \text{ volt}$$

$$i = 10 \sin (314\,t + 315°) \text{ amperes.}$$

(*a*) **Identify the circuit element.**

(*b*) **Find its value.**

(*c*) **Obtain expression for power.**

4. Verify maximum power transfer theorem for d.c. network with its statement. Find the value of R in the given circuit such that the maximum power transfer takes place.

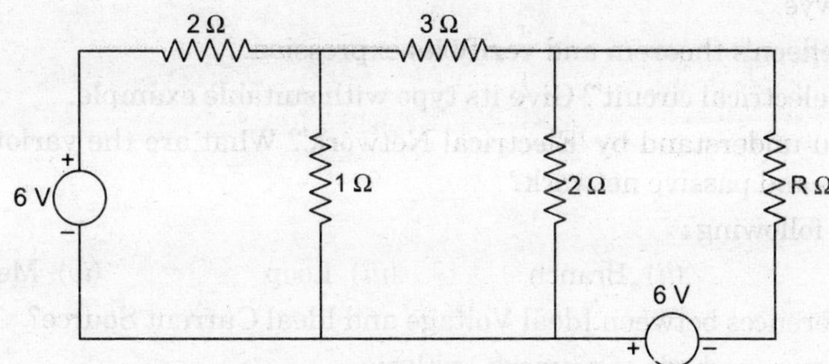

5. Using Wye-Delta transformation to obtain the resistance betwen the terminal a-b of the circuit shown below. All the resistances are in ohms.

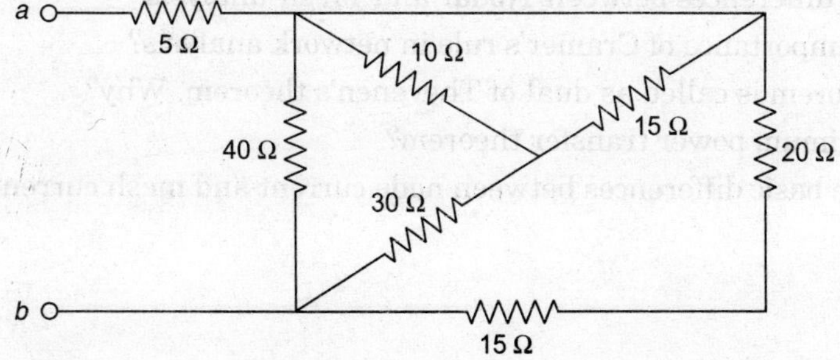

6. Derive Norton's equivalent circuit and find the current through 15 ohm resistance shown in figure. Also state Norton's theorem.

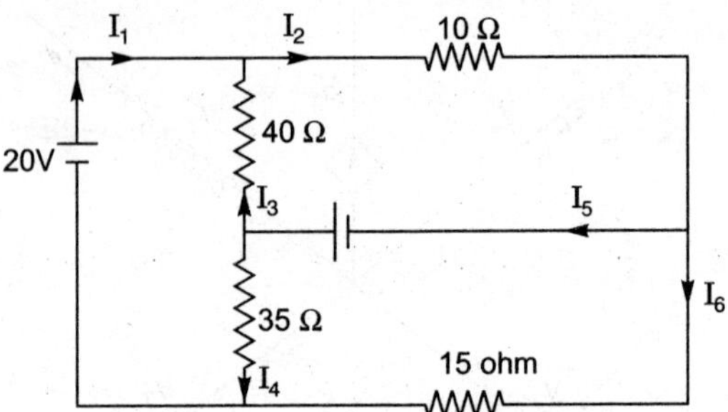

7. State Millman's theorem and also reduce the circuit shown in figure to a single current generator. Where all the resistances are in ohms?

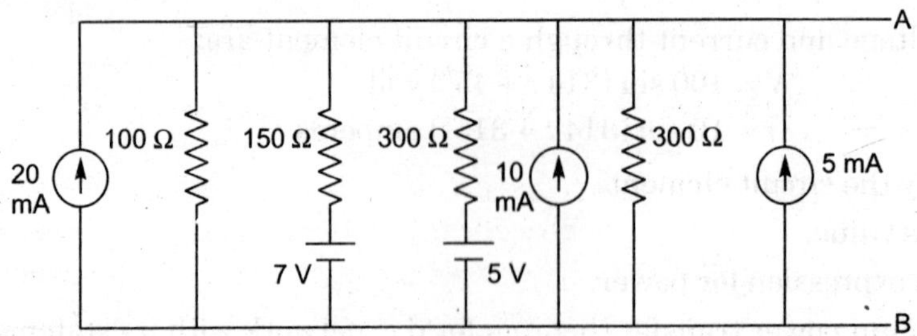

8. A delta network has the resistance values of 10 Ω, 20 Ω and 30 Ω. Find the parameters of equivalent wye.

9. State the Telleen's theorem and verify its expression.

10. What is an 'electrical circuit'? Give its type with suitable example.

11. What do you understand by 'Electrical Network'? What are the various differences between active and passive network?

12. Explain the following :

(*i*) Node (*ii*) Branch (*iii*) Loop (*iv*) Mesh

13. What is differences between Ideal Voltage and Ideal Current Source?

14. Explain the two methods of network analysis.

15. Explain Kirchhoff's Current Law and verify that Incoming Current = Outgoing Current.

16. What are the two laws of Kirchhoff?

17. Give various differences between 'Nodal' and 'Mesh' analysis.

18. What is the importance of Cramer's rule in network analysis?

19. Norton's theorem is called as dual of Thevenen's theorem. Why?

20. What is maximum power transfer theorem?

21. What are the basic differences between node current and mesh current?

3

Basic Concepts of Alternating Current and Voltage

3.1 INTRODUCTION

In an electric circuit, direct current flows continuously in one direction only. If the applied voltage and circuit parameters are kept constant, the magnitude of current flowing through this circuit remains constant over time. While for an alternating voltage or current its magnitude and polarity with respect to time varies periodically. Therefore an alternating quantity either current or e.m.f. is one, which periodically passes through a definite cycle, each consisting of two half cycles during one of which the current or e.m.f. around the circuit varies in one direction and during the other cycle it is in the opposite direction.

3.2 GENERATION OF ALTERNATING VOLTAGE

When a conductor is rotated in a uniform magnetic field with constant angular velocity then alternating e.m.f. will generate. The e.m.f. generated will depend upon the strength of the magnetic field, number of turns in the coil and the speed at which the coil or magnetic field rotates.

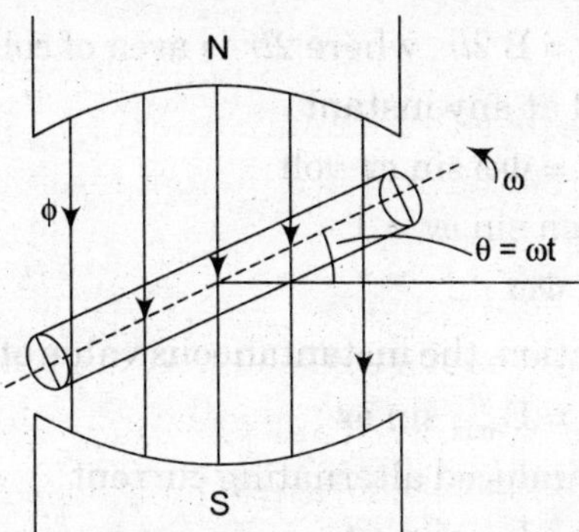

Fig. 3.1 *Generation of alternating voltage.*

Let, B = Flux density of magnetic field (tesla)

l = Length of the coil (m)
b = Breadth of coil (m)
ω = Angular velocity (rads)
v = Linear velocity (m/s)
A = Area of the coil (m^2)
r = Radius of path (m)
f = Frequency of rotation of coil (Hz).

Considering a rectangular coil rotating with constant angular velocity of ω; radian per second in a uniform magnetic field, its axis of revolution being perpendicular to the magnetic lines of force. At any instant the coil rotates through an angle θ.

It can be written as

$$\text{Angular velocity} = \frac{\text{Angular distance}}{\text{Time}}$$

$$\omega = \frac{\theta}{t}$$

or, $$\theta = \omega t, \text{ radian}$$

Linear velocity of coils sides $v = \omega r, \frac{m}{s}$

Linear velocity component perpendicular to magnetic field

$$= v \sin \theta \approx v \sin \omega t$$

Flux cut per second by coil $Blv \sin \omega t$

Thus e.m.f. generated in the coil side at time t

$$= Blv \sin \omega t$$

l

2r

area = 2lr

e.m.f. generated in the coil at time t

$$= 2Blv \sin \omega t$$

$$= 2Bl\omega r \sin \omega t$$

Maximum flux linking the coil

$$\Phi = B.2lr, \text{ where } 2lr \text{ is area of coil}$$

Thus e.m.f. generated at coil at any instant

$$l = \Phi\omega \sin \omega t \text{ volt}$$

e.m.f. will be maximum, when $\sin \omega t = 1$

So $$E_{max} \approx \Phi\omega$$

Substituting it in above equation, the instantaneous value of e.m.f. generated at any time t is

$$e = E_{max} \sin \omega t$$

Similarly the expression for induced alternating current

$$i = I_{max} \sin \omega t$$

$$\omega = 2\pi f$$

So, instantaneous e.m.f. generated is given as

$$e = E_{max} \sin (2\pi ft)$$

$$= E_{max} \sin (2\pi/T)t$$

Instantaneous value of induced current

$$I = I_{max} \sin (2\pi ft)$$

$$= I_{max} \sin \left(\frac{2\pi}{T}\right)t$$

where T is called the periodic time *i.e.* the time taken to complete one cycle

i.e. $$f = \frac{1}{T}$$

If the e.m.f. or current value as given by equations, from instant to instant are plotted along y-axis and time along x-axis, the graph will be as shown in Fig. 3.2.

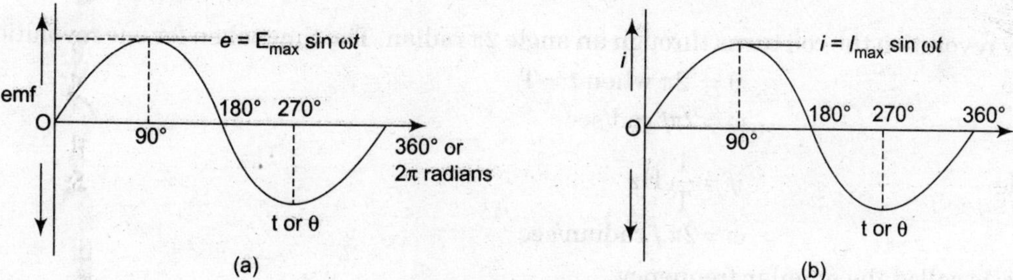

Fig. 3.2 *Plot of alternating voltage and current.*

Such a graph of e.m.f. or current is called a sinusoidal alternating voltage or current. The trace OPQRS of the graph completes one cycle and consists of two alternations, one positive and other negative. In practice horizontal axis is marked out in radians or degrees instead of seconds.

3.3 CYCLE

One complete alternation of a set of values of e.m.f. or current is called a cycle. Each complete cycle consists of the positive half-cycle above the horizontal axis and the negative half-cycle below the horizontal axis. Hence in the positive half-cycle all the values of the current are positive, while in the negative half-cycle all the values of the current are negative.

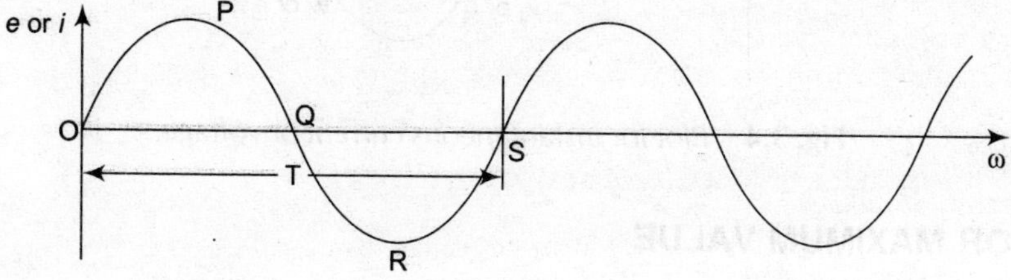

Fig. 3.3 *Wave cycle.*

3.4 FREQUENCY

Number of cycle completed in one second is called the frequency. It is denoted by f and is expressed in hertz (Hz) or cycle per second (1 Hz =1 c/s). Standard frequency generation adopted 50 Hz or 60 Hz (in India 50 Hz).

Period

The time required to complete one cycle its called periodic time or period. It is devoted by T and is expressed in second. The relationship between frequency and periodic time is given by

$$T = \frac{1}{f} \text{ sec}$$

3.5 ANGULAR FREQUENCY

In every revolution the coil turns through an angle 2π radian. The time taken for one revolutions.

Since, $\theta = 2\pi$ when $t = T$

$$\omega = 2\pi f \text{ rad/sec}$$

While $$f = \frac{1}{T} \text{ Hz}$$

$$\omega = 2\pi . f \text{ radian/sec}$$

where ω is called the angular frequency.

3.6 INSTANTANEOUS VALUE

The value of the alternating quantity at particular instant (time) is called the instantaneous value. It varies from time to time. Instantaneous values are denoted by small (lower case) letters. Thus i denotes the instantaneous current as shown in Fig. 3.4, and v denotes the instantaneous voltage.

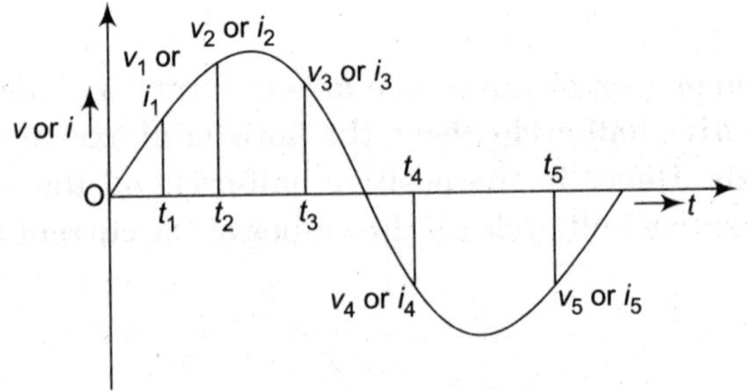

Fig. 3.4 *Plot for instantaneous current or voltage.*

3.7 PEAK OR MAXIMUM VALUE

The maximum value of alternating quantity attained in a cycle or highest instantaneous value is called peak or maximum, crest value or the amplitude. It is denoted by a capital letter with subscript m. So I_m or V_m denotes the maximum value of current or voltage, as shown in Fig. 3.5.

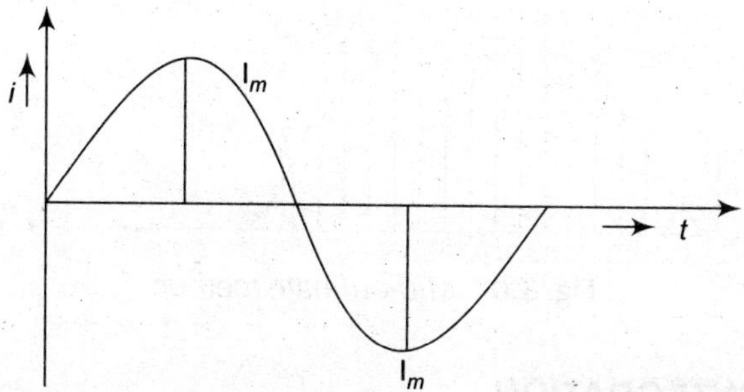

Fig. 3.5 *Plot for maximum value of current.*

Average or Mean Value

The average is mean value the current or voltage of an alternative quantity over a given interval is sum of all instantaneous values divided by the number of values taken over that interval. The average value (arithmetic mean) over the complete cycle will be zero for sinusoidal as well as non-sinusoidal wave; provided the wave shape is symmetrical.

3.8 DETERMINATION OF AVERAGE OR MEAN VALUE

Basically average value of curve is the average height of the curve. The average height of the curve is found by dividing the area under the curve by the length of interval of the curve over which the curve extends.

$$\text{Average value} = \frac{\text{Area under the curve}}{\text{Length of the interval of the curve}}$$

The average value can be determined by:

1. The mid-ordinate method
2. The method of integration.

3.9 MID-ORDINATE METHOD

In this method the area under the curve is divided into n number of strips of equal width by vertical lines. At the centre of each strip a vertical line in drawn up to the curve such vertical lines are called the mid-ordinates of the strips. At the equal intervals mid-ordinates are shown by full lines. Let the height of each be $i_1, i_2, i_3, \ldots, i_n$. Each mid-ordinates is taken as the average height of its strip as shown in Fig. 3.6. Hence, the average height of the curve is determined by taking the average value of these n mid-ordinates. Number of the strips must be as more as possible for better accuracy,

$$\text{Average height of the curve} = \frac{(i_1 + i_2 + i_3 + \ldots + i_n)}{n}$$

This method is convenient for non-sinusoidal waves.

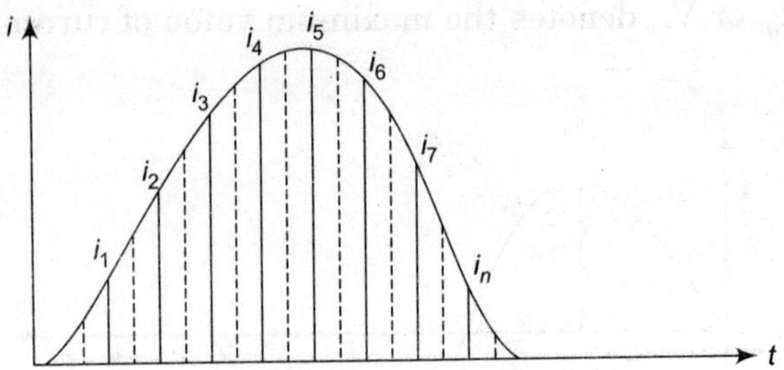

Fig. 3.6 Mid-ordinate method.

3.10 METHOD OF INTEGRATION

In this method an infinite number of mid-ordinates are used. For a given curve a shaded strip of width $d\theta$ is considered as shown in Fig. 3.7(*a*).

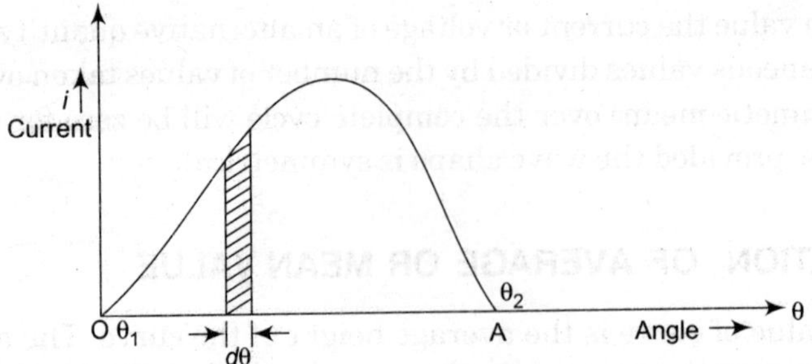

Fig. 3.7 (a) Method of integration.

The area of strip = $id\theta$. The area under the curve and the horizontal axis will be the sum of all such strips. The curve has the limits from θ_1 to θ_2 as shown in Fig. 3.7(*b*). The length of the base of the curve is therefore,

$$OA = (\theta_2 - \theta_1)$$

$$\text{Average length of the curve} = \frac{\text{Area under the curve}}{\text{Base of the curve}}$$

$$= \frac{1}{(\theta_2 - \theta_1)} \int i d\theta$$

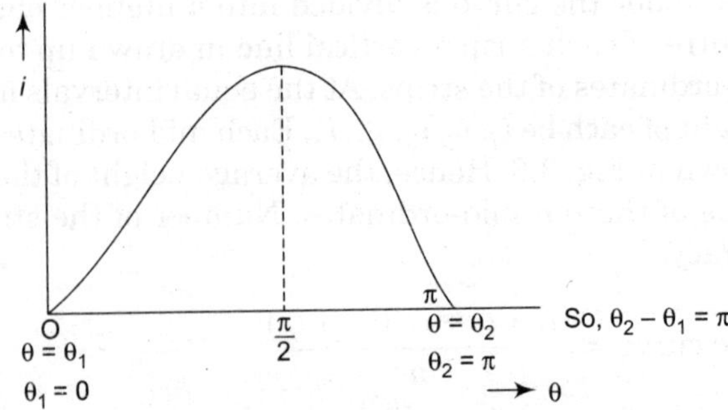

Fig. 3.7 (b) Method of integration.

3.11 AVERAGE VALUE OF SINE WAVE

Sine wave may be expressed as

$$i = I_{max} \sin \theta$$

$$\text{Average value over one half cycle} = \frac{1}{\pi} \int_0^{\pi} I_{max} \sin \theta . d\theta$$

$$= \frac{I_{max}}{\pi} [-\cos \theta]_0^{\pi}$$

$$= \frac{2\, I_{max}}{\pi} = \frac{I_{max}}{\frac{\pi}{2}} = 0.637\, I_{max}$$

Thus the average value of current for a sine wave

$$= 0.637 \times \text{maximum value of current.}$$

Root Mean Square or Effective Value: The alternating voltage or current varies from instant to instant. It becomes maximum at a certain instead only. In specifying such a varying quantity, its maximum or peak value is rarely used. Therefore, it is necessary to determine an equivalent direct current, which will produce the same amount of work or heat in the same time of interval in the same the resister. Usually the values of alternating voltage or current are taken as rms or effective values; unless specified. Thus, effective value is the d.c. equivalent of a.c.

Determination of r.m.s. Value by Mid-ordinate Method

Consider the positive half-cycle of a non-sinusoidal alternating current wave shape in Fig. 3.8. Let this cycle of periodic time t be divided into n strips, each of width $\frac{t}{n}$, with mid-ordinate i_1, i_2, i_3, ..., i_n. Suppose that this current flows through a resistor R. Consider each mid-ordinate to be the mean value of the current over the interval of time $\frac{t}{n}$ represented by its own strip.

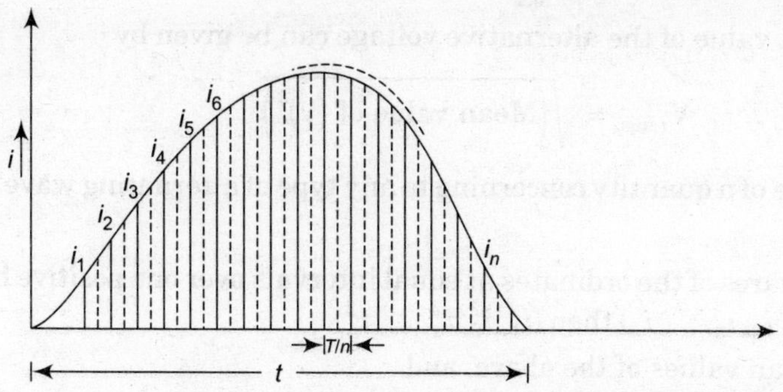

Fig. 3.8 *RMS value of AC wave 1.*

Then work done or heat produced in different interval will be

By $\quad i_1 \quad W_1 \text{ or } H_1 = i_1^2 \frac{R \cdot t}{n}$

By $\qquad i_2 \quad W_2$ or $H_2 = i_2^2 \dfrac{R \cdot t}{n}$

By $\qquad i_n \quad W_n$ or $H_n = i_n^2 \dfrac{R \cdot t}{n}$

Thus total work done or heat produced in t seconds on applying alternating current wave to a resistance R.

$$H_{ac} \text{ or } W_{ac} = \left[\frac{Rt \cdot (i_1 + i_2 + i_3 + \ldots + i_n)}{n}\right]; \text{joules}$$

Heat produced by the equivalent direct current I in resistor R in time t seconds is given by

$$H_{dc} \text{ or } W_{dc} = I^2Rt$$

If $\qquad H_{dc} = H_{ac}$

Then,

$$I^2Rt = \frac{(i_1^2 + i_2^2 + i_3^2 + \ldots + i_n^2) R \cdot t}{n}$$

$$I^2 = \frac{(i_1^2 + i_2^2 + i_3^2 + \ldots + i_n^2)}{n}$$

$$I = \sqrt{\left[\frac{(i_1^2 + i_2^2 + i_3^2 + \ldots + i_n^2)}{n}\right]}$$

= Square root of the mean of the squares of the current
= Root-mean-square (r.m.s.) value of the current
= $I_{r.m.s.}$

Similarly, r.m.s. value of the alternative voltage can be given by

$$V_{r.m.s.} = \sqrt{(\text{Mean value of } (v)^2)}$$

Hence rms value of a quantity concerning to any type of alternating wave shape can be found as following:

1. Take the squares of the ordinates at equal intervals over one positive half-cycle. Let these values be $(i_1, i_2, i_3, \ldots, i_n)$ then $(i_1^2, i_2^2, i_3^2, \ldots, i_n^2)$.
2. Take the mean values of the above, and
3. Take the square root of the obtained mean value.

Determination of r.m.s. Value by Method of Integration

Work done or heat produced by an alternating of instantaneous value i in resistor R in time dt is i^2Rdt. Therefore total heat produced in one cycle is given by

$$W_{ac} \text{ or } H_{ac} = \int_0^t i^2 R dt$$

Heat produced is equivalent to direct current I in resistance R in time t is given by:

$$H_{dc} = I^2 Rt$$

If, $\quad H_{dc} = H_{ac}$, then

$$I^2 Rt = \int_0^t i^2 R dt$$

or,
$$I^2 = \frac{1}{t}\int_0^t i^2 dt$$

or,
$$I = \int_0^t \int_0^t i^2 dt$$

R.m.s. value of Sine Wave

Analytical method

Let
$$i = I_{max} \sin\theta$$

Mean value of (i^2) over one complete cycle

$$= \frac{1}{2\pi}\int_0^{2\pi} i^2 d\theta$$

$$= \frac{1}{2\pi}\int_0^{2\pi} (I_{max} \sin\theta)^2 d\theta = \frac{I_{max}}{2\pi}\int_0^{2\pi} \sin^2\theta d\theta$$

But
$$\sin^2\theta = \frac{(1-\cos 2\theta)}{2}$$

Thus mean value of (i^2) over one complete cycle is expressed as

$$= \frac{I_{max}^2}{2\pi}\int_0^{2\pi}\left[\frac{(1-\cos 2\theta)}{2}\right] d\theta$$

$$= \frac{I_{max}^2}{4\pi}\left[\theta - \frac{\sin 2\theta}{2}\right]_0^{2\pi}$$

$$= \frac{I_{max}^2}{4\pi}\left[2\pi - \frac{\sin 4\pi}{2} - 0 + \sin 0\right]$$

Hence r.m.s. value of current $= \sqrt{\text{mean value of } (i^2)}$

$$= \frac{I_{max}^2}{4\pi}[2\pi - 0 - 0 + 0]$$

$$= \frac{I_{max}^2}{2}$$

$$= \sqrt{\frac{I_{max}^2}{2}} = \frac{I_{max}}{\sqrt{2}} = 0.707\, I_{max}$$

Thus for sinusoidal wave form, r.m.s. value of current will be 0.707 times the maximum value of current. Similarly for voltages.

3.12 FORM FACTOR

The form factor of an alternating wave shape is defined as the ratio of its r.m.s. value to the average value of a half-cycle.

$$\text{Form factor } (K_f) = \frac{(\text{r.m.s. value of current})}{\text{Average value of current}}$$

For sine wave,
$$K_f = \frac{0.707\, I_{max}}{0.637\, I_{max}}$$
$$K_f = 1.11.$$

3.13 PEAK FACTOR

The peak factor of an alternating wave shape is defined as the ratio its maximum value to the r.m.s. value.

$$\text{Peak factor } (K_f) = \frac{\text{Maximum value of current}}{\text{r.m.s. value of current}}$$

This is also called crest factor, or amplitude factor

For a sine wave

$$\text{Peak factor } K_p = \frac{I_{max}}{\left(\frac{I_{max}}{\sqrt{2}}\right)} = \sqrt{2} = 1.414$$

The form factor and peak factor gives the idea about the shape of a wave form. The more pointed the peak of a wave, the higher are the value of these factors.

For rectangular wave form,

Average value ≈ maximum value ≈ r.m.s. value

So, both, $K_f = K_p = 1.$

3.14 PHASE AND PHASE DIFFERENCE

Alternating wave shapes are normally represented by their maximum values and the frequency of alternations. Their instantaneous values are different at different interval of time. When two alternating quantity have the same phase angle, they are said to be in phase with each other or when they reach their maximum and zero values at the same time, though their maximum values may be different in magnitude as shown in Fig. 3.9.

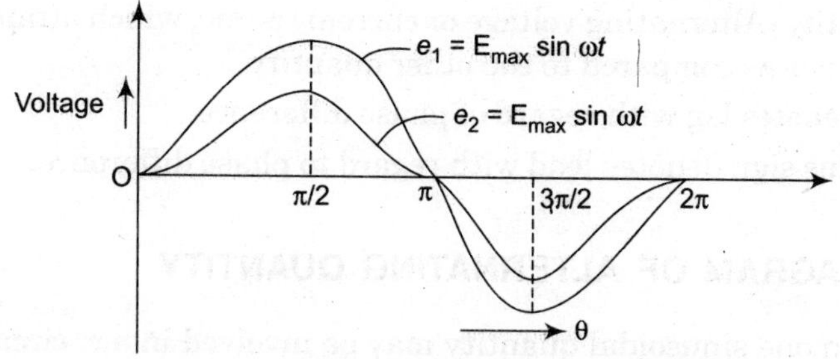

Fig. 3.9 *Voltage wave shapes in phase.*

Now consider the sinusoidal alternating voltage shape have the same maximum value and the same frequency. However, there may be an important difference in these wave shapes. These waves become zero and maximum at different instants. Suppose the voltage wave shape e_1, e_2, e_3 are shown in Fig. 3.10.

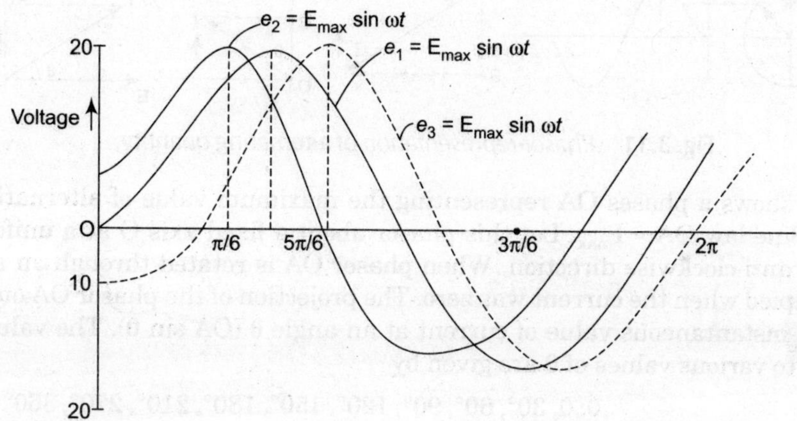

Fig. 3.10 *Phase difference between voltage waves.*

Wave form	*Maximum point*	*Minimum point*
e_1	$\theta = \dfrac{\pi}{2}$	$\theta = 0$
e_2	$\theta = 60° \text{ or } \dfrac{\pi}{3}$	$\theta = 150° \text{ or } \dfrac{5\pi}{6}$
e_3	$\theta = \dfrac{2\pi}{3} \text{ or } 120°$	$\theta = 210° \text{ or } \dfrac{7\pi}{6}$

The phase difference between e_1 and e_2 is clearly 30° and between e_1 and e_3 in also 30° but e_3 lags voltage wave e_1.

Therefore, following conclusions can be drawn with regard to the phase difference between the various alternating waves shapes quartities.

1. A lagging quantity (Alternating voltage or current) is one, which attains its maximum, or zero value latter than the other quantity with which it is being compared.

2. A leading quantity (Alternating voltage or current) is one, which attains its maximum, or zero values earlier as compared to the other quantity.
3. A minus sign denotes lag with regard of phase difference.
4. A positive or plus sign denotes lead with regard to phase difference.

3.15 PHASOR DIAGRAM OF ALTERNATING QUANTITY

In practice more than one sinusoidal quantity may be involved in a.c. circuit calculations. All these quantities may be represented on the same diagram by drawing phasor inclined at different angles to the reference line. If these phasors have one same frequency, they will rotate with the same angular speed ω. Their phases will change with time, but the between any two phasors will remain constant with time whatever be their position. It is the relative position of the phasors or the phase difference, which is important in a.c. calculations.

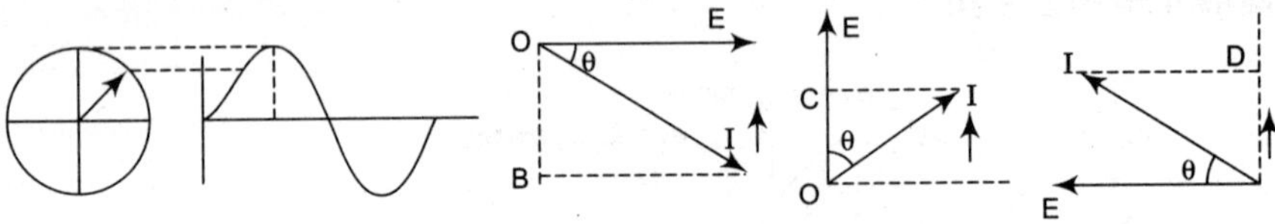

Fig. 3.11 *Phasor representation of alternating quantity.*

Figure 3.11 shows a phases OA representing the maximum value of alternating quantity following the Sine law OA ≈ I_{max}. Let this phasor about a fixed axis O at a uniform angular velocity in the anti-clockwise direction. When phasor OA is rotated through an angle θ from position it occupied when the current was zero. The projection of the phasor OA on the vertical axis gives the instantaneous value of current at an angle θ (OA sin θ). The value of current corresponding to various values of θ are given by

$$\theta: 0, 30°, 60°, 90°, 120°, 150°, 180°, 210°, 270°, 360°$$

Thus, the phasor is straight line of fixed magnitude rotating about one of its ends at a uniform angular velocity. From figure drawn above, it is clear that when the instantaneous value of voltage E is zero and becoming positive as instant latter, whereas the instantaneous value of current I is negative (equals OB) and diminishing. After quarter of a period, the instantaneous value of voltage becomes positive maximum, whereas current is positive (equal to OC) and increasing. The phase angle θ between voltage and current phasor remains constant in all the cases discussed above. Hence the relative position of the voltage phasor with current phasor at any instant is same.

3.16 ADDITION AND SUBTRACTION OF PHASORS

In phasor diagrams for a.c. circuits as shown in Fig. 3.12, the addition and subtraction of phasors representing their r.m.s. values, have to be carried out very often. In such circuits, the addition/subtraction of phasors becomes essential to find out the resultant voltage or current. The addition/subtraction of phasor follows the rules, which are used for vectors. Hence, the phasor addition may be performed either by graphically or analytically. The phasor addition can be done more quickly and easily as compared with the wave form addition or analytical method.

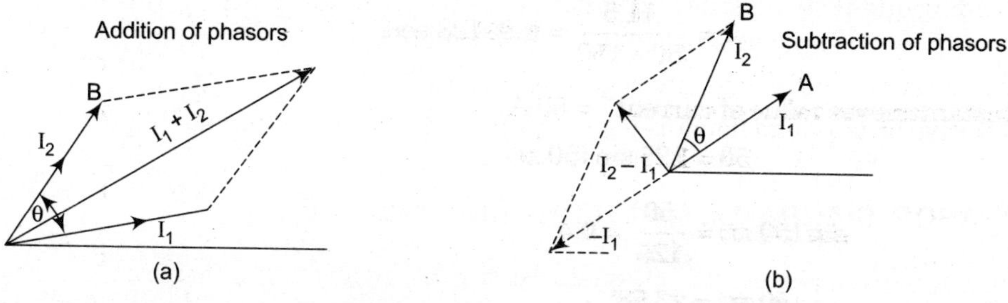

Fig. 3.12 *Addition and subtractions of phasors.*

The phasors E_1(OA) and E_2 (OB) represents the r.m.s. values of the two alternating current having the same frequency. The current phasor I_2 is leading the phasor I_1 by an angle θ. The resultant voltage can be found by adding these two phasors. The addition these two phasors is carried out according to the parallelogram law. The parallelogram OACB is completed and the diagonal OC represents the resultant voltage *i.e.* the phasor sum of the phasors E_1 and E_2.

Similarly, in subtraction if E_1 phasor is to be subtracted from E_2. Phasor OA representing E_1 is extended in reverse direction so that OC is equal OA in magnitude. Parallelogram OBDC is then completed with the diagonal OD of the parallelogram representing the phasor difference of two phasors E_2 and E_1.

SOLVED NUMERICAL PROBLEMS

Example 1. *A sinusoidal alternating current of frequency 25 Hz has a maximum value of 125 A. How long will it take for the current to attain values of 25 A, 50 A and 100 A?*

Solution: The alternating current wave following sinusoidal low is expressed as:

$$i = I_{max} \sin(\omega t)$$

$$= I_{max} \times \sin(2\pi f t)$$

$$I_{max} = 125 \text{ A}$$

$$f = 25 \text{ Hz}$$

Thus instantaneous value of current

$$i = 125 \sin(2\pi \times 25t)$$

(*a*) Instantaneous value of current

$$i = 25 \text{ A}$$

Thus, $$25 = 125 \sin(50\,\pi t)$$

$$\sin(50\,\pi t) = \frac{25}{125} = 0.2$$

$$= \sin 11.5°$$

Hence $$50 \times 180° \times t = 11.5$$

$$t = \frac{11.5}{50 \times 180} = 0.00128 \text{ sec.}$$

(*b*) Instantaneous value of current $i = 50$ A

$$50 = 125 \sin (50\, \pi t)$$

$$\sin (50\, \pi t) = \frac{50}{125} = 0.4$$

$$(50\, \pi t) = 23.58°$$

$$50 \times 180 \times t = 23.58°$$

$$t = \frac{23.58}{50 \times 180}$$

$$t = 2.61 \times 10^{-3} \text{ sec.}$$

$$t = 2.61 \text{ m-sec.}$$

(*c*) Instantaneous current $i = 100$ A

$$100 = 125 \sin (50\, \pi t)$$

$$\frac{100}{125} = \sin (\pi t)$$

$$\sin (50\, \pi t) = 0.8$$

$$50\, \pi t = 53.13$$

$$t = \frac{53.13}{50 \times 180°}$$

$$t = 5.90 \text{ m-sec.}$$

Example 2. *Find the rate of change of voltage per second of an emf wave given by*

$$e = 125 \sin (2\pi \times 50t)$$

0.005 and 0.01 sec after the voltage wave passes through zero and is increasing in a positive direction.

Solution:

$$e = 125 \sin (2\pi \times 50t)$$

Rate of change of voltage per second

$$\frac{de}{dt} = 125 \times 100\pi \cos (100 \times \pi \times \text{t})$$

$$= 125000\, \pi \cos (100 \times 180° \times t)$$

(*i*) At $t = 0.005$ sec

$$\frac{de_1}{dt} = 12500\, \pi \cos (100 \times 180 \times 0.005)$$

$$= 0 \text{ V/s.}$$

(*ii*) At $t = 0.01$ sec.

$$\frac{de_2}{dt} = 12500\,\pi \cos(100 \times 180 \times 0.01)$$

$$= -12500\,\pi$$

$$= -39250 \text{ V/s.}$$

Example 3. *A sinusoidal current wave is given by*

$i = 30 \sin(120\,\pi t)$ then calculate the following:

(*i*) *The highest rate of change of current,*

(*ii*) *The average value,*

(*iii*) *Root mean square value (RMS), and*

(*iv*) *The time interval between a maximum value and the next zero value.*

Solution:

(*i*) $$i = 30 \sin(120\pi \times t)$$

Differentiating it w.r.t. time

$$\frac{di}{dt} = 30 \times 120\pi \cos(120\pi \times t)$$

The rate of change of current will be highest when $\cos(120\pi \times t)$ becomes equal to 1.

Therefore, highest rate of change of current $= 30 \times 120 \times \pi$

$$= 30 \times 120 \times 3.14$$

$$= 11314.28 \text{ A/s.}$$

(*ii*) $$i = 30 \sin(120\pi \times t)$$

The frequency of given alternating quantity, $\phi = 50$ Hz.

Time period, $$T = \frac{1}{f} = \frac{1}{50} = 0.025$$

Thus, the positive half of the wave be completed in 0.01 s.

Mean value of current over half-cycle

$$= \frac{1}{0.01}\int_0^{0.01} 30 \sin(120\,\pi t)\,dt$$

$$= \frac{30}{0.01} \times \frac{1}{120\pi}\left[-\cos(120 \times 180° \times t)\right]_0^{0.01}$$

$$= \frac{30}{0.01 \times 120 \times 3.14}[2.5] = \frac{30 \times 2.5}{3.768} = 19.90 \text{ A.}$$

The average value of the given current wave = 19.90 A.

(*iii*) $$i = 30 \sin(120\pi t) = 30 \sin\theta$$

Mean value of i^2

$$= \frac{1}{2\pi}\int_0^{2\pi} i^2\, d\theta = \frac{1}{2\pi}\int_0^{2\pi} \left[30 \sin (120\pi t)^2\right] d\theta$$

$$= \frac{1}{2\pi}\int_0^{2\pi} (30 \sin \theta)^2\, d\theta = \frac{900}{2\pi}\int_0^{2\pi} \left(\frac{1-\cos 2\theta}{2}\right)^2 d\theta$$

$$= \frac{900}{2\pi}\left[\frac{\theta}{2} - \frac{\sin 2\theta}{4}\right]_0^{2\pi} = \frac{900}{2\pi} \times \pi = 450$$

Hence, r.m.s. value of current

$$I = \sqrt{\text{Mean value of } i^2}$$

$$= \sqrt{450} = 21.22 \text{ A}$$

(*iv*) $i = 30 \sin (120\pi t)$

$$= 30 \sin (120 \times 180° \times t)$$

This wave shape will have maximum value of current of 50 A that will occur when sin (120 × 180° × *t*) becomes 1. *i.e.*,

$$\sin (120 \times 180° \times t) = 1$$

$$120 \times 180° \times t = \sin^{-1} (1)$$

$$120 \times 180° \times t = 90°$$

$$t = \frac{90°}{120 \times 180°} = \frac{90°}{21600} = 0.004 \text{ sec.}$$

Thus, current becomes maximum after 0.004 s from its earlier zero value.

Therefore,

$$\sin (120 \times 180 \times t) = 0$$

$$120 \times 180 \times t = 180°$$

$$t = \frac{180}{120 \times 180} = 0.01 \text{ s.}$$

Interval will be 0.004 s to 0.01 sec.

Example 4. *The minimum value of a sinusoidal alternating current having frequency 60 Hz is 20 A. Determine the equation for instantaneous value of the alternating current. Calculate i at 0.002 s and 0.015 s. Also obtain the time at which current is 15 A.*

Solution: We know that,

$$\omega = 2\pi f = 2\pi \times 60$$

$$= 120\ \pi \text{ rad./s.}$$

The equation for instantaneous value alternating current is

$$i = 20 \sin 120\ \pi t$$

Now, at 0.002 s

$$i = 20 \sin (120\pi \times 0.002)$$

$$i = 0.2630 \text{ A}$$

and AT 0.015 s

$$i = 20 \sin (120\pi \times 0.015)$$

$$= 1.9697 \text{ A}$$

Now, time at current 15 A will be

$$15 = 20 \sin 120\pi t_1$$

$$t_1 = \frac{15}{20 \sin 120\pi} = \frac{15}{5.7806} = 2.594 \text{ m-sec.}$$

Example 5. *If a half wave rectified sinusoidal current having $I_m = 15$ A delayed by an angle $q = 45°$, calculate the average r.m.s. value and form factor.*

Solution:

$$I_m = \frac{15}{2\pi}\int_{\pi/4}^{\pi} \sin\theta \, d\theta = \frac{15}{2\pi}[-\cos\theta]_{\pi/4}^{\pi}$$

$$= \frac{15}{2\pi}\left[-\cos\pi + \cos\frac{\pi}{4}\right] = \frac{15}{2\pi}[1 + 0.707]$$

$$= \frac{15 \times 1.707}{2 \times 3.14} = \frac{25.605}{6.28} = 4.077 \text{ A}$$

$$I_{rms}^2 = \frac{(15)^2}{4\pi}\int_{\pi/4}^{\pi}(-\cos 2\theta)\, d\theta = \frac{225}{4\pi}\left[\pi - \frac{\pi}{4} - \frac{1}{2}\left(\sin 2\pi - \sin\frac{\pi}{2}\right)\right]$$

$$= \frac{56.25}{3.14}\left[\frac{3\pi}{4} + \frac{1}{2}\right] = 17.91\,[2.355 + 0.5]$$

$$I_{rms}^2 = 51.133 \text{ A}$$

Thus,

$$\text{Form factor} = \frac{\text{r.m.s. value}}{\text{Average value}} = \frac{51.133}{4.077} = 12.541.$$

Example 6. *An alternating current of peak value 45 A has the following wave forms in turn:*

(*a*) *Sinusoidal,*

(*b*) *Full wave rectified sinusoidal,*

(*c*) *Rectangular, and*

(*d*) *Triangular.*

Solution: (*a*) Peak value of current wave = 45 A

Instantaneous value of the current wave can be exposed by the equation,

$$i = 45 \sin \omega t = 45 \sin \theta$$

Such a wave shape can be shown below:

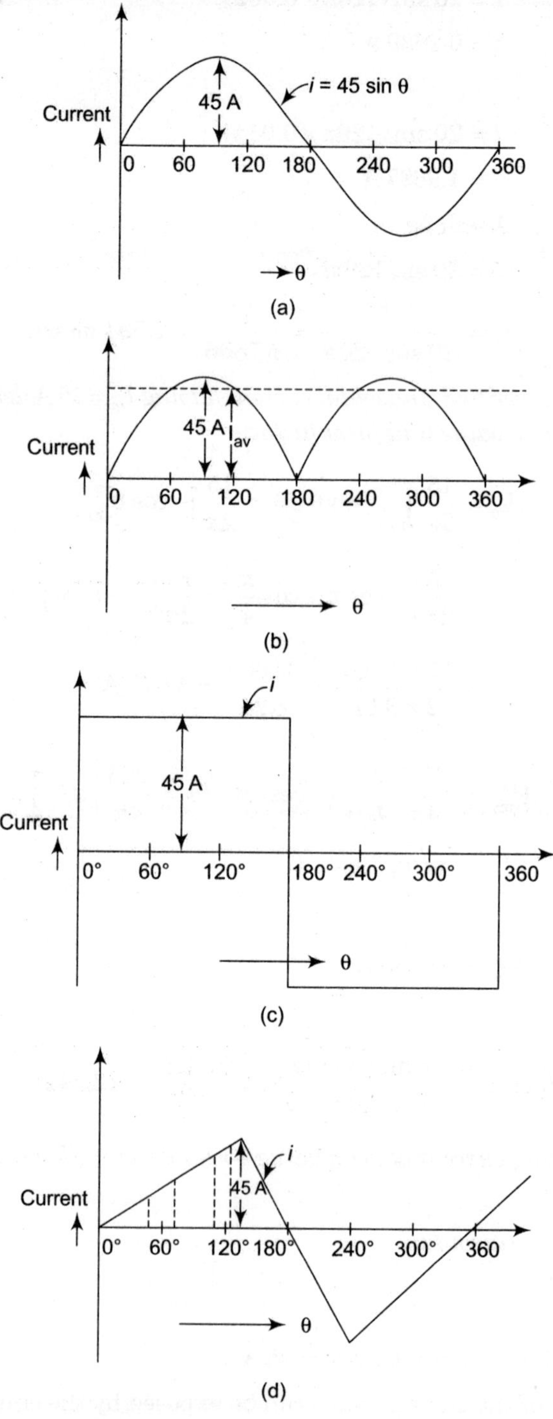

Fig. 3.13

When sinusoidal current is passed through a moving coil ameter connected in series, it reads the average value of the current wave over the complete cycle. So, average value of current over one complete cycle.

$$I_m = \frac{1}{2\pi}\int_0^{2\pi} 45 \sin\theta \, d\theta$$

$$= \frac{45}{2\pi}[-\cos\theta]_0^{2\pi} = 0$$

Hence, ammeter reads zero in case of moving coil instruments.

r.m.s. value of this current wave will be

$$I_{r.m.s.} = \sqrt{\text{mean value of } (i)^2}$$

Mean value of

$$(i^2) = \frac{1}{2\pi}\int_0^{2\pi} (45 \sin\theta)^2 \, d\theta$$

$$= \frac{2025}{2\pi}\int_0^{2\pi}\left(\frac{1-\cos 2\theta}{2}\right) d\theta$$

$$= \frac{2025}{2\pi}\left[\frac{\theta}{2} - \frac{\sin 2\theta}{4}\right]_0^{2\pi} = \frac{2025}{2} = 1012.5 \text{ A}$$

r.m.s. value of this current wave will be

$$I_{r.m.s.} = \sqrt{1012.5} = 31.819 \text{ A.}$$

Similarly for all the wave forms shown, currents can be calculated in similar fashion.

Example 7. *Determine the phase difference between the respective emf and current. Also state the condition of frequency in each case.*

(*i*)
$$e = 300 \sin\left(350.14\, t + \frac{\pi}{12}\right)$$
$$i = 30 \sin (350.14t)$$

(*ii*)
$$e = E_{max} \sin\left(\omega t + \frac{\pi}{3}\right); \; i = I_{max} \sin \omega t$$

(*iii*)
$$e = E_{max} \sin (\omega t + a)$$
$$i = \frac{E_{max}}{\sqrt{R^2 + X^2}} \sin\left(\omega t - \tan^{-1}\frac{X}{R}\right)$$

Solution: (*i*) Voltage and current wave shapes are given by

$$e = 300 \sin\left(350.14t + \frac{\pi}{12}\right)$$

$$e = 300 \sin (350.14t + 15°)$$

$$= 300 \sin (2\pi \times 60t + 15°)$$

$$i = 30 \sin (350.14t) = 30 [\sin (2\pi \times 60t)]$$

Phase difference between voltage and current wave shapes

$$\Rightarrow \quad (350.14t + 15°) - (350.14t) = 15°$$

Hence, voltage wave shape leads the current by 15°.

The frequency of the wave shape,

$$\phi = 60 \text{ Hz}$$

(ii) $e = E_{max} \sin (\omega t + 60°)$

$$= E_{max} \sin \left(2\pi \times \frac{\omega}{2\pi} t + 60°\right)$$

$$i = I_{max} \sin \omega t = I_{max} \sin \left(2\pi \times \frac{\omega}{2\pi} t\right)$$

Phase difference between the voltage and current wave shapes

$$= (\omega t + 60°) - \omega t$$

$$= 60°$$

Thus, voltage leads current by an angle of 60°.

(iii) $e = E_{max} \sin (\omega t + \alpha) = E_{max} \sin \left(2\pi \times \frac{\omega}{2\pi} t + \alpha\right)$

$$i = \frac{E_{max}}{\sqrt{R^2 + x^2}} \sin \left(\omega t - \tan^{-1} \frac{x}{R}\right)$$

$$= \frac{E_{max}}{\sqrt{R^2 + x^2}} \sin \left(2\pi \times \frac{\omega}{2\pi} t - \tan^{-1} \frac{x}{R}\right)$$

Phase difference between current and voltage wave shapes is

$$= \left(\omega t - \tan^{-1} \frac{x}{R}\right) - (\omega t + \alpha)$$

$$= -\tan^{-1} \frac{x}{R} - \alpha = -\left(\tan^{-1} \frac{x}{R} + \alpha\right)$$

Hence, current wave lags the voltage wave shape by an angle

$$\left(\tan^{-1} \frac{x}{R} + \alpha\right)$$

Frequency of the wave shapes,

$$f = \frac{\omega}{2\pi}.$$

EXERCISES

1. Derive an expression for instantaneous value of alternating current in terms of its maximum value, angular velocity and time.
2. Explain the method of calculating r.m.s. value for the non-sinusoidal and sinusoidal alternating quantity.
3. Why are the root mean square values of an alternating quantities more important than their average values?
4. Explain the terms frequency, cycle and periodic time in connection with a.c. circuits.
5. Derive an expression for the r.m.s. value of the half wave rectified alternating current in terms of its maximum value.
6. What is average value of sine wave over a complete cycle? Justify your answer.
7. Define the root mean square value of an alternating current. Explain why this value is more generally employed in a.c. measurement than either the average or peak value.
8. An alternating current is represented by the following equation:
$$i = 50 \sin (100\ \pi t)$$
How long will it take the current to attain values of 10, 20, 30, 40, 50 amps?
9. Draw the following alternative waves extending over one period:
(*i*) $e_1 = 150 \cos \left(\frac{\omega t - \pi}{6}\right)$, (*ii*) $e_2 = 150 \cos \left(\frac{\omega t - \pi}{3}\right)$,
(*iii*) $e_3 = 100 \sin 2\ \omega t$.
10. The current wave is represented by the equation $i = 50 \sin (100\ \pi t)$. Calculate the maximum and r.m.s. values of the current and its frequency.
11. The voltages of a three-phase, 50 Hz supply are given by the following expressions:
Phase R; $e_R = 100 \sin \pi t$, Phase Y; $e_y = 100 \sin \left(\frac{\omega t - 2\pi}{3}\right)$, Phase B; $e_B = 100 \sin \left(\frac{\omega t + 2\pi}{3}\right)$.
Find the instantaneous values of e_R, e_y, e_B at the following instants after e_R passes through zero in positive direction.
12. Three voltages represented by the following equations:
$e_1 = 10 \sin \omega t$, $e_2 = 5 \sin \left(\frac{\omega t + \pi}{6}\right)$, $e_3 = 10 \cos \omega t$ act together in an a.c. circuit. Represent these voltages by phasor and calculate an expression for the resultant voltage. Check the result so obtained graphically.
13. An alternating current flowing through a circuit has maximum value of 150 A and lags the applied voltage by 60°.The maximum value of voltage is 250 V. Both current and voltage wave shapes in their correct relationship for one complete cycle. What is the value of current when the voltage is at its maximum value?
14. Three alternating currents expressed by, $i_1 = 10 \sin \omega t$, $i_2 = 20 \sin \left(\frac{\omega t - \pi}{4}\right)$, $i_3 = 30 \sin \left(\frac{\omega t + \pi}{4}\right)$ act together in an a.c. circuit. Represent these voltages by Phasor and calculate an expression for the resultant voltage. Check the result so obtained graphically.

4

Single Phase AC Series Circuits

4.1 INTRODUCTION

In case of d.c. supply applied voltage and current flowing are constant with respect to time. Solution of the circuit can be found simply by applying Ohm's law as following:

$$\text{Current,} \qquad I = \frac{\text{Applied voltage}}{\text{Resistance offered by the circuit}}$$

While in case of a.c. circuits voltage applied to the circuit and the current flowing through it changes from instant to instant. Thus the above simple relationship will not hold good in these circuits. The variation of current with respect to time sets up magnetic effect and variation of e.m.f. sets up electrostatic effect. Magnetic effects will be appreciably large with low voltage, heavy current circuits. Electrostatic effects are usually appreciable with high voltage circuits.

4.2 RESISTIVE CIRCUIT ONLY

Figure 4.1 shows a circuit consisting only a pure resistance R which is connected to which an alternating voltage source and can be given by

$$v = V_{max} \sin \omega t$$

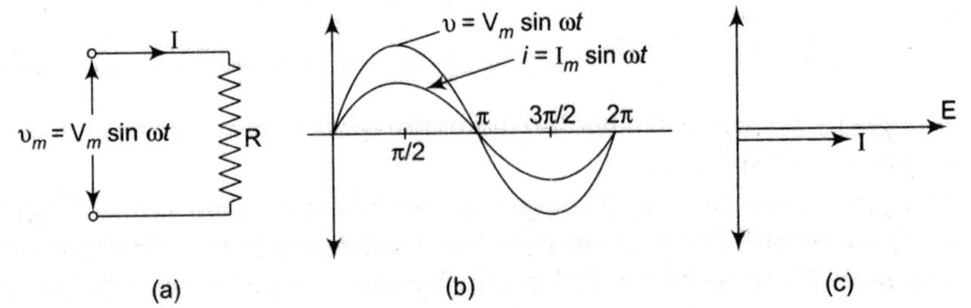

Fig. 4.1 *(a) Circuit diagram (b) Wave form (c) Phasor diagram.*

BY Ohm's law the instantaneous value of current in the circuit will be

$$i = \frac{V}{R} = \left(\frac{V_m}{R}\right) \sin \omega t$$

$$i = I_m \sin \omega t$$

where, $$I_m = \frac{V_m}{R}$$

Comparison of the voltage and current equation shows that applied voltage and current are in phase and follow the sine law. Figure 4.1 (*b*) and (*c*) shows the wave form and phasor diagrams of the voltage and current respectively in a circuit containing only a resistance.

4.3 POWER IN RESISTIVE CIRCUIT

Power drawn by the circuit at any instant is product of instantaneous voltage and instantaneous current *i.e.*

$$P = v \cdot i = (V_m \sin \omega t)(I_m \sin \omega t)$$

$$= V_m I_m \sin^2 \omega t$$

$$= \frac{V_m \cdot I_m}{2} (1 - \cos 2\omega t)$$

$$= \frac{V_m I_m}{2} - \frac{V_m I_m}{2} (\cos 2\omega t)$$

The above expression consists of a constant part of value $\frac{V_m I_m}{2}$ and fluctuating part $\frac{(V_m I_m \cos 2\omega t)}{2}$. The frequency of the fluctuating power is twice the applied voltage frequency and its average value over one complete cycle is zero.

Hence average power in this circuit is

$$P = \frac{V_m I_m}{2} = \left(\frac{V_m}{\sqrt{2}}\right) \cdot \left(\frac{I_m}{\sqrt{2}}\right) = V \cdot I \text{ watts.}$$

Thus the power in a purely resistive circuit is equal to the product of the r.m.s. values of voltage and current.

4.4 A.C. CIRCUITS CONTAINING INDUCTANCE

An a.c. circuit having purely inductive circuit containing only an inductance L Henry. A sinusoidal alternative voltage is applied to this circuit as a result of which alternating currents flows in the circuit. This current produces a self-induced e.m.f. e_L in the circuit given in Fig. 4.2.

$$e_L = \frac{-L di}{dt}$$

By KVL in circuit,

$$v + e_L = 0$$

$$v - L \frac{di}{dt} = 0$$

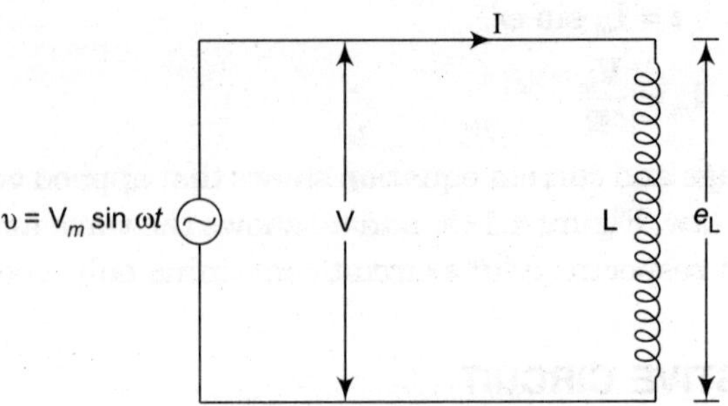

Fig. 4.2 *Circuit diagram of Inductive circuit.*

$$v = \frac{Ldi}{dt}$$

Let, $$v = V_m \sin \omega t$$

So $$v_m \sin \omega t = \frac{Ldi}{dt}$$

or $$di = \frac{1}{L}(V_m \cdot \sin \omega t \cdot dt)$$

Integrating both side of the equation,

$$i = \int \left(\frac{V_m}{L}\right) \cdot \sin \omega t \cdot dt = \left(\frac{-V_m \cos \omega t}{\omega L}\right)$$

$$i = \frac{V_m}{\omega L} \cdot \sin\left(\omega t - \frac{\pi}{2}\right)$$

$$i = i_m \sin\left(\omega t - \frac{\pi}{2}\right)$$

where, $$i_m = \frac{V_m}{\omega L}$$

So it is observed that the current attains its maximum value when sin $(\omega t - \pi/2)$ becomes unity, *i.e.* the current lags behind the applied voltage by an angle $\frac{\pi}{2}$ or 90°. Hence in a purely inductive circuit the current flowing in the circuit lags the voltage applied to the circuit by an angle of 90° as shown in Fig. 4.3.

The quantity ωL is called inductive reactance and it is denoted by X_L.

$$X_L = \omega L = \frac{V_m}{I_m} = \frac{V \cdot \sqrt{2}}{I \cdot \sqrt{2}} = \frac{V}{I} \text{ ohm}$$

$$X_L = \frac{V}{I} \text{ ohm } V = I \cdot X_L$$

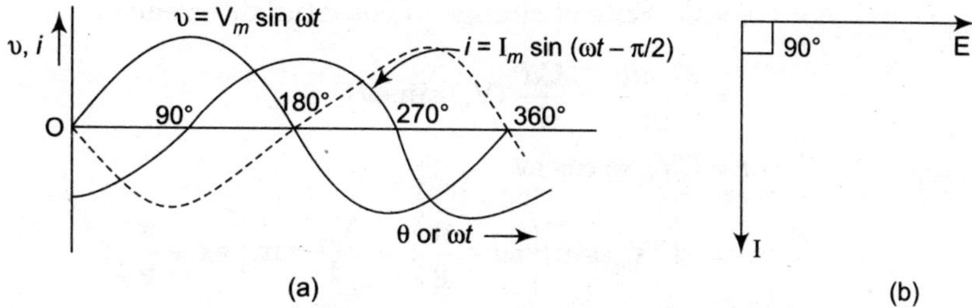

Fig. 4.3 *Phasor diagram and its waveform.*

4.5 POWER IN PURELY INDUCTIVE CIRCUIT

Instantaneous power in an a.c. circuit is given by

$$P = v \cdot i = V_m \sin \omega t \, I_m \sin\left(\omega t - \frac{\pi}{2}\right)$$

$$= -V_m I_m \sin \omega t \cos \omega t$$

$$= -\left(\frac{V_m I_m}{2}\right)(2 \sin \omega t \cos \omega t)$$

$$= -\left(\frac{V_m}{\sqrt{2}}\right)\left(\frac{I_m}{\sqrt{2}}\right) \sin 2\,\omega t$$

$$P = -V \cdot I \sin 2\,\omega t$$

Average power for one complete cycle

$$P = -V \cdot I \text{ Average of } (\sin 2\,\omega t) \approx 0$$

Hence the total power consumed by a purely inductive circuit is zero.

4.6 A.C. CIRCUIT CONTAINING CAPACITANCE

Figure 4.4 shows an a.c. circuit containing a capacitor of capacitance C farads. Let the supply voltage given to circuit.

$$v = V_m \sin \omega t$$

Charging current in the capacitance circuit is given by applied voltage v.

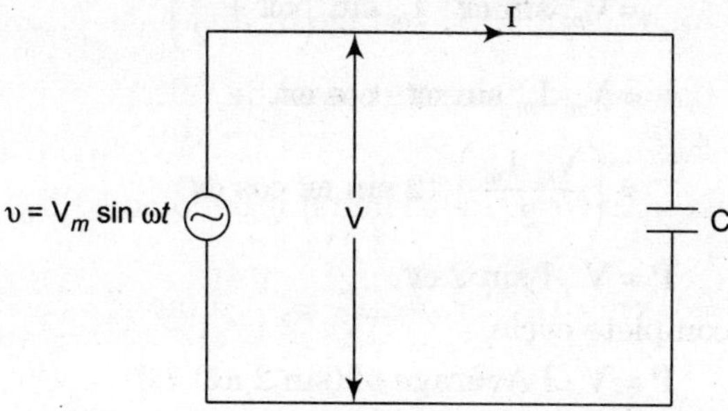

Fig. 4.4 *Circuit diagram of capacitive circuit.*

$$i = \text{C} \cdot \text{Rate of change of potential difference}$$

$$= \frac{\text{C} \cdot dv}{dt} = \frac{\text{C}d}{dt}(\text{V}_m \sin \omega t)$$

$$i = \text{CV}_m \, \omega \cos \omega t$$

$$= \omega\text{C}\,\text{V}_m \sin\left(\omega t + \frac{\pi}{2}\right) = \frac{V_m}{\frac{1}{\omega\text{C}}} \sin\left(\omega t + \frac{\pi}{2}\right)$$

where, $$\text{I}_m = \frac{\text{V}_m}{\frac{1}{\omega\text{C}}}$$

$$\text{I}_m = \frac{\text{V}_m}{\text{X}_\text{C}}$$

Since $$\text{X}_\text{C} = \frac{1}{\omega\text{C}}$$

where X_C is called capacitive reactance, its unit is ohm. Comparing voltage and current equation it is found that current leads the voltage by an angle 90° or $\frac{\pi}{2}$ as shown in Fig. 4.5.

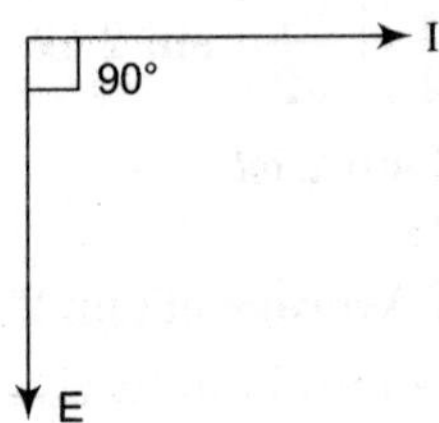

Fig. 4.5 *Phasor diagram for capacitive circuit.*

4.7 POWER IN PURELY CAPACITIVE CIRCUIT

Instantaneous power is an a.c. circuit is given by

$$\text{P} = v \cdot i$$

$$= \text{V}_m \sin \omega t \cdot \text{I}_m \sin\left(\omega t + \frac{\pi}{2}\right)$$

$$= \text{V}_m \, \text{I}_m \sin \omega t \cdot \cos \omega t.$$

$$= \left(\frac{\text{V}_m \, \text{I}_m}{2}\right)(2 \sin \omega t \cos \omega t)$$

$$\text{P} = \text{V} \cdot \text{I} \sin 2\,\omega t.$$

Average power for one complete cycle

$$\text{P} = \text{V} \cdot \text{I Average of} (\sin 2\,\omega t) \approx 0$$

Thus, the total power consumed by a purely capacitive circuit is zero.

Series Resistive and Inductive Circuit

A coil can be represented by its resistance and inductance connected in series as shown in Fig. 4.6.

Let, R = Resistance in Ohms.

L = Inductance in Henry

V = Applied voltage

I = RMS current flowing in circuit

V_R = Voltage drop across resistance

V_L = Voltage drop across inductance

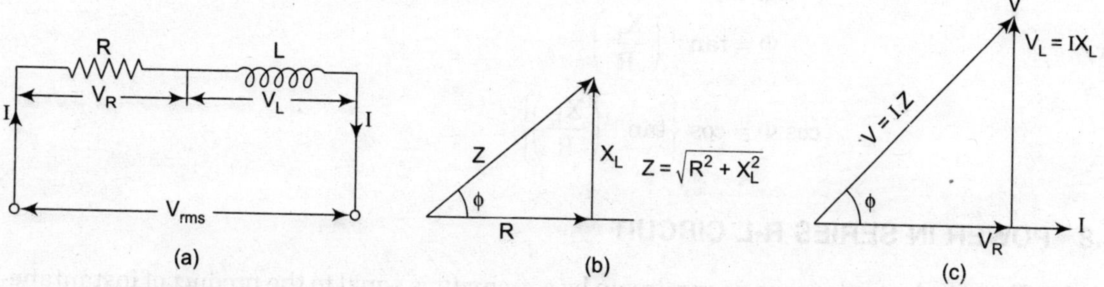

Fig. 4.6 *(a) Series R-L circuit, (b) Phasor diagram for series R-L circuit.*

Since two components of the circuit are connected in series so that the same current will flow in both of them and current is used as reference phasor. The voltage V_R is in phase with the current and V_L leads by $\frac{\pi}{2}$ or 90°. The voltage V is the phasor sum of V_R and V_L

$$V = V_R + V_L$$

Applied voltage, $$V = \sqrt{(V_R^2 + V_L^2)}$$

$$= \sqrt{\left[(I \cdot R)^2 + (I \cdot X_L)^2\right]} = I^2 \sqrt{(R^2 + X_L^2)}$$

So, current flowing through the circuit

$$I = \frac{V}{Z} \approx \frac{V}{\sqrt{R^2 + X_L^2}}$$

$Z = \sqrt{(R^2 + X_L^2)}$ is usually called impedance of circuit and is denoted by symbol Z, its unit is Ohms.

or $$I = \frac{V}{Z}$$

For the wave form are shown in Fig. 4.7.

$$\tan \Phi = \frac{V_L}{V_R} = \frac{IX_L}{IR} = \frac{X_L}{R} \text{ (from Fig. 4.6)}$$

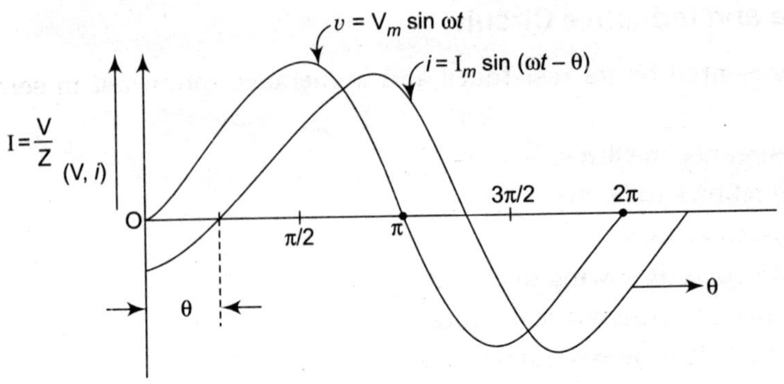

Fig. 4.7 *Wave shapes of R-L series circuit.*

$$\Phi = \tan^{-1}\left(\frac{X_L}{R}\right)$$

or,

$$\cos \Phi = \cos\left\{\tan^{-1}\left(\frac{X_L}{R}\right)\right\}$$

4.8 POWER IN SERIES R-L CIRCUIT

Active Power : Instantaneous power drawn by a.c. circuit is equal to the product of instantaneous value of voltages and currents.

Instantaneous power $= v \cdot i$

where v = Instantaneous value of voltage $= V_m \cdot \sin \omega t$

i = Instantaneous value of current $= I_m \cdot \sin(\omega t - \phi)$

$$P = V_m \sin \omega t \, I_m \sin(\omega t - \phi)$$

$$= I_m \cdot V_m \sin \omega t \cdot \sin(\omega t - \phi)$$

$$= \left(\frac{V_m I_m}{2}\right)(\cos \phi - \cos(2\omega t - \phi))$$

$$= \left(\frac{V_m I_m}{2}\right)\cos \phi - \left(\frac{V_m I_m}{2}\right) \cdot \cos(2\omega t - \phi)$$

The above equation consists of two terms:

1. $\left(\frac{V_m I_m}{2}\right)$ cos ϕ, remains constant irrespective of time.
2. $\left(\frac{V_m I_m}{2}\right)$ cos $(2\omega t - \phi)$, power varies at twice the supply frequency. So average value of power over one complete cycle is zero. Hence it does not contribute to average value of power drawn from the supply.

So, average power over one cycle

$$P = \frac{1}{2} \cdot V_m I_m \cdot \cos \phi$$

$$= \left(\frac{V_m}{\sqrt{2}}\right)\left(\frac{I_m}{\sqrt{2}}\right) \cos\phi$$

$$P = V \cdot I \cos\phi$$

where, V and I are r.m.s. value of voltage and current.

where cos ϕ is called power factor.

Thus the average power drawn by circuit is find out by multiplying the r.m.s. value of voltage and current by cos ϕ. Its units is watt.

Apparent Power or Volt-Amperes

The product of r.m.s. values of voltage and current in a.c. circuit is called volt-amperes. It is also known as apparent power and denoted by S; which is measured in volt-amperes (VA) or kilo volt-amperes (KVA)

$$S = V \cdot I; \text{ volt-amperes}$$

Reactive Power

In a.c. circuits, current lags or leads the applied voltage by an angle ϕ, thus the current can be resolved into active and reactive components. The reactive component of circuit is equal to I sin ϕ as shown in Fig. 4.8. Therefore power drawn of reactive component of current is called reactive power.

Hence, reactive power = VI sin ϕ

It is expressed in Var or KVAR or MVAR

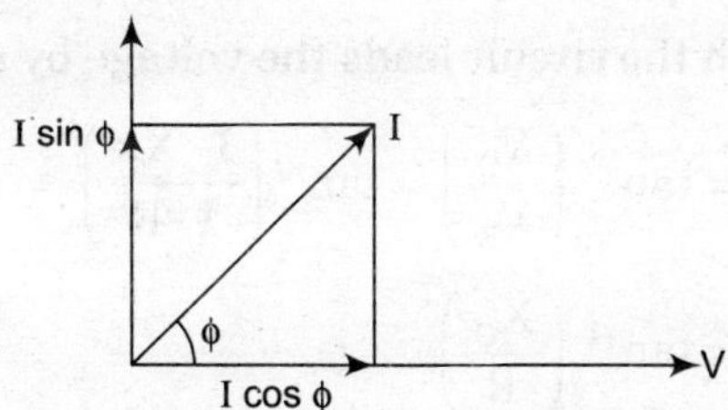

Fig. 4.8 *Phasor diagram for reactive and active current.*

Circuit Containing Resistor and Capacitor

Figure 4.9 shows an A.C. circuit containing resistance and capacitance in series and connected to a single phase A.C. supply of frequency *f*, Hertz.

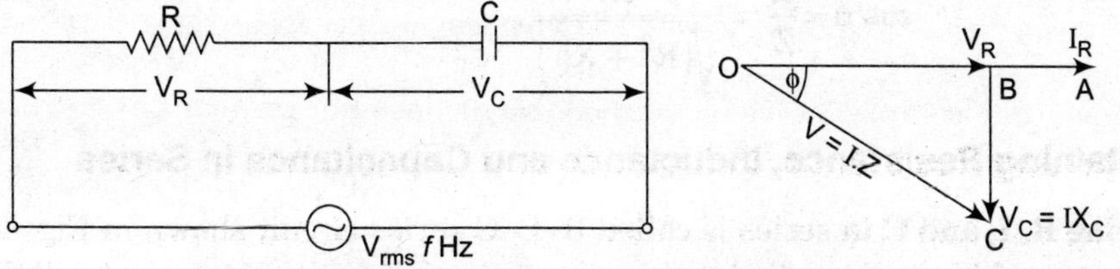

Fig. 4.9 *Diagram for R-C series circuit and its phasor.*

Let,

V is the r.m.s. value of voltage applied to the circuit.

I is the r.m.s. value of current flowing in the circuit.

V_R, voltage drop across resistance = $I \cdot R$

V_C voltage drop across capacitance = $I \cdot X_C$

The phasor diagram for circuit can be drawn by taking the current as reference (OA). The voltage drop V_R across the resistance is in phase with current (OB). The voltage across the capacitance V_C lags the current by 90°(BC). The OC is the phasor sum of the two voltages V_R and V_C, hence, the phasor OC represents the applied voltage. Thus, in a capacitive circuit current leads the voltage by angle ϕ.

Therefore applied voltage is given by

$$V = \sqrt{(V_R^2 + V_C^2)}$$

$$V = \sqrt{\{(IR)^2 + (-IX_C)^2\}}$$

$$V = I\sqrt{(R^2 + X_C^2)}$$

$$\frac{V}{I} = \sqrt{(R^2 + X_C^2)}$$

or

$$I = \frac{V}{\sqrt{R^2 + X_C^2}} = \frac{V}{Z},$$

$$Z = \sqrt{R^2 + X_C^2}\text{; is the impedance of the circuit.}$$

Also, Current flowing through the circuit leads the voltage by an angle ϕ which is given by

$$\phi = \tan^{-1}\left(\frac{V_C}{V_R}\right) = \tan^{-1}\left(\frac{I \cdot X_C}{I \cdot R}\right)$$

$$= \tan^{-1}\left(\frac{X_C}{R}\right)$$

$$\tan\phi = \frac{X_C}{R}$$

and

$$\sin\phi = \frac{X_C}{Z}$$

Power factor of the circuit

$$\cos\phi = \frac{R}{Z} = \frac{R}{\sqrt{(R^2 + X_C^2)}}.$$

Circuit Containing Resistance, Inductance and Capacitance in Series

A circuit having R, L and C in series is called R–L–C series circuit shown in Fig. 4.10. An a.c. supply of frequency *f*, Hertz is applied to this circuit since current is common to all the elements of circuit, so used as reference phasor.

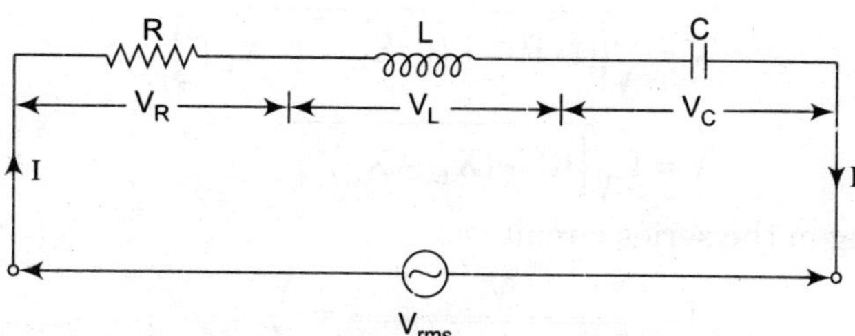

Fig. 4.10 *Series R-L-C circuit.*

Let V be the r.m.s. value of the voltage applied to the circuit.

V_R, the r.m.s. value of voltage across the resistance R.

V_L, the r.m.s. value of the voltage across the inductance L.

V_C, the r.m.s. value of the voltage across the capacitance C.

I, the r.m.s. value of current flowing through the circuit.

The current phasor is represented by OA. The voltage drop V_R across the resistance R is in phase with the current and is represented by OB. The voltage V_L across the inductance L leads the current phasor by 90°(OC). The voltage V_C across the capacitance lags the current by 90° and is shown by OF. Phasor OC and OF are in direct opposition and hence their resultant is given by

$$OD = (OC - OF); \text{ (Assuming OC to be greater than OF)}$$

$$OD = (OF - OC); \text{ (Assuming OF to be greater than OC)}$$

Hence the applied voltage will be the phasor sum of the phasors OB and BD. Referring to the following Fig. 4.11.

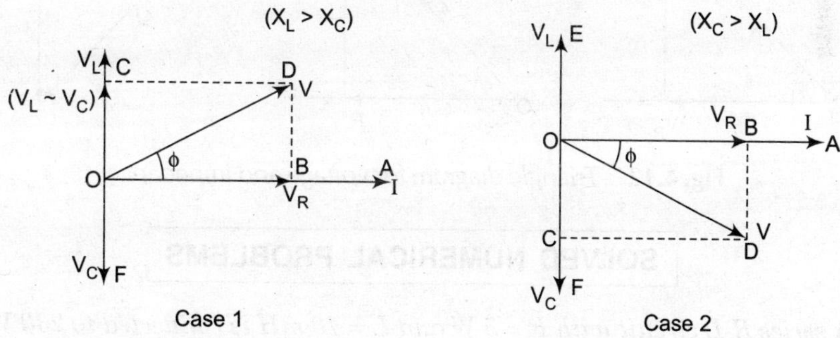

Fig. 4.11 *Phasor diagram for R-L-C series circuit case 1 and case 2.*

Case 1: From the phasor diagram drawn above,

$$OD = \sqrt{\{(OB)^2 + (BD)^2\}}$$

$$V = \sqrt{\{(OB)^2 + (OC - OF)^2\}}$$

$$V = \sqrt{(V_R^2 + (V_L - V_C)^2)}$$

$$V = \sqrt{\left\{(I \cdot R)^2 + (I \cdot X_L - I \cdot X_L)^2\right\}}$$

$$V = I\sqrt{\left\{R^2 + (X_L - X_C)^2\right\}}$$

Or, current flowing in the series circuit is

$$I = \frac{V}{\sqrt{R^2 + (X_L - X_C)^2}} = \frac{V}{Z}$$

Case 2: From the above phasor diagram

$$OD = \sqrt{\left(OB^2 + OC^2\right)} = \sqrt{\left\{OB_f^2 + (OF - OE)^2\right\}}$$

$$V = \sqrt{\left\{V_R^2 + (V_C - I \cdot X_L)^2\right\}}$$

$$= \sqrt{\left\{(I \cdot R)^2 + (I \cdot X_C - I \cdot X_L)^2\right\}} = I\sqrt{\left\{R^2 + (X_C - X_L)^2\right\}}$$

Or, current flowing in the series circuit is

$$I = \frac{V}{\sqrt{\left\{R^2 + (X_C - X_L)^2\right\}}}$$

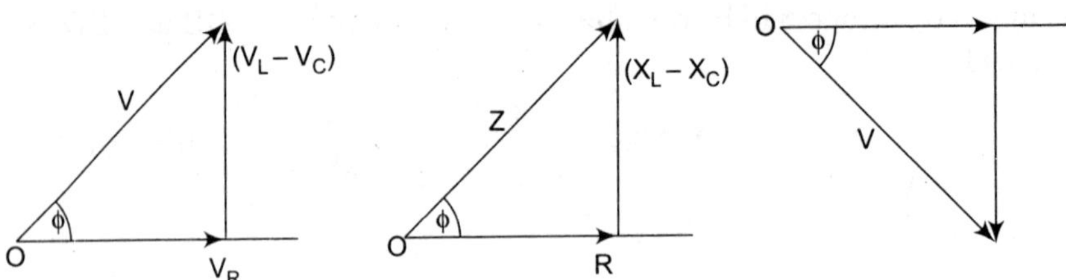

Fig. 4.12 *Triangle diagram for voltage and impedance.*

SOLVED NUMERICAL PROBLEMS

Example 1. *A series R-L circuit with R = 5 W and L = 10 mH is connected to 230 V, 50 Hz single phase supply. Calculate : (i) reactance ; (ii) impedance ; (iii) current drawn by the circuit; and (iv) P.f. of the circuit.*

Solution:

$R = 5$ W, $L = 10$ mH, $V = 230$ V, $f = 50$ Hz

$= 10 \times 10^{-3}$ H

(*i*) Reactance $\quad X_L = 2\pi f l$

$= 2 \times 3.14 \times 50 \times 10 \times 10^{-3}$ H

$= 3.14$ W

(*ii*) Impedance $\quad Z = \sqrt{R^2 + X_L^2}$

$$= \sqrt{(5)^2 + (3.14)^2}$$

$$= 5.90 \text{ W}$$

(*iii*) Current drawn $\quad I = \frac{V}{Z} = \frac{230}{5.90} = 38.98$ amp.

(*iv*) Power factor $\quad \cos\phi = \frac{R}{Z} = \frac{5}{5.90}$

$$\cos\phi = 0.85.$$

Example 2. *A series R-C circuit has R = 5 W and C = 100 mF is connected to 230 V, 50 Hz single phase supply. Calculate (i) impedance; (ii) reactance; (iii) current drawn by circuit; (iv) power factor.*

Solution: $\quad$ R = 5 W, C = 100 mF, V = 230 V, *f* = 50 Hz

(*i*) Reactance $\quad X_C = \frac{1}{WC} = \frac{1}{2\pi fc} = \frac{1}{2 \times 3.14 \times 50 \times 100 \times 10^{-6}}$

$$X_C = 31.85 \text{ W}$$

(*ii*) Impedance $\quad Z = \sqrt{R^2 + X_L^2}$

$$= \sqrt{(5)^2 + (31.85)^2}$$

$$Z = 32.24 \text{ W}$$

(*iii*) Current $\quad I = \frac{V}{Z}$

$$= \frac{230}{32.24} = 7.13 \text{ amp.}$$

(*iv*) Power factor $\quad \cos\phi = \frac{R}{Z} = \frac{5}{32.24}$

$$\text{Pf} = \cos\phi = 0.155 \text{ (reading).}$$

Example 3. *A series R-L-C circuit is connected to a 230 V, 50 Hz single phase a.c. supply. The value of R = 6 W, L = 15 mH and C = 125 mF. Find (i) total reactance (ii) impedance ; (iii) current drawn by the circuit ; (iv) P.f. of the circuit.*

Solution:

(*i*) Total reactance

$$X_L = W_L = 2 \times 3.14 \times 50 \times 15 \times 10^{-3}$$

$$X_L = 4.71 \text{ W}$$

$$X_C = \frac{1}{WC} = \frac{1}{2 \times 3.14 \times 50 \times 125 \times 10^{-6}}$$

$$X_C = 25.48 \text{ W}$$

Since $X_C > X_L$, so the circuit is capacitive

$$X = X_C - X_L = 25.48 - 4.71$$

$$X = 20.77 \text{ W}$$

(*ii*) Impedance $Z = \sqrt{R^2 + X^2}$

$$= \sqrt{(6)^2 + (20.77)^2}$$

$$Z = 21.62\ W$$

(*iii*) Current drawn $I = \frac{230}{21.62}$

$$I = 10.64\ \text{Amp.}$$

(*iv*) Power factor $\cos\phi = \frac{R}{Z}$

$$= \frac{6}{21.62}$$

$$\text{P.f.} = \cos\phi = 0.28\ \text{(leading).}$$

Example 4. *A capacitor of 100 mF is connected across a 200 V, 50 Hz single phase supply calculate (i) the reactance of the capacitor; (ii) r.m.s. value of current and (iii) the maximum current.*

Solution: $C = 100\ \text{mF} = 100 \times 10^{-6}\ \text{F}$

$$V = 200\ \text{V},\ \phi = 50\ \text{Hz}$$

(*i*) Reactance $X_C = \frac{1}{WC} = \frac{1}{2\pi fc}$

$$= \frac{1}{2 \times 3.14 \times 50 \times 100 \times 10^{-6}} = 31.85\ W$$

(*ii*) r.m.s. value of applied voltage = 200 V

r.m.s. value of current drawn:

$$I_{rms} = \frac{V}{X_C} = \frac{200}{31.85} = 6.28\ \text{amp.}$$

(*iii*) Maximum current

$$I_{max} = \sqrt{2}\ I_{rms} = \sqrt{2} \times 6.28 = 8.88\ \text{Amp.}$$

Example 5. *A non-inductive load takes 20 A at 200 V. Calculate the inductance of the reactor to be connected in series in order that some current be supplied from 240 V, 50 Hz supply. Also determine the phase angle between the 230 V supply and the current. Neglect the resistance of reactor.*

Solution: Voltage applied to non-inductive load = 200 V

Current flowing through the load = 20 A

Resistance of non-inductive load $= \frac{200}{20} = 10$ ohm.

Now a reactor of inductance L is connected in series with the non-inductive resistance of 10 W.

Impedance of the circuit

$$= \sqrt{R^2 + X_L^2} = \sqrt{(10)^2 + (2\pi \times 50 \times L)^2}$$

Voltage applied to the circuit = 240 V

Current flowing in the circuit = 20 A

Impedance $= \frac{V}{I}$

or $\sqrt{(10)^2 + (100\pi L)^2} = \frac{240}{20}$

$= 12$

$$(10)^2 + (100\pi L)^2 = (12)^2$$

$$(100\pi L)^2 = (12)^2 - (10)^2$$

$$100\pi L = \sqrt{44}$$

$$L = \frac{6.63}{100\pi}$$

$$L = 0.021 \text{ H.}$$

Phase angle between applied voltage and current

$$\phi = \tan^{-1}\left(\frac{X_L}{R}\right) = \tan^{-1}\left(\frac{6.63}{10}\right)$$

$$\phi = 33.54°.$$

Example 6. *A capacitor of 25 mF is connected in series with a variable resistor. The circuit is connected across a 50 Hz mains. Find the value of the resistor for a particular condition when the voltage across the capacitor is half the supply voltage.*

Solution: $C = 25 \text{ mF} = 25 \times 10^{-6} \text{ F}, f = 50 \text{ Hz}$

Capacitive reactance $X_C = \frac{1}{2\pi fc}$

$$= \frac{1}{2 \times 3.14 \times 50 \times 25 \times 10^{-6}} = 127 \text{ W}$$

Let the resistance of the variable resistor for the condition mentioned be R ohms.

Impedance $Z = \sqrt{R^2 + X_C^2}$

$$Z = \sqrt{(R)^2 + (127)^2}$$

Let the current drawn be I Amp.

$$V = I \cdot Z = I\sqrt{R^2 + (127)^2}$$

Voltage across the capacitor

$$V = I \cdot X_C = I \cdot (127)$$

As per given condition,

voltage across the capacitor $= \frac{1}{2}$ (Voltage applied)

$$I \cdot X_C = \frac{1}{2}(I \cdot Z)$$

$$I(127) = \frac{1}{2}\left(I \cdot \sqrt{R^2 + (127)^2}\right)$$

$$254 = \sqrt{R^2 + (127)^2}$$

$$R^2 = (254)^2 - (127)^2$$

$$R = \sqrt{(254)^2 - (127)^2}$$

$$R = 219 \text{ W}.$$

Example 7. *A two element series circuit is connected across an a.c. source V = $100\sqrt{2}$ sin (314t + 25°) V. The current in the circuit then is found to be i = $10\sqrt{2}$ sin (314t + 45°) amp. Determine the parameters of circuit.*

Solution:

$$V = 100\sqrt{2}\,\sin(314t + 25°)\text{ V}$$

$$i = 10.\sqrt{2}\,\sin(314t + 45°)\text{ A}$$

Phase angle between voltage and current (45 – 25°) = 20° *i.e.*, current leads

$$\text{P.f.} = \cos 20° = 0.94$$

Impedance
$$Z = \frac{V}{i} = \frac{100}{10} = 10 \text{ W}$$

so,
$$R = Z\cos\phi$$

$$R = 10\cos 20° = 9.4 \text{ W}$$

$$X_C = Z\sin\phi$$

$$= 10\,.\sin 20°$$

$$X_C = 3.42 \text{ W}$$

$$C = \frac{1}{2\pi f X_C} = \frac{1}{2 \times 3.14 \times 50 \times 3.42}$$

$$C = 9.31 \times 10^{-4}\text{ F}$$

$$C = 9.31 \text{ mF}.$$

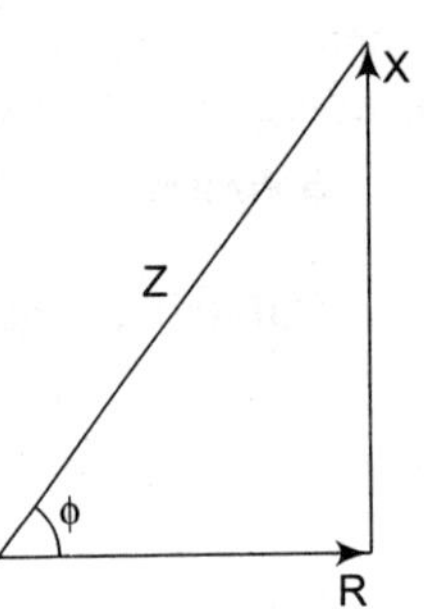

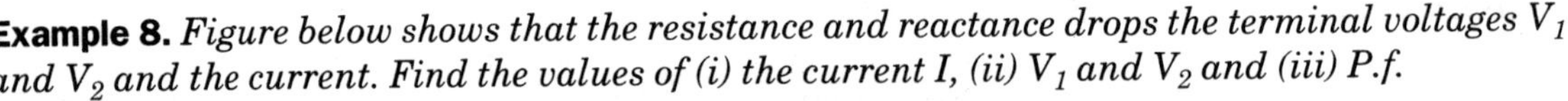

Example 8. *Figure below shows that the resistance and reactance drops the terminal voltages V_1 and V_2 and the current. Find the values of (i) the current I, (ii) V_1 and V_2 and (iii) P.f.*

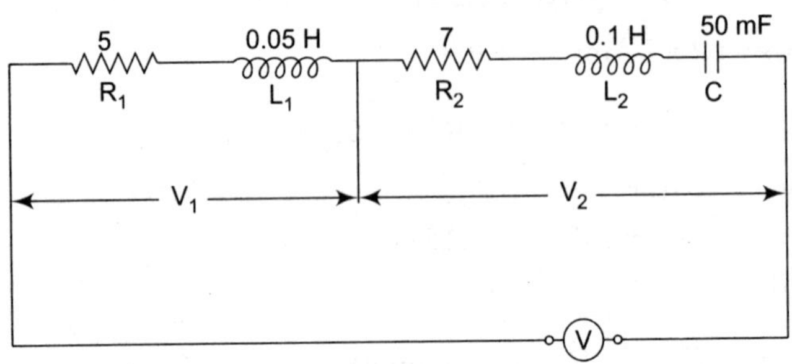

Solution:

$$R = 5 + 7 = 12 \text{ W}$$

$$L = 0.05 + 0.1 = 0.15 \text{ H}$$

$$C = 50 \text{ mF} = 50 \times 10^{-6}\text{ F}$$

So,

$$X_L = 2\pi f L$$
$$= 2 \times 3.14 \times 50 \times 0.15$$
$$= 47.1 \text{ H}$$

$$X_C = \frac{1}{2\pi fc} = \frac{1}{2 \times 3.14 \times 50 \times 50 \times 10^{-6}}$$

$$X_C = 63.70 \text{ W}$$

$$Z = \sqrt{R^2 + (X_L - X_C)^2} = \sqrt{(12)^2 + (47.1 - 63.70)^2}$$

$$Z = 20.48 \text{ W}$$

Since $X_C > X_L$ so circuit is capacitive in nature.

(*i*) Current $$I = \frac{V}{Z} = \frac{230}{20.48} = 11.23 \text{ Amp.}$$

(*ii*) Terminal voltages $$V_1 = I \cdot Z_1 = 11.23 \cdot \sqrt{R_1^2 + X_{L_1}^2}$$

$$= 11.23\sqrt{(5)^2 + (2\pi \times 50 \times 0.05)^2}$$

$$V_1 = 185.04 \text{ volt}$$

$$V_2 = I \cdot Z_2$$

$$= 11.23\sqrt{R_2^2 + (X_{L_2} - X_C)^2}$$

$$= 11.23\sqrt{(7)^2 + \left(2\pi \times 50 \times 0.1 - \frac{1}{2\pi \times 50 \times 50 \times 10^{-6}}\right)^2}$$

$$V_2 = 11.23\sqrt{49 + (31.4 - 63.69)^2}$$

$$V_2 = 371 \text{ volt}$$

(*iii*) P.f. of circuit $$\cos\phi = \frac{R}{Z} = \frac{12}{20.48} = 0.586$$

$$\text{P.f.} = \cos\phi = 0.586 \text{ leading.}$$

Example 9. *Branch A of a parallel circuit of an inductive coil R = 50 W, L = 0.1 H and the branch B consists of a resistor of 45 W in series with a capacitor of 100 mF. Calculate the current, power and power factor of the circuit, when the supply voltage is 230 V at 50 Hz.*

Solution:

(*i*) Inductive reactance of branch A,

$$X_A = 2\pi f l$$
$$= 2 \times 3.14 \times 50 \times 0.1$$
$$= 31.4 \text{ W}$$

Impedance of branch A

$$Z_A = \sqrt{(50)^2 + (3.14)^2}$$
$$= 59.04 \text{ W}$$

Current in branch A

$$I_A = \frac{V}{Z_A} = \frac{230}{59.04}$$

$$I_A = 3.9 \text{ Amp.}$$

Phase angle $$f_A = \tan^{-1}\left(\frac{31.4}{59.04}\right)$$

$$f_A = 32.13°$$

Capacitive reactance of branch B, X_B

$$X_B = \frac{1}{2\pi fc}$$

$$= \frac{1}{2\pi \times 50 \times 100 \times 10^{-6}} = 31.83 \text{ W}$$

Impedance $$Z_B = \sqrt{(45)^2 + (31.83)^2} = 55.12 \text{ W}$$

$$I_B = \frac{V}{Z_B} = \frac{230}{55.12} = 4.17 \text{ Amp.}$$

Angle of lead

$$f_B = \tan^{-1}\left(\frac{X}{R}\right)$$

$$= \tan^{-1}\left(\frac{31.83}{45}\right) = 35.27°.$$

(ii) Sum of active components of branch current

$$= I_A \cos\phi_A + I_B \cos\phi_B$$

$$= 3.9 \cos 32.14 + 4.17 \cos 35.27°$$

$$= 6.7 \text{ A}$$

Sum of reactive components of branch current

$$= -I_A \sin\phi_A + I_B \sin\phi_B$$

$$= -3.9 \sin 32.14 + 4.17 \sin 35.27°$$

$$= 0.33 \text{ A}$$

Total amount drawn by the parallel

$$\text{circuit } I = \sqrt{(6.7)^2 + (0.33)^2}$$

$$I = 6.71 \text{ A.}$$

(iii) Power factor of the current

$$\cos\phi = \frac{6.7}{6.71} = 0.998 \text{ (leading)}$$

(iv) Power taken by the parallel current

$$= VI\cos\phi = 230 \times 6.71 \times 0.998$$

$$= 1541 \text{ W.}$$

Example 10. *A resistance of 5 W and inductance of 0.03 H are connected in parallel and fed from a since having V = 250 sin (314t + 50°). Determine (i), the total current (ii) the power factor, equivalent resistance and inductance of the circuit.*

Solution:

$$R = 4\text{ W}, L = 0.03\text{ H}, W = 3.14\text{ rad./s}$$

$$X_L = WL$$

$$= 3.14 \times 0.03 = 9.42\text{ W}$$

and

$$Z_L = R + jX_L$$

$$= 5 + j9.42 = 10.66\ \angle 6.204$$

$$Z_{eq} = \frac{R\,(j \times L)}{R \times j \times L}$$

$$= \frac{5\,(9.42\ \angle 96°)}{10.66\ \angle 62.04} = 4.42\ \angle 27.96°$$

(*i*) Total current

$$i = \frac{V}{Z_{eq}} = \frac{250\angle 50°}{4.42\angle 27.96}$$

$$i = 56.56\ \angle 22.04$$

$$i = 56.56\sin(314\,t + 22.04)\text{ A}$$

(*ii*) Power factor, equivalent resistance and inductance:

Power factor = cos φ = cos 27.96 = 0.883 (lassing). **Ans.**

$$Z = 4.42\ \angle 27.96$$

$$Z = 3.90 + j2.07$$

∴

$$R_{eq} = 3.9\ \Omega$$

$$X_{eq} = 2.07\ \Omega$$

$$L = \frac{X_{eq}}{W} = \frac{2.07}{314}$$

$$L = 6.60 \times 10^{-3}\text{ H}$$

$$L = 6.6\text{ mH.}$$

Example 11. *A voltage of 230 ∠53.8° is applied to two impedance in parallel. The value of impedance are (10 + j12) n and (9 – j16) W, determine P, Q and S in kW and kVA in each branch and power factor of the circuit.*

Solution:

$$V = 230\ \angle 53.8°$$

Impedances

$$Z_1 = (10 + j12)$$

$$Z_2 = (9 - j16)$$

$$Z_1 = \frac{1}{Z_1} = \frac{1}{(10 + j12)}$$

$$= \frac{10 - j12}{10^2 + 12^2} = \frac{10}{244} - \frac{j12}{244}$$

$$Z_1 = (0.04 - 10.049)\text{ W}$$

$$= 0.063\ \angle -50.77$$

$$Y_2 = \left(\frac{1}{9 - j16}\right) = \frac{9}{337} \times \frac{j16}{337}$$

$$Y_2 = (0.026 + j0.047)\ \text{W}$$

$$Y_2 = 0.054 \angle 61.05$$

$$V = 230 \angle 53.8$$

$$V = 135.84 + j185.6$$

$$I_1 = VZ_1$$

$$= (230 \angle 53.8)(0.063 \angle -50.77)$$

$$= 14.49 \angle 3.303$$

$$I_1 = (14.47 + j0.76)\ \text{amp.}$$

$$I_2 = VZ_2$$

$$= (230 \angle 53.8)(0.054 \angle 61.05)$$

$$= 12.42 \angle 114.85$$

$$= (-5.2 + j11.27)\ \text{amp.}$$

For branch 1,

$$S_1 = VI_1$$

$$= (230 \angle 53.8)(14.49 \angle -3.03)$$

$$= 333.27 \angle 50.77°$$

$$S_1 = 2107.72 + j2581$$

$$P_1 = 2107.72\ \text{W}$$

$$= 2.11\ \text{kW}$$

$$Q_1 = 2581\ \text{VA}$$

$$= 2.581\ \text{kVA}$$

$$S_1 = 3.33\ \text{kVA}$$

For branch 2,

$$S_2 = VI_2$$

$$= (230 \angle 53.8)(12.48 \angle -114.85)$$

$$= 2870.4 \angle -61.05$$

$$= 1389.41 - j2511.72$$

$$P_2 = 1389.41\ \text{W}$$

$$P_2 = 1.4\ \text{kW}$$

$$Q_2 = -2511.72\ \text{VA}$$

$$Q_2 = -2.5\ \text{kVA}$$

$$S_2 = 2.87\ \text{kVA}$$

Current drawn from source $I = I_1 + I_2$

$$I = (14.47 + j0.76) + (-5.2 + j11.27)$$

$$I = (9.27 + j12.03)$$

$$I = 15.19 \angle 52.83$$

Current is lagging by an angle

$$= (53.80 - 52.83)$$
$$= 0.97°$$
$$\cos \phi = \cos (0.97°)$$
$$\cos \phi = 0.999.$$

Example 12. *Two circuits having the same numerical ohmic impedance are joined in parallel. The power factor of one circuit is 0.8 and the other is 0.75. What is the power factor of the combination?*

Solution: Let Z be impedance of each circuit

$$\cos \phi_1 = 0.8$$
$$Z_1 = Z \angle 36.86Z \qquad \phi_1 = 36.86°$$
$$Z_1 = Z\,(0.8 + j0.6) \qquad \cos \phi_2 = 0.75$$
$$Z_2 = Z \angle 41.41° \qquad \phi_2 = 41.41°$$
$$Z_2 = Z\,(0.75 + j0.66)$$

Since $Z_1 \,||\, Z_2$

So,
$$Z = \frac{Z_1 . Z_2}{Z_1 + Z_2} = \frac{(Z \angle 36.86)(Z \angle 41.41)}{Z\,(0.8 + j0.6 + 0.75 + j0.66)}$$
$$= \frac{Z \angle 78.27}{(\angle 1.55 + j1267)} = \frac{Z \angle 78.27}{2 \angle 39.26}$$
$$= \frac{Z}{2} \angle 39.01$$

Power factor of combination is

$$\cos \phi = \cos 39.01$$
$$\text{P.f.} = 0.777.$$

EXERCISES

1. Define the term inductive reactance of a coil and explain its variation with frequency.
2. What is the phase angle between the applied voltage and the current in a purely inductive circuit? Justify your answer analytically.
3. Define the term capacitive reactance and explain its variation with frequency. Also derive an expression for the instantaneous current in a purely capacitive circuit, when a sinusoidal voltage given by $e = E_{max} \sin \omega t$ is applied to the circuit.
4. An inductor having a reactance of 15 Ω and negligible resistance is connected across a 240 V, 50 Hz single phase supply. Find the current drawn by the coil. What will be the effect on the current drawn if (*i*) frequency is reduced to half and (*ii*) frequency is three time of fundamental? The supply voltage is kept constant.
5. A choke coil having a fixed resistance of 4 ohm and an inductive reactance of 15 ohm is connected in series with a variable resistor. A voltage of 230 V is applied to this circuit. For what value of variable resistor is the power consumed in it maximum? Also determine its value when the power consumed by the complete circuit is maximum.

6. An inductance of 50 mH is connected in series with a resistance of 10 ohm. The voltage applied to the circuit is 230 V, 50 Hz. Calculate (*i*) impedance, (*ii*) current, (*iii*) power absorbed and (*iv*) power factor.
7. A choke coil having a resistance of 10 Ω and inductance of 0.05 H is connected in series with a resistance of 20 Ω, an inductance of 0.06 H and a capacitance of 100 μF. The final circuit is energized from 230 V, 50 Hz supply. Calculate (*i*) Current taken from the supply, (*ii*) Voltage across the choke coil (*iii*) Voltage across the remaining part of the circuit.
8. A coil having resistance of 15 Ω and inductance of 0.05 H is connected in series with a condenser of 100 μF.The whole circuit has been connected to 230 V, 50 Hz supply. Calculate (*i*) impedance, (*ii*) current, (*iii*) power factor and absorbed power.
9. A resistor of 15 Ω and capacitance of 0.05 μF is connected in series. The circuit has been connected to 230 V, 50 Hz supply. Calculate (*i*) impedance, (*ii*) current, (*iii*) power factor and absorbed power. Also draw the phasor diagram with proper scaling.
10. A series circuit consisting of resistance of 100 Ω, inductance 0.05 H and a capacitance of 50 μF is energized fro a 230 V, 50 Hz mains. Determine (*i*) impedance (*ii*) the current in the circuit, (*iii*) power factor, (*iv*) the phase angle, (*v*) voltage across the resistance and (*vi*) voltage across the capacitance.
11. A resistor and a capacitor are connected in series across a 230 V a.c. supply. The current taken by the circuit is 6 A when the frequency of supply is 50 Hz. The current is reduced to 5 A when the frequency of the supply is decreased to 25 Hz. Determine (*i*) the value of resistor and (*ii*) the value of capacitor.
12. A coil having a resistance of 20 Ω and inductive reactance 25 Ω is connected in series with a capacitor of reactance 30 V. The circuit as a whole is connected across 230 V, 50 Hz supply. Determine (*i*) current drawn (*ii*) angle of phase difference between the applied voltage and the current, (*iii*) active component of current, (*iv*) reactive component of current, (*v*) power factor of the circuit and (*vi*) voltage across the coil.

5

Single Phase AC Parallel Circuit

5.1 INTRODUCTION

A parallel circuit consists of two or more branches connected in parallel. The voltage is same for each branch, but the current may differ in magnitude and phase depending upon the branch impedance. Each branch of the circuit is analyzed separately; the voltage is taken as the reference phasor. The magnitude and phase of each current is determined. The total current in the circuit is equal to the phasor sum of the branch current. Figure 5.1(*a*) shows an AC circuit possessing resistance R Ω, inductance L Henry and capacitance C farad all are connected in parallel across a supply of V volts and f Hertz. Current flowing in the resistive, inductive and capacitive branches are respectively represented by I_R, I_L and I_C. The current taken by this circuit from the supply is the phasor sum of the branch currents *i.e.* the phasor sum of I_R, I_L and I_C.

Current through the resistive branch R, $I_R = \dfrac{V}{R}$ and is in phase with applied voltage.

Current through the inductive branch L, $I_L = \dfrac{V}{X_L} = \dfrac{V}{(2\pi f L)}$

This current is lagging behind the applied voltage by 90°.

Current through the capacitive branch C, $I_C = \dfrac{V}{X_C} = 2\pi f CV$ ampere.

The current in the capacitive branch leads the applied voltage by 90°.

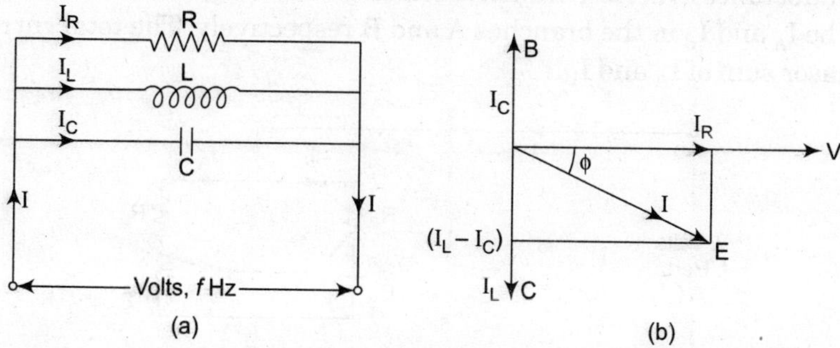

Fig. 5.1 *Fundamental parallel circuit and its phasor representation.*

In parallel circuits, voltage across the various branches of the circuit is the same and as such the applied voltage should be taken as reference phasor to draw the phasor diagram of parallel circuits. The phasor diagram of this circuit with applied voltage as reference phasor is shown in Fig. 5.1(*b*). Current in the inductive branch has been taken greater than that of the current in the capacitive branch.

Thus, the resultant of I_L and I_C is $(I_L - I_C)$.

Current drawn from the supply,

$$I = \text{Phasor sum of } I_R \text{ and } (I_L - I_C)$$

$$= \sqrt{\{I_R^2 + (I_L - I_C)^2\}}$$

The phasor difference between the applied voltage and the resultant current I is

$$\phi = \tan^{-1}\left[\frac{(I_L - I_C)}{I_R}\right]$$

$$\tan\phi = \frac{(I_L - I_C)}{I_R}$$

The resultant current lags the applied voltage, if tan ϕ is negative and leads when it is positive. The above circuit is basic one and its solution is quite simple. However parallel circuit may consist of a large number of branches connected in parallel and each branch might be consisting of more than one element connected in series.

5.2 METHODS TO SOLVE PARALLEL CIRCUITS

When a number of series circuits are connected in parallel across an a.c. supply, the total current taken by the circuit as a whole is the phasor sum of the currents drawn by various branches of the circuit. The total current in such cases can be calculated by either of three methods. (1) Phasor diagram method, (2) Admittance method and (3) Symbolic method.

5.2.1 Phasor Diagram Method

A parallel circuit consisting of two branches has been shown in Fig. 5.2. Branch A consists of a resistance and inductance in series, branch B consists of a resistance and capacitance in series. Let the current be I_A and I_B in the branches A and B respectively. The total current I drawn by the circuit is phasor sum of I_A and I_B.

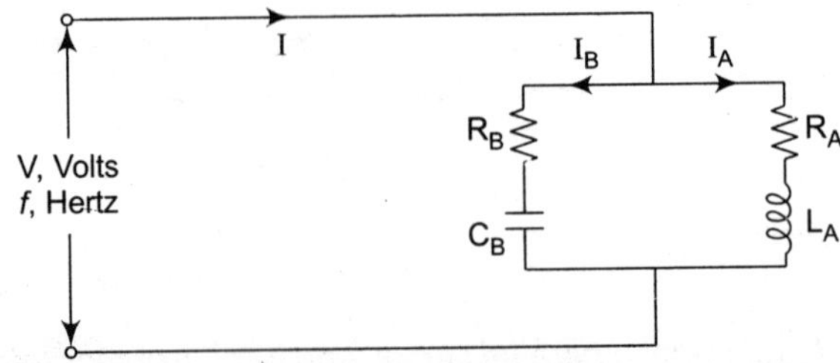

Fig. 5.2 *Parallel circuit with capacitive and inductive branch.*

Branch A

Voltage across branch A $= V$, volts

Impedance of branch A, $Z_A = \sqrt{\{(R_A)^2 + (X_{LA})^2\}}$

Thus current taken by branch A, $I_A = \dfrac{V}{Z_A}$

$$= \frac{V}{\{\sqrt{R_A^2 + X_{LA}^2}\}}$$

Phase difference of this current with respect to the applied voltage is given by

$$\phi_A = \tan^{-1}\left(\frac{X_{LA}}{R_A}\right)$$

This current will lag the applied voltage by an angle ϕ_A.

Branch B

Branch B is a capacitive branch and thus the current I_B will lead the applied voltage

Impedance of branch B, $Z_B = \sqrt{\{R_B^2 + X_{CB}{}^2\}}$

Current drawn by branch B, $I_B = \dfrac{V}{Z_B}$

$$= \frac{V}{\{\sqrt{R_B^2 + X_{CB}^2}\}}$$

The branch current, I_B leads the applied voltage V by an angle ϕ_B given by

$$\phi_B = \tan^{-1}\left(\frac{X_{CB}}{R_B}\right)$$

Now, we can drawn the phasor diagram for above currents and phase angle with voltage as reference vector or phasor (OC).

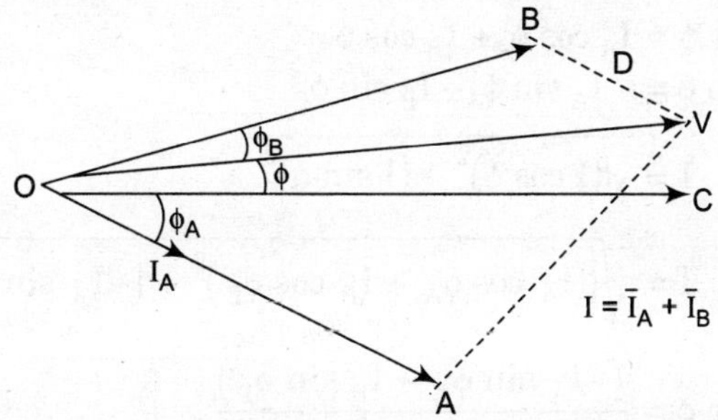

Fig. 5.3 *Phasor diagram of parallel circuit.*

Currents in the branches A and B have been represented by OA and OB respectively. Phasor sum of OA and OB is given by OD, which represents total current, I. The power factor of the circuit can also be obtained from the phasor diagram. This method of finding the resultant current by drawing a phasor diagram, gives clear insight into the problem. This method can be used for complicated circuit also.

5.2.2 Analytical Method

The resultant current can also be obtained by resolving the branch currents, I_A and I_B into their X- and Y-components as shown in Fig. 5.4 and then combining these properly.

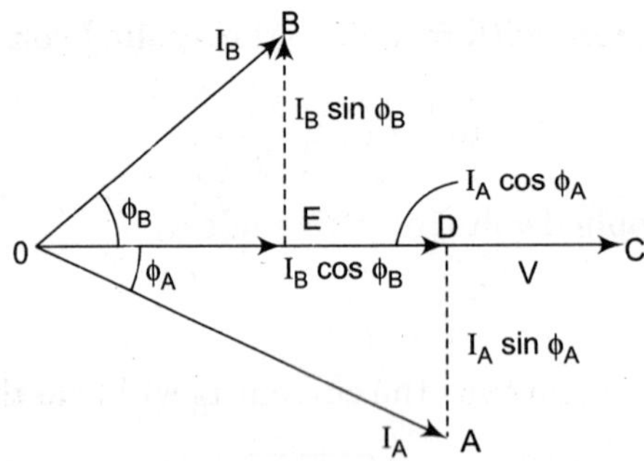

Fig. 5.4 *Resolution of branch current.*

From the above phasor diagram,

$$\text{X-component of } I_A(OD) = I_A \cos \phi_A$$
$$\text{X-component of } I_B(OE) = I_B \cos \phi_B$$

Sum of X-component (active components) of branch currents

$$= I_A \cos \phi_A + I_B \cos \phi_B$$

Now,

$$\text{Y-component of } I_A(AD) = -I_A \sin \phi_A$$
$$\text{Y-component of } I_B(BE) = I_B \sin \phi_B$$

Therefore the sum of Y components (reactive components) of branch currents

$$= -I_A \sin \phi_A + I_B \sin \phi_B$$

Active components of resultant current = $I \cos \phi$

Reactive component of resultant current = $I \sin \phi$

Hence,

$$I \cos \phi = I_A \cos \phi_A + I_B \cos \phi_B$$
$$I \sin \phi = -I_A \sin \phi_A + I_B \sin \phi_B$$

The resultant current $\quad I = \sqrt{(I \cos \phi)^2 + (I \sin \phi)^2}$

$$I = \sqrt{\left\{(I_A \cos \phi_A + I_B \cos \phi_B)^2 + (-I_A \sin \phi_A + I_B \sin \phi_B)^2\right\}}$$

$$\tan \phi = \frac{(-I_A \sin \phi_A + I_B \sin \phi_B)}{(I_A \cos \phi_A + I_B \cos \phi_B)}$$

$$= \frac{\text{Sum of reactive components of branch currents}}{\text{Sum of active components of branch currents}}$$

Power factor of the circuit as a whole is,

$$\cos \phi = \frac{(I_A \cos \phi_A + I_B \cos \phi_B)}{I}$$

$$= \frac{\text{Sum of active components of branch currents}}{\text{Resultant current}}$$

The analytical method of finding the resultant current is quick and more convenient but does not give clear insight into the problem.

5.2.3 Admittance Method

The admittance Y is defined as reciprocal of impedance

$$Y = \frac{1}{Z} = \frac{I}{V}$$

Unit of admittance is siemens (S)

Pure resistance : For a circuit containing pure resistance R, the admittance is

$$Y_R = \frac{1}{Z_R} = \frac{1}{R} = G$$

where G is the conductance, it is the reciprocal of resistance.

Pure inductance : For a circuit containing pure inductance L, the admittance in

$$Y = \frac{1}{Z_L} = \frac{1}{jX_L} = -\frac{j}{X_L} = -jB_L$$

where B_L is called the inductive susceptance. It is the reciprocal of the inductive reactance X_L.

Inductive susceptance

$$B_L = \frac{1}{X_L} = \frac{1}{\omega L}$$

Pure capacitance : For a circuit containing pure capacitance C, there admittance is

$$Y_C = \frac{1}{Z_C} = \frac{1}{-jX_C} = \frac{j}{X_C} = j\omega C = j\,B_C$$

where B_C is called the capacitive susceptance, it is the reciprocal of capacitive reactance X_C.

Capacitive susceptance

$$B_C = \frac{1}{X_C} = \omega C$$

A circuit in which $X_L > X_C$ is called an inductive circuit. Similarly a circuit in which $X_C > X_L$ is called capacitive circuit.

Impedance Z of the circuit can be resolved into two components, a real components R (Resistance) and an imaginary component X(reactance) as shown in the impedance triangle Fig. 5.5(*a*).

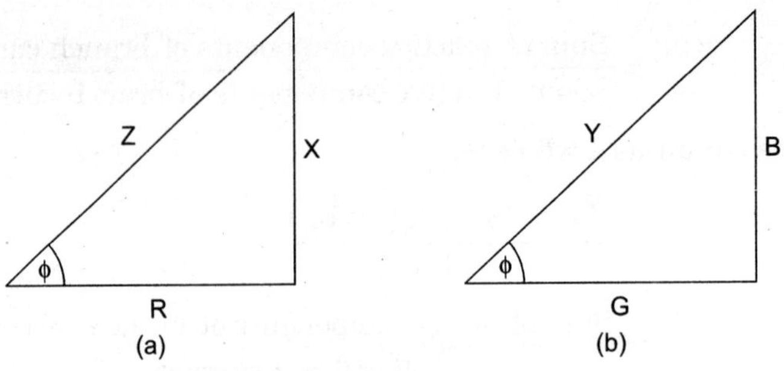

Fig. 5.5 *(a) Impedance triangle (b) Admittance triangle.*

Similarly the admittance may be resolved into two components, a real component called conductance G and an imaginary component termed susceptance B as shown in the admittance triangle in Fig. 5.5(*b*).

$$\tan\phi = \frac{X}{R} = \frac{B}{G}$$

$$Y^2 = G^2 + B^2$$

Admittance

$$Y = \sqrt{(G^2 + B^2)}$$

Resultant current, $I = V \cdot Y$

Power factor of the circuit,

$$\cos\phi = \frac{G}{Y}$$

5.3 ADMITTANCE OF A SERIES CIRCUIT

$$Y = \frac{1}{Z} = \frac{1}{(R + jX_L)}$$

$$= \frac{(R - jX_L)}{\{(R + jX_L) \cdot (R - jX_L)\}}$$

$$= \frac{(R - jX_L)}{R^2 + X_L^2} = \frac{R}{(R^2 + X_L^2)} - \frac{jX_L}{(R^2 + X_L^2)}$$

$$= G - jB_L = \sqrt{(G^2 + B_L^2)}$$

Conductance of this circuit

$$G = \frac{R}{(R^2 + X_L^2)} = \frac{R}{Z^2}$$

Inductive susceptance of this circuit,

$$B_L = \frac{X_L}{(R^2 + X_L^2)} = \frac{X_L}{Z^2}$$

A series circuit consisting of resistance R and capacitive reactance X_C is considered, Impedance of this circuit

$$Z = \sqrt{(R^2 + X_C^2)} = R - jX_C$$

$$= \frac{(R + jX_C)}{(R - jX_C)\cdot(R + jX_C)} = \frac{R + jX_C}{(R^2 + X_C^2)}$$

$$= \frac{R}{(R^2 + X_C^2)} + \frac{jX_C}{(R^2 + X_C^2)} = G + jB_C$$

$$= \sqrt{(G^2 + B_C^2)}$$

Conductance $$G = \frac{R}{(R^2 + X_C^2)} = \frac{R}{Z^2}$$

Capacitive susceptance $$B_C = \frac{X_C}{(R^2 + X_C^2)} = \frac{X_C}{Z^2}$$

Hence for series circuit, the conductance is the ratio of its resistance to the square of its impedance and the susceptance is the ratio of its reactance to square of impedance.

5.4 ADMITTANCE OF A PARALLEL CIRCUIT

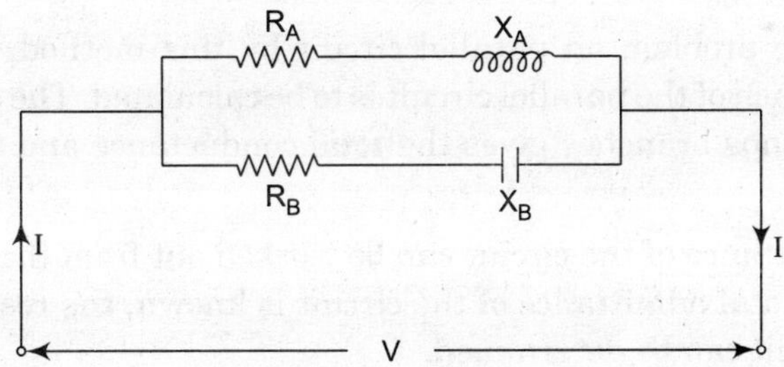

Fig. 5.6 *Parallel circuit.*

Conductance of the branch A, $$G_A = \frac{R_A}{(R_A^2 + X_A^2)} = \frac{R_A}{Z_A^2}$$

Inductive susceptance of branch A, $$B_A = \frac{X_A}{(R_A^2 + X_A^2)} = \frac{X_A}{Z_A^2}$$

Admittance of branch A, $$Y_A = G_A - jB_A$$

$$= \frac{R_A}{Z_A^2} - \frac{j.X_A}{Z_A^2}$$

Conductance of branch B, $$G_B = \frac{R_B}{(R_B^2 + X_B^2)} = \frac{R_B}{Z_B^2}$$

Capacitive susceptance of branch B, $B_B = \dfrac{X_B}{(R_B^2 + X_B^2)} = \dfrac{X_B}{Z_B^2}$

Admittance of branch B, $Y_B = G_B + jB_B$

$$= \frac{R_B}{Z_B^2} + \frac{j.X_B}{Z_B^2}$$

Total admittance of the parallel circuit

$$Y = Y_A + Y_B \text{ (Phasor sum)}$$
$$= (G_A - jB_A) + (G_B + jB_B)$$
$$= (G_A + G_B) - j.(B_A + B_B)$$

$$Y = \frac{R_A}{Z_A^2} + \frac{R_B}{Z_B^2} - j\left(\frac{X_A}{Z_A^2} - \frac{X_B}{Z_B^2}\right)$$

Total admittance, $Y = \sqrt{\left[\left(\dfrac{R_A}{Z_A^2} + \dfrac{R_B}{Z_B^2}\right)^2 + \left(\dfrac{X_A}{Z_A^2} - \dfrac{X_B}{Z_B^2}\right)^2\right]}$

Total current drawn by the parallel circuit I = VY

Power factor of the circuit as whole

$$\cos\phi = \frac{G}{Y} = \frac{(G_A + G_B)}{Y}$$

Hence in solving the problem on parallel circuit by this method, first conductance and susceptance of each branch of the parallel circuit is to be calculated. The sum of the conductance and susceptance of various branches gives the total conductance and susceptance of parallel circuits respectively.

Therefore total admittance of the circuit can be worked out from the total conductance and susceptance. Once the total admittance of the circuit is known, the resultant current and the power factor of the circuit can be determined.

EXERCISES

1. A parallel circuit consisting of the following three branches is connected to a 220 V a.c. supply.

   Branch A: purely resistive, resistance, 10 ohm.

   Branch B: inductive, resistance 5 Ω, reactance, 8 Ω

   Branch C: capacitive, resistance, 6 Ω reactance 16 Ω

   Calculate (*i*) current in each branch, (*ii*) total current taken from the supply and (*iii*) power of the combined circuit.

2. Two inductive coils of resistançes 10 and 20 Ω, and inductive reactance's of 7.5 and 15 Ω are connected in parallel across a 230 V, a.c. mains. Calculate (*i*) current in each coil (*ii*) total current drawn by the combination and (*iii*) the power factor of the combination.

3. An inductive circuit having a resistance of 5 Ω and inductance of 10 μH and a capacitive circuit with resistance with resistance 8 Ω and capacitance 314 μF are connected in parallel across a 100 V, 50 Hz supply. Calculate (*i*) current through the inductive and capacitive circuit, (*ii*) angle of phase difference of branch currents with respect to applied voltage and (*iii*) total current drawn by the combination.
4. An inductive coil of resistance 15 Ω and inductive reactance 45 Ω is connected in parallel with a capacitor of capacitive reactance 50 Ω. The combination is charged with a 230 V, 50 Hz a.c. supply. Find the total current drawn by the circuit and its power factor. Draw the phasor diagram of the circuit with proper scaling.
5. A 200 W discharge lamp takes a current of 2 amps at unity power factor. Calculate the inductance of choke required to enable the lamp to work from 230 V, 50 Hz supply. Find also the capacitance of the condenser to be connected across the mains. Find also the capacitance of condenser to be connected across the mains to bring the resultant power factor unity.
6. Two impedances, $Z_1 = 8 + j6\ \Omega$ and $8 - j6\ \Omega$ are connected in parallel across a single phase, 50 Hz supply. Find the power taken by each branch if the total current drawn is 20 Amp.
7. A coil having resistance of 20 Ω and an inductance of 0.1 H is connected in parallel with 10 μF condenser across 230 V variable frequency supply. Calculate (*i*) the frequency and (*ii*) current through the coil, when the total current drawn by the parallel combination is in phase with the voltage.
8. A series -parallel circuit is arranged as follows:

   Branch A—resistance 8 Ω, Inductive reactance 20 Ω

   Branch B—resistance 10 Ω, Inductive reactance 5 Ω

   These two branches are connected in parallel. A third branch C is now connected in series with the parallel combination, whose resistance is 5 Ω and capacitive reactance, 2 Ω. Calculate (*i*) equivalent impedance of the entire circuit and (*ii*) phase angle between applied voltage and total current.
9. Three impedances $Z_A = (20 - j15)\ \Omega$, $Z_B = -j50\ \Omega$ and $Z_C = (12 + j13)\ \Omega$ are connected in parallel across a 230 V, 50 Hz supply. Using the symbolic method, find out the values of three branch currents and the total current in complex notation.
10. A series parallel circuit consists of three branches A, B, C in parallel. The parameters of the various branches are given below:

    Branch A: resistance 5 Ω, inductance 0.015 H

    Branch B: resistance 5 Ω, capacitance 0.015 μF

    Branch C: resistance 10 Ω.

    The entire circuit is connected across a voltage of 230 V, 50 Hz.Calculate (*i*) the impedance of the circuit, (*ii*) current flowing in each branch and (*iii*) power consumed in each branch.
11. Two impedances $Z_A = (10 + j15)\ \Omega$ and $Z_B = (10 + R) + j\,(25 - X_C)$ are connected in parallel across a single-phase a.c. supply. The current taken by the two impedance branches A and B are equal in magnitude and the phase angle between them 90°.Calculate (*i*) the values of R and X_C and (*ii*) the phase difference of the branch currents with respect to the applied voltage.
12. Three impedances, $Z_A = (11 + j4)\ \Omega$, $Z_B = (3 + j4)\ \Omega$ and $Z_C = (6 - j8)\ \Omega$ are in parallel. Find the conductance, susceptance and admittance of each branch and for total circuit.

6

Electric Resonance

6.1 INTRODUCTION

Resonance is defined as the condition in a circuit containing at least one inductor and one capacitor, when the supply voltage and the supply current are in phase. Thus, at resonance the equivalent impedance of the circuit is purely resistive. Since the supply voltage and supply current are in phase, the power factor of a circuit is unity. At resonance the circuit impedance Z and admittance Y are real quantities.

6.2 EFFECT OF FREQUENCY VARIATION IN R, L AND C SERIES CIRCUIT

An a.c. circuit containing resistance, inductance and capacitance in series has been shown in Fig. 6.1.

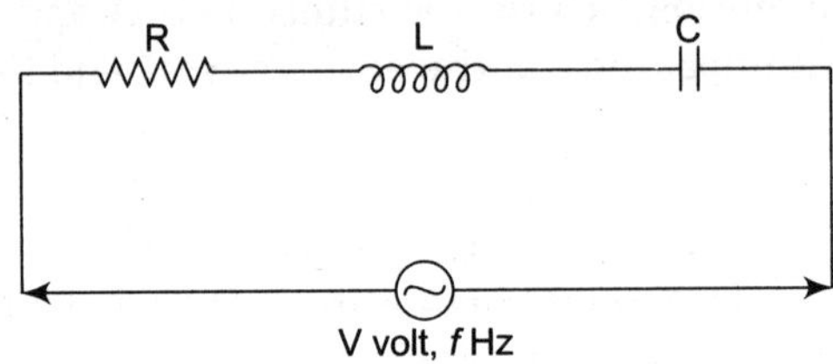

Fig. 6.1 *R-L-C Series circuit.*

The relationship of various parameters of this circuit with the frequency of supply is given by,

Inductive reactance of the circuit, $X_L = \omega L = 2\pi f L$

Capacitive reactance of the circuit, $X_C = \dfrac{1}{\omega C} = \dfrac{1}{2\pi f C}$

Resultant reactance $X = X_L - X_C = \left(2\pi f L - \dfrac{1}{2\pi f C}\right)$

Resultant reactance will be positive if inductive reactance is greater than the capacitive reactance, in case capacitive reactance is greater than the inductive reactance of the circuit, the resultant will be negative.

Impedance of the circuit, $Z = \sqrt{\{R^2 + (X_L - X_C)^2\}}$

$$= \sqrt{\left\{R^2 + \left(2\pi fL - \frac{1}{2\pi fC}\right)^2\right\}}$$

The above equation indicates that the various parameters of this circuit except resistance R are functions of frequency. Inductive reactance is directly proportional to the supply frequency, whereas capacitive reactance is inversely proportional to frequency. Thus on a plot of frequency vs. parameters of the circuit, inductive reactance is represented by the straight line and capacitive reactance by a rectangular hyperbola. Moreover capacitive reactance has been shown in the fourth quadrant, because of its negative nature. The curve of resultant reactance is obtained from the curves of X_L and X_C. In a similar way, the impedance curve has been plotted from resistance line and the effect of frequency upon various parameters of this circuit as discussed above.

From Fig. 6.2, it is observed that the resultant reactance is negative for all parameters below OA, and positive for frequencies greater than OA for a particular value of inductance and capacitance. It is equal to zero at the frequency OA. Hence the impedance of the circuit is minimum at the instant the resultant reactance becomes zero, *i.e.* at frequency f_o. At all other frequencies, the impedance is higher than this value. The r.m.s. value of current flowing in such a circuit is given by,

$$I = \frac{V}{Z}$$

$$I = \frac{V}{\sqrt{\{R^2 (X_L - X_C)^2\}}} = \frac{V}{\sqrt{\left\{R^2 \left(2\pi fL - \frac{1}{2\pi fC}\right)^2\right\}}}$$

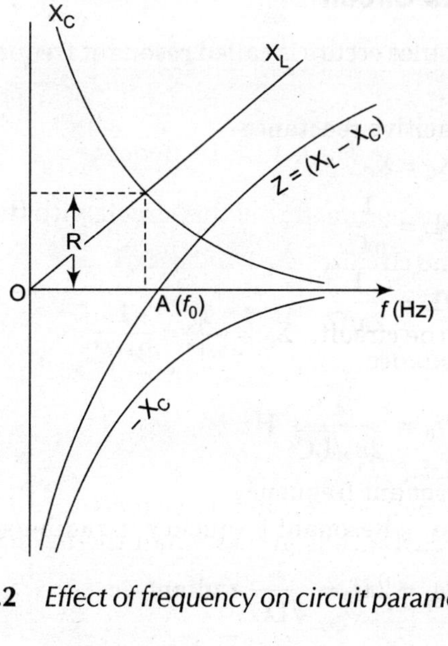

Fig. 6.2 *Effect of frequency on circuit parameters.*

Thus, the current is maximum when the impedance of the circuit is minimum, *i.e.* at frequency f_0.At all other frequencies other than f_0, impedance increases and therefore the current decreases Fig. 6.3 shows the effect of frequency variation upon the current drawn by the circuit and the voltages across the various parameters of the circuit. It is observed that the voltages across the inductance are equal to the voltage across the capacitance at the frequency f_0. Current is maximum and the impedance is minimum at this frequency. Moreover, the whole circuit behaves as a purely resistive circuit and thus the current is in phase with applied voltage at this frequency. This condition is said to be the condition for electrical resonance, an effect that is extremely important in radio work. Under the condition of resonance, the voltages across the inductor at capacitor are equal and may be many times greater than the applied voltage.

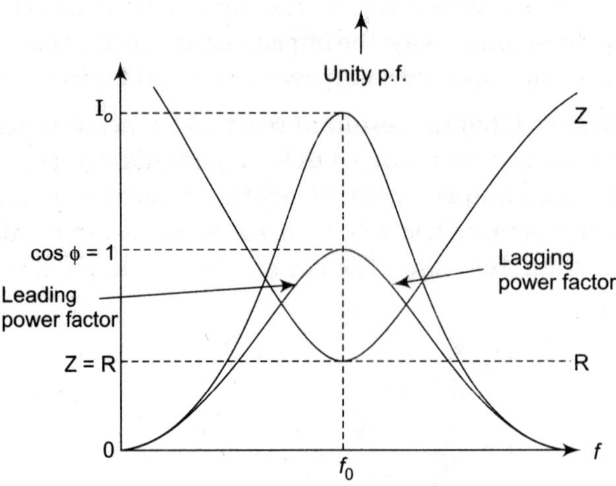

Fig. 6.3 *Effect of frequency on voltage, current and power factor.*

Resonance in R-L-C Series Circuit

The frequency at which resonance occurs is called resonant frequency, denoted by f_0. Resonance occurs when,

Inductive reactance = Capacitive reactance

$$X_L = X_C$$

$$\omega L = \frac{1}{\omega C}$$

$$2\pi f L = \frac{1}{2\pi f C}$$

Thus the frequency at resonance

$$f_0 = \frac{1}{2\pi\sqrt{LC}} \text{ Hz}$$

Fig. 6.4 *(a) When $V_L = V_C$, then $f = f_0$.*

This frequency is called resonant frequency.

If ω_0 = Resonant frequency in radian/second.

$$\omega_0 = 2\pi f_0 = \frac{1}{\sqrt{LC}} \text{ radian/sec.}$$

The series resonance effect may be provided by the following methods:

1. Varying frequency, keeping L and C constant.
2. Varying either L or C (or both) for a given frequency shown in Fig. 6.4.

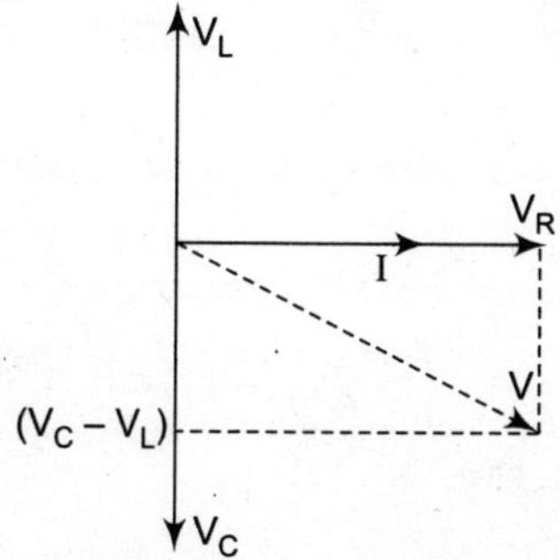

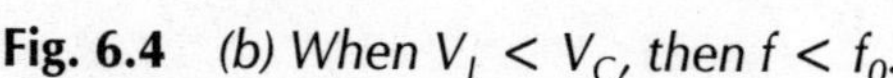

Fig. 6.4 *(b) When $V_L < V_C$, then $f < f_0$.*

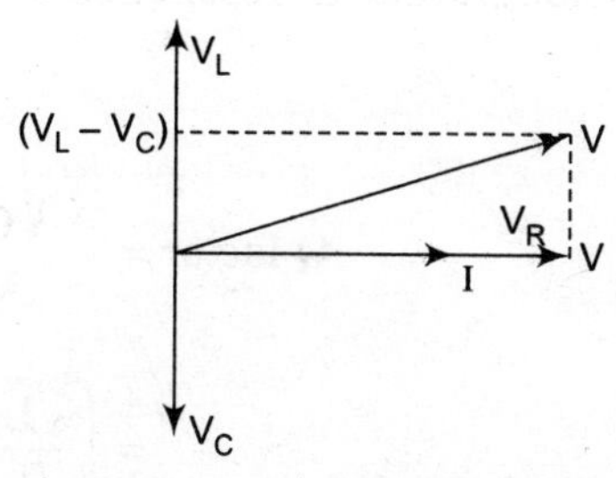

Fig. 6.4 *(c) When $V_L > V_C$, then $f > f_0$.*

6.3 VOLTAGE MAGNIFICATION

Voltage across the inductance at resonant frequency,

$$V_L = I \cdot X_L = I \cdot \omega L$$

$$= I \cdot (2\pi f_0)\, L$$

$$= I\left(\frac{2\pi \times 1}{2\pi\sqrt{LC}}\right) L = I \cdot \left(\sqrt{\frac{L}{C}}\right)$$

Current flowing in the circuit at resonance, $I = \dfrac{V}{R}$

$$V_L = \left[\frac{V}{R}\right]\left[\sqrt{\frac{L}{C}}\right]$$

$$V_L = V\left[\sqrt{\frac{L}{CR^2}}\right]$$

Voltage across the capacitance at resonance frequency is given by,

$$V_C = I \cdot X_C = I \cdot \left(\frac{1}{\omega C}\right)$$

$$= \frac{I}{2\pi f_0 C} = \frac{I}{2\pi \times \dfrac{1}{2\pi\sqrt{LC}}} \times C \qquad \left(f_0 = \frac{1}{2\pi\sqrt{LC}}\right)$$

$$= \frac{1}{2\pi C} \times \frac{2\pi\sqrt{LC}}{1} = 1\sqrt{\frac{L}{C}}$$

$$= \frac{V}{R}\left(\frac{1}{\sqrt{\dfrac{C}{L}}}\right) = \frac{V}{R}\left(\sqrt{\frac{L}{C}}\right) = V \times \sqrt{\frac{L}{CR^2}} \qquad \left(I = \frac{V}{R}\right)$$

Thus at resonant frequency f_0, both the voltages are equal and each is greater than applied voltage, hence a voltage magnification occurs at the resonance condition.

The voltage magnification at resonance is given by,

$$\text{Voltage magnitude at resonance} = \frac{\text{Voltage across L or C}}{\text{Supply voltage}}$$

$$\text{Q-factor} = \frac{V\sqrt{\dfrac{L}{CR^2}}}{V}$$

$$= \left(\frac{\sqrt{L}}{R}\right)\left(\frac{1}{\sqrt{C}}\right)$$

Multiplying neumerator and denominator by

$$\frac{\sqrt{L}}{R} \times \frac{\sqrt{L}}{\sqrt{LC}} = \frac{L}{R} \times \frac{1}{\sqrt{LC}}$$

Putting value of $\dfrac{1}{\sqrt{LC}} = 2\pi f_0$ so equation becomes

$$= \frac{L}{R} \times 2\pi f_0 = \frac{2\pi f_0 L}{R}$$

This is generally termed as Q-factor or quality factor of the circuit.

$$\text{Q-factor} = \frac{2\pi f_0 L}{R}$$

$$\text{Q-factor} = \frac{\omega_0 L}{R}$$

Here the current is maximum at resonance. Such a series resonant circuit is often referred as an acceptor circuit.

6.4 BANDWIDTH OF SERIES RESONANT CIRCUIT

The bandwidth of series resonant circuit is given by the band of frequencies, which lie between two points on either side of resonant frequency where the current falls to $\dfrac{1}{\sqrt{2}}$ or 0.707 of its maximum value at resonance. When the resistance of the circuit is zero *i.e.* curve 1, the current rises to infinity at resonant frequency f_0. The maximum current is comparatively high for low value of the resistance R and current curve exhibits very sharp peak as shown in curve 2. The maximum value of current corresponding to resonant frequency diminishes with a large value of circuit resistance and the current curves becomes flatter as indicated by the curves 3 and 4.

If a horizontal line AB at a height of $\frac{V}{R\sqrt{2}}$ is drawn, so as to cut any particular current - frequency (resonant) curve at f_1 and f_2 then $(f_2 - f_1)$ shown by the shaded strip in Fig. 6.5 is called the bandwidth. f_1 is called the lower cut-off frequency and f_2 is called the upper cutoff frequency point.

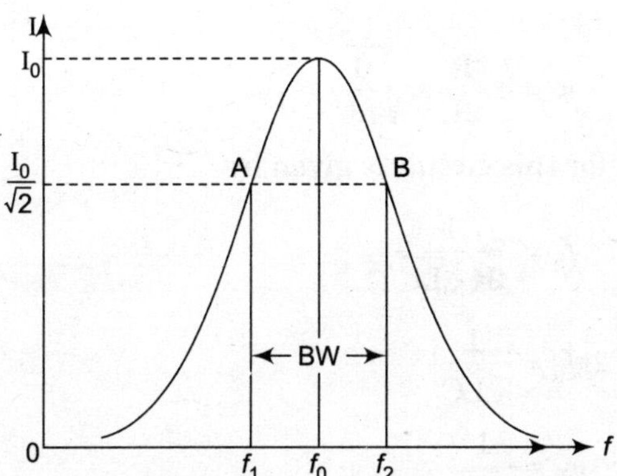

Fig. 6.5 *Bandwidth of resonant circuit.*

The current drawn by the circuit is given by

$$I = \frac{V}{\sqrt{R^2 + \left(\omega L - \frac{1}{\omega C}\right)^2}}$$

At point A and B shown in the Fig. 6.5, current

$$I = \frac{V}{R\sqrt{2}}$$

Thus to find f_1 or ω_1 and f_2 or ω_2 corresponding to points A and B

$$\frac{V}{R\sqrt{2}} = \frac{V}{\sqrt{R^2 + \left(\omega L - \frac{1}{\omega C}\right)^2}}$$

$$R^2 + \left(\omega L - \frac{1}{\omega C}\right)^2 = 2R^2$$

$$\left(\omega L - \frac{1}{\omega C}\right) \pm R = 0$$

$$(\omega^2 LC - 1 \pm R\omega C) = 0$$

$$\omega^2 - \frac{1}{LC} \pm \frac{R\omega}{L} = 0$$

or $$\omega^2 - \frac{R\omega}{L} - \frac{1}{LC} = 0$$

$$\omega = \pm \frac{R}{2L} \pm \sqrt{\frac{R^2}{4L^2} + \frac{1}{LC}}$$

For low values of R, the term $\left(\frac{R^2}{4L^2}\right)$ can be neglected in comparison with the term $\frac{1}{LC}$. Then ω is given by

$$\omega = \pm \frac{R}{2L} \pm \sqrt{\frac{1}{LC}}$$

The resonant frequency for this circuit is given by

$$f_0 = \frac{1}{2\pi\sqrt{LC}}$$

$$2\pi f_0 = \frac{1}{\sqrt{LC}}$$

$$\omega_0 = \frac{1}{\sqrt{LC}}$$

Substituting

$$\omega = \pm \frac{R}{2L} \pm \omega_0$$

$$\omega = \omega_0 \pm \frac{R}{2L}$$

$$\omega_1 = \omega_0 - \frac{R}{2L}$$

$$\omega_2 = \omega_0 + \frac{R}{2L}$$

$$\omega_2 - \omega_1 = \frac{R}{2L} + \frac{R}{2L}$$

$$\omega_2 - \omega_1 = \frac{R}{L}$$

$$f_2 - f_1 = \frac{R}{2\pi L}$$

Hence bandwidth of series resonant circuit is

$$f_2 - f_1 = \frac{R}{2\pi L}.$$

6.5 RESONANCE IN PARALLEL CIRCUITS

The basic condition of resonance, *i.e.* power factor of the entire circuit being unity, remains the same for parallel circuit also. So resonance will occur in a parallel circuit, when the power factor of entire circuit becomes unity.

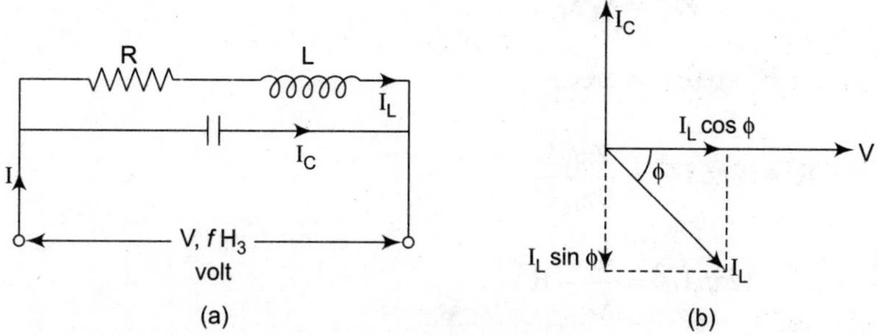

Fig. 6.6 *Parallel resonant circuit.*

The Fig. 6.6(*a*) shows a parallel circuit consisting of an inductive coil in parallel with a capacitor. The phasor diagram of the circuit with applied voltage as reference phasor has been drawn in Fig. 6.6(*b*). The current drawn by the inductive branch lags the applied voltage by an angle ϕ. It can be resolved into two components, an active component of current and reactive component as shown in Fig. 6.6(*b*). The current drawn by the capacitive branch leads the applied voltage by 90 degree. The power factor of the circuit becomes unity; when the total current drawn by entire circuit is in phase with the applied voltage. This will happen only when the current drawn by the capacitive branch I_C equal reactive component of current of inductive branch.

Hence for resonance in parallel circuit,

$$I_C = I_L \sin \phi$$

Current in capacitive branch, $I_C = \dfrac{V}{X_C}$

Current in inductive branch, $I_L = \dfrac{V}{Z_L}$

where Z_L is the impedance of inductive branch,

Impedance of inductive branch $Z_L = R + jX_L$

$$= \sqrt{\left(R^2 + X_L^2\right)}$$

angle of lag of inductive branch current with respect to applied voltage, as shown in Fig. 6.7

$$\phi = \tan^{-1}\left(\frac{X_L}{R}\right)$$

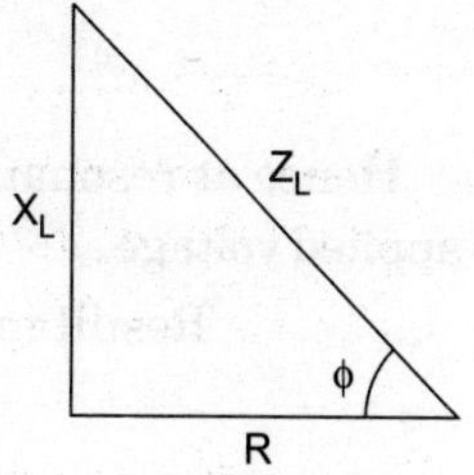

Fig. 6.7

Thus $$\sin \phi = \frac{X_L}{Z_L}$$

$$I_L \sin \phi = \left(\frac{V}{Z_L}\right) \cdot \left(\frac{X_L}{Z_L}\right)$$

$$= \left(\frac{VX_L}{Z_L^2}\right)$$

Resonance in parallel circuit will occur when

$$\frac{V}{X_C} = \frac{VX_L}{Z_L^2}$$

or
$$Z_L^2 = X_L X_C$$

$$R^2 + (\omega L)^2 = \omega L\left(\frac{1}{\omega C}\right)$$

$$R^2 + (2\pi f_0 L)^2 = \frac{2\pi f_0 L}{2\pi f_0 C}$$

$$(2\pi f_0 L)^2 = \frac{L}{C} - R^2$$

or
$$(2\pi f_0)^2 = \frac{1}{CL} - \frac{R^2}{L^2}$$

$$2\pi f_0 = \sqrt{\left(\frac{1}{LC} - \frac{R^2}{L^2}\right)}$$

$$f_0 = \left(\frac{1}{2\pi}\right)\left(\sqrt{\frac{1}{LC} - \frac{R^2}{L^2}}\right) \text{ Hz}$$

6.6 CURRENT MAGNIFICATION

At resonance in a parallel circuit, the branch current may be many times greater than the supply. Thus by means of a parallel resonant circuit, the current taken from the supply can be greatly magnified. Hence the type of resonance is called current resonant. This current magnification in parallel circuit is normally termed as Q-Factor of the circuit.

$$\text{Q-Factor of the circuit} = \frac{I_C}{I_L \cos\phi}$$

Putting value of I_C

$$= \frac{I_L \sin\Phi}{I_L \cos\Phi} = \tan\phi$$

$$= \frac{X_L}{R} = \frac{2\pi f_0 L}{R}$$

Hence at resonance the resultant current drawn by the parallel circuit is in phase with the applied voltage.

$$\text{Resultant current } I' = I_L \cos\phi$$

$$= \left(\frac{V}{Z_L}\right)\left(\frac{R}{Z_L}\right) = \frac{VR}{Z_L^2}$$

Under the condition of resonance,

$$Z_L^2 = X_L X_C$$

$$= \frac{\omega L \times 1}{\omega C} = \frac{L}{C}$$

Substituting for Z_L^2, resultant current

$$I' = \frac{VR}{\left(\frac{L}{C}\right)} = \frac{V}{\left(\frac{L}{CR}\right)}$$

Thus the impedance offered by a resonant parallel circuit, $Z_0 = \frac{L}{CR}$.

Z_0 is called the dynamic impedance of the circuit. This impedance is purely resistive because it is independent of frequency. It is seen that lower the resistance of the coil, the higher the value of Z_0. Hence the value of impedance at resonance is maximum and the resultant is minimum. A parallel resonant circuit is also called a rejector circuit. Since the current at resonance is minimum. In other words, a tank circuit almost rejects the current at resonance. Figure 6.8 shows the characteristics of the parallel circuit consisting of an inductance L and capacitance C in parallel plotted against frequency, the applied voltage being constant.

$$\text{Inductive susceptance} = \frac{1}{X_L} = \frac{1}{2\pi f L}$$

Thus inductive susceptance is inversely proportional to the frequency and it is represented by a rectangular hyperbola.

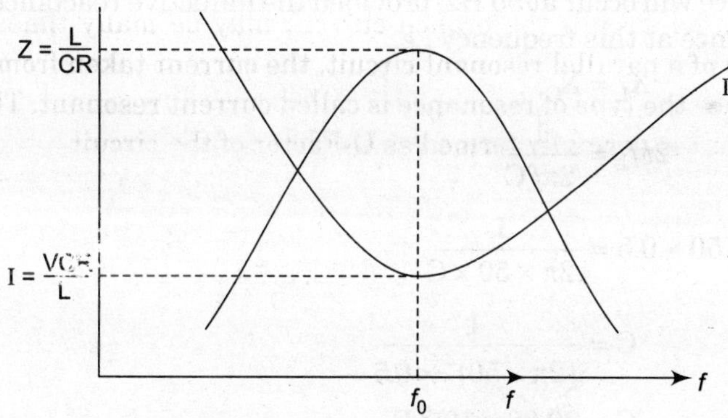

Fig. 6.8 *Resonance characteristic.*

Capacitive susceptance is directly proportional to frequency and is represented by a straight line passing through the origin. Total susceptance is the sum of the inductive and capacitive susceptance. As the conductance in the circuit is zero, the total susceptance is also equal to the total admittance of the circuit. The total current is zero (minimum) when the total admittance of the circuit becomes zero. The frequency at which total current becomes minimum is the resonant frequency f_0. Below the resonant frequency the inductive susceptance predominates thus making the circuit current to be lagging whereas beyond f_0 capacitive susceptance predominates and the current leads the applied voltage. At resonant frequency f_0, the current is in phase with the applied voltages.

SOLVED NUMERICAL PROBLEMS

Example 1. *The series R-L-C circuit is composed of components having R = 1.0 Ω, L = 100 mH and C = 50 μF. Calculate the resonance frequency and corresponding input current at 25 V.*

Solution: $R = 1.0, L = 100\ m\text{H and } C = 50\ \mu F = 5 \times 10^{-6}\ F$

$$V = 25\ V, L = 100 \times 10^{-3}\ H$$

Resonance frequency, $f_0 = \dfrac{1}{2\pi\sqrt{LC}}$

$$= \frac{1}{2 \times 3.14\sqrt{100 \times 10^{-3} \times 50 \times 10^{-6}}}$$

$$= 71.21\ Hz$$

At resonance $Z \approx R = 1.0\ \Omega$

$$I_0 = \frac{V}{R} = \frac{25}{1} = 25 \text{ amp.}$$

Example 2. *A circuit having a resistance of 10 Ω, an inductance of 0.5 H and a variable capacitance in series, is connected across a 110 V, 50 Hz supply. Calculate (i) the value of capacitance is give resonance, (ii) current, (iii) voltage across the inductance, (iv) voltage across the capacitance, and (v) Q-factor of the circuit.*

Solution : Resonance will occur at 50 Hz, provided the inductive reactance of the circuit equals the capacitive reactance at this frequency *i.e.*,

At resonance, $X_L = X_C$

$$2\pi f_L = \frac{1}{2\pi f C}$$

$$2\pi \times 50 \times 0.5 = \frac{1}{2\pi \times 50 \times C}$$

$$C = \frac{1}{(2\pi \times 50)^2 \times 0.5}$$

$$C = 20.28 \times 10^{-6}\ F$$

$$C = 20.28\ \mu F.$$

(*ii*) At resonance, $X_L = X_C$. Thus the impedance of the circuit,

$$Z = R = 10\ \Omega$$

Current drawn by the circuit $= \dfrac{110}{10} = 11\ A$

(*iii*) Inductive reactance of the circuit $X_L = 2\pi f L$

$$X_L = 2\pi \times 50 \times 0.5$$

$$= 157\ \Omega$$

Voltage across inductance, $V_L = I \cdot X_L$

$$= 11 \times 157$$

$$= 1727 \text{ volt.}$$

(*iv*) Capacitive reactance, $X_C = X_L = 157\ \Omega$

Voltage across the capacitor $= I \cdot X_C$

$$= 11 \times 157$$
$$= 1727 \text{ volt.}$$

(*v*) Q-factor of the circuit $= \dfrac{\text{Voltage across L or C}}{\text{Supply voltage}}$

$$= \frac{1727}{110} = 15.7.$$

Example 3. *A large coil of inductance 1.5 H and resistance 50 Ω is connected in series with a capacitor of capacitance 25 μF. Calculate the frequency at which the circuit resonates. If a voltage of 110 V is applied to the circuit at resonant condition, calculate drawn current from the supply and voltage across the coil and capacitor.*

Solution: $R = 50\ \Omega$, $L = 1.5$ H, $C = 25\ \mu F = 25 \times 10^{-6}$ F

$V = 110$ V

$$f_0 = \frac{1}{2\pi\sqrt{LC}} = \frac{1}{2\pi\sqrt{1.5 \times 25 \times 10^{-6}}} = 26 \text{ Hz}$$

$$I_0 = \frac{V}{R} = \frac{110}{50} = 2.2 \text{ A}$$

Reactance of coil at resonance

$$X_{L_0} = 2\pi f_0 L$$
$$= 2\pi \times 26 \times 1.5 = 244.92\ \Omega$$

Reactance of capacitor at resonance

$$X_{C_0} = X_{L_0} = 244.92\ \Omega$$

Impedance of the coil at resonance

$$Z_{coil} = \sqrt{R^2 + X_{L_0}^2} = \sqrt{(50)^2 + (244.92)^2}$$
$$= 250\ \Omega$$

Voltage across the coil $= Z_{coil} \cdot I_0$

$$= 250 \times 2.2 = 550 \text{ volt.}$$

Voltage across the capacitor $= X_{C_0} \cdot I_0$

$$= 244.92 \times 2.2 = 538.82 \text{ volt.}$$

Example 4. *A choke coil is connected in series with a 25 μF capacitor. With a constant supply voltage of 220 V the circuit takes its maximum current of 45 amp, when the supply frequency is 50 Hz. Calculate (a) the resistance and inductance of the choke coil; (b) the voltage across the capacitor.*

Solution: $f_0 = 50$ Hz, $V = 220$ V, $I = 45$ amp, $C = 25\ \mu F = 25 \times 10^{-6}$ F

(*a*) At resonance,

$$I_0 = \frac{V}{R}$$

$$R = \frac{V}{I_0} = \frac{220}{45} = 4.89\ \Omega.$$

$$f_0 = \frac{1}{2\pi\sqrt{LC}}$$

$$50 = \frac{1}{2 \times 3.14\sqrt{L \times 25 \times 10^{-6}}}$$

$$25 \times 10^{-6}\,L = \frac{1}{(314)^2}$$

$$L = \frac{1}{(314)^2 \times 25 \times 10^{-6}}$$

$$L = 0.405\text{ H.}$$

(*b*) Reactance of the capacitor

$$X_{C_0} = \frac{1}{2\pi f_0 C} = \left(\frac{1}{2\pi \times 50 \times 25 \times 10^{-6}}\right)$$

$$X_{C_0} = 127.39\ \Omega$$

Voltage across the capacitor

$$= X_{C_0} . I_0$$

$$= 127.39 \times 45$$

$$= 5732.48\text{ volts.}$$

Example 5. *In measurement a series R-L-C circuit resonance occur at 1.50 MHz if capacitance is 200 pF and Q-factor was 50. Determine the inductance and effective series resistance of the inductor; neglect resistance of capacitor.*

Solution: $f_0 = 1.5$ MHz, $C = 200$ pF

$$= 200 \times 10^{-12}\text{ F}$$

At resonance,

$$X_L = X_C \quad Q = 50$$

$$2\pi f L = \frac{1}{2\pi f C}$$

$$L = \frac{1}{(2\pi f)^2 C}$$

$$L = \frac{1}{(2\pi \times 1.5 \times 10^6)^2 \times 200 \times 10^{-12}}$$

$$L = 5.63 \times 10^{-5}\text{ H}$$

$$L = 56.3\ \mu\text{H}$$

$$Q = \frac{W_0 L}{R} = \frac{2\pi f_0 L}{R}$$

$$50 = \frac{2 \times 3.14 \times 1.5 \times 10^6 \times 56.3 \times 10^{-6}}{R}$$

$$R = 10.61\ \Omega.$$

Example 6. *If the bandwidth of a resonant circuit is 10 kHz and lower half-power frequency is 120 kHz, what is the value of the upper half power frequency? What is the value of the quality factor?*

Solution: Lower half power frequency = Resonant frequency – $\frac{1}{2}$ bandwidth

$$f_1 = f_0 - \frac{1}{2}\,\Delta f$$

$$120 = f_0 - \frac{1}{2} \times 10$$

$$f_0 = 120 + 5 = 125 \text{ kHz}$$

Upper half-power frequency = resonant frequency + $\frac{1}{2}$ bandwidth

$$f_2 = f_0 + \frac{1}{2}\,\Delta F$$

$$= 125 + \frac{1}{2} \times 10 = 130 \text{ kHz}$$

$$\text{Q-factor} = \frac{\text{Resonant frequency}}{\text{Bandwidth}}$$

$$Q = \frac{f_0}{\Delta f} = \frac{125}{10} = 12.5.$$

Example 7. *In the take circuit components are 500 μF capacitor and coil of resistance of 50 Ω and inductance 250 mH. Calculate the frequency of resonance and Q-factor.*

Solution:

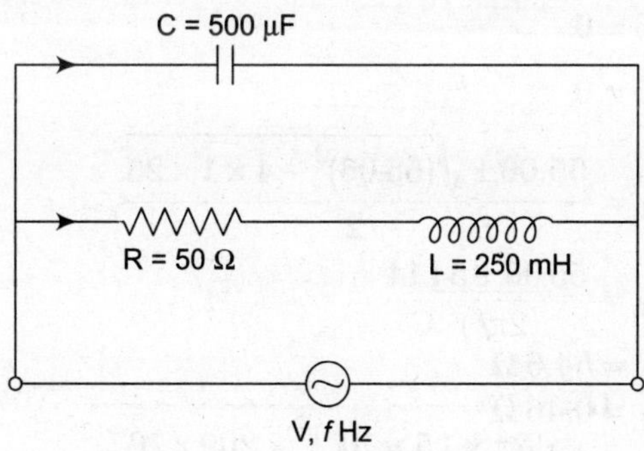

$$L = 250 \text{ mH} = 250 \times 10^{-3} \text{ H},\ C = 500\ \mu\text{F} = 500 \times 10^{-9}\ R = 50\ \Omega$$

Resonant frequency,

$$f_0 = \frac{1}{2\pi}\sqrt{\frac{1}{LC} - \frac{R^2}{L^2}}$$

$$= \frac{1}{2 \times 3.14}\sqrt{\frac{1}{250 \times 10^{-3} \times 500 \times 10^{-9}} - \left(\frac{50}{250 \times 10^{-3}}\right)^2}$$

$$= \frac{1}{2 \times 3.14} \sqrt{8{,}00{,}000 - 40{,}000} = 449.26 \text{ Hz.}$$

Example 8. *Find the value of L for the circuit drawn below in figure :*

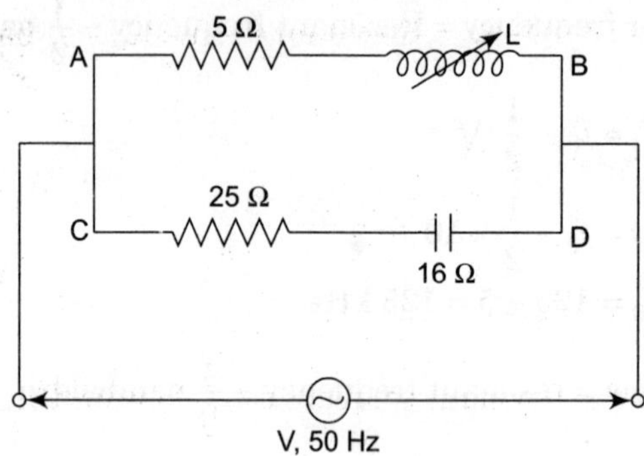

Solution: Let the admittance of branch AB be Y_1 and that of CD be Y_2.

$\therefore$
$$Y_1 = \frac{1}{5 + jX_L} = \frac{5 - jX_L}{25 + X_L^2} = \frac{5}{25 + X_L^2} - \frac{jX_L}{25 + X_L^2}$$

and
$$Y_2 = \frac{1}{25 - j16} = \frac{25 + j16}{625 + 256} = \frac{25}{881} + \frac{j16}{881}$$

$$Y_2 = (0.028 + j0.018)$$

At resonance,

$$\frac{X_L}{25 + X_L^2} = \frac{16}{881}$$

$$881\, X_L = 16\, X_L^2 + 25 \times 16$$

$$16\, X_L^2 - 881\, X_L + 25 \times 16 = 0$$

$$X_L^2 - 55.06\, X_L + 25 = 0$$

$$X_L = \frac{55.06 \pm \sqrt{(55.06)^2 - 4 \times 1 \times 25}}{2}$$

$$X_L = \frac{55.06 \pm 54.14}{2}$$

$$X_L = 54.6\ \Omega$$

$$X_L = 0.46\ \Omega$$

When $X_L = 54.6 \quad X_L = 2\pi f L$

$$L = \frac{54.6}{2 \times 3.14 \times 50}$$

$$L = 0.173 \text{ H}$$

and $X_L = 0.46$

$$L = \frac{0.46}{100\pi}$$

$$L = 1.46 \text{ mH.}$$

Example 9. *A parallel R-L-C circuit has inductance of 10 mH and resistance of 5Ω. What value of the capacitance will produce resonance frequency 800 Hz? Calculate also the quality factor of the circuit.*

Solution: $R = 5\ \Omega,\ L = 10\ \text{mH},\ f_0 = 800\ \text{Hz}$

$$L = 10 \times 10^{-3}\ \text{H}$$

Let the value of capacitance to be C farad

$$f_0 = \frac{1}{2\pi\sqrt{LC}}$$

$$800 = \frac{1}{2\pi\sqrt{10 \times 10^{-3} \times C}}$$

$$C = \frac{1}{(800 \times 2\pi)^2 (10 \times 10^{-3})}$$

$$C = 3.96 \times 10^{-6}\ \text{F}$$

$$C = 3.96\ \mu\text{F}$$

$$\text{Q-factor} = \omega_0 RC = 2\pi f_0 RC$$

$$= 2 \times 3.14 \times 800 \times 5 \times 3.96 \times 10^{-6}$$

$$= 0.099.$$

Example 10. *A series R-L-C circuit resonates at a frequency of 1500 Hz and consumes 75 W power for 50 V a.c. source at resonant frequency. The bandwidth is 0.75 kHz, calculate R, L and C.*

Solution: $f_0 = 1500\ \text{Hz},\ P = 75\ \text{W},\ V = 50\ \text{V},\ BW = 0.75\ \text{kHz} = 750\ \text{Hz}$

We know that,

$$P = \frac{V^2}{R}$$

or

$$R = \frac{V^2}{P} = \frac{(50)^2}{75} = 33.33\ \Omega$$

$$BW = \frac{R}{2\pi L}$$

or

$$L = \frac{R}{2\pi\ BW} = \frac{33.33}{2 \times 3.14 \times 750}$$

$$L = 7.08 \times 10^{-3}\ \text{H}$$

$$L = 7.08\ \text{mH}$$

Resonant frequency

$$f_0 = \frac{1}{2\pi\sqrt{LC}}$$

$$1500 = \frac{1}{2 \times 3.14\sqrt{7.08 \times 10^{-3}\ C}} = 1.56 \times 10^{-6}\ \text{F}$$

$$C = \frac{1}{(62.8 \times 1500)^2 \times 7.08 \times 10^{-3}} = 1.56\ \text{mF}.$$

EXERCISES

1. Explain briefly the phenomena of electrical resonance in a.c. circuits.
2. What happens when resonance takes place in an a.c. circuit? When the frequency of the supply is varied above and below the resonant frequency, explain what happens to the power factor of the circuit?
3. Discuss the effect of varying the frequency over a wide range upon the parameters of an RLC series circuit.
4. Discuss the effect of varying the frequency upon the current drawn and the power factor in RLC series circuit.
5. Draw the phasor diagram showing the voltage drop across the various elements of series RLC circuit. Explain the term resonance as applicable to this circuit and establish a relationship between L, C and the frequency for the resonance to, take place.
6. An inductive coil of resistance 2 ohm an inductance 0.05 H is connected to the following values of capacitance of the condenser in turn:(1) 250 μF, (2) 1000 μF and (3) 3000 μF. Find in each case frequency of the supply at which current drawn by the circuit will be maximum. Determine also the current and the voltage across the condenser in each case when voltage of 230 V, 50 Hz is applied to the circuit.
7. What are the features of resonance in parallel circuit? Calculate the value of C which in resonance for the circuit shown below when frequency is 1,000 Hz and find Q-factor for each branch. *(U.P.Technical. University, EE-2001)*

   4 Ω   5 Ω

   J8 Ω   C

8. Draw resonance curve for a series R-L-C series circuit. Locate LCF, UCF and BW. Write an expression for Q.
9. Find the half-power frequencies for R-L-C series circuit which has Q = 50 and f_0 = 50 kHz.
10. Show that the condition for resonance in a parallel R-L-C circuit is same as that in a series R-L-C circuit. State the application of resonance.

7

Polyphase Circuits

7.1 INTRODUCTION

Why sinusoidal function?

1. First reason is that the natural response of an under damped second-order system is a damped sinusoid, and if in losses are present it is a pure sinusoid. So sinusoid appears natural signal.
2. Second reason is that by Fourier theorem it proves that most of the useful mathematical functions of time which repeat themselves to times per second by the sum of an infinite number of sinusoidal time functions with frequency that are integral multiples of to; the given periodic function $f(t)$ can be approximated as closely as we wish by the sum of finite number of such terms.

   Therefore, decomposition of a periodic function into a number of approximately chosen sinusoidal forcing function is a very powerful analytical method, it enables us to superimpose the partial responses produced in any linear circuit by the respective sinusoidal components and thereby obtain the desired response caused by the given periodic forcing function so response to a sinusoidal forcing function is found is the dependence of other forcing function on Sinusoidal Analysis.
3. Sinusoidal function derivatives and integrals are also sinusoidal. Since the forcing response takes on the form of the forcing function, its integral and its derivatives, the sinusoidal forcing function will produce a sinusoidal forced response throughout a linear circuit. The sinusoidal forcing function thus allows a much earlier mathematical analysis than does almost every other forcing function.
4. It is an easy function to generate as the most household and industry utilizes the alternating current.

The entire output of the electric power industry world made is generated and distributed as polyphase power at 50 Hz or 60 Hz frequencies. Most common polyphase system is balanced three-phase system. For special application more number of phases are used.

The use of polyphase systems having higher numbers of phases, such as 6- and 12-phase system is limited to the supply to large rectifiers. Here, the rectifier converts AC into DC. As the number of phases increases, the ripple in the output of the rectifiers decreases.

The system is said to be balanced when the various voltages are equal in magnitude and displaced from one another by equal angles and draws power equally from the three phases but when one of voltages is instantaneously zero the phase relationship shows that the other two must be at half amplitude. So at any instant total instantaneous power remains constant. This is an advantage for rotating machinery to keep its torque on rotor much more constant than the single-phase source were used; thereby less vibration.

There are other advantages over single-phase system

1. Size of three-phase machine is smaller than single-phase machine of same capacity.
2. Conductor required by three-phase transmission system is less compared to equivalent single-phase system.
3. Voltage regulation is better in a three-phases system.
4. The instantaneous power in three-phase system is constant rather than pulsating in single phase system.

7.2 GENERATION OF A THREE-PHASE SUPPLY

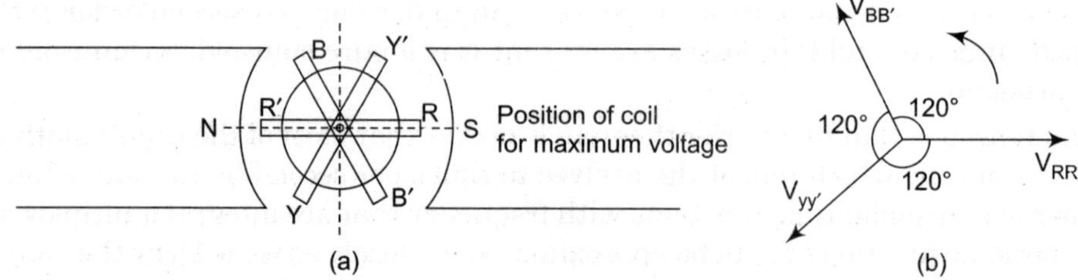

Fig. 7.1 *Three-phase power generation.*

When three identical coils RR', YY' and BB' are placed 120° apart and rotated in the uniform magnetic field, a sinusoidal voltage is generated across each coil (according the Faraday's law) In the Fig. 7.1(*a*) shows a three-phase two pole alternator. Its three sets of coils are symmetrically mounted on a rotor such that their axes are at 120° apart each other. When the rotor is rotated in the anticlockwise direction at constant angular velocity ω radians per second a sinusoidal voltage is generated across each coil. Since the coils are rotated at the same velocity ($\omega = 2\pi f$), the generated voltages have the same frequency. Coils are identical so generated voltages be of same magnitude but phase difference of 120° between each of voltages.

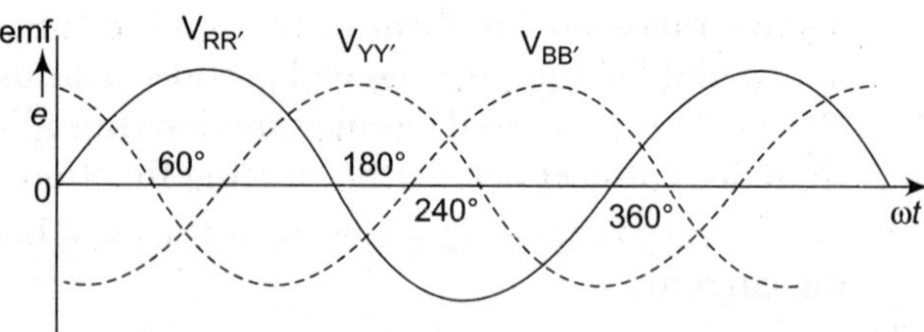

Fig. 7.1 *(c) Wave form of 3 – φ emfs.*

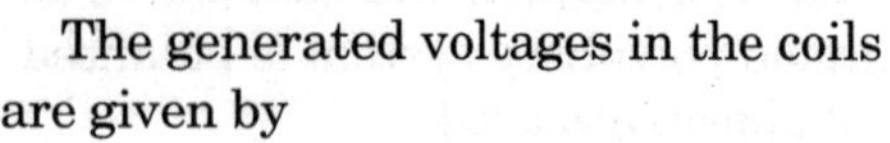
The generated voltages in the coils are given by

$$V_{RR'} = V_m \sin \omega t$$

$$V_{YY'} = V_m \sin (\omega t - 120°)$$

$$V_{BB'} = V_m \sin (\omega t - 240°)$$

In polar form

$$V_{RR'} = V_m\angle 0°, V_{YY'} = V_m\angle -120°, V_{BB'} = V_m\angle -240° \text{ or } V_m\angle 120°$$

It is seen that $V_{RR'}$ leads $V_{YY'}$ by 120° and $V_{YY'}$ leads $V_{BB'}$ by 120° also these three voltages reach their maximum positive values in the order of $V_{RR'}$, $V_{YY'}$, $V_{BB'}$. The order in which the phase voltages reach their maximum values is called phase sequence. For Fig. 7.2(*b*) shown coils are rotating in anticlockwise direction, the phase sequences of the rotor is rotated in clockwise direction the voltages reach their maximum positive values in the order $V_{RR'}$, $V_{YY'}$, $V_{BB'}$ to phase sequence in RYB. Thus the phase sequence determines the direction of rotation.

7.3 CONNECTION OF THREE-PHASE SUPPLIES

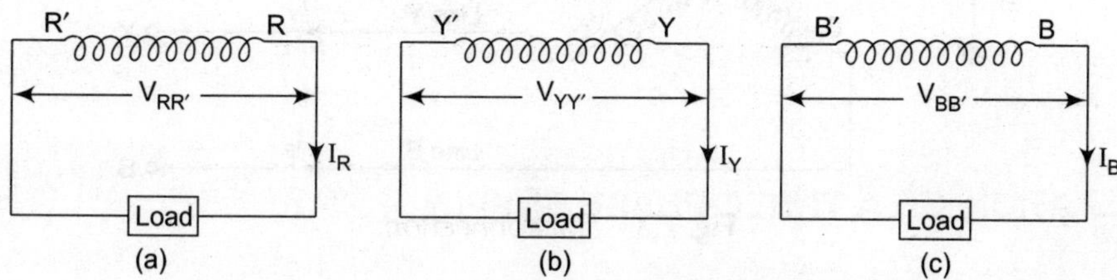

Fig. 7.2 *Coils of generator.*

Since voltage is generated in each coil so it would supply separate loads if isolated from each other. The arrangement is shown in the above Fig. 7.2.

From the Fig. 7.2 it requires six wires to carry energy from the coils to the loads. This is equivalent to three separate single phase system, such system is called a three-phase. If six wire-system of the coils are interconnected properly the number of connecting wires may be reduced. There are two methods of interconnection *i.e.* star and delta connections. It is preferred that the algebraic sum of ordinates of e.m.f. wave shapes of the three-phase is always zero, irrespective of the position of the ordinates. Hence the sum of instantaneous values of e.m.f. in a three-phase system is always zerom *i.e.*

$$V_{RR'} + V_{YY'} + V_{BB'} = 0$$

$$V_{RR'} + V_{YY'} + V_{BB'} = V_m \sin\theta + V_m \sin(\theta - 120°) + V_m \sin(\theta - 240°)$$

$$= V_m [\sin\theta + \sin(\theta - 120°) + \sin(\theta - 240°)]$$

$$= V_m [\sin\theta + \sin\theta \cos 120° - \cos\theta \sin 120° + \sin\theta \cos 240° - \cos\theta \sin 240°]$$

$$= V_m \left[\sin\theta + \sin\theta\left(-\frac{1}{2}\right) - \cos\theta\left(\sqrt{\frac{3}{2}}\right) + \sin\theta\left(-\frac{1}{2}\right) - \cos\theta\left(\sqrt{\frac{3}{2}}\right)\right]$$

$$= V_m \left[\sin\theta - \frac{1}{2}\sin\theta - \frac{1}{2}\sin\theta\right] = 0$$

Similarly for currents flowing through the three-phase circuit.

7.4 STAR (γ) CONNECTION

In star connection R′Y′B′ are connected together to form a common point N while R, Y and B to external circuit though three conductors called lines. The point N is called neutral or star point.

A wire brought out from star point is called neutral wire. The three line conductors and a neutral wire system is called three-phase four wires system. While three line conductors is called three-phase system. The neutral point is connected to ground; Fig. 7.3.

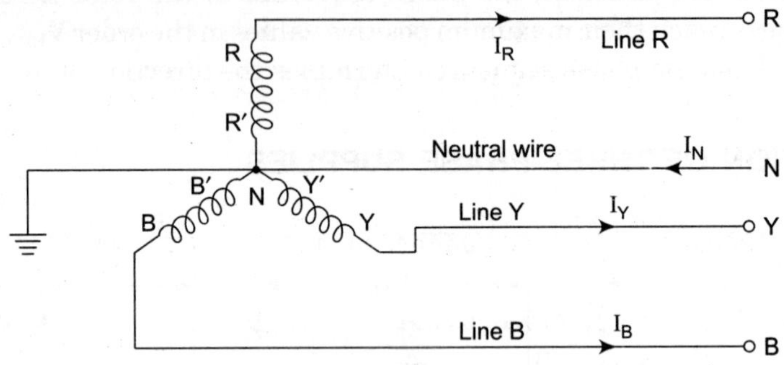

Fig. 7.3 *Star connection.*

7.5 DELTA (Δ) CONNECTION

If three isolated coils are connected to form a closed delta by starting end of one phase being joined the finishing end of another phase as shown in Fig. 7.4.

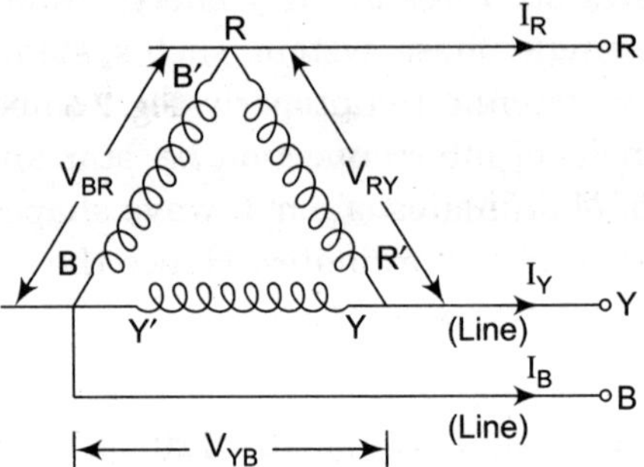

Fig. 7.4 *Delta connection.*

As earlier shown that algebraic summation of V_{RY}, V_{YB} and V_{BR} is zero so there will not be circulating current in the mesh.

7.6 LINE VOLTAGE, PHASE VOLTAGE AND CURRENT IN STAR CONNECTION

The voltage between two lines is called the line-to-line voltage or simply the line voltage. The line voltages are V_{RY} ,V_{YB} ,V_{BR} between the pairs of lines R-Y, Y-B and B-R respectively. In a symmetrical system these voltages are equal in magnitude and each is designated as V_L

$$|V_{RY}| = |V_{YB}| = |V_{BR}| = V_L$$

The voltage between each line and neutral is called the phase voltage (V_{Ph}). The three-phase voltages are V_{RN}, V_{YN} and V_{BN}. In symmetrical system these voltages are equal in magnitude

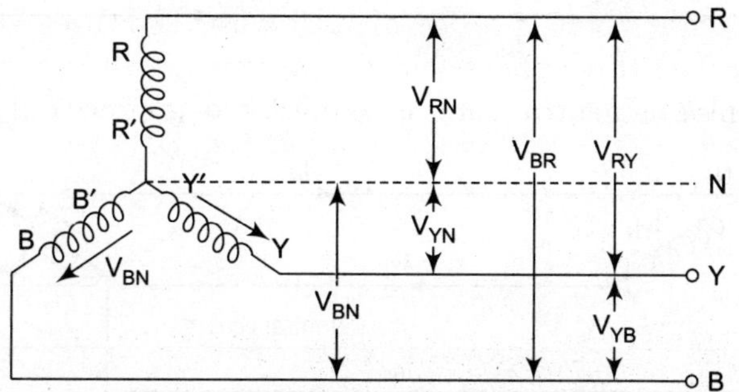

Fig. 7.5 *Star connection showing line and phase voltage.*

$$|V_{RN}| = |V_{YN}| = |V_{BN}| = V_{Ph}$$

$$V_{RN} = V_{Ph} \angle 0°$$

$$V_{YN} = V_{Ph} \angle -120°$$

$$V_{BN} = V_{Ph} \angle +120°$$

Line voltage $V_{YB} = V_{YN} - V_{BN} = 2V_{YN} \cos 30°$

where $V_{YN} = V_{Ph}$

so $$V_{YB} = 2V_{Ph}\left(\sqrt{\frac{3}{2}}\right) = \sqrt{3}\, V_{Ph}$$

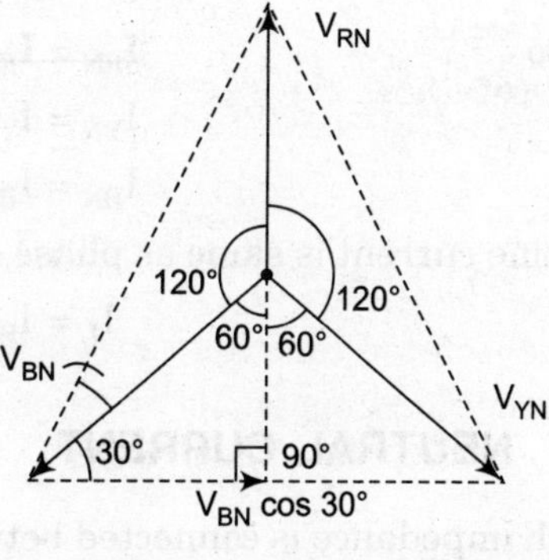

Fig. 7.6 *Star connection phasor diagram.*

Similarly, line voltage

$$V_{BR} = V_{BN} - V_{RN}$$

$$= 2V_{BN} \cos 30°$$

$$= 2V_{Ph}\left(\sqrt{\frac{3}{2}}\right) = \sqrt{3}\, V_{Ph}$$

Line voltage $V_{RY} = V_{RN} - V_{YN}$

$$= 2V_{RN} \cos 30° = \sqrt{3}\, V_{Ph}$$

Hence in star connected three-Phase system line voltage $V_L = \sqrt{3}$. Phase voltage (V_{Ph}).

While, the current in the line is equal to the phase current.

Hence in star connection

$$V_L = \sqrt{3}\, V_{Ph}$$

$$I_L = I_{Ph}$$

The angle between the line currents and the corresponding line voltage is $(30 + \phi)$ the case of lagging load and $(30 - \phi)$, in case of leading loads.

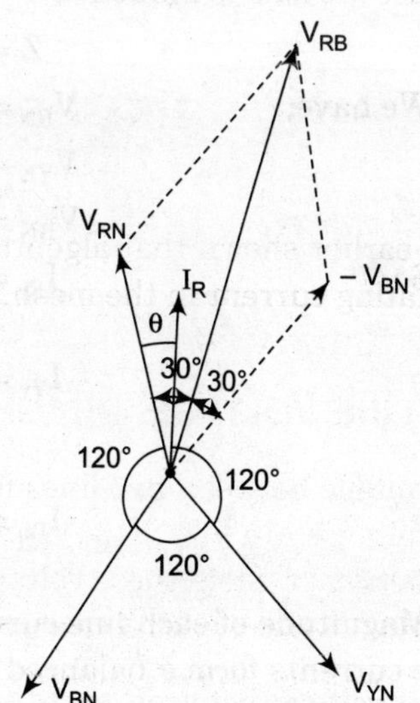

Fig. 7.7 *Showing angle of lead or lag between voltage and current.*

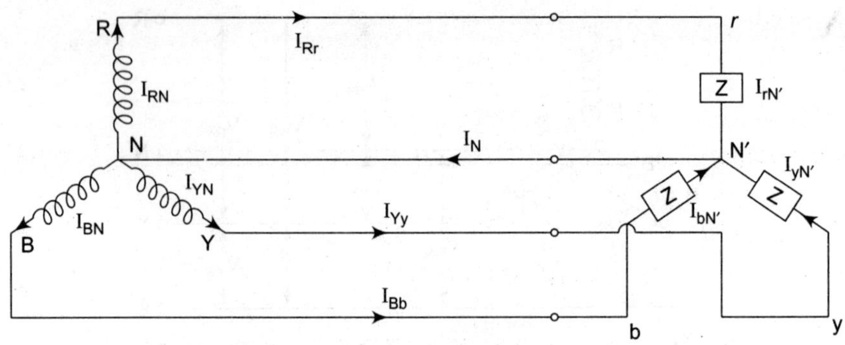

Fig. 7.8 *Three-phase, four-wire supply system.*

So

$$I_{RN} = I_{Rr} = I_{rN'}$$
$$I_{YN} = I_{Yy} = I_{yN'}$$
$$I_{BN} = I_{Bb} = I_{BN'}$$

Line current is same as phase current both in magnitude and phase.

$$I_L = I_{Ph}$$

7.7 NEUTRAL CURRENT

Each impedance is connected between a line and neutral.

The line currents are given by

$$I_{Rr} = \frac{V_{RN}}{Z},\ I_{Yy} = \frac{V_{YN}}{Z},\ I_{Bb} = \frac{V_{BN}}{Z}$$

Let the load is inductive

$$Z = |Z_{Ph}| \angle \phi$$

We have,

$$V_{RN} = V_{Ph} \angle 0°$$
$$V_{YN} = V_{Ph} \angle -120°$$
$$V_{BN} = V_{Ph} \angle -240° \text{ or } V_{Ph} \angle +120°$$

So

$$I_{Rr} = V_{Ph} \angle 0°/Z_{Ph}\, \phi$$

$$I_{Yy} = V_{Ph} \angle -120°/Z_{Ph} \angle \phi = \frac{V_{Ph} \angle -120 - \phi}{Z_{Ph}}$$

$$I_{Bb} = V_{Ph} \angle 120°/Z_{Ph} \angle \phi = \frac{V_{Ph} \angle 120 - \phi}{Z_{Ph}}$$

Magnitude of each line current is same and phase is 120° apart from each other. Since the line currents form a balanced system, so the phasor sum of the current must be zero.

By, KCL at neutral point N'

$$I_N = I_{Rr} + I_{Yy} + I_{Bb} = 0 \text{ [balanced system]}$$

7.8 LINE VOLTAGE, PHASE VOLTAGES AND CURRENT IN DELTA CONNECTION

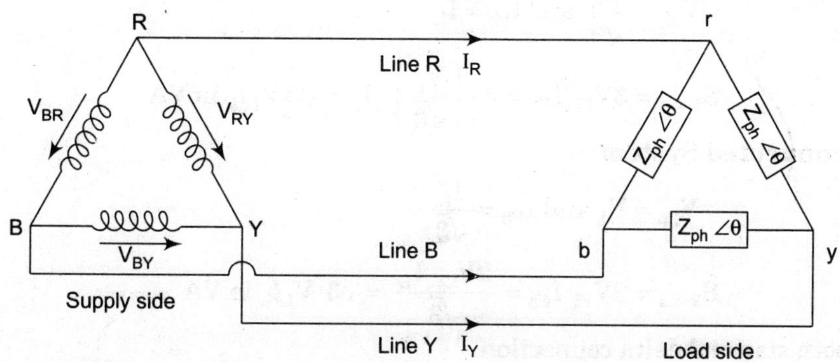

Fig. 7.9 *Delta connection, showing line and phase quantity.*

The voltage between any two lines is the line voltage V_L. In delta connection the phase coils are connected between lines. Therefore the phase voltage is equal to the line voltage.

$$|V_{RY}| = |V_{YB}| = |V_{BR}| = V_L = V_{Ph}$$

7.8.1 Phase and Line Currents in Delta Connection

If V_{RY} is considered as the reference voltage

The phase current can be found as

$$I_{RY} = V_{RY}/Z_{Ph} = V_L \angle 0°/Z_{Ph} \angle\phi = V_L/Z_{Ph} \angle -\phi$$

$$I_{YB} = V_{YB}/Z = V_L \angle -120°/Z_{Ph} \angle\phi = V_L/Z_{Ph} \angle -120 - \phi$$

$$I_{BY} = V_{BY}/Z = V_L \angle 120°/Z_{Ph} \angle\phi = V_L/Z_{Ph}\, \phi \angle 120 - \phi$$

The three-phase currents therefore have equal magnitudes

$I_{Ph} = V_L/Z_{Ph}$ and separated by 120° phase angle.

Line current

$$I_R = I_{ry} - I_{br} = 2\, I_{Ph} \cos 30° = \sqrt{3}\, I_{Ph}$$

$$I_Y = I_{yb} - I_{ry} = 2I_{Ph} \cos 30° = \sqrt{3}\, I_{Ph}$$

$$I_B = I_{br} - I_{yb} = 2I_{Ph} \cos 30° = \sqrt{3}\, I_{Ph}$$

7.9 VOLT-AMPERES, POWER AND REACTIVE VOLT-AMPERES IN THREE-PHASE SYSTEM

Volt-Amperes. In single-phase volt-amperes is given by the expression:

$$S_{1-\phi} = V \cdot I, \text{ volt-amp}$$

In three phase whether balanced or unbalanced is given by sum of the volt-amperes in each phase.

If the load is balanced, total volt-ampere will be three times of volt-ampere per phase.

$$S_{3-\phi} = 3V_{Ph}I_{Ph}, \text{ volt-amperes}$$

Where, V_{Ph} and I_{Ph} are the r.m.s. phase voltage and r.m.s. phase current.

In term of line values for star connection

$$V_{Ph} = \frac{V_L}{\sqrt{3}} \text{ and } I_{Ph} = I_L$$

$$S_{3-\phi} = 3V_{Ph}\, I_{Ph} = 3 \cdot \left(\frac{V_L}{\sqrt{3}}\right) \cdot I_L = \sqrt{3}\, V_L I_L \text{ in VA.}$$

For delta connected System

$$V_{Ph} = V_L \text{ and } I_{Ph} = \frac{I_L}{\sqrt{3}}$$

$$S_{3-\phi} = 3V_{Ph}\, I_{Ph} = \frac{2V_L \cdot I_L}{\sqrt{3}} = \sqrt{3}\ V_L I_L \text{ in VA}$$

Hence for both star and delta connection

$$S_{3-\phi} = \sqrt{3}\ V_L I_L \text{ VA.}$$

7.9.1 Active Power

In single phase

$$P_{1-\phi} = V \cdot I \cos\phi \text{ W}$$

Power consumed by a three-phase load, whether balanced or unbalanced is equal to the sum of the power of each phase. For balanced load the total power is equal to three times the power per phase

$$P_{3-\phi} = 3P_{1-\phi} = 3V_P I_P \cos\phi \text{ W}$$

In terms of line values for both star and delta connections

$$P_{3-\phi} = \sqrt{3}\, V_L I_L \cos\phi, \text{ W}$$

7.9.2 Reactive Power

In a single phase circuits are given by

$$Q_{I-\phi} = V \cdot I \sin\phi \text{ var.}$$

Reactive power of three-phase load either balanced or unbalanced are equal to the sum of reactive volt-amperes of each phase.

$$Q_{3-\phi} = 3V_P I_P \sin\phi \text{ var.}$$

In term of line values for star and delta connections

$$Q_{3-\phi} = \sqrt{3}\, V_L I_L \sin\phi \text{ var.}$$

7.9.3 Apparent Power

The product of r.m.s. values of voltage and current in a circuit is called volt-amperes or apparent power. It is denoted by S and is measured is volt-amperes (VA).

$$S = VI$$

$$V = ZI$$

$$S = I^2Z$$

The unit of S are volt-ampere (VA), Kilo voltmeter (kVA) and megavoltamperes (MVA).

$$1 \text{ kVA} = 10^3 \text{ VA}$$

$$1 \text{ MVA} = 10^6 \text{ VA or } 10^3 \text{ kVA}$$

7.10 MEASUREMENT OF POWER

7.10.1 Measurement of Power by Single-wattmeter Method

If the load on each phase is balanced then the power consumed by load can be measured with the help of only single-wattmeter with its current coil in one line and the pressure coil between the line and the neutral point as shown in Fig. 7.10. The reading of wattmeter thus connected, gives the power per phase. So

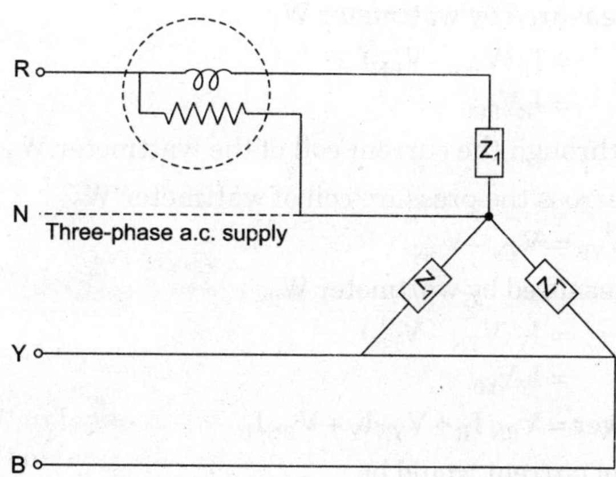

Fig. 7.10 *Circuit diagram of power measurement by single-wattmeter.*

$$\text{Total power} = 3 \times \text{power per phase}$$

or $$= 3 \times \text{wattmeter reading.}$$

7.10.2 Measurement of Power by Two-wattmeter Method

By two-wattmeter method power can be measured whether the load is balanced or unbalanced. The load may be star or delta connection form. The currents coils of the two wattmeters are connected in any of two lines and the pressure coils are connected between these lines and third line as shown in Fig. 7.11.

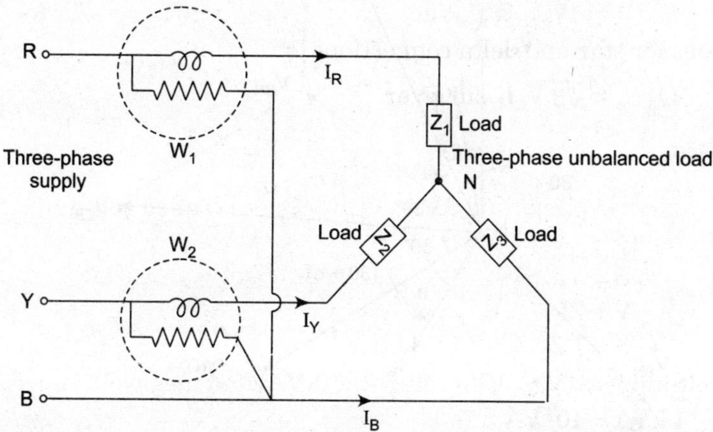

Fig. 7.11 *(a) Circuit diagram of power measurement by two wattmeter.*

Let V_{RN},V_{YN} and V_{BN} be the voltages across the three phases of the load and I_R,I_Y and I_B the currents flowing in the three lines.

Total instantaneous power in the load = $V_{RN} I_R + V_{YN} I_Y + V_{BN} I_B$

Instantaneous current through the current coil of the wattmeter, $W_1 = I_R$

Instantaneous voltage across the pressure coil of wattmeter W_1

$$V_{RB} = V_{RN} - V_{BN}$$

Instantaneous power measured by wattmeter W_1

$$= I_R (V_{RN} - V_{BN})$$
$$= I_R V_{RB}$$

Instantaneous current through the current coil of the wattmeter, $W_2 = I_Y$

Instantaneous voltage across the pressure coil of wattmeter W_2

$$V_{YB} = V_{YN} - V_{BN}$$

Instantaneous power measured by wattmeter W_2

$$= I_Y (V_{YN} - V_{BN})$$
$$= I_Y V_{YB}$$

Total instantaneous power = $V_{RN} I_R + V_{YN} I_Y + V_{BN} I_B$

By KCL algebraic sum of current would be

$$I_R + I_Y + I_B = 0$$

or

$$I_B = -(I_R + I_Y)$$

Putting the values of I_B in equation

Total instantaneous power = $V_{RN} I_R + V_{YN} I_Y + V_{BN} I_B$

$$= V_{RN} I_R + V_{YN} I_Y + V_{BN} (-I_R - I_Y)$$

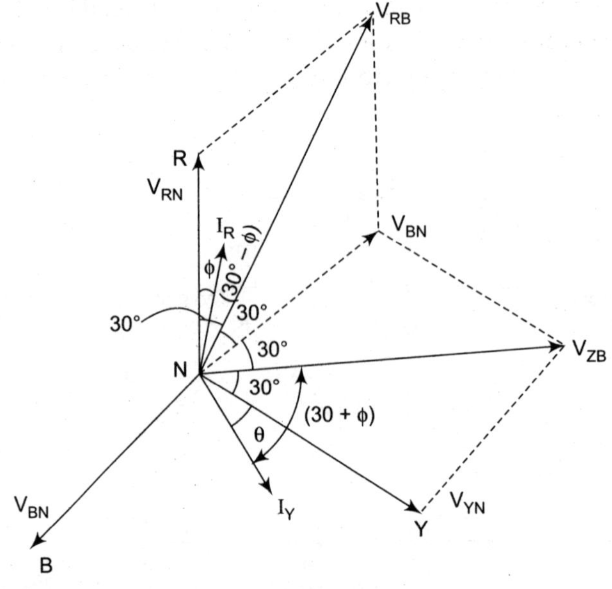

Fig. 7.11 *(b) Phasor diagram of measurement of power.*

$= I_R(V_{RN} - V_{BN}) + I_Y(V_{YN} - V_{BN})$

= Instantaneous power measured by W_1 + Instantaneous power measured by W_2

So at any instant power measured by two-wattmeter is equal to total power consumed in load. This is valid for whether load is balanced or unbalanced.

7.10.3 Measurement of Power and Power Factor in Three-Phase System with Balanced Load using Two Wattmeter Method

Let I_R, I_Y and I_B be the rms values of the currents in the lines and V_{RN}, V_{YN} and V_{BN} the rms values of Voltages across the phases. The Phasor diagram shown in Figure 7.11 for lagging current. The load being balanced, currents I_R, I_Y and I_B will be equal in magnitude and lagging by an angle Φ with respect to its phase voltage. So the phasor voltages V_{RN}, V_{YN} and V_{BN} will also be equal in magnitude but displaced by 120°, the phase sequence being RYB.

Current through the current coil of wattmeter $W_1 = I_R$

Voltage across the pressure coil of wattmeter $W_1 = (V_{RN} - V_{BN}) = V_{RB}$

While phase difference between I_R and $V_{RB} = (30 - \Phi)$

Thus wattmeter reading $W_1 = I_R, V_{RB} \cos(30 - \Phi)$

Current through the current coil of wattmeter $W_2 = I_Y$

Voltage across the Pressure coil of the wattmeter $W_2 = (V_{YN} - V_{BN}) = V_{YB}$

While phase difference between the current I_Y and Voltage $V_{YB} = (30 + \Phi)$

Thus reading of wattmeter $W_2 = I_Y \cdot V_{YB} \cos(30 + \Phi)$

Since the load is balanced

$$I_R = I_Y = I_B = I_L$$

$$V_{RY} = V_{YB} = V_{BR} = V_L$$

So,

$$W_1 = V_L \cdot I_L \cos(30 - \Phi)$$

$$W_2 = V_L \cdot I_L \cos(30 + \Phi)$$

Adding these equations

$$W_1 + W_2 = V_L \cdot I_L \cos(30 - \Phi) + V_L \cdot I_L \cos(30 - \Phi)$$

$$= V_L . I_L [\cos 30 \cos \Phi + \sin 30 \sin \Phi + \cos 30 \cos \Phi - \sin 30 \sin \Phi]$$

$$= V_L \cdot I_L \left[\sqrt{\frac{3}{2}} \cos \Phi + \sqrt{\frac{3}{2}} \cos \Phi\right]$$

$$= \sqrt{3}\, V_L I_L \cos \Phi$$

= Total power input to a balanced load.

Subtracting

$$\begin{aligned} W_1 - W_2 &= V_L \cdot I_L \cos(30 - \Phi) - V_L \cdot I_L \cos(30 + \Phi) \\ &= V_L \cdot I_L [\cos 30 \cos \Phi + \sin 30 \sin \Phi - \cos 30 \cos \Phi + \sin 30 \sin \Phi] \\ &= V_L \cdot I_L [2 \sin 30 \sin \Phi] \\ &= V_L \cdot I_L \sin \Phi \end{aligned}$$

Dividing

$$\frac{(W_1 - W_2)}{(W_1 + W_2)} = \frac{V_L \cdot I_L \sin \Phi}{\sqrt{3} V_L I_L \cos \Phi}$$

Or

$$\tan \Phi = \sqrt{3} \frac{(W_1 - W_2)}{(W_1 + W_2)}$$

$$\Phi = \tan^{-1} \left[\sqrt{3} \frac{(W_1 - W_2)}{(W_1 + W_2)} \right]$$

$$\cos \Phi = \cos \left[\tan^{-1} \left\{ \sqrt{3} \frac{(W_1 - W_2)}{(W_1 + W_2)} \right\} \right]$$

Wattmeter reading at different power factor

1. When power factor is unity

$$\cos \Phi = 1 \text{ then } \Phi = 0$$

The readings of the two watt-meters are

$$\begin{aligned} W_1 &= \sqrt{3} V_L \cdot I_L \cos(30 - \Phi) \\ &= \sqrt{3} V_L \cdot I_L \cos(30 - \Phi) \\ &= \left(\sqrt{3} V_L \cdot I_L\right)\left(\frac{\sqrt{3}}{2}\right) \\ &= \frac{3 V_L \cdot I_L}{2} \end{aligned}$$

$$\begin{aligned} W_2 &= \sqrt{3} V_L \cdot I_L \cos(30 + \Phi) \\ &= \sqrt{3} V_L \cdot I_L \cos(30 + \Phi) \\ &= \left(\sqrt{3} V_L \cdot I_L\right)\left(\frac{\sqrt{3}}{2}\right) \end{aligned}$$

$$= \frac{3V_L \cdot I_L}{2}$$

So $\qquad W = W_1 + W_2 = 3V_L \cdot I_L$

So at unity power factor, total power $W = 3V_L \cdot I_L \cos \Phi$

$$= 3V_L I_L \qquad \{\text{Since } \cos \Phi = 1\}$$

Hence at unity power factor, the readings of the two wattmeters are equal and each wattmeter measures half of the total power.

2. When the power factor is 0.5.

$$\cos \phi = 0.5$$

$$\Rightarrow \qquad \phi = 60°$$

The readings of the two wattmeters are

$$W_1 = \sqrt{3}\, V_L \cdot I_L \cos (30 - \phi)$$

$$= \sqrt{3}\, V_L \cdot I_L \cos (30 - 60) = \sqrt{3}\, V_L \cdot I_L \cos 30°$$

$$= \frac{3V_L \cdot I_L}{2}$$

$$W_2 = \sqrt{3}\, V_L \cdot I_L \cos (30 + \phi)$$

$$= \sqrt{3}\, V_L \cdot I_L \cos (30 + 60) = \sqrt{3}\, V_L \cdot I_L \cos 90° = 0$$

so $\qquad W = W_1 + W_2 = \dfrac{3V_L \cdot I_L}{2}$

Hence when the power factor is 0.5 one of wattmeter reads zero while other measures total power.

3. When power factor is zero. We have $\phi = 90°$

$$\cos \phi = 0$$

$$\Rightarrow \qquad \phi = 90°$$

$$W_1 = \sqrt{3}\, V_L \cdot I_L \cos (30 - \phi)$$

$$= \sqrt{3}\, V_L \cdot I_L \cos (30 - 90) = \frac{\sqrt{3} V_L \cdot I_L}{2}$$

$$= \frac{\sqrt{3} V_L \cdot I_L}{2}$$

$$W_2 = \sqrt{3}\, V_L \cdot I_L \cos (30 + 90) = \sqrt{3}\, V_L \cdot I_L \cos 120 = \sqrt{3}\, V_L \cdot I_L \left(-\frac{1}{2}\right)$$

$$= -\frac{3}{2} V_L \cdot I_L$$

Total power $W = W_1 + W_2 = 0$

Therefore, with zero power factor, the reading of two wattmeters are equal but of opposite sign.

4. If the power factor is below 0.5, one of wattmeters will give negative indication. So to read the wattmeter, we must either reverse the current coil or pressure coil connections. The wattmeters will give positive reading but this must be taken as negative for calculating the total power.

7.11 POWER FACTOR

Power factor is defined as the ratio of active power (kW) to the total apparent power (kVA):

$$Pf = \frac{kW}{kVA} \text{ or } kVA \times Pf = kW$$

Trigonometrically the PF

$$\frac{kW}{kVA} = \text{cosin } \phi$$

Power factor can also be defined as the factor to multiply apparent power is order to obtain active power. For example, assume a load on a 440 V, three-phase system. The ammele indicates 200 amp and the watt meter reads 120 kW. What is the power factor of the load?

The apparent power for a three-phase circuit is given by the expression.

$$kVA = \frac{\sqrt{3} \times V \times I}{1000}$$

$$= \frac{\sqrt{3} \times 440 \times 200}{1000} = 152.24 \text{ kVA}$$

$$Pf = \frac{kW}{kVA} = \frac{120}{152.24} = 0.788 \text{ or } 78.8\%$$

Power factor is often stated as a percentage, but since it is a ratio its better expressed a decimal form.

The real power can be expressed as

$$kW = \frac{\sqrt{3} \times V \times I \times Pf}{1000}$$

and reactive power

$$kVAR = \frac{\sqrt{3} \times V \times I \times \sqrt{1 - Pf^2}}{1000}$$

As shown in Fig. 7.12 φ is angle by which the current lags the voltage and can be expressed trigonometrically as

$$\cos\phi = \frac{\text{kW}}{\text{kVA}} = \text{Pf}$$

This means the power factor of a circuit can be expressed by the cosine of the angle by which the current lags (or leads) the voltage is that circuit.

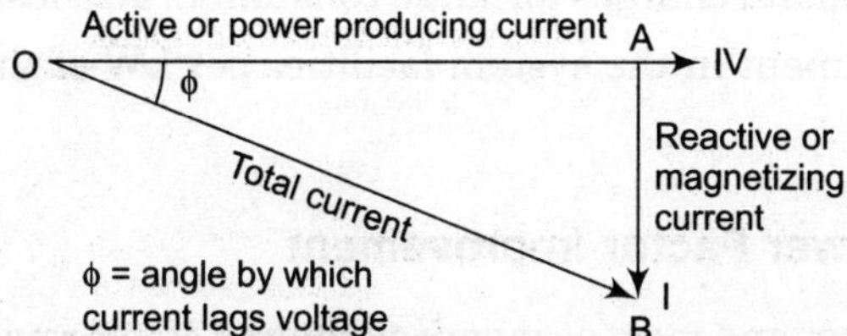

Fig. 7.12 *Currents component in AC inductive circuit.*

7.11.1 Causes of Low Power Factor

(*a*) In the industries as well as in domestic application the most commonly used motors are induction motors. At full load the three phase induction motor operates at a power factor of around 0.80 lagging. At less than the full load the power factor is poor. The power factor of single phase induction motors is about 0.5 to 0.6 lagging.

(*b*) The transformer draws a magnetizing current from the supply. This current is at a power factor of zero lagging.

(*c*) Different type of special device/equipment like arc lamps, welding machine etc. also contribute to low power factor to the system.

(*d*) The transmission and distribution line usually behaves like inductive circuit thereby contributes to low power factor to the system.

7.11.2 Problem of Low Power Factor

(*a*) To meet the load requirement at a low power factor, the capacity of power plant, transmission & distribution equipment has to be more than that which would be necessary if the load was demanded at unity power factor.

(*b*) For the same active power, operation of an existing power system at a lower p.f. means overloading the equipment at times of full load.

(*c*) For the same active power, a low power factor means a greater current and hence higher energy loss.

(*d*) Low power factor causes the voltage regulation to be poor.

7.11.3 Advantages of Power Factor Improvement

Generally the power factor improvement of any system is performed by placement of Shunt capacitor-bank, Synchronous condenser, to improve the power factor towards, results the following advantages:

(*a*) Reduced circuit current.

(*b*) Improved voltage profile.

(*c*) Reduction in the copper losses in the system due to reduced current.

(*d*) Improvement in power factor of the generators.

(*e*) Reduction in kVA loading of the generators and circuits. This reduction in kVA loading may relieve an over load condition or release capacity for addition growth of the load.

(*f*) Reduced in kVA demand charges for large consumers also less possibility of penalty.

(*g*) Reduction in investment in the system facilities per kW of the load supplied, in case for utilities.

7.11.4 Concept of Power Factor Improvement

Using capacitors is simplest and most economical method of improving power factor in plants or of premises that do not require additional large motor drives. When properly applied to the system, capacitors supply the reactive magnetizing current and remove the reactive current from the plant circuit. This improves the overall power factor. Capacitors also improve a plant's efficiency by releasing electrical system capacity (kVA); raising the voltage level and reducing system losses so that additional loads can be added to the same system.

Here's how improving the power factor can save money. Assume that a plant's load is 1500 kW at 0.75 power factor and the utility has a cost-rate schedule based on ₹ 5/kVA, with a maximum billed power factor of 0.9.

The billed kVA is

$$\frac{1500 \text{ kW}}{0.75 \text{ PF}} = 2{,}000 \text{ kVA}$$

The utility demand charge is:

$$2000 \text{ kVA} \times ₹\ 5/\text{kVA} = ₹\ 10000 \text{ a month}$$

The minimum kVA on which the plant's 1,500 kW demand cost can be based is:

$$\frac{1500 \text{ kW}}{0.90 \text{ PF}} = 1667 \text{ kVA}$$

The utility demand charge is:

$$1667 \text{ kVA} \times ₹\ 5/\text{kVA} = ₹\ 8335 \text{ a month}$$

This means that by improving the power factor from 0.75 to 0.9, the demand charge will be reduced ₹ 1665 a month (₹ 10,000 – ₹ 8335 = ₹ 1665)

At 1500 kW with 0.75 power factor, the reactive kVA (kVAR) is 1500 kW × 0.882 = 1323 kVAR. At 1500 kW with 0.9 power factor the kVAR is 1500 kW × 0.484 = 726 kVAR. The required corrective capacity is 597 kVAR (1323 kVAR – 726 kVAR).

SOLVED NUMERICAL PROBLEMS

Example 1. *Three impedances each having a resistance of 20 Ω and an inductive reactance of 18 Ω are connected in star across a 400 V, three-phase supply. Calculate (i) the line current, (ii) the power factor, (iii) total power in kW.*

Solution: Resistance/Phase $R_F = 20\ \Omega$

Reactance/Phase $X = 18\ \Omega$

$V_L = 400\ V$

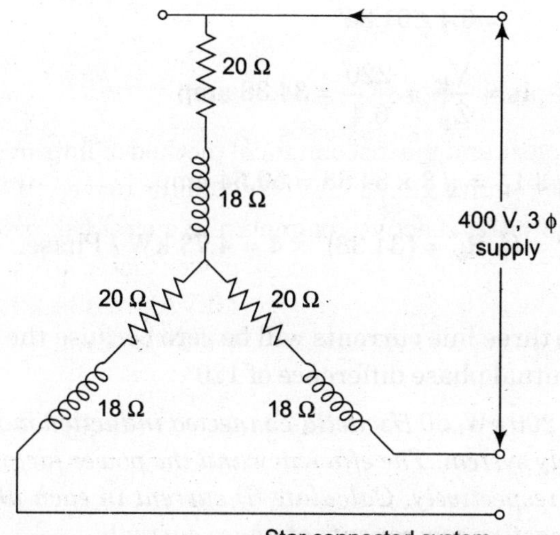

Fig. 7.13

Phase voltage $V_P = \dfrac{V_L}{\sqrt{3}} = \dfrac{400}{\sqrt{3}} = 231\ V$

Impedance/Phase $= \sqrt{R^2 + X^2}$

$$Z_P = \sqrt{(20)^2 + (18)^2} = 26.91\ \Omega$$

(*i*) Line current, for star connection $I_L = I_P$

$$= \frac{V_P}{Z_P} = \frac{231}{26.91} = 8.58 \text{ amp}$$

(*ii*) Power factor $\cos\phi = \dfrac{R}{Z} = \dfrac{20}{26.91} = 0.74$ lagging

(*iii*) Total power

$$P = \sqrt{3}\, V_L I_L \cos\phi$$

$$= \sqrt{3} \times 400 \times 8.58 \times 0.74 = \mathbf{4.40\ kW.}$$

Example 2. *A 220 V, three-phase voltage is applied to a balanced delta connected three-phase load by phase impedance (4 + j5) Ω.*

(i) Find the line current in each line.

(ii) What is the power consumed per phase?

(iii) What is the phasor sum of the three line currents why does it have this value?

Solution: For delta connected system $V_L = V_P$

$$V_L = V_P = 220 \text{ V}$$

$$Z_P = (4 + j5)$$

$$= 6.4 \angle 51.34°$$

Phase current $$I_P = \frac{V_P}{Z_P} = \frac{220}{6.4} = 34.38 \text{ amp}$$

(i) Line current $I_L = \sqrt{3}\, I_P = \sqrt{3} \times 34.38 = 59.54$ amp.

(ii) Power consumed $P = I_P^2\, R_P = (34.38)^2 \times 4 = 4.73$ kW / Phase.

(iii) Phasor sum.

The phasor sum of the three line currents will be zero because the line currents are equal in magnitude and have a mutual phase difference of 120°.

Example 3. *A three-phase 200 kW, 50 Hz, delta-connected induction motor is supplied from three-phase, 400 V, 50 Hz supply system. The efficiency and the power factor of three-phase induction motor are 90% and 0.85 respectively. Calculate (i) current in each phase of phase, (ii) the line current, (iii) active and reactive components of phase currents.*

Solution:

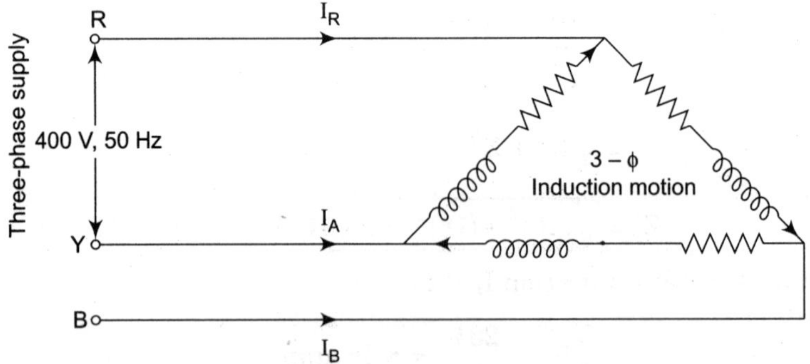

Fig. 7.14 Phase delta connected induction motor.

(i) Line voltage $V_L = 400$ V

Voltage across the motor phase, V_{Ph} = line voltage, V_L

$$= 400 \text{ V}$$

Output of three-phase I.M. = 200 kW

Input to the motor $= \dfrac{\text{Output}}{\text{Efficiency}} = \dfrac{200 \times 10^3}{0.90}$

$= 222222.22 \text{ W}$

Input to the motor $= 3 \, V_{Ph} \, I_{Ph} \cos \phi$

$$222222 = 3 \times 400 \times I_{Ph} \times 0.85$$

$$I_{Ph} = 217.86 \text{ amp}$$

Current drawn by each phase would be same in magnitude.

(*ii*) Line current,

$$I_L = \sqrt{3} I_{Ph}$$

$$= \sqrt{3} \times 217.86$$

$$= 377.35 \text{ amp.}$$

(*iii*) Active component of phase current $= I_{Ph} \cos \phi$

$$= 217.86 \times 0.85$$

$$= 185.18 \text{ amp}$$

Reactive component of phase current $= I_{Ph} \sin \phi$

$$= 217.86 \times 0.53$$

$$= 114.76 \text{ amp.}$$

Example 4. *A balanced three-phase star connected load of 100 kW takes a leading current of 75 amp, when connected across a three-phase, 1.1 kV, 50 Hz supply. Find the circuit constants of the load per phase.*

Solution: Line voltage across the load, $V_L = 1.1 \text{ kV} = 1100 \text{ V}$

Phase voltage, $V_{Ph} = \dfrac{1100}{\sqrt{3}} = 635 \text{ V}$

Current takes by the load = 75 amp.

For star connection, $I_{Ph} = I_L$

Thus phase current, $I_{Ph} = 75 \text{ amp}$

Impedance per phase, $Z_{Ph} = \dfrac{V_{Ph}}{I_{Ph}}$

$$= \frac{635}{75} = 8.47 \, \Omega$$

Total power drawn by the load = 100 kW

$$= 100 \times 10^3 \, \Omega$$

$$P = \sqrt{3} \, V_L \cdot I_L \cos \phi$$

$$100 \times 10^3 = \sqrt{3} \times 1100 \times 75 \times \cos \phi$$

$$\cos \phi = \frac{100 \times 10^3}{\sqrt{3} \times 1100 \times 75} = 0.699$$

We know that, $$\cos \phi = \frac{R_{Ph}}{Z_{Ph}}$$

or $$R_{Ph} = Z_{Ph} \cos \phi$$

$$= 8.47 \times 0.699$$

$$= 5.92\ \Omega$$

As the load power factor is leading so nature would be capacitive,

$$Z_{Ph} = \sqrt{R_{Ph}^2 + X_C^2}$$

$$X_C = \sqrt{Z_{Ph}^2 - R_{Ph}^2} = \sqrt{(8.47)^2 - (5.92)^2}$$

$$X_C = 6.06\ \text{W}$$

$$X_C = \frac{1}{WC} = \frac{1}{2\pi fc}$$

$$C = \frac{1}{X_C \cdot 2\pi f} = \frac{1}{2\pi \times 50 \times 6.06}$$

$$C = 5.26 \times 10^{-4}\ \text{Farad}$$

$$C = 526\ \mu f.$$

Example 5. *Three non-inductive loads of 2 kW, 3 kW, 24 kW are connected between the three line conductors and neutral of a three-phase four wires system with line voltage of 400 V as shown in Fig. 7.14. Find out (i) the current in each line and (ii) the current in the neutral conductor.*

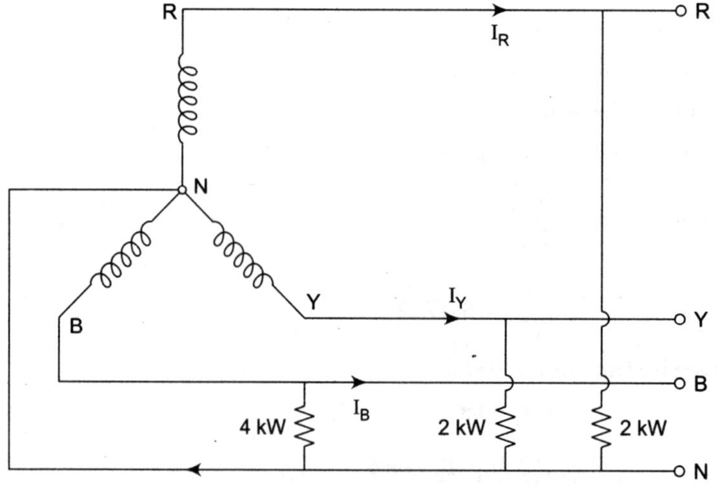

Fig. 7.15 *Circuit diagram.*

Solution:

(*i*) Line voltage, $V_L = 400$ V

Phase voltage $\quad V_P = \dfrac{V_L}{\sqrt{3}} = \dfrac{400}{\sqrt{3}} = 231$ V

Current in R-phase, $I_R = \dfrac{2 \times 10^3}{231} = 8.61$ amp

Current in Y-phase, $I_Y = \dfrac{3 \times 10^3}{231} = 12.99$ amp

Current in B-phase, $I_B = \dfrac{4 \times 10^3}{231} = 17.30$ amp.

(*ii*) All the loads are resistive in nature, therefore power factor will be unity *i.e.*, current in phase with respective voltage. Assuming D.C. phase sequence RZB,

$$\begin{aligned} I_R &= 8.65 \angle 0° \\ &= (8.65 + j0°) \text{ amp} \\ I_Z &= 12.99 \angle -120° \\ &= (-6.50 - j11.25) \text{ amp} \\ I_B &= 17.30 \angle -240° \\ &= (-8.65 + j14.98) \text{ amp} \end{aligned}$$

Current in the neutral conductor

$$\begin{aligned} I_N &= I_R + I_Z + I_B \\ &= 8.65 - 6.50 - j11.25 - 8.65 + j14.98 \\ &= -6.50 + j3.73 \\ I_N &= 7.49 \angle 150°. \end{aligned}$$

Example 6. *The power input to a three-phase induction motor is read by two wattmeters. The readings are 1000 W and 500 W respectively. Calculate the power factor of the motor.*

Solution: $\quad W_1 = 500$ W, $W_2 = 1000$ W

We know that,

$$\tan \phi = \frac{\sqrt{3}\,(W_2 - W_1)}{W_2 + W_1} = \frac{\sqrt{3}\,(1000 - 500)}{1000 + 500}$$

$$\begin{aligned} \tan \phi &= 0.577 \\ \phi &= \tan^{-1}(0.577) \\ \phi &= 30° \\ \cos \phi &= \cos 30° \\ \cos \phi &= 0.866 \end{aligned}$$

P.f. of the motor $\quad = \cos \phi = 0.866.$

Example 7. *In the three-phase circuit power measured by two wattmeter gives 16.2 kW and – 8.2 kW. The line voltage was 400 V. Determine (i) total active power drawn by the load; (ii) power factor; and (iii) line current.*

Solution: $W_2 = 16.2 \text{ kW}, W_1 = -8.2 \text{ kW}, V_L = 400 \text{ V}$

(*i*) Total active power,

$$P = W_2 + W_1 = 16.2 - 8.2 = 8.0 \text{ kW}$$

(*ii*) Power factor,

$$\tan\phi = \frac{\sqrt{3}\,(W_2 - W_1)}{W_2 + W_1} = \frac{\sqrt{3}\,(16.2 + 8.2)}{16.2 - 8.2} = 5.283$$

$$\phi = \tan^{-1}(5.283)$$

$$\phi = 79.28$$

$$\cos\phi = \cos 79.28 = 0.186 \text{ lagging.}$$

(*iii*) Line current, $P = \sqrt{3}\, V_L \cdot I_L \cos\phi$

$$8 \times 10^3 = \sqrt{3} \times 400 \times I_L \times 0.186$$

$$I_L = \frac{8 \times 10^3}{128.86} = 62.08 \text{ amp.}$$

Example 8. *A balanced star connected load is supplied from a symmetrical, three-phase, 440 V, 50 Hz supply system the current in each phase is 15 A and lags behind its phase voltage by an angle of 36°. Calculate (i) phase voltage, (ii) load parameters, (iii) total power and (iv) readings of two wattmeters, connected in the load circuit to measure the total power.*

Solution:

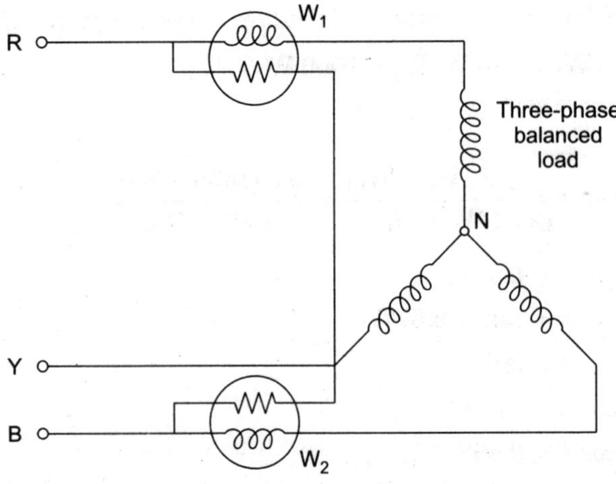

Fig. 7.16 *Two wattmeter connections.*

(*i*) Line Voltage, $V_L = 440$ V

Phase voltage in star connected Ckt $V_{Ph} = \dfrac{V_L}{\sqrt{3}}$

$$= \frac{440}{\sqrt{3}} = 254 \text{ Volt.}$$

(*ii*) Current in each phase = 15 A

Impedance of the load per phase.

$$Z_{Ph} = \frac{V_{Ph}}{I_{Ph}} = \frac{254}{15} = 16.93\ \Omega$$

$$Z_{Ph} = \sqrt{R^2 + X^2} = 16.93\ \Omega$$

Current in each phase lags by its voltage 36°.

Hence, $\tan 36° = \dfrac{X}{R}$

$$X = R \tan 36° = 0.726\ R$$

Substituting the value of X in equation,

$$R^2 + (0.726\ R)^2 = (16.93)^2$$

$$1.52\ R^2 = (16.93)^2$$

$$R = \frac{16.93}{\sqrt{7.52}} = 13.73\ \Omega$$

So, $X = 0.726\ R = 0.726 \times 13.73$

$$X = 9.97\ \Omega$$

(*iii*) Power consumed by each phase = $V_{Ph} \cdot I_{Ph} \cos \phi$

$$= 254 \times 15 \times \cos 36°$$

$$= 3082.35 \text{ W}$$

$$= 3.08 \text{ kW}$$

So, total power $= 3 \times P_{Ph}$

$$= 3 \times 3.08$$

$$= 9.24 \text{ kW}$$

(*iv*) Total power = $W_1 + W_2$ (reading of two wattmeters)

Thus, $W_1 + W_2 = 9.24$ kW

$$\tan 36° = \frac{\sqrt{3}\,(W_2 - W_1)}{W_2 + W_1}$$

$$W_2 - W_1 = \frac{9.24 \times 10^3 \times 0.726}{\sqrt{3}}$$

$$W_2 - W_1 = 3873 \text{ W}$$

$$W_2 + W_1 = 9.24$$

$$W_2 - W_1 = 3.87$$

So, solving these equations

$$W_1 = 2.69 \text{ kW}$$

$$W_2 = 6.56 \text{ kW}.$$

Example 9. *Two wattmeters connected to read the total power in a three-phase system supplying a balanced load of 12 kW and – 3 kW, respectively. Calculate the total power and power factor. Also explain the Significance of (i) equal wattmeter readings and (ii) A zero reading on one wattmeter.*

Solution: $W_2 = 12 \text{ kW}, W_1 = -3 \text{ kW}$

Total power $W = W_1 + W_2$

$$= -3 + 12 = 9 \text{ kW}$$

$$\tan\phi = \frac{\sqrt{3}\,(W_2 - W_1)}{W_2 + W_1} = \frac{\sqrt{3}\,(12+3)}{12-3} = \frac{\sqrt{3} \times 15}{9}$$

$$\tan\phi = 2.89$$

$$\phi = \tan^{-1}(2.89) = 70.89°$$

Power factor, $\cos\phi = \cos 70.89° = 0.33.$

(*i*) Equal wattmeter reading cases:

For reading of the two wattmeters to be equal,

$$\sqrt{3}\, V_L \cdot I_L \cos(30 - \phi) = \sqrt{3}\, V_L \cdot I_L \cos(30 + \phi)$$

$$\cos(30 - \phi) = \cos(30 + \phi)$$

or $$\theta = 0°$$

$$\cos\phi = 1.0$$

This suggest unity power factor, *i.e.*, readings of two wattmeters are equal.

(*ii*) Zero reading on one wattmeter

As reading of wattmeters are

$$W_2 = \sqrt{3}\, V_L \cdot I_L \cos(30 - \phi)$$

$$W_1 = \sqrt{3}\, V_L \cdot I_L \cos(30 + \phi)$$

For zero reading on one of the wattmeter, ϕ must be 60°, which makes

$$W_1 = \sqrt{3}\, V_L \cdot I_L \cos 90° = 0.$$

EXERCISES

1. Deduce the relationship between the phase and the line voltages in a three-phase star connected circuit.
2. Derive an expression for the total power input for a balanced three-phase load in terms of line voltage, line current and power factor.
3. Show that the power intake by a three-phase circuit can be measured by two wattmeters connected properly in a circuit.
4. Draw suitable phasor diagram to explain the significance of (*i*) equal wattmeters readings (*ii*) zero reading on one wattmeters.
5. Each phase of a three-phase, Δ-connected load consists of an impedance $Z = 10\angle 60°$ ohm. The line voltage is 440 volts at 50 Hz. Compute the total power.
6. Each phase of a three-phase, star-connected load consists of an impedance $Z = 10\angle 60°$ ohm. The line voltage is 440 volts at 50 Hz. Compute the total power.
7. Three identical impedances are connected in delta to a three-phase supply of 415 V 50 Hz; the line current is 25 amps. The total power taken from the supply is 10 kW.Calculate the resistance and inductance of the impedances.
8. Three identical impedances are connected in delta to a three-phase supply of 415 V 50 Hz; the line current is 25 amps. The total power taken from the supply is 10 kW. Calculate the resistance and inductance of the impedances.
9. A balance load of power factor 0.80 lagging is supplied from a three-phase 415 V supply. The input as shown by the sum of the readings of two wattmeter connected in the supply leads in is 30 kW. Find the readings of each wattmeter.
10. Three similar coils each having a resistance of 10 Ω and an inductance of 0.50 H are connected in (*a*) star and (*b*) delta across a 415 V, three phases, 50 Hz supply. Calculate the current and the power absorbed in each case.
11. Three similar coils resistances are connected in star across a 440 V, three phase supply. The line current is 10 A. Calculate the value of each resistor. If the resistors were delta connected to get the same value of the line current, what should be the value of supply?
12. A balanced mesh-connected load of $(4 + j3)$ Ω per phase is connected to a three-phase, 50 Hz, 240 V supply system. Calculate (*i*) line current, (*ii*) power factor, (*iii*) power (*iv*) reactive volt-ampere and (*v*) total volt-ampere.
13. Two wattmeters connected to measure the total power in a balanced three-phase load read 10 kW and 2 kW respectively. The later reading has been taken by reversing the connections of the current coil. Calculate the total power.
14. A three-phase 415 V, 50 Hz induction motor has a full load output of 15 kW at which efficiency and power factor are 85% and 0.85 respectively. Find the readings on the two wattmeters connected to measure the power input to the motor. What is full load current?
15. A three-phase 415 V, 50 Hz induction motor has a full load output of 50 hp at which efficiency and power factor are 80% and 0.80 respectively. Three delta-connected condensers are connected across the motor to improve the power factor to 0.90 lag. Calculate the capacitance of each condenser and its kVA rating.

8

Electromagnetism

8.1 INTRODUCTION

An iron core called magnetic (Fe_3O_4) possesses the property of attracting small pieces of iron. The bodies, which exhibit the property of attracting small pieces of iron. The bodies, which exhibit the property of attracting iron, are generally termed as magnets. These are classified in two types: (1) Natural magnets (2) Artificial magnets. Artificial magnet can be prepared either by rubbing the iron bar with magnet or by passing an electric current through the wire wound round a iron piece. When a magnet is suspended and free to rotate in a horizontal plane, it comes to rest, pointing north and south. The end of the magnet-facing north is called North Pole, whereas that faces south is known as a south pole. A line joining these two poles is termed as magnetic axis.

8.2 MAGNETIC FIELD

When there is an electric current in a conductor magnetic forces are produced surrounding the conductor. The area around a magnetic pole or a magnet with its influence is perceptible is called its magnetic field. A magnetic field is a condition of space; it is shown by hypothetical magnetic lines. These lines of force travel from the north pole to south pole as shown in Fig. 8.1. The magnetic lines always forms close loops.

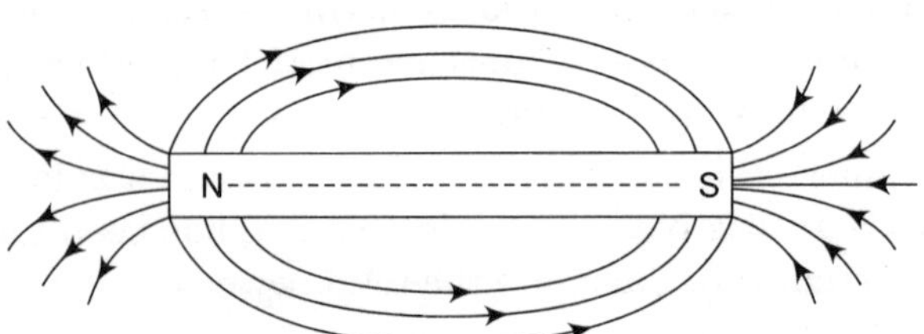

Fig. 8.1 *External lines of force due to small bar magnet.*

8.3 MAGNETIC FLUX

The total number of lines of force in the magnetic field is called the magnetic flux. It is denoted by a symbol Φ. The unit of magnetic flux is Weber (Wb). The lines of magnetic flux have no physical existence. The concept of these flux lines is purely hypothetical and was introduced by Faraday as a pictorial method of representing the distribution and density of a magnetic field. Lines of magnetic flux posses the following properties:

1. They form closed loops.
2. They always start form the North Pole and end in South Pole and are then continuous through the body of magnet.
3. They never intersect each other.
4. Lines of forces are like stretched lash cords, tending to contract lengthwise.
5. Lines of force exert lateral pressure *i.e.* they lend to bulge outside ways.
6. Lines of magnetic flux that are parallel and in the same direction repel each other.

8.4 MAGNETIC FLUX DENSITY

Magnetic flux density is defined as the magnetic flux per unit area of a surface at right angle to the magnetic field. This is also known as magnetic induction. Its symbol is B and measured in Weber per square meter or Tesla.

$$\Phi = \text{B.A (Wb)}$$

or

$$\text{B} = \frac{\Phi}{\text{A}} \frac{\text{Wb}}{m^2} \text{ or Tesla}$$

Φ = Magnetic field, in Wb

B = Flux density, Wb/m^2

A = Surface area, m^2.

8.5 DIRECTION OF CURRENT IN A CONDUCTOR

It is convenient to show the cross-section of a conductor instead by showing the conductor as a whole. The current direction is represented by an arrow. Consider an arrow AB as shown in Fig. 8.2. The current is coming towards point B.

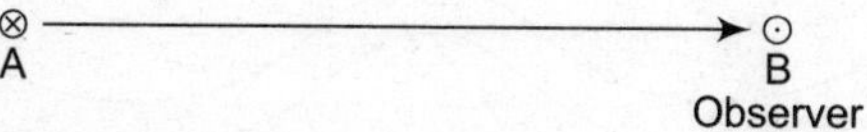

Fig. 8.2 *Direction of current in a conductor.*

⨀ Conductor with no current
⊙ Conductor carrying current towards the observe
⊗ Conductor carrying current away from the observer.

Hence a dot within the conductor represents the current coming towards the observer. An observer towards end A shows the tail or crossed features of the arrow. The current is going away from the observer; hence the cross within a conductor represents the current going away from observer.

8.6 MAGNETIC FIELD DUE TO A CURRENT CARRYING CONDUCTOR

When a conductor carries an electric current a magnetic field is produced all along its length, the lines of magnetic flux being concentric circles in planes at right angles to the conductor. The magnetic field associated with a current carrying conductor depends upon the magnitude of current and also on the direction of flow of current as shown in Fig. 8.3.

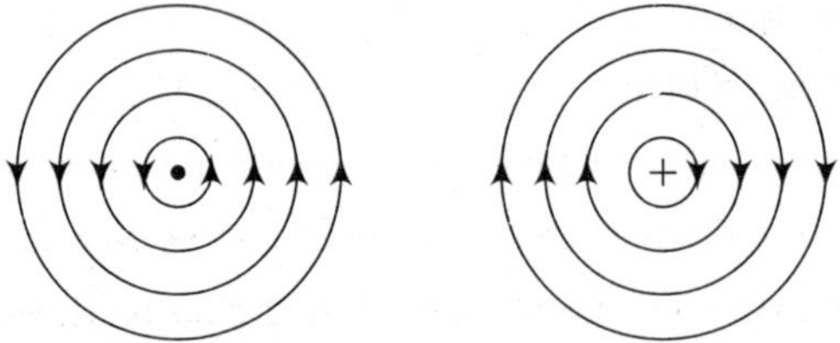

Fig. 8.3 *Magnetic field around a current carrying conductor.*

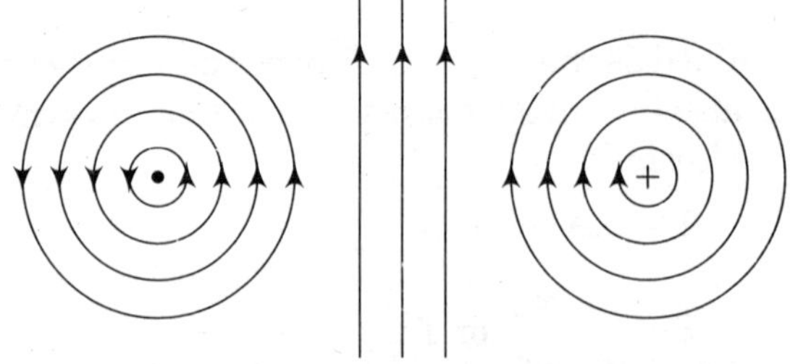

Fig. 8.4 *Resultant flux distribution (opposite direction).*

The two conductors shown in Fig. 8.4. Carry the same current in opposite directions and their fields so produced have different directions. If these two conductors are laid side by side, the two fields so produced will cancel each other thereby resulting in zero magnetic effect however, if they carry the same current in the same direction the magnetic effect will be twice compared to one such conductor alone as shown in Fig. 8.5.

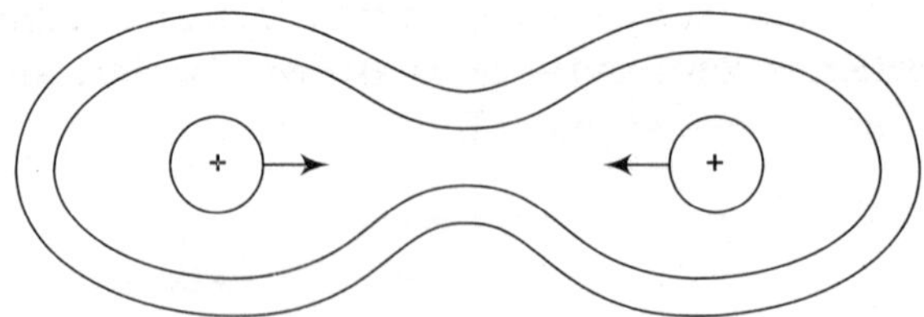

Fig. 8.5 *Resultant flux distribution (same direction).*

Thus we conclude that if there are two conductors carrying current in the same direction there is a force of attraction between them. If the two conductors carrying current in opposite directions there is force of repulsion between them.

8.7 DIRECTION OF MAGNETIC FIELD/FLUX

The direction of magnetic field can be found either of the following rules:

8.7.1 Right Hand Rules

Assume that the conductor is be gripped by the right hand in such a way that the thumb points in the direction of current, then the fingers encircling the conductor point in the direction of the magnetic field or flux as shown in Fig. 8.6.

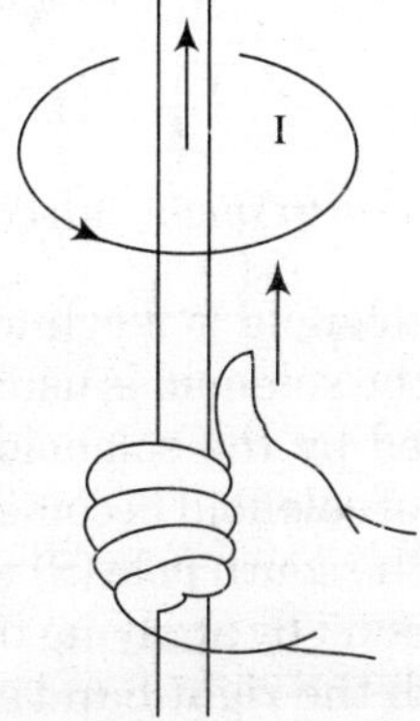

Fig. 8.6 *Right hand rule showing direction of field.*

8.7.2 Right hand screw rule

Assume that a right-handed screw along the side of the current carrying conductor. Advance the screw in the direction of current then the direction of the magnetic flux is in the direction; in which the screw has to be turned to move it in the forward direction as shown in Fig. 8.7.

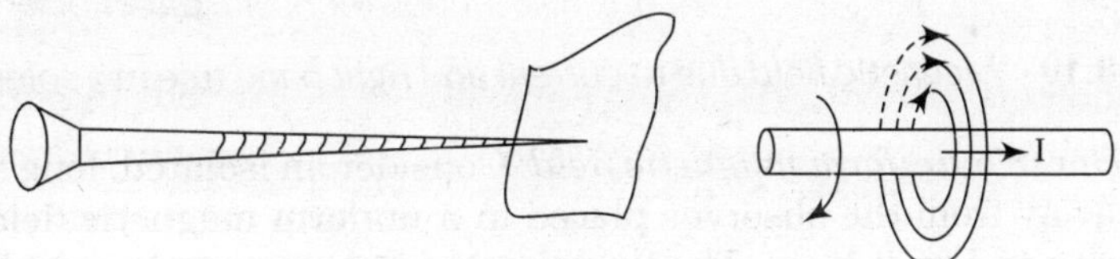

Fig. 8.7 *Right hand screw rule.*

8.8 REPRESENTING OF UNIFORM MAGNETIC FIELD

A uniform magnetic field is represented by equally spaced lines as given in Fig. 8.8.

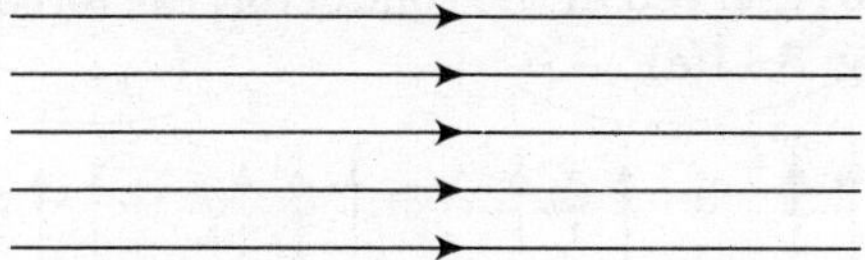

Fig. 8.8 *Uniform magnetic field.*

8.9 FLUX DISTRIBUTION OF A SINGLE TURN COIL

If a conductor is made into loop and current is passed through it, the magnetic flux lines are concentric circles all along its length as shown in Fig. 8.9. The direction of field can be determined by right hand rule or right hand screw rule.

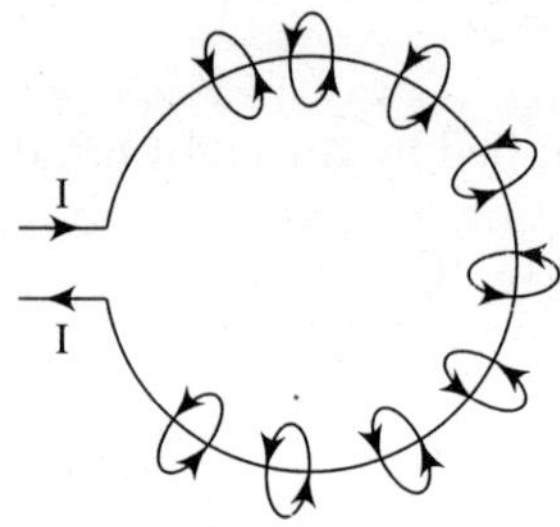

Fig. 8.9 *Magnetic field due to single turn conductor carrying current.*

Flux distribution of a solenoid: A solenoid is a cylindrical coil wound with large number of turns of insulated wire. The length of the solenoid is usually much larger in comparison with its diameter. The magnetic field produced by the solenoid resembles more or less that of a bar magnet as shown in Fig. 8.10 one-end of solenoid becomes a north pole (N) where the flux leaves the solenoid; while other end becomes the south pole (S) where the flux enters in it. The direction of the field inside the solenoid may be found by applying the right hand rule for solenoids this rule states that if a solenoid is gripped with the right hand such that the fingers point in the directions of the current in the wire, then the thumb will point in the directions of magnetic flux.

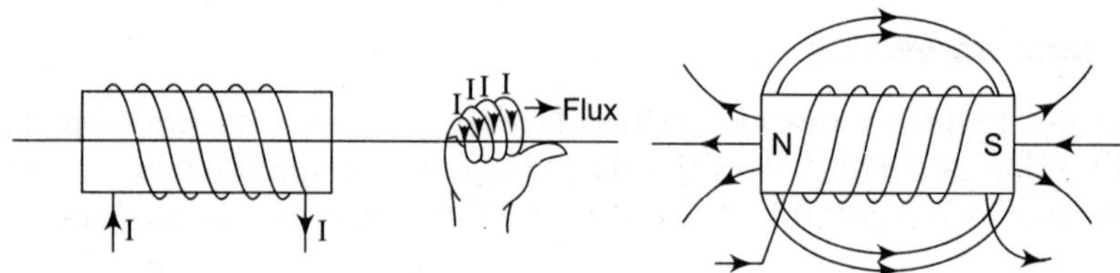

Fig. 8.10 *Magnetic field due to current and right hand rule in a solenoid.*

Isolated conductor in a uniform magnetic field: Consider an isolated, long straight conductor carrying current away from the observer placed in a uniform magnetic field acting vertically downwards as shown in Fig. 8.11(*a*). The field due to conductor on its right hand side is in the direction of the main magnetic field. On the left hand side of the conductor the field due to conductor is in a direction opposite to that of main field. Therefore the resultant field is stronger on the right hand side of the conductor and it is weaker on the left hand side of the conductor as shown in Fig. 8.11(*b*). The non-uniformity of the magnetic flux on the two sides of the conductor pushes the conductor form stronger flux side to weaker flux side.

If the direction of current is reserved in the conductor, the force produced on the conductor is also reversed, as shown in Fig. 8.11(*c*).

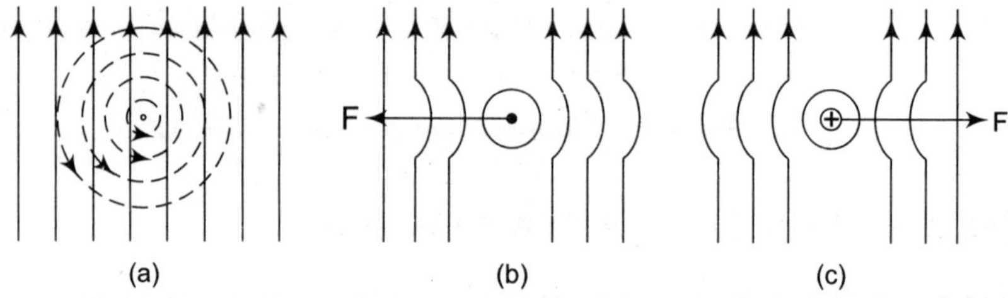

Fig. 8.11 *Magnetic field due to an isolated long straight current carrying conductor placed in a uniform field.*

8.10 PERMEABILITY

A magnetic material when placed in a magnetic field acquires magnetism due to induction. The measure of the degree to which the lines of force of the magnetizing field can penetrate or permeate the medium is called the absolute permeability of the medium. It is denoted by a symbol μ.

The permeability of all non-magnetic material including air represented by μ_0 equal to $4\pi \times 10^{-7}$ H/m.

The absolute permeability μ of a medium or magnetic material can also be expressed in terms of its relative permeability μ_r and the permeability of free space or air μ_0 *i.e.*

$$\mu = \mu_0 \times \mu_r$$

The relative permeability of a magnetic material may be defined as the ratio of the flux density produced in the material to the flux density produced in vacuum or in a non-magnetic core, provided the magnetic field strength is same in both the cases. It is denoted by μ_r.

8.11 RELATION BETWEEN MAGNETIC FLUX DENSITY AND FIELD INTENSITY

At any point in a magnetic field, field strength or field intensity H is the force maintaining the magnetic flux and producing a particular value of flux density B at that point. Hence the field intensity H is cause and the flux density B the effect. Thus the flux density can be assumed proportional to field intensity in the magnetic field *i.e.* in free space,

$$B = \mu_0 H$$

or

$$\frac{B}{H} = \mu_0$$

where μ_0 is called permeability of free space or magnetic space constant. Its value is $4\pi \times 10^{-7}$ H/m.

The flux density B also depends upon the nature of the medium. Thus the relative permeability of medium is μ_r, flux density at any point in a magnetic field is given by

$$B = \mu_0 \mu_r H$$

8.12 FORCE ON A CONDUCTOR IN A MAGNETIC FIELD

If a conductor carrying a current of I is placed in a uniform magnetic field of flux density B Wb/m^2, it experiences a mechanical force that depends upon the magnitude of current, the length of the conductor and flux density of the magnetic field as shown in Fig. 8.12(*a*).

$$F = BIl$$

where, F = Force on the conductor in newton (N)

B = Flux density in $\frac{\text{Wb}}{\text{m}^2}$ or Tesla (T)

I = Current in the conductor in amperes (A)

l = Effective length of conductor perpendicular to the field in meters (m).

The mechanical force experienced by the conductor is directly proportional to

1. Current flowing in the conductor I.
2. Length of conductor l.
3. Flux density of the uniform magnetic field B.

As shown in Fig. 8.12(*b*).

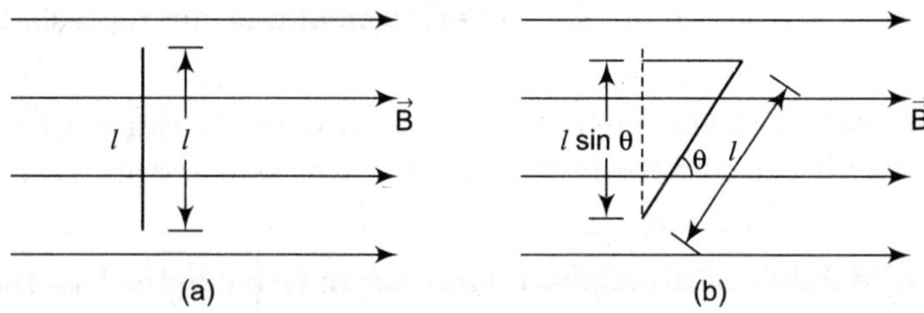

Fig. 8.12 *Current carrying conductor placed in uniform magnetic field.*

If the angle between the conductor and the field is θ then the effective length of the conductor perpendicular to the field is $l \sin \theta$ and the force on conductor is given by

$$F = BIl \sin \theta$$

If $\theta = 0°$ *i.e.* the conductor is placed parallel to the field, then there is no force exerted on the conductor.

The principle of force production on a current carrying conductor placed in a magnetic field forms the basis of the operation of the electric motor. That is why it is known as the motor principle.

8.13 FLEMING'S LEFT HAND (OR MOTOR) RULE

The thumb, the fore finger and the middle finger of the left hand are held mutually at right angle as shown in Fig. 8.13.

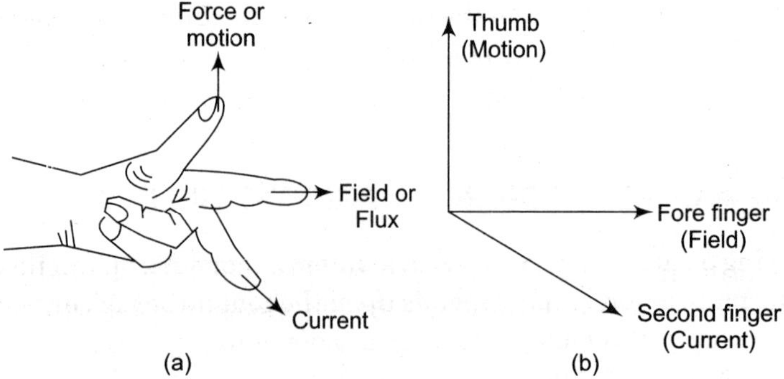

Fig. 8.13 *Left hand rule representing mutual direction of motion, field and current.*

8.14 ELECTROMAGNETIC INDUCTION

Oersted at Copenhagen in 1820; discovered a very important phenomenon giving the relationship between magnetism and electricity. As per this relationship a conductor carrying a current

I is surrounded all along its length by a magnetic field, the lines of magnetic flux being concentric circles in planes at right angles to the conductor. This phenomenon of a magnetic field being associated with a current carrying conductor lead to a question whether the converse of the above is possible, *i.e.* can a magnetic field generate a current? Michael Faraday on 29 August 1831, succeed in generating an electric current with the aid of magnetic flux.

From his experiments Faraday concluded that a current was generated in a coil so long as the magnetic lines of force being through the conductor changed. The current thus generated is called the induced current and e.m.f. that gives rise to this induced current is called the induced e.m.f. This phenomenon of generating an induced current in a closed circuit by changing the magnetic field through it is called electromagnetic induction.

8.15 FARADAY'S LAWS OF ELECTROMAGNETIC INDUCTION

There are two principal laws of electromagnetic induction known as Faraday's laws.

***Faraday's first law*:** It states that whenever the magnetic flux associated or linked with a closed circuit is changed or alternatively when a conductor cuts or is cut by the magnetic flux; an e.m.f. is induced in the circuit resulting in an induced current. This e.m.f. is induced so long as the magnetic flux changes.

***Faraday's second law*:** It states that the magnitude of the induced e.m.f. generated in a coil is directly proportional to the rate of change of magnetic flux.

The change of flux as discussed in the Faraday's laws can be produced in two different ways:

(*i*) By the motion of the conductor or the coil in a magnetic field *i.e.* the magnetic field is stationary and the moving conductors cut through it. The e.m.f. generated in this way is normally called dynamically induced e.m.f.

(*ii*) By changing the current in a circuit thereby changing the flux linked with stationary conductors *i.e.* The conductor or coils remain stationary and flux linking with the conductors is changed. The e.m.f. is termed statically induced e.m.f. can be further subdivided into (*a*) self-induced e.m.f. (*b*) mutually induced e.m.f.

(*iii*) The concept of dynamically induced e.m.f. gives rise to the development of generators whereas statically induced e.m.f. was helpful in developing transformers.

8.16 LENZ'S LAW

The direction of statically induced e.m.f. can be obtained with the help of Lenz's law, which states: "The direction of the induced e.m.f. would always be such that its tends to setup a current opposing the change of flux responsible for producing that e.m.f."

When the north pole of the magnet is inserted in the coil, an e.m.f. is induced in it due to the motion of the magnet thereby generating induced current. According to Lenz's law the direction of the induced current generated in the coil should be such that the motion of the magnet is opposed; which is possible only when the upper end of the coil behaves as a north pole. For this is to happen the current generated in the coil should be in the anticlockwise direction as was observed by Faraday as shown in Fig. 8.14.

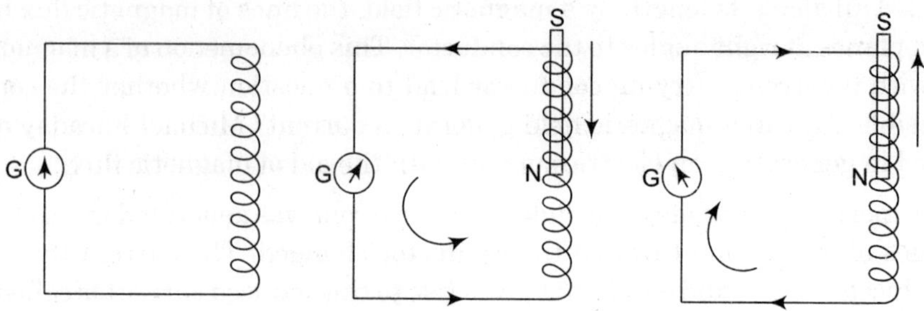

Fig. 8.14 *Circuit demonstrating Lenz's law.*

8.17 FLEMING'S RIGHT HAND RULE

The thumb, fore finger and the middle finger of the right hand are held mutually at right couples as shown Fig. 8.15. If the fore finger points in the direction of the magnetic flux, the thumb points in the direction of motion of the conductor relative to the magnetic field. Then the middle finger represents the direction of the induced e.m.f.

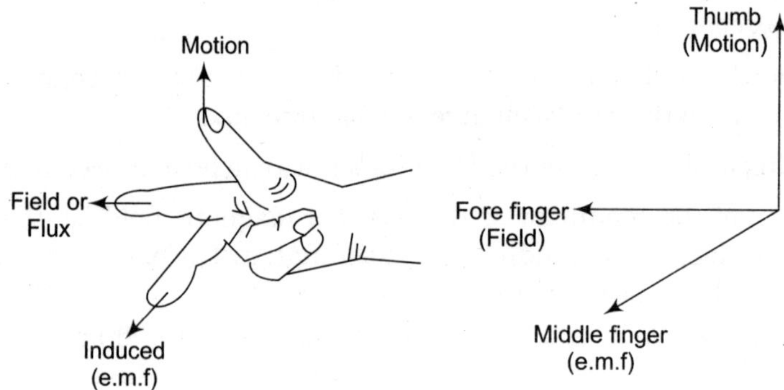

Fig. 8.15 *Right hand rule representing motion, field and induced e.m.f.*

8.18 MAGNITUDE OF INDUCED E.M.F. IN A COIL

Let a coil consist of N number of turns. Assume that the flux through the coil changes from its initial value ϕ_1 to ϕ_2 in an interval of t seconds.

Then,

$$\text{Initial value of flux linkages} = \text{N}\ \phi_1$$

$$\text{Value of flux linkages after time } t \text{ seconds} = \text{N}\ \phi_2$$

$$\text{Change of flux linkages in time } t \text{ seconds} = (\text{N}\ \phi_1 - \text{N}\ \phi_2)$$

As per Faraday's laws of electromagnetic induction the induced e.m.f. in the above coil due to a change of flux is given by

$$\text{Induced e.m.f.} = \frac{\text{N}\left(\phi_1 - \phi_2\right)}{t}$$

The instantaneous value of e.m.f. induced in the coil can be represented as

$$e = -\frac{d}{dt}(N\phi)$$

$$e = -N\,d\frac{\phi}{dt} \text{ volt}$$

The negative sign signifies that the induced e.m.f. generates a current tending to oppose the increase of flux through the coil.

8.19 DYNAMICALLY INDUCED E.M.F.

Dynamically induced e.m.f. is produced by the movement of the conductor in magnetic field. Figure 8.16 shows a uniform magnetic field of flux density B Tesla. In which the conductor is moving in the direction shown and cuts the flux at right angles?

If l = Length of the conductor *ii* metre.

v = Velocity of melon of conductor in m/s.

dx = Distance moved by the conductor in time dt.

Then area swept by moving conductor = $l \cdot dx$

Hence the changes in flux; when the conductor moves a distance dx in time dt is given by

P ⟶ u

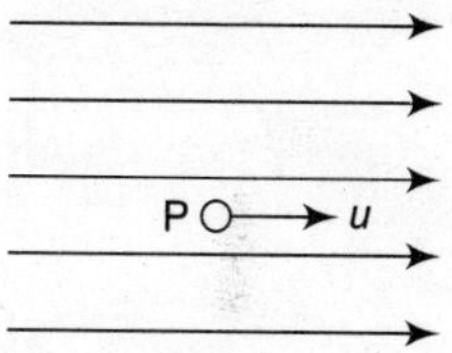

Fig. 8.16 *Uniform magnetic field.*

$$d\phi = Bl dx \text{ (Wb)}$$

The dynamically induced e.m.f. is rate of change of flux linkages *i.e.*

$$\text{Dynamically induced e.m.f. } e = \frac{d\phi}{dt} = \frac{Bldx}{dt}$$

As $$\frac{dx}{dt} = v \text{ (velocity)}$$

Thus dynamically induced e.m.f. = Blv volts.

Suppose that the conductor moves at an angle θ to the direction of field as shown in Fig. 8.17. The component of velocity which is perpendicular to the direction of field will be responsible for inducing voltage in the conductor. The component of velocity perpendicular to the field in (v sinθ). Therefore the induced voltage when the conductor moves in the direction making an angle θ with the direction of field is

$$e = Blv \sin\theta \text{ volts.}$$

The conductor forms part of a closed circuit; there will be a current in free conductor. By Lenz's law the current will be in such a direction as to produce a force that tends to oppose the motion of the conductor.

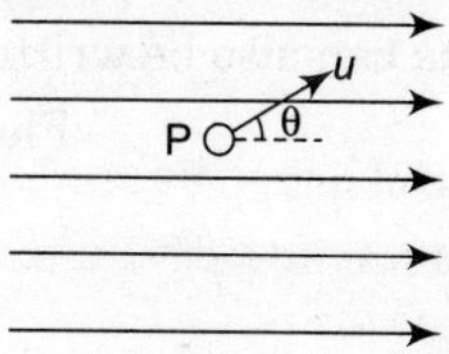

P u θ

Fig. 8.17 *Conductor moving in uniform magnetic field at angle θ.*

8.20 STATICALLY INDUCED E.M.F.

When the conductor or coil remains stationary and flux linking with these conductors undergo a change; an e.m.f. is induced in the conductors. Such an induced e.m.f. is termed as statically induced e.m.f. Statically induced e.m.f. can be further classified as (*i*) self-induced e.m.f. (*ii*) mutual induced e.m.f.

8.20.1 Self-induced e.m.f.

Any electrical circuit in which the change of current is accompanied by the change of flux and thereby an induced e.m.f. is said to be inductive or to posses self-inductance. Thus the property of the coil that enables to induce an e.m.f. in it whenever the current changes is called self-induction.

Consider a coil having N turns carrying a current of I amperes and let ϕ be the resulting flux linking in the coil. The magnetic flux forms complete loops as shown in Fig. 6.18.

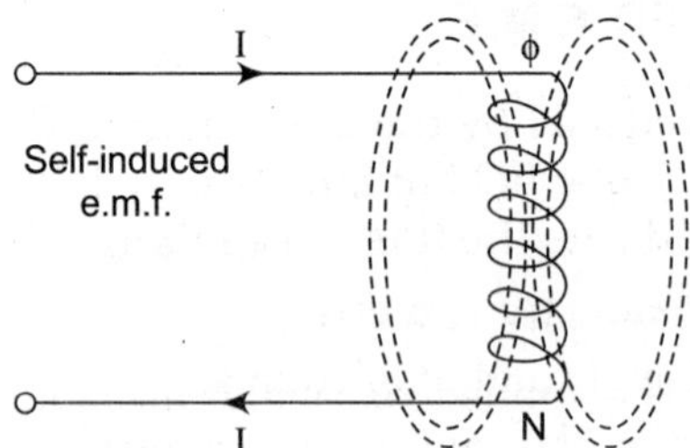

Fig. 8.18 *Statically induced e.m.f.*

The product $N\phi$ is normally termed as flux linkages. Now if the current flowing in coil is changed; then the number of coil linking the flux also changes such e.m.f. is induced in the according to the Faraday's laws of electromagnetic induction. This e.m.f. is termed as statically self-induced e.m.f. or the e.m.f. of self-induction.

As per Faraday's laws of electromagnetic induction this induced e.m.f. is given by,

$$e = \frac{-Nd\phi}{dt} \text{ volt}$$

Assuming iron core permeability to be constant. Thus the flux is proportional to the current through the coil *i.e.*

$$\Phi \propto I$$

$$\frac{\Phi}{I} = \text{Constant}$$

Now flux can also be written as

$$\text{Flux} = (\text{flux/current}). \text{Current}$$

i.e.

$$\phi = \left(\frac{\Phi}{I}\right) \cdot i$$

Now if current is changed at a certain rate the flux also changes at the same rate

$$\text{Thus the rate of change of flux} = \left(\frac{\phi}{I}\right) \times \text{Rate of change of current}$$

$$\frac{d\Phi}{dt} = \left(\frac{\phi}{I}\right) \cdot \frac{di}{dt}$$

Substituting this equation

$$e = -N\left(\frac{\phi}{I}\right) \times \text{Rate of change of current}$$

$$e = -N\left(\frac{\phi}{I}\right)\cdot\frac{di}{dt}$$

The term $\left(\frac{N\phi}{I}\right)$ *i.e.* flux linkages/amperes is generally called the self-inductance of the coil or coefficient of self-induction and is denoted by a symbol L.

Therefore, $e = -L\, di/dt$

Where $L = N\frac{\phi}{I}$ Henry

The negative sign in equation indicates that it is an e.m.f. opposing the change *i.e.* if the current is increasing then this e.m.f. will oppose the increase in current, in case the current is decreasing the induced e.m.f. tends to prevent the decrease of current and its direction is therefore the same as that of current or the applied voltage. It also indicates in that energy is being absorbed from the electric circuit and stored as magnetic energy in the coil.

8.20.2 Mutually Induced e.m.f.

The phenomenon of generation of induced e.m.f. in a circuit by changing the current in a neighbouring circuit is called mutual induction.

Consider two coils Q_1 and Q_2 such that Q_1 is connected to a battery through key K and Q_2 to a galvanometer as shown in Fig. 8.19. When the key is closed suddenly then current I start flowing in the coil Q_2, the galvanometer gives a sudden 'Kick' in one direction. Now when K is opened; the galvanometer shows a deflection but in the opposite direction. The above observations indicate clearly that an induced current is setup in the coil Q_2 when the current is changed in the coil Q_1 though the coil Q_2 is not connected physically to coil Q_1. Two coils possessing this property are said to be mutual inductance. The unit of mutuai inductance is also Henry. It is denoted by M.

Let ϕ_1 be the flux in coil Q_1 due to current I flowing in it and ϕ_2; the flux induced in Q_2 due to flux ϕ_1 in coil Q_1.

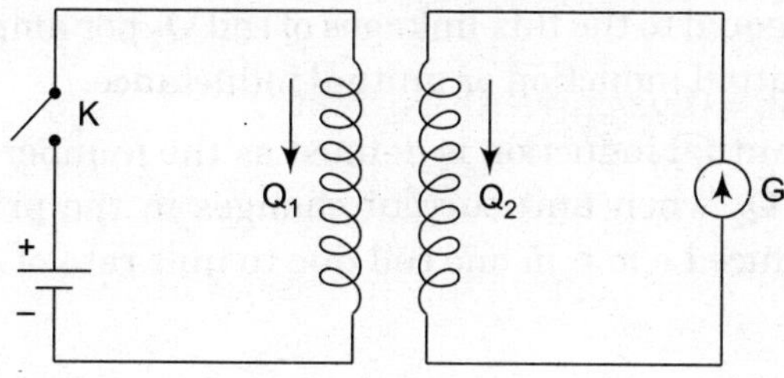

Fig. 8.19 *Mutral.*

The ratio $\frac{\phi_2}{\phi_1}$ is denoted by K.

Thus, $\frac{\phi_2}{\phi_1} = K$

or $$\phi_2 = K\,\phi_1$$

Also $$\phi_1 \propto I$$

or $$\frac{\phi_1}{I} = \text{Constant}$$

Now, $$\phi_2 = \left(\frac{\phi_2}{I}\right) \cdot I$$

Putting the value of ϕ_2

$$\phi_2 = \left(\frac{K\phi_1}{I}\right) \cdot \text{Current}$$

When current is changed at a certain rate ϕ_2 also changes at the same rate. Thus rate of change $$\phi_2 = K\,\frac{\phi_1}{I} \times \text{Rate of change of current}$$

$$= \frac{d\phi_2}{dt} = \left(\frac{K\,\phi_1}{I}\right)\frac{di}{dt}$$

According Faraday's law of electromagnetic induction the e.m.f. induced in coil Q_2 is given by

$$e_s = N_2 \cdot (\text{rate of change of flux } \phi_2)$$

where; N_2 is the number of flux in coil Q_2

$$e_s = N_2 \cdot \left(\frac{K\,\phi_1}{I}\right) \cdot \frac{di}{dt}$$

$$e_s = M \cdot \frac{di}{dt}$$

where $$M = \frac{N_2\,K\,\phi_1}{I} = \frac{N_2\phi_2}{I}$$

$$= \frac{\text{Flux linkages of coil } Q_2}{\text{Current in coil } Q_1}$$

The constant M; which is equal to the flux linkages of coil Q_2 per ampere of current in coil Q_1; is called the coefficient of mutual induction or mutual inductance.

Hence the coefficient of mutual induction is defined as the number of lines of force passing through the secondary coil Q_2 when unit current changes in the primary coil Q_1. It is also numerically equal to the induced e.m.f. in one coil due to unit rate of change of current in the other coil.

8.21 COEFFICIENT OF COUPLING

Consider two coils Q_1 and Q_2 having N_1 and N_2 turns respectively and coupled magnetically. Mutual inductance between the coils Q_1 and Q_2 is given by

$$M = \frac{N_1\phi_1}{I_2}$$

or $$M = \frac{N_1 K_1 \phi_2}{I_2}$$

where ϕ_2 is the flux produced by current I_2 flowing in the coil Q_2. Mutual inductance between the coils Q_1 and Q_2 can also be written as,

$$M = \frac{N_2 \phi_2}{I_1}$$

or $$M = \frac{N_2\ K_2\ \phi_1}{I_1}$$

where ϕ_1 is the flux produced by current I_1 flowing in coil Q_1; however only a fraction $K_2\ \phi_1$ links the coil Q_2.

Multiplying above eq.

$$M^2 = \frac{(N_1\ N_2\ K_1\ K_2\ \phi_1\ \phi_2)}{I_1\ I_2}$$

$$= K_1 K_2 \left(\frac{N_1\ \phi_1}{I_1}\right)\left(\frac{N_2\ \phi_2}{I_2}\right)$$

$$M^2 = K_1 K_2\ (L_1)\ (L_2)$$

where L_1 and L_2 are self-inductance of coil Q_1 and Q_2 respectively.

$$M = \left(\sqrt{K_1\ K_2}\right)\left(\sqrt{L_1\ L_2}\right)$$

If, $$K_1 = K_2 = K$$

Then $$M = K\left(\sqrt{L_1\ L_2}\right)$$

or $$K = \frac{M}{\left(\sqrt{L_1\ L_2}\right)}$$

The constant K is called; the coefficient of coupling. Hence coefficient of coupling depends upon the mutual inductance between the two coils and also upon the self-inductance of these coils.

8.22 INDUCTANCES OF INDUCTIVELY COUPLED COILS CONNECTED IN SERIES

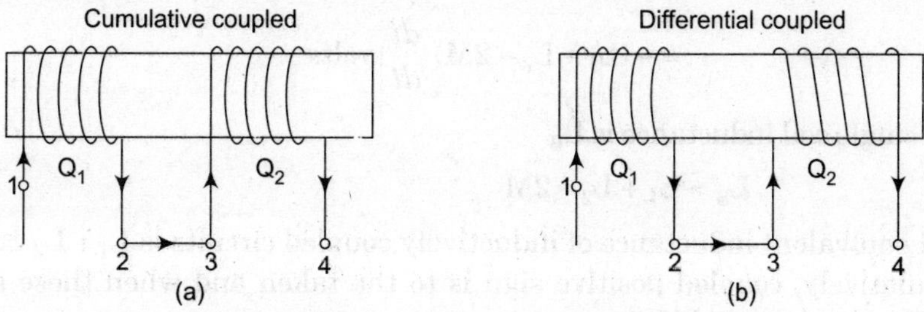

Fig. 8.20 *Inductively coupled coils.*

Two inductive coils Q_1 and Q_2 connected in series so that the currents through the two coils are in the same direction in order to produce flux in the same direction. Coils connected in this manner are said to be cumulative coupled. As shown in Fig. 8.20, let L_1 and L_2 be self-inductances of coils Q_1 and Q_2 respectively and M the mutual inductance between them.

If the current in the coils increases by *di* amperes in *dt* second, then

e.m.f. induced in coil Q_1 due to its self-inductance $= -L_1\frac{di}{dt}$

e.m.f. induced in coil Q_2 due to it self-inductance $= -L_2\frac{di}{dt}$

e.m.f. induced in coil Q_1 due to increase of current in coil $Q_2 = -M \cdot \frac{di}{dt}$

e.m.f. induced in coil Q_2 due to increase of current in coil $Q_1 = -M \frac{di}{dt}$

Hence that induced e.m.f. in coils Q_1 and Q_2

$$= -\left(\frac{L_1 di}{dt} + \frac{L_2 di}{dt} + \frac{2M di}{dt}\right)$$

or

$$= -(L_1 + L_2 + 2M)\frac{di}{dt} \text{ volts}$$

Equivalent single coil having inductance L_A

So, $\qquad L_A = L_1 + L_2 + 2M$

Now if the coil are differentially coupled the induced e.m.f. in coil Q_1 due to an increase of current *di* amperes in *dt* second in coil Q_2 (*i.e.* $M\frac{di}{dt}$) is in the same direction as the current.

Similarly e.m.f. induced in the coil Q_2 due to a change of current in coil Q_1 is also in the same direction as the current, as shown in Fig. 9.19(*b*).

Hence total e.m.f. induced in coils Q_1 and Q_2

$$= -L_1\frac{di}{dt} - L_2\frac{di}{dt} + 2M\frac{di}{dt}$$

or

$$= -(L_1 + L_2 - 2M)\frac{di}{dt} \text{ volts}$$

Equivalent single coil inductance is L_B

So, $\qquad L_B = L_1 + L_2 - 2M$

Hence total equivalent inductance of inductively coupled circuits is $L_1 + L_2 \pm 2M$. When the coils are cumulatively, coupled positive sign is to the taken and when these are differently coupled negative sign is considered.

SOLVED NUMERICAL PROBLEMS

Example 1. *Calculate the magnetising force along the axis of a solenoid wound with 800 turns of wire when a current of 5A is flowing through it. The length of solenoid is 2.5 metre.*

Solution: Number of turns, N = 800

Current $i = 5$ amp

Length $l = 2.5$ m

$$H = \frac{Ni}{l} = \frac{800 \times 5}{2.5} = 1600 \text{ AT/m}.$$

Example 2. *Calculate the magnetic field strength and flux density if flux $\phi = 10^{-3}$ weber and area of cross-section $a = 4 \times 10^{-4}$ m^2. The permeability of the medium may be taken as of air.*

Solution: $\phi = 10^{-3}$ weber

$a = 4 \times 10^{-4}$ m^2

$\mu = \mu_0 = 4\pi \times 10^{-7}$ H/m line $\mu_r = 1$

Flux density $$B = \frac{\phi}{a} = \frac{10^{-3}}{4 \times 10^{-4}} = \frac{10}{4} = 2.5 \text{ WB/m}^2$$

Field strength $$H = \frac{B}{\mu_0} = \frac{2.5}{4\pi \times 10^{-7}} = 1.99 \times 10^6 \text{ AT/m}.$$

Example 3. *The field coils of a two pole generator has 500 turns each and are connected in series. The magnetic flux per pole is 0.020 when field coils are excited. If the field circuit is opened in 0.02 second and residual magnetism is 0.0020 Wb/pole, determine the average voltage which is induced across field coils.*

Solution: Number of turns, N = 2 × 500 = 1000

Total initial flux = 0.020 × 2 = 0.04 Wb

Total residual flux = 0.0020 × 2 = 0.0040 Wb

∴ Change in flux ($d\phi$) = (0.04 – 0.0040) Wb

= 0.036 Wb

Change in time dt = 0.02 seconds

∴ $$e = N\frac{d\phi}{dt}$$

$$= 1000 \times \frac{0.036}{0.020} = 1800 \text{ volt}.$$

Example 4. *A conductor 2.0 m long carried a current of 55 amp at right angles to a magnetic field of density 1.4 Tesla. Calculate the force on the conductor.*

Solution: $l = 2.0$ m, I = 55 amp, B = 1.4 T

Force $F = Bil$

= 1.4 × 55 × 2.0 = 154 N.

Example 5. *In a loudspeaker, the moving coil consists of 150 turns of fine wire wound on a circular former so that the coil has an effective diameter of 25 mm. The coil is situated in a radial field of the speaker magnet so that the whole of its length is perpendicular to a field which has a flux density of 1.2 T. Calculate the force on the coil when it is carrying a current of 100 mA.*

Solution: *The coil may be considered as one conductor of length = Number of turns × Circumference of the former*

$$l = 150 \times \pi \times 25 \times 10^{-3} = 11.78 \text{ m}$$

$$i = 100 \text{ mA} = 100 \times 10^{-3} \text{ A}$$

$$B = 1.2 \text{ T}$$

$$F = Bil$$

$$= 1.2 \times 100 \times 10^{-3} \times 11.78 = 1.41 \text{ N}.$$

Example 6. *A conductor of length 0.6 m moves in a uniform magnetic field of density 1.2 T at a velocity of 25 m/s. Calculate the induced voltage in the conductor when the direction of motion is (a) perpendicular to the field (b) inclined at 45° to the direction of the field.*

Solution: $l = 0.6$ m, B = 1.2 T, $u = 25$ m/s.

(*a*) Induced voltage in inductor when perpendicular to field

$$e = \text{Blu}$$

$$= 1.2 \times 0.6 \times 25 = 18 \text{ V}$$

(*b*) Induced voltage when conductor is at 45° to the direction of the field.

$$e = \text{Blu} \sin \theta$$

$$= 1.2 \times 0.6 \times 25 \times \sin 45° = 12.73 \text{ volt}.$$

Example 7. *A conductor 0.25 m long carries a current of 100 A and lies at right angle to a magnetic field of density of 0.45 T, calculate the force acting on the conductor. If the force causes the conductor to move at a velocity of m/s. Calculate the voltage generated in it and the power developed by it.*

Solution: $l = 0.25$ m, $i = 100$ A, B = 0.45 T

Force $F = B \, . \, i \, . \, l$

$$= 0.45 \times 100 \times 0.25$$

$$= 11.25 \text{ N}$$

Induced voltage $e = B \, . \, l \, . \, u$

$$= 0.45 \times 0.25 \times 10$$

$$= 1.13 \text{ volt}$$

Power developed = Force × Velocity

$$= 11.25 \times 10 = 112.5 \text{ W}.$$

Example 8. *A coil of 300 turns has a flux of 0.45 mWb linking with it when carrying a current of 2A. Determine the inductance of the coil.*

$$N = 300, \phi = 0.45 \text{ mWb} = 0.45 \times 10^{-3} \text{ Wb}$$

$$i = 2.0 \text{ amp}$$

$$L = \frac{N\phi}{i} = 300 \times \frac{0.45 \times 100^{-3}}{2} = 0.0675 \text{ H}$$

$$= 67.50 \text{ mH.}$$

Example 9. *A coil consists of 600 turns carrying current of 8 amp in gives rise to a magnetic flux of 1.1 mWb. Calculate the inductance of the coil. If the current in coil is reversed in 0.01 second, determine the average voltage induced in the coil.*

Solution: $N = 600, i = 8 \text{ amp}, \phi = 1.1 \text{ mWb},$

$$\phi = 1.1 \times 10^{-3} \text{ Wb}$$

$$L = \frac{N\phi}{i}$$

$$= 600 \times \frac{1.1 \times 10^{-3}}{8} = 0.0825 \text{ H}$$

Initial current = 8 A

Final current (in reverse direction) = – 8 A

Change in current = Final current – Initial current

$$di = -8 - 8 = -16 \text{ amp}$$

Change in time $dt = 0.01 \, s$

Induced voltage $e = -L \dfrac{di}{dt}$

$$= -0.0825 \times \frac{(-16)}{0.01} = 132 \text{ volt.}$$

Example 10. *The number of turns in two coupled coils are 600 and 1500 respectively. When a current of 5 A flows in coil 2, the total flux in this coil is 0.75 mWb and the flux linking the first coil is 0.45 mWb. Calculate L_1, L_2, M and K.*

Solution: $N_1 = 600, N_2 = 1500, \phi_2 = 0.75 \text{ mWb} = 0.75 \times 10^{-3} \text{ Wb}$

$$\phi_{21} = 0.45 \text{ mWb} = 0.45 \times 10^{-3} \text{ Wb}$$

$$i_2 = 5 \text{ amp}$$

So, $L_2 = \dfrac{N_2\phi_2}{i_2} = \dfrac{1500 \times 0.75 \times 10^{-3}}{5} = 0.0225 \text{ H}$

$$K = \frac{\phi_{21}}{\phi_2} = \frac{0.45}{0.75} = 0.60$$

We know that

Self-inductance $\propto N^2$

$$L_1 \propto N_1^2 \text{ and } L_2 \propto N_2^2$$

$$\frac{L_1}{L_2} = \frac{N_1^2}{N_2^2}$$

$$L_1 = \frac{N_1^2}{N_2^2} L_2$$

$$L_1 = \left(\frac{600}{1500}\right)^2 \times 0.0225 = 3.6 \times 10^{-3} \text{ H or } 3.6 \text{ mH}$$

Mutual inductance

$$M = K\sqrt{L_1 L_2}$$

$$= 0.60\sqrt{3.6 \times 10^{-3}} \times 0.0225 = 5.4 \times 10^{-3} \text{ H.}$$

Example 11. *The coefficient of coupling between two coils is 0.70. There are 200 turns in coil 1. The total flux of coil 1 is 0.4 mWb, when the current in this coil is 3.5 A. When the current is changed from 3.5 A to zero in 3 m sec., the voltage induced in coil is 70 V. Calculate L_1, L_2, M and N_2.*

Solution:

$$K = 0.70, N_1 = 200, \phi_1 = 0.4 \text{ mWb} = 0.4 \times 10^{-3} \text{ Wb}$$

$$i_1 = 3.5 \text{ A}$$

$$L_1 = N_1 \frac{\phi_i}{i_1} = 200 \times \frac{0.4 \times 10^{-3}}{3.5} = 22.85 \text{ mH.}$$

$$V_2 = M\frac{di}{dt}$$

$$70 = M\frac{3.5}{3 \times 10^{-3}}$$

$$M = \frac{70 \times 3 \times 10^{-3}}{3.5} = 0.06 \text{ H}$$

$$M = K\sqrt{L_1 L_2}$$

$$0.06 = 0.70\sqrt{22.85 \times 10^{-3} \times L_2}$$

$$\frac{0.06}{0.70} = \sqrt{22.85 \times 10^{-3} \times L_2}$$

$$L_2 = \left(\frac{0.06}{7}\right)^2 \times \frac{1}{22.85 \times 10^{-3}}$$

$$L_2 = 0.321 \text{ H}$$

$$\frac{L_2}{L_1} = \left(\frac{N_2}{N_1}\right)^2$$

or $$N_2 = N_1 \sqrt{\frac{L_2}{L_1}}$$

$$= 200 \sqrt{\frac{0.0321}{22.85 \times 10^{-3}}}$$

$$N_2 = 749.62 \approx 750. \textbf{ Ans.}$$

Example 12. *Two coils with a coefficient of coupling of 0.70 between them, are connected in series to as to magnetize (i) in the same direction (ii) in the opposite direction. The corresponding values of equivalent inductances are (i) 2.0 H and for (ii) 0.75 H. Find the self-inductance of two coils and the mutual inductance between them.*

Solution : Given $K = 0.70$

We know, $L = L_1 + L_2 \pm 2M$

(*i*) For the same direction, L = 2.0 H

$$2 = L_1 + L_2 + 2M \quad \text{... (1)}$$

(*ii*) For the opposite direction, L = 0.75 H

$$0.75 = L_1 + L_2 - 2M \quad \text{... (2)}$$

Substracting equation (2) from equation (1), we get

$$1.25 = 4M$$

or $$M = 0.3125 \text{ H.}$$

Now adding equation (1) and equation (2)

$$2(L_1 + L_2) = 2.75$$

$$L_1 + L_2 = 1.375 \quad \text{... (3)}$$

$$M = K\sqrt{L_1 L_2}$$

$$L_1 L_2 = \frac{M^2}{K^2} = \frac{(0.3125)^2}{(0.70)^2} = 0.199$$

Since $$(L_1 - L_2)^2 = (L_1 + L_2)^2 - 4 L_1 \cdot L_2$$

$$= (1.375)^2 - 4 \times 0.199$$

$$L_1 - L_2 = 1.05 \text{ H} \quad \text{... (4)}$$

Solving equations (3) and (4), we get

$$L_1 = 1.21 \text{ H}$$

$$L_2 = 0.163 \text{ H. } \textbf{Ans.}$$

EXERCISES

1. State the following :
   (*a*) Magnetic flux and its properties.
   (*b*) Flux density.
   (*c*) Fleming's right-hand rule.
   (*d*) Fleming's Left-hand rule.
   (*e*) Lenz's law.
2. Write a short note on electromagnetic induction. State and explain Faraday's laws of electromagnetic induction.
3. Explain how is an e.m.f. induced (*i*) statically, (*ii*) dynamically?
4. Derive the expressions :
   $$e = Blv \sin \theta$$
   and $$F = Bil \sin \theta.$$
   where the symbols have their usual meanings.
5. A coil of 100 turns gives rise to a magnetic flux of 2.0 mWb, when carrying a certain current. If this is completely reversed in 0.2 seconds, what is average voltage induced in the coil?
6. A conductor of length 150 cm moves at rightangles to a unit for *m* magnetic field of flux density 1.2 T with a velocity of 50 m/s. Calculate the e.m.f. induced in it. Find also the e.m.f. induced if the conductor moves at an angle of 30° to direction of field.
7. A conductor of effective length 0.4 m moves with a velocity of 6 m/s perpendicular to a magnetic field of density 0.75 T. Calculate the (*a*) voltage induced in the conductor; (2) force acting on the conductor when it carries a current of 25 A and (*b*) power required to derive the conductor.
8. Calculate the force on a conductor of length 2.5 m carrying a current of 25 A, if it is (*a*) induced at 45° (*b*) parallel to a magnetic field of flux density 0.75 T.
9. The current flowing through a coil of 300 turns is suddenly arranged so that the flux linking the turns is increased by 5 m Wb in 0.2 *s*. Calculate the induced voltage.
10. Two coils are wound side by side on a former. An e.m.f. of 0.30 V is induced in coil A when the flux linking it changes at rate of 1 mWb/sec. A current of 2.5 amp in coil B causes a flux of 10^{-5} Wb to link coil A. What is the mutual inductance between the coils?
11. A coil of inductance 200 mH is magnetically coupled to another coil of inductance 800 mH. The coefficient of coupling between the coils is 0.75. Calculate the equivalent inductance of (*i*) series aiding, (*ii*) series opposing.
12. The total inductance of two coils when connected in series are 0.75 H and 0.35 H depending on their relative directions of current in the coils. Coil 1 when isolated from other coil has a self-inductance of 0.30 H. Calculate (*i*) the mutual inductance between the two coils, (*ii*) the self-inductances of coils, (*iii*) the coupling factor between the coils and (*iv*) the two possible value of induced e.m.f. in coil A when the current is decreasing at 1000 A per second in series combination.

9

Magnetic Circuits

9.1 INTRODUCTION

Magnetic flux lines always form closed loops. The complete closed path followed by the flux lines is called a magnetic circuit. Thus, a magnetic circuit provides a path for magnetic flux, just as an electric circuit provides a path for flow of electric current. There are lots of similari-ties between the magnetic and electric circuits. The study of magnetic circuit concepts is essential in the design, analysis and application of electromagnetic devices like transformers, rotating machines electromagnetic relays etc.

9.2 MAGNETO-MOTIVE FORCE (M.M.F.)

Flux is produced around any current carrying coil. Magneto-motive force can be produced when current flows in a coil of single or more turns *i.e.* product of the current and number of turns; is defined as the magneto-motive force (m.m.f.).

$$\text{m.m.f. } F = NI \text{ Amp-Turn}$$

where N = Number of turns in the coil

I = Current through the coil (amp.)

Since N is dimensionless then m.m.f. is taken as a.m.p.

9.3 MAGNETIC FIELD STRENGTH OR INTENSITY

If the magnetic circuit of a magnetic material is homogeneous and of uniform cross-section area, the magneto-motive force per metre length of the magnetic circuit is called the magnetic field strength. It is represented by H. Hence the magnetic field strength is

$$H = \frac{\text{m.m.f.}}{l} = \frac{NI}{l}\ \frac{A}{m}$$

where l is effective length of the magnetic flux path in metre.

$$H = \frac{F}{l}$$

or

$$F = H \cdot l$$

9.4 RELUCTANCE (S)

The opposition offered by a magnetic circuit to the establishment of magnetic flux is called the reluctance of the magnetic circuit. Let the iron ring have a mean circumference of l meter, cross-sectional area of a m^2 and N turns carrying current of I ampere then the total flux flowing in the dotted path is given by, as shown is Fig. 9.1.

$$\text{Flux} = \text{Flux density} \times \text{Cross-section area}$$

$$\phi = B \cdot a$$

Also m.m.f. = Magnetic field strength × Length of the magnetic flux path

$$\text{m.m.f. (F)} = H \cdot l$$

Dividing above equations

$$\frac{\phi}{F} = \frac{B \cdot a}{H \cdot l}$$

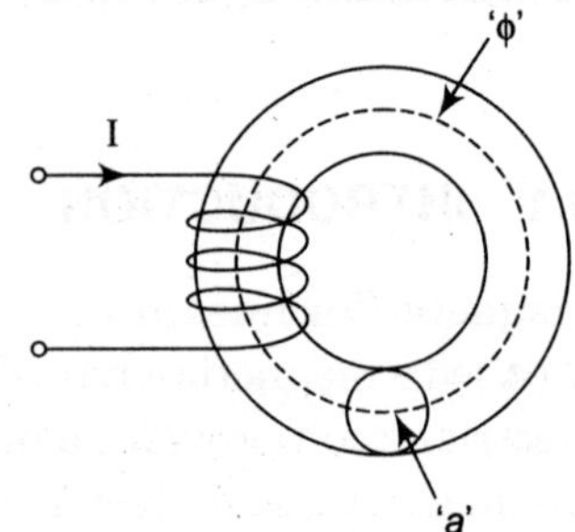

Fig. 9.1 *Magnetic flux path.*

We know that,

$$\frac{B}{H} = \mu_0 \mu_r$$

$$\frac{\phi}{F} = \frac{\mu_0 \mu_r \cdot a}{l}$$

or

$$\frac{F}{\phi} = \frac{l}{\mu_0 \mu_r \cdot a}$$

or

$$\frac{\text{m.m.f.}}{\text{flux}} = \frac{l}{\mu_0 \mu_r \cdot a} = S$$

Similar in electric circuit

$$\frac{\text{e.m.f.}}{\text{current}} = R = \frac{\rho \cdot l}{a}$$

Thus the reluctance is the property of magnetic material, which opposes the flow of magnetic flux through it.

Reluctance 'S' $= \dfrac{l}{\mu_0 \mu_r \cdot a}$ for magnetic material

Reluctance 'S' $= \dfrac{l}{\mu_0 a}$ for non-magnetic material *i.e.* $\mu_r = 1$

Hence the reluctance offered by the magnetic circuit or a part of magnetic circuit depends upon

1. Nature of magnetic material *i.e.* $\mu_0 \mu_r$.
2. Length of magnetic flux path in the part of magnetic circuit *i.e.* l.
3. Cross-section area of the material through which flux is passing *i.e.* 'a'.

The reluctance is expressed in Ampere/Webber and is denoted by 'S'. The reciprocal of reluctance is termed as permeance, which is analogous to conductance in electric circuits.

9.5 LAWS OF MAGNETIC CIRCUITS

All the laws applicable to electric circuit such as Ohm's law, Kirchhoff's laws etc. can also be applied to magnetic circuits simply by replacing the electrical terms with analogous magnetic terms.

Ohm's law for electrical circuits is

$$\text{e.m.f.} = \text{Current} \times \text{Resistance}$$

or $$E = I \cdot R$$

Current $$I = \frac{E}{R}$$

Ohm's law for magnetic circuit will be

$$\text{m.m.f.} = \text{Flux} \times \text{Reluctance}$$

$$F = \phi \cdot S$$

Flux $$\phi = \frac{F}{S}$$

So we can compare these two relations, it is found that the flux is analogous to current, m.m.f. is analogous to e.m.f. and reluctance is analogous to resistance but reluctance is not a energy-loss component.

Also

$$\text{Resistance} = \frac{\rho \cdot l}{a} = \frac{\text{Resistivity} \times \text{Length}}{\text{Cross-section area}}$$

$$= \left(\frac{1}{\text{conductivity}}\right) \cdot \left(\frac{\text{length}}{\text{area}}\right)$$

Similarly for magnetic circuit

$$\text{Reluctance} = \left(\frac{1}{\text{permeability}}\right) \cdot \left(\frac{\text{length}}{\text{area}}\right)$$

$$= \left(\frac{1}{\mu_0 \mu_r}\right) \cdot \frac{l}{a}$$

In series electric circuit, the total resistance of the circuit is equal to the sum of all the resistances in series. Similarly when the flux has to permeate a number of portions of a magnetic circuit in series, the total reluctances of the complete magnetic circuit will be equal to the sum of the reluctances of the various portions *i.e.*

$$S = S_1 + S_2 + S_3 + \dots + S_n$$

In parallel magnetic circuits the same m.m.f. is applied to each of the parallel paths and the total flux divides between the paths is inversely proportional to their reluctances

$$\phi = \phi_1 + \phi_2 + \phi_3 + \dots \phi_n$$

$$\frac{1}{S} = \frac{1}{S_1} + \frac{1}{S_2} + \frac{1}{S_3} + \dots + \frac{1}{S_n}$$

In a series electric circuit the total voltage drop is equal to the sum of the voltage drop in various elements of the circuits. Similarly, the total m.m.f. required to establish a given flux in the magnetic circuit is equal to the sum of the m.m.f. necessary to establish the flux through the various parts of the circuit. Thus the total m.m.f. for the complete magnetic circuit consisting of a number of homogeneous is given by

$$\text{Total m.m.f.} = F_1 + F_2 + F_3 + \ldots + F_n$$

$$F = H_1 l_1 + H_2 l_2 + H_3 l_3 + \ldots + H_n l_n$$

$$\text{Total m.m.f. } F = \frac{B_1 l_1}{\mu_1} + \frac{B_2 l_2}{\mu_2} + \frac{B_3 l_3}{\mu_3} + \ldots + \frac{B_n l_n}{\mu_n}$$

where l_1, l_2, l_3 ... etc. are the magnetic flux path length in the various parts of the magnetic circuits and μ_1, μ_2, μ_3 etc. are the absolute permeability of the medium of various of magnetic circuit.

9.6 AMPERE TURNS FOR A MAGNETIC CIRCUIT

The m.m.f. acting around a complete magnetic circuit is equal to the total ampere turns required to force the given flux through the magnetic circuit *i.e.*

$$\text{Total ampere turns} = \frac{B_1}{\mu_1 \cdot l_1} + \frac{B_2}{\mu_2 \cdot l_2} + \frac{B_3}{\mu_3 \cdot l_3} + \ldots + \frac{B_n}{\mu_n \cdot l_n}$$

Above equation clearly indicates that the m.m.f. per unit length or the ampere turns per unit length of each part of a magnetic circuit depends upon the working flux density and the absolute permeability of the material of that part. The absolute permeability depends on the nature of magnetic material and also upon the working value of flux density. Hence ampere turns per unit length of the magnetic flux path for a particular part of the magnetic material can be found out standard curves of the material, plotted as flux density vs ampere turns per unit length.

9.7 CALCULATIONS OF AMPERE TURNS

Ampere turns for varies path of the magnetic circuit will be calculated separately. To calculate the ampere-turns of a particular part, the following procedures are followed in general:

1. Cross-sectional area of the part is calculated from the given dimensions.
2. The reluctance is calculated using equation reluctance, $S = \dfrac{l}{\mu_0 \mu_r a}$.
3. The magnetic flux established in that part is calculated using $\phi = \dfrac{\text{m.m.f.}}{S}$.
4. Magnetic flux density is found by using $B = \dfrac{\phi}{a}$.
5. Ampere turns per meter of the magnetic flux path length in that part at the flux density calculated above is found by using the magnetization curve of the magnetic material of that part.

6. Length of the magnetic flux path in that part is estimated from the given dimensions.
7. Total ampere-turns for the part are obtained by multiplying ampere-turns per meter by the length of flux path.
8. General procedure is now applied to a various path of magnetic circuit.
9. Total ampere-turns for the complete magnetic circuit can now be formed by adding algebraically the ampere turns needed by the various parts of the magnetic circuit.

9.8 LEAKAGE FLUX

The part of total magnetic flux that has its path wholly within the magnetic circuit is called the useful magnetic flux. The part of magnetic flux which is in air is called the leakage magnetic flux as shown in Fig. 9.2. Thus, the total magnetic flux produced is equal to the sum of the useful magnetic flux and the leakage flux. The ratio of the total flux produced to the useful flux is called the leakage factor or leakage coefficient. It is denoted by λ.

$$\text{Leakage factor `}\lambda\text{'} = \frac{\text{Total flux produced}}{\text{Useful flux}}$$

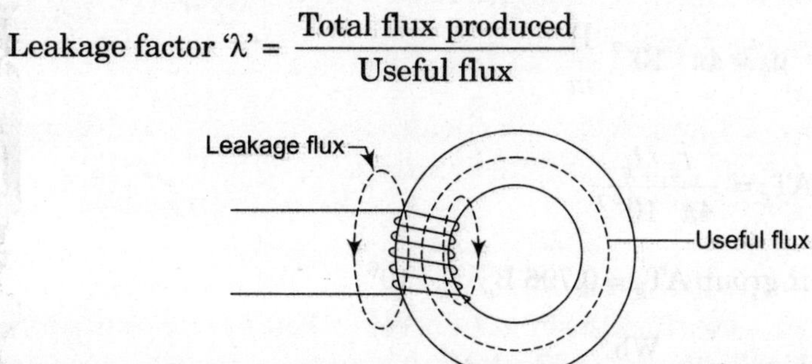

Fig. 9.2 *Useful and leakage flux.*

The value of λ is always greater than unity. Typical values of leakage factor are from about 1.12 to 1.25. Leakage is a characteristic of all magnetic circuits and can never be completely eliminated.

9.9 FRINGING

Air gaps are provided in many practical magnetic circuits. Consider a ring as shown in Fig. 9.3 provided with an air gap. When the flux line crosses the air gaps they tend to bulge out across the edges of the air gap. This effect is called fringing as shown in Fig. 9.3. The effect of fringing to make the effective gap area larger than that of the ring, consequently the flux density in the air gap is reduced. The effective increase depends upon the length of air gap. The air gap length is kept as small as possible so that the reduction in the flux density is the minimum. If the air gap length is small compared with the gap width, the effect of fringing can be neglected. In calculations, the effect of fringing is allowed by introducing an empirical factor.

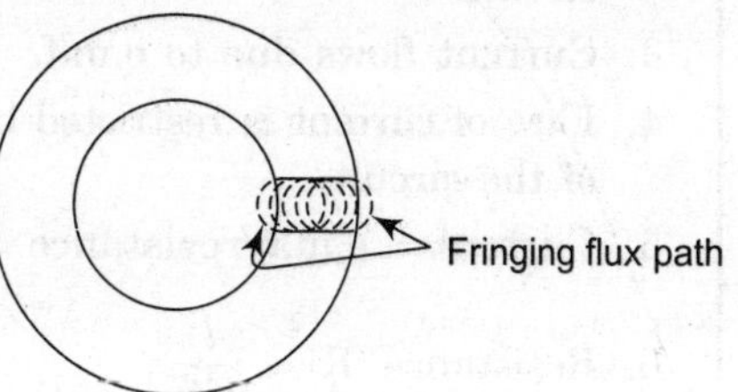

Fig. 9.3 *Fringing flux.*

9.10 CALCULATION OF AMPERE TURNS FOR THE AIR GAP

Total ampere turns for the air gap is given by

$$\text{m.m.f.} = AT_g = \text{Flux* Reluctance}$$

Reluctance for air gap (for which $\mu_r = 1$), $S = \dfrac{l_g}{\mu_0} \cdot A_g$

Thus, $$AT_g = \phi \cdot S = \frac{\phi \cdot l_g}{\mu A_g} \qquad \left(B = \frac{\phi}{A}\right)$$

$$= \left(\frac{\phi}{A_g}\right) \cdot \left(\frac{l_g}{\mu_0}\right) = \frac{B_g . l_g}{\mu_0}$$

where, $$\mu_0 = 4\pi \cdot 10^{-7} \frac{H}{m}$$

so $$AT_g = \frac{l_g \cdot l_g}{4\pi \cdot 10^{-7}}$$

or ampere turns for the air group $AT_g = 0.796\ B_g \cdot l_g \cdot 10^6$

where, B_g = Gap flux density in $\dfrac{Wb}{m^2}$.

l_g = Length of magnetic flux path in air gap in metre.

9.11 COMPARISON OF THE ELECTRIC AND MAGNETIC CIRCUIT

SIMILARITIES

Electric circuit	*Magnetic circuit*
1. Current flows in the circuit.	Flux is assumed to flow.
2. The path of current is called electric circuit.	Path of flux is called magnetic circuit.
3. Current flows due to e.m.f.	Flux flows due to m.m.f.
4. Flow of current is restricted by resistance of the circuit.	Flows of flux is restricted by reluctance of the circuit.
5. Current = e.m.f./resistance	Flux = e.m.f./reluctance
6. Resistance (R) = $\dfrac{l}{\sigma A}$	Reluctance (S) = $\dfrac{l}{\mu A}$.

DISSIMILARITIES

Electric circuit	*Magnetic circuit*
1. Current actually flows in the circuit.	Flux does not flow, it is only assumed to flow for finding out certain effects.
2. Energy is needed till the current flows.	Energy is needed only to create the magnetic flux.
3. Resistance of the circuit is independent current.	Reluctance of the circuit changes with the magnetic flux.

9.12 CALCULATION OF AMPERE TURNS OF SERIES-PARALLEL MAGNETIC CIRCUIT

In as shown in Fig. 9.4, the magnetic frame, in which magnetic circuit exists in parallel symmetrical pairs. The common section generally known as central limbs is wound with a number of turns, establishing the total flux ϕ Wb. The frame consist of two outer limbs which are mainly to provide a path to the magnetic flux. The total flux ϕ then divides into two parts and follows the different paths which recombine at the other end of the common section.

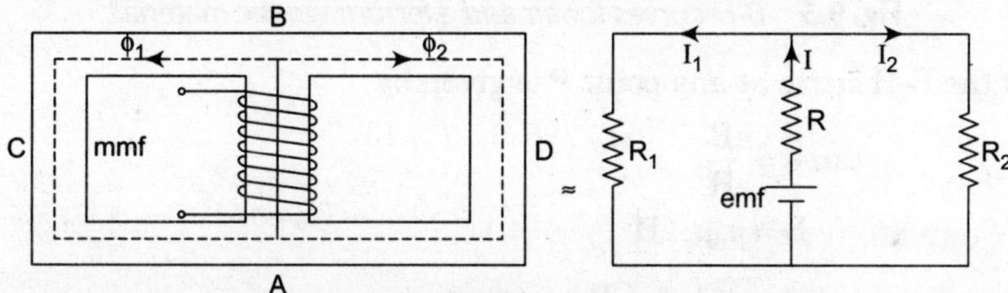

Fig. 9.4 *Series-parallel magnetic circuit.*

Let the flux in limb C be ϕ_1 and that in limb D be ϕ_2

Then $\phi = \phi_1 + \phi_2$

Electrical equivalent is

$$I = I_1 + I_2$$

where S = Reluctance of the path AB

S_1 = Reluctance of the path BCA and

S_2 = Reluctance of the path BDA

Then, $\quad$ m.m.f. $= \phi \cdot S + \phi_1 S_1$

and $\quad$ m.m.f. $= \phi \cdot S + \phi_2 S_2$

Electrical equivalent equations can be given as

$$\text{e.m.f.} = IR + I_1R_1$$

$$\text{e.m.f.} = IR + I_2R_2$$

Hence, total ampere turns required = Ampere turns required for the common section + Ampere turns required for one parallel path.

These type of magnetic frames are needed for shell type of transformers and for most of the rotating electrical machines.

9.13 MAGNETIC CURVE OR B-H CURVE

The graph between the flux density B and field intensity H of a magnetic material is called the magnetization curve or B-H curve. In air or a non-magnetic material the flux density B is proportional to magnetic field intensity H. Therefore, the graph of B versus H is a straight line through the origin as shown in Fig. 9.5. In the characteristic diagram except for a small portion near the origin the curve is linear from A to K. Over this portion the flux density B is proportional to the field intensity H and μ_r is constant. For higher values of H, there is no further increase in B with H. The curve becomes almost horizontal after the point K. The material is then said to be saturated and the point K is called the point of saturation for the material.

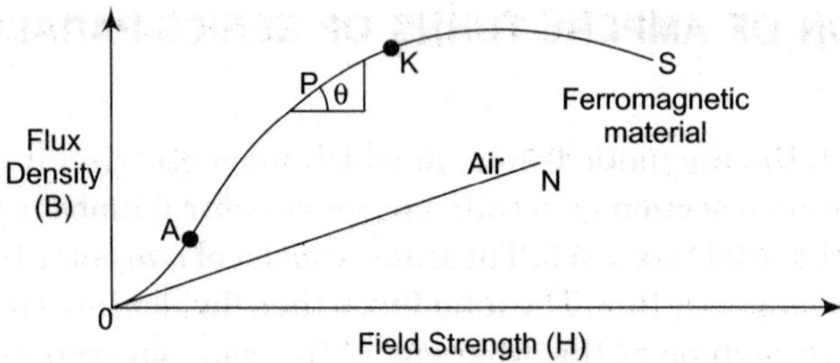

Fig. 9.5 *B-H curves for air and a ferromagnetic material.*

The slope of the B-H curve at any point P is given by

$$\tan \theta = \frac{B}{H}$$

$$B = \mu_0 \mu_r \cdot H$$

$$\mu_r = \left(\frac{1}{\mu_0}\right) \cdot \left(\frac{B}{H}\right) = \frac{\tan \theta}{\mu_0}$$

Thus, μ_r is proportional to slope of the B-H curve at any point starting from a definite value at the origin. The slope increases as B increases until its becomes maximum. It then gradually decreases as B increases further. The slope becomes almost zero in the saturation region when the curve becomes horizontal. The B-H curves shows that permeability μ_r of a magnetic material changes with the flux density B.

9.14 EXPLANATION OF THE SHAPE OF B-H CURVE

The shape of B-H curve can be explained on the basis of domain theory. Initially the specimen is magnetically neutral because the domains are randomly oriented so that their magnetic effects cancel each another. If an external magnetic field is applied, the domains having their magnetic axis close to the direction of the applied field which grow in size at expense of their neighbring domains. This growth is reversible and if the applied field is removed, the flux becomes zero. The portion OA on the B-H curve is due to this initial alignment. As H is increased further, more domains align themselves with the field then grow in size. This growth is irreversible and if the applied field removed the flux will still be there due to the aligned domains. The building of domains at a steady state is represented by the linear portion AK of the B-H curve. At higher values of H, rotation of the remaining domains taken place. When most of the domains are aligned, the material is said to be magnetically saturated. It is shown by the portion KS of the B-H curve. The point K where the saturation begins is called the knee of the curve.

9.15 HYSTERESIS

Hysteresis is the name given to the lagging of flux density B behind the magnetizing force H, when a specimen of ferromagnetic material is taken through a cycle of magnetization. If the specimen has been completely demagnetized and the magnetizing force H is increased in steps from zero. The relationship between flux density B and H is represented by curve OC; which is normal magnetization curve. If the value of H is now decreased the trace of B is higher than OC and follows the curve CD until H in reduced to zero. Thus when H reaches zero there is a residual flux density referred to as remnant flux density denoted by B_r. In order to reduce B to be zero, negative field strength OE must be applied. The magnetic field intensity OE required to wipe out the residual magnetism Br is called coercive force. As H is further increased in negative direction, the specimen becomes magnetized with the opposite polarity as shown by curve EF. If H varied backwards, the flux density curve follows a path FGC that is similar to the curve CDEF. The closed loop CDEFGC thus traced out is called the Hysteresis loop of the specimen. The term remnant flux density Br is called retentivity and the term coercive force is often called coercitivity.

The shape of the Hysteresis loop will depend upon the nature of magnetic material. Steel alloyed with 4% silicon has a very narrow Hysteresis loop. Hysteresis in magnetic material results in dissipation of energy, which is proportional to the area of the Hysteresis loop. Hence the following conclusion can be drawn:

1. Flux density B always lags with respect to the magnetizing force H.
2. An expenditure of energy is essential to carry the specimen through a complete cycle of magnetization.
3. Energy loss is proportional to the area by Hysteresis loop and depends upon the quality of the magnetic material.

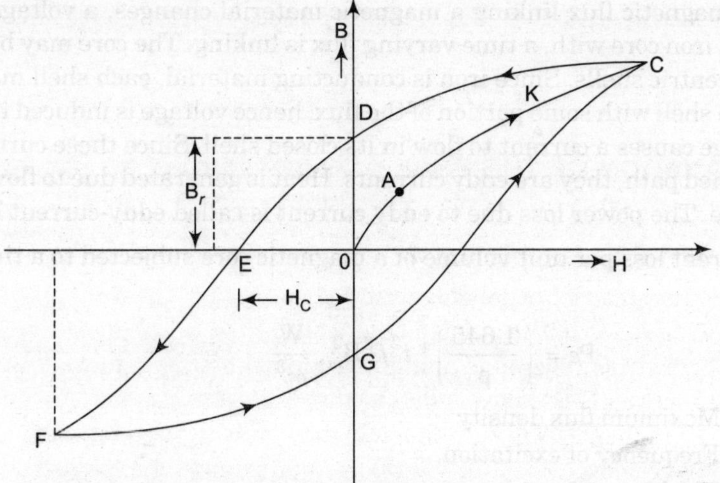

Fig. 9.6 *Hysteresis loop.*

9.16 CORE LOSSES OR IRON LOSSES

Time varying fluxes produce losses in ferromagnetic materials, known as core loss or iron losses. The core losses consist of hysteresis and eddy current losses. That is $P_{core} = P_h + P_e$. Both Hysteresis loss and eddy current loss produce heat in the magnetic circuit.

9.16.1 Hysteresis Loss

When a material is magnetized its domains are aligned in the direction of applied field. If the material is magnetized in the reverse direction the domains swing around to align themselves in the opposite direction. If there is cyclic reversal of the applied field, the domain rotate to and fro. In this process, there is loss of power called Hysteresis loss. This loss appears in the form of the heat and raises the temperature of the magnetic material.

It is not convenient to find out the area of the Hysteresis loop because B-H curve is non-linear, multivalued and no simple mathematical expression can describe the loop. Several empirical formulae for computing the Hysteresis loss P_h have been developed. The following empirical formula was evolved by Stein Metz and is widely used.

$$P_h = vK_h B^n{}_m f$$

where v = Volume of core.

K_h = Constant whose value depends upon the ferromagnetic material.

B_m = Maximum value of flux density.

f = Frequency of variation of the current i.

The exponent n varies in the range 1.5 to 2.5 depending upon the material. However, it is important to note that the Hysteresis loss varies directly as the frequency for a given B_m.

Using good quality silicon steel for core material reduces the Hysteresis loss in a magnetic circuit. The Hysteresis loop for such a material should be narrow.

9.16.2 Eddy Current Loss

Whenever the magnetic flux linking a magnetic material changes, a voltage is induced in it. Consider a solid iron core with, a time varying flux is linking. The core may be considered to be made up of concentric shells. Since iron is conducting material, each shell may be treated as a closed coil. Each shell with some portion of the flux, hence voltage is induced in each shell. Each induced a voltage causes a current to flow in its closed shell. Since these current do not flow in any clearly defined path, they are eddy currents. Heat is generated due to flow of eddy currents through the core. The power loss due to eddy current is called eddy-current loss.

The eddy current loss per unit volume of a magnetic core subjected to a time varying flux is given by

$$Pe = \left(\frac{1.645}{\rho}\right) * t^2 f^2 B_m^2, \frac{W}{m^3}$$

where B_m = Maximum flux density

f = Frequency of excitation.

t = Thickness of core.

ρ = Resistivity of the core material

The eddy current losses increases the power loss and produce additional temperature rise. Making the core of a number of thin sheets called laminations reduces the eddy current loss. Varnish layer tightly insulates the laminations from one another. Thus the closed paths along which eddy currents flow are reduced in length. The lamination thickness varies from 0.5 to 5 mm in electrical machine and from 0.01 to 0.5 mm in devices used in electronic circuits operating at higher frequency.

9.17 APPLICATION OF EDDY CURRENT

1. Eddy current heating is used for heating metals, for example melting, hardening and other heat treatment processes.
2. Eddy current damping is used in permanent magnet moving coil instruments.
3. Eddy current braking is used in induction energy meters.

9.18 STACKING FACTOR

Magnetic circuits are usually constructed by laminations which are coated with insulating varnish layer, thus there is a small space present between the successive laminations. Consequently, the effective magnetic cross-sectional area is less than the overall area of stack. The rates of effective area of the overall area are called the stacking factor. It is also defined is the ratio of volume occupied by magnetic material to total volume of the core. This factor is important in calculating flux densities in magnetic parts. Stacking factor is usually less than 1.0, it approaches 1.0 as the lamination thickness increases.

SOLVED NUMERICAL PROBLEMS

Example 1. *An iron ring of mean circumference 1.2 m is uniformly wound with 400 turns of wire. When a current of 1.5 amp is passed through the coil, a flux density of 1.25 Wb / m² is produced in the iron. Find the relative permeability of iron under these conditions:*

Solution: Number of turns in the coil, N = 400

Current through the coil, i = 1.5 amp

Thus total ampere turns provided on the ring = 400 × 1.5 = 600

Mean circumference of the ring = 1.2 m

Thus, the length of mean flux path in the ring = 1.2 m

Hence ampere turns per metre of flux length,

$$H = \frac{Ni}{l}$$

$$= \frac{600}{1.2} = 500, \frac{\text{Amp-Turns}}{\text{metre}}$$

Flux density produced in the ring = 1.25 Wb/m^2

Thus $$\mu_0\mu_r = \frac{B}{H} = \frac{1.25}{500} = 2.5 \times 10^{-3}$$

Relative permeability of iron ring $$\mu_r = \frac{2.5 \times 10^{-3}}{4\pi \times 10^{-7}} = 1990. \textbf{ Ans.}$$

Example 2. *An iron ring with mean diameter of 50 cm has an air gap of 0.9 mm and a winding of 210 turns. If the permeability of the iron is 450 when a current of 1.20 amp flows through the coil, find the flux density.*

Solution: Let the flux density through the iron ring as well as air gap be B Wb/m^2.

$$l = 50 \text{ cm or } 0.5 \text{ m}, \mu_r = 450, i = 1.2 \text{ amp}, N = 210$$

Ampere turns required for iron ring is

$$AT_i = Hi \times l = \frac{B}{\mu_0 \cdot \mu_r} l = \frac{0.5\,B}{450\,\mu_0}$$

Similarly, for air gap $AT_g = H_g \times l$

$$= \frac{B \times 0.9 \times 10^{-3}}{\mu_0}$$

Thus total ampere turns required is given by

$$AT_{total} = AT_i + AT_g$$

$$= \frac{0.5\,B}{450\,\mu_0} + \frac{0.9 \times 10^{-3} B}{\mu_0}$$

Ampere turns supplied = 210 × 1.20 = 252.

$$\frac{0.5\,B}{450\,\mu_0} + \frac{0.9 \times 10^{-3} B}{\mu_0} = 252$$

$$\frac{B}{\mu_0}\left[\frac{0.5}{450} + 0.9 \times 10^{-3}\right] = 252$$

$$\frac{B}{\mu_0}\left[2.01 \times 10^{-3}\right] = 252$$

$$B = \frac{252 \times \mu_0}{2.01 \times 10^{-3}}$$

$$B = \frac{252 \times 4\pi \times 10^{-7}}{2.01 \times 10^{-3}} = 0.158 \text{ Wb/m}^2.$$

Example 3. *A ring shaped electromagnet has an air gap of 6 mm long and 25 cm^2 in area, the mean length of the core being 50 cm and its cross-section is 9 cm^2. Calculate the ampere turns required to produce a flux density of 0.75 Wb / m^2 in the air gap. Assume the permeability of iron as 1600.*

Solution: In air gap, $$H = \frac{B}{\mu_0} = \frac{0.75}{4\pi \times 10^{-7}} = 5.97 \times 10^5 \text{ AT/m}.$$

AT required $= H \times l$

$= 5.97 \times 10^5 \times 6 \times 10^{-3} = 3580.98$

In core, core flux

$= \text{Air gap flux}$

$= 0.75 \times 25 \times 10^{-4} = 1.8 \times 10^{-3}$ Wb

Flux density in case $= \dfrac{\phi}{A} = \dfrac{1.8 \times 10^{-3}}{9 \times 10^{-4}}$

$= 2$ Wb/m^2.

$$H = \frac{B}{\mu_r \cdot \mu_0} = \frac{2}{1600 \times 4\pi \times 10^{-7}}$$

$= 994.72$ AT/m

Here length $= (50 - 0.6) = 49.4$ cm

$= 0.494$ m

$\therefore$ AT required $= 994.72 \times 0.494$

$= 491.39$

Total AT required is given $AT_{total} = AT_{core} + AT_{air\ gap}$

$= 491.39 + 3580.98$

$= 4072.37$. **Ans.**

Example 4. *A ring has a mean diameter of 25 cm and a cross section area of 12 cm^2. The ring is made up of semi-circular sections of cost iron and cast steel, with each joint having a reluctance equal to an air gap of 0.25 mm. Find the ampere turns required to produce a flux of 0.75×10^{-4} Wb. The relative permeabilities of cast steel and cast iron are 800 and 175, respectively, neglect fringing and leakge effects.*

Solution: $\phi = 0.75 \times 10^{-4}$ Wb, $a = 12$ cm^2 $= 12 \times 10^{-4}$ m^2

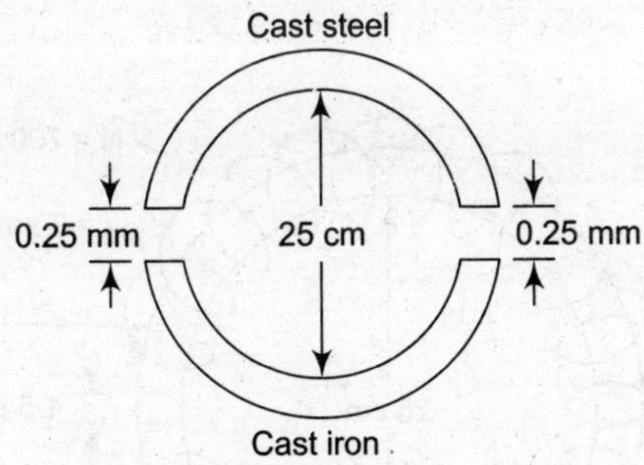

Fig. 9.7

$$B = \frac{\phi}{A} = \frac{0.75 \times 10^{-4}}{12 \times 10^{-4}} = 0.0625 \text{ Wb/m}^2$$

Air gap : $H = \dfrac{B}{\mu_0} = \dfrac{0.0625}{4\pi \times 10^{-7}} = 4.97 \times 10^4$ AT/m.

Total air gap length $= 2 \times 0.25 = 0.50 \text{ mm} = 5 \times 10^{-4}$ m.

AT required $= \text{H} \times l$

$= 4.97 \times 10^4 \times 5 \times 10^{-4} = 24.86$

Cast steel, $$\text{H} = \frac{\text{B}}{\mu_0 \cdot \mu_r} = \frac{0.0625}{4\pi \times 10^{-7} \times 800} = 62.17 \text{ AT/m.}$$

Length of cast steel path $$= \frac{\pi d}{2} - 25 = \frac{3.14 \times 25}{2} - 0.25 = 39.25 - 0.25 = 39.00 \text{ cm.}$$

AT required $= \text{H} \times l$

$= 62.17 \times 0.39 = 24.25$

Cast iron path

$$\text{H} = \frac{0.0625}{4\pi \times 10^{-7} \times 175} = 284.2 \text{ AT/m.}$$

Cast iron path length $= 0.39$

AT required $= 284.2 \times 0.39$

$= 110.84$

Total AT required is given by

$$\text{AT}_{\text{total}} = 24.86 + 24.25 + 110.84$$
$$= 159.95. \text{ AT.}$$

Example 5. *A steel ring of 25 cm mean diameter and of circular section 3 cm in diameter has an air gap of 1.5 mm length. It is wound uniformly with 700 turns of wire carrying a current of 2 amp. Calculate: (i) Magneticmotive force, (ii) Flux density, (iii) Magnetic flux (iv) Reluctance and (v) Relative permeability of steel ring.*

Neglect magnetic leakage and assume that iron path takes about 35 per cent of total magnetomotive force.

Solution:

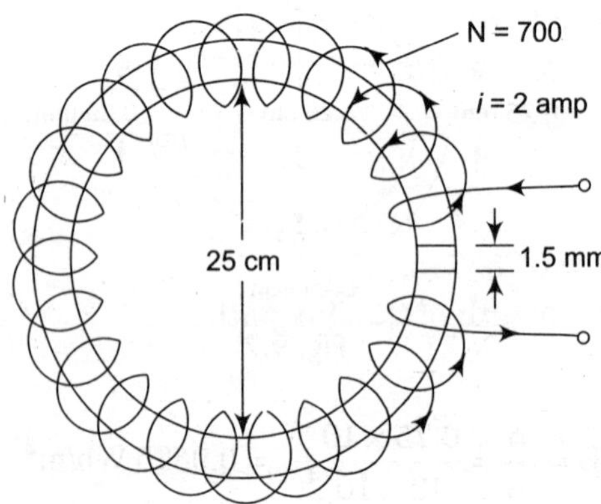

Fig. 9.8

(*i*) Number of turns provided on steel ring, N = 700

Current carried by the winding in the ring, I = 2.0 amp

Total ampere turns provided on the ring = 700 × 2 =1400

m.m.f. produced = NI = 1400 AT.

(*ii*) Total ampere turns or m.m.f. = 1400 AT

Iron portion takes 35% of the total m.m.f. Thus m.m.f. for iron portion

$= 0.35 \times 1400 = 490$ AT

Hence m.m.f. of air gap = 1400 – 490

= 910 AT

Ampere turns for the air gap is given by

$$AT_g = 0.796\, B_g A_g \times 10^6$$

Length of air gap, $l_g = 1.5 \text{ mm} = 1.5 \times 10^{-3}$ m

Thus flux density in air gap

$$B_g = \frac{AT_g}{0.796 \times l_g} \times 10^{-6}$$

$$= \frac{910 \times 10^{-6}}{0.796 \times 1.5 \times 10^{-3}} = 0.762 \text{ Wb/m}^2.$$

(*iii*) Diameter of circular section of the ring = 3 cm

Sectional area $= \frac{\pi}{4}(3)^2 = 7.07 \text{ cm}^2$

$= 7.07 \times 10^{-4} \text{ m}^2$.

Magnetic flux = Flux density × Sectional area

$= 0.762 \times 7.07 \times 10^{-4} = 0.538$ mWb

(*iv*) m.m.f. = Reluctance × Flux

or Reluctance $= \frac{\text{m.m.f.}}{\text{flux}} = \frac{1400}{0.538 \times 10^{-3}}$

$= 2.6 \times 10^6$ AT/Wb.

(*v*) Total ampere turns, NI = 1400

Length of the mean flux path in the steel ring = $\pi \times D_m$

Mean diameter of steel ring D_m = 25 cm.

Thus length of mean flux path in the steel ring

$= \pi \times 25 = 78.57$ cm

= 0.7857 m.

Ampere turns per metre length of the flux path *i.e.*,

$$H_i = \frac{\text{Ampere turn for iron part}}{0.7857}$$

$$= \frac{490}{0.7857} = 623.6$$

Now $\frac{B_i}{H_i} = \mu_0 \cdot \mu_r$

$$\mu_r = \frac{B_i}{H_i \cdot \mu_0} = \frac{0.762}{623.6 \times 4\pi \times 10^{-7}} = 972.$$

Example 6. Determine the total m.m.f. acting in the magnetic circuit shown in Fig. 9.9.

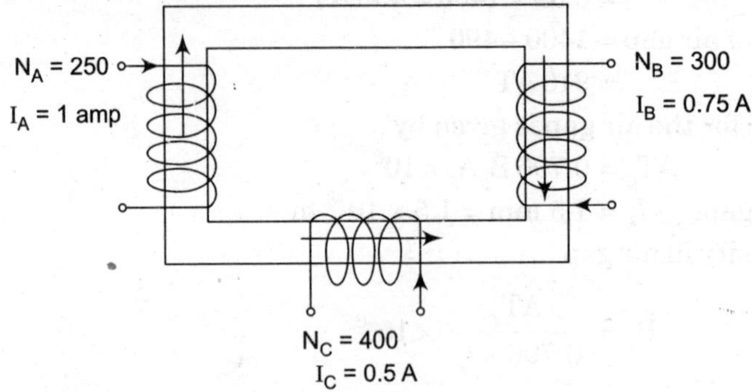

Fig. 9.9

$N_A = 300$
$I_A = 0.75$ A
$N_C = 400$
$I_C = 0.5$ A

Solution: Total m.m.f. acting in the circuit $= I_A N_A + I_B N_B - I_C N_C$

$$= 1 \times 250 + 0.75 \times 300 - 0.5 \times 400$$
$$= 475 - 200 = 275 \text{ AT}.$$

Example 7. *A total flux of 0.0006 Wb is required in the air gap of an iron ring of cross-section 5.0 cm² and mean length 2.5 m with an air gap of 4.5 mm. Find the number of ampere turns required. Points on the B-H curve for the material of the ring are as follows:*

H (AT/m) :	*200*	*400*	*500*	*600*	*800*	*1000*
B (Wb/m²) :	*0.4*	*0.8*	*1.00*	*1.09*	*1.17*	*1.19*

Solution: Total ampere turns required, AT = Ampere, turns the iorn portion, AT_i + Ampere turns for gap, AT_g. Drawing B-H curve.

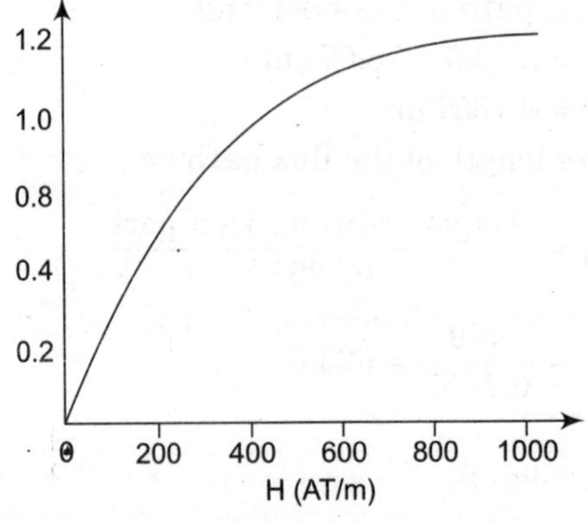

Fig. 9.10

(*i*) Flux $\phi = 0.0006$ Wb

Cross-sectional area of iron ring = 5.0 cm^2

$= 5.0 \times 10^{-4} \text{ m}^2$

$$\text{Flux density in iron portion} = \frac{0.0006}{5.0 \times 10^{-4}} = 1.2 \text{ Wb/m}^2$$

From the B-H curve for above flux density (1.2 Wb/m^2) the ampere turns required would be 1000.

Mean length of iron ring = 250 cm

Length of air gap, $l_g = 4.5$ mm

$= 0.45$ cm

Thus length of mean flux path in the iron portion of the ring,

$l_i = 250 - 0.45 = 249.55$ cm

$= 2.4955$ m.

Thus ampere turns required for iron portion

$AT_i = 1000 \times 2.4955 = 2495.5$

(*ii*) Ampere turns required for air gap = $0.796\ B_g\ l_g \times 10^6$

$= 0.796 \times 1.2 \times 4.5 \times 10^{-3} \times 10^6$

$= 4298.4$

Total ampere turns required = $2495.5 + 4298.4$

$= 6793.9.$

EXERCISES

1. Define m.m.f., Magnetic field intensity, reluctance and permeability.

An iron ring has a mean length c.f 1.5 m and a cross-section area of 20 cm^2. It has a radial air gap of 5 mm. The ring is uniformly wound with 250 turns. What direct current would be needed in the coil to produce a flux of 1.0 mWb in the air gap? Assume relative permeability of iron as 400 and leakge negligible.

2. Explain with help of suitable diagram (*i*) leakage flux and (*ii*) fringing.

An iron magnetic circuit has a uniform cross-sectional area of 5 cm^2 and a length of 16 cm. A coil of 100 turns is wound uniformly over the magnetic circuit. When the current in the coil is 1.4 amp, the total flux is 0.3 mWb; when the current is 5 A, the total flux is 0.5 mWb. For each value of current, calculate (*a*) the magnetising force (*b*) the relative permeability of the iron.

3. (*a*) Explain with the help of B-H curve. The meaning of the following terms:

Relative permeability, remanence and coerctivity.

(*b*) What is meant by magnetic hysteresis?

(*c*) What information can be derived from B-H loop.

4. (*a*) Compare the electric and magnetic circuits.

(*b*) What are different types of magnetic losses? How can they be minimised?

(c) Deduce an expression for the force between two parallel inductors.

5. (*a*) Draw and explain the B-H curves for magnetic material:
   (*b*) What is eddy current loss and under what conditions does it occur ? How can they be minimised? Mention some application of eddy currents.
6. (*a*) A steel ring has a mean diameter of 16 cm and a cross-sectional area of 4 cm^2. Calculate the ampere turns required for producing a flux of 0.75 mWb in the ring.
   (*b*) If the sawtooth cut of 1 mm width is now made in the above ring, determine the flux produced assuming the m.m.f. to remain constant. Assume the relative permeability of the mild steel to remain constant at 1000.
7. An iron ring of cross-sectional area 9 cm^2 and mean diameter of 40 cm. It is uniformly wound with a coil of 450 turns. When the coil is carrying a current of 2.5 amp, the flux set up in the ring is found to be 0.009 Wb. Find out the relative permeability of iron at this flux density.
8. An iron ring of mean circumference 50 cm has an air gap of 1 mm. It is uniformly wound with a coil having 250 turns. If the relative permeability of iron is 400, when a current of 1.2 A flows through the coil, calculate the flux density.
9. A cast steel electromagnet has an iron path of mean length 40 cm and an air gap of 1.5 mm. It is desired to produce a flux density of 1.0 Wb/m^2 in the gap by putting a coil on the electromagnet. Assuming negligible leakage and fringing, find the total ampere turns required. B-H curve may be plotted as per the datas given below:

H (AT/m) :	500	1000	2000	3000	4000
B (Wb/m^2) :	0.6	1.05	1.38	1.5	1.58

10. An electromagnet has mean length of iron path of 48 cm with the area of cross-section of core 10 cm^2. It has an air gap of 0.5 cm width. The electromagnetic is excited by two coils each having 450 turns. When the current in the coil is 0.8 amp. The resulting flux density gives a relative permeability of 1200. Neglect leakage and fringing. Calculate (*i*) reluctance of magnetic circuit, (*ii*) reluctance of air gap, (*iii*) total reluctance, (*iv*) total flux and (*v*) flux density in the gap.

10

Transformer

10.1 INTRODUCTION

A transformer is a static device consisting of two or more coils coupled through a magnetic circuit. The transfer of energy from one circuit to another takes place at different voltage levels (normally) without a change in frequency.

Transformers are so widely used as electrical apparatus ranging from low-power, low-current electronics and control circuits to extra high voltage power systems. Transformers in which ferrites or ferromagnetic magnetic cores are used to provide tight magnetic coupling and high flux densities such transformers are known as iron core transformers. While in which air as coupling medium is called air core transformers but have poor coupling and sometimes used in low power, electronics circuit. Generally work performed by transformers are: changing voltage and current level in electric power system. Impedance matching for maximum power transfer, Isolation of one circuit from another. Also used to measure voltage and currents in conjunction with measuring instrument. These are known as instrument transformers. Normally two types of core construction are used; core types and shell types Fig. 10.1.

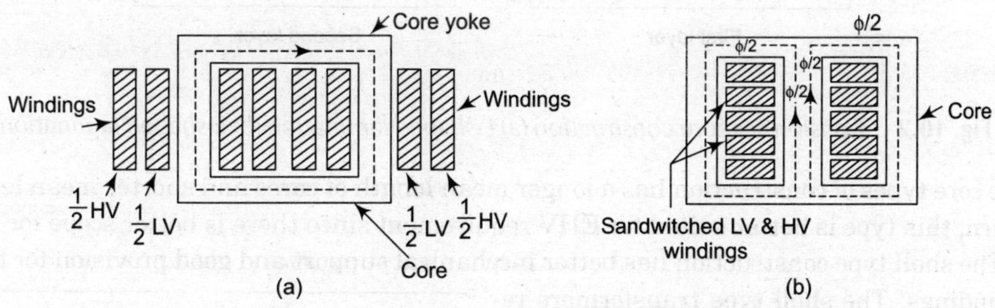

Fig. 10.1 *(a) Core type transformer, (b) Shell type transformer.*

In the core type the windings are wound around two legs of a magnetic core of rectangular shape while in the shell type, the windings are wound around the center leg of a three-legged magnetic core. To reduce core losses, the magnetic core are formed of a stack of thin laminations *i.e.*, silicon steel of 0.35 mm thickness material allows the use of high flux densities (1-1.5 T) while L Shaped laminations are used for shell type construction. To avoid continuous air gap lamination are stacked alternatingly as shown a Fig. 10.2.

For low power and high frequencies, compressed powdered ferromagnetic alloys known as Permalloy are used. Pulse transformer and high frequency electronic transformer cores made of soft ferrites.

A schematic representation of a two-winding transformer is shown as in Fig. 10.3. The two verticals lines are used to signify tight magnetic coupling between the windings. One winding, which is connected to a.c. supply, is referred as the primary windings while other windings connected to an electrical load is referred as the secondary windings.

The windings with the higher numbers of turns will have a high voltage and it is called the High voltage (HV) or high tension (HT) winding. The windings with the lower number of turns are called the low voltage (LV) or low tension (LT) windings.

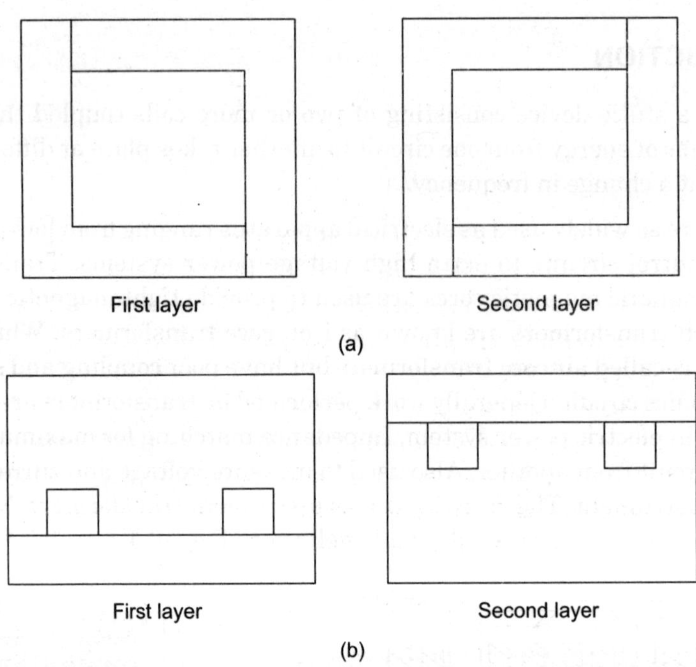

Fig. 10.2 *Transformer core construction (a) L-shaped lamination (b) E-shaped lamination.*

The core types of construction has a longer mean length of cores and shorter mean length of coil turn; this type is better suited for EHV requirement since there is better scope for insulation. The shell type construction has better mechanical support and good provision for bracing the windings. The shell type transformers requires more specialized fabrication facilities than core types while the latter offers the additional advantage of permitting visual inspection of coils in the case of a fault and ease of repair at sub-station site. For these reasons, the present practice is to use the core type transformers in large high voltage installation.

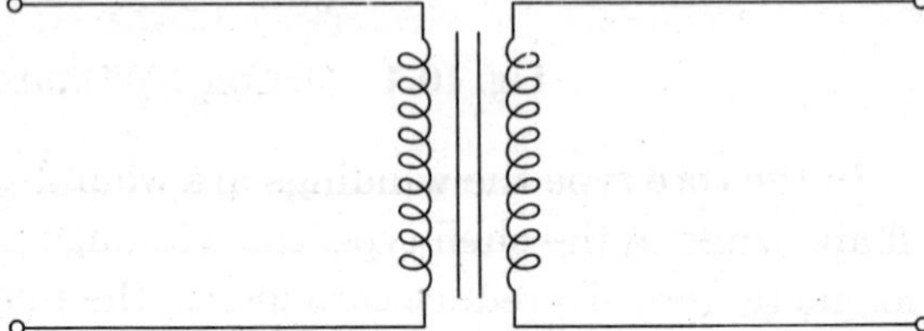

Fig. 10.3 *Schematic representation of two-winding transformer.*

10.2 IDEAL TRANSFORMER

In order to see the effect of flow of secondary current in transformers, certain assumptions will be made which are close approximations for a practical transformer. A transformer having ideal properties is hypothetical one. Assumptions for transformer to be ideal are as following:

1. The primary and secondary windings have zero resistance *i.e.* no copper loss (ohmic loss) and no resistive voltage drop in the ideal transformers. While in practical transformers finite but small windings resistances.
2. There is no leakage fluxes so all the flux is combined to the core and links both the winding while a practical transformer have a small amount of leakage fluxt *i.e.* magnetization curve should be linear.
3. The permeability of core is infinite so that exciting current required to establish flux in core is zero (negligible).
4. Losses in the core (eddy current losses and hysteresis losses) are considered to be zero.

Figure 10.4 shows an ideal transformer having a primary of N_1 turns and secondary of N_2 turns on a common magnetic core. The voltage of the source to which primary is connected is

$$v_1 = \sqrt{2}\, V_1 \cos \omega t$$

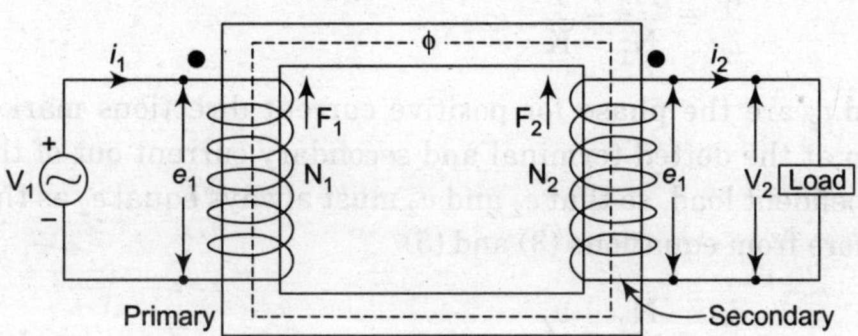

Fig. 10.4 *Ideal transformer on load.*

While the secondary is initially assumed to be an open circuited. As a consequence, flux ϕ is established in the core such that

$$e_1 = v_1 = N_1 \frac{d\phi}{dt} \qquad ...(1)$$

But the exciting current drawn from the source is nil (*i.e.* zero) by virtue of assumption that core has infinite permeability. The core flux ϕ which is completely mutual (no leakage) flux that causes an e.m.f. *i.e.* $e_2 = N_2 \dfrac{d\phi}{dt}$ to be induced in the secondary of polarity marked on the diagram for the windings direction indicated. The dots marked at one-end of each winding indicate the windings ends which simultaneously have the same polarity due to e.m.fs. induced.

So that $$\frac{e_1}{e_2} = \frac{N_1}{N_2} = K \qquad ...(2)$$

Since K, the transformation ratio is a constant. While e_1 and e_2 are in phase. The secondary terminal is

$$v_2 = e_2$$

Hence $$\frac{v_1}{v_2} = \frac{e_1}{e_2} = \frac{N_1}{N_2} = K \qquad ...(3)$$

Therefore an ideal transformer changes voltage in direct ratio of the number of turns in the two-windings. In terms of r.m.s. values equation (3) will be written as

$$\frac{V_1}{V_2} = \frac{E_1}{E_2} = \frac{N_1}{N_2} = K \qquad ...(4)$$

Now let the secondary be connected to a load of impedance Z_2 so that the secondary feeds a sinusoidal current of instantaneous value i_2 to the load. Due to flow of current (i_2) the secondary creates m.m.f. $F_2 = i_2 N_2$ opposing the flux ϕ. However, the mutual flux ϕ cannot change as otherwise the (v_1, e_1) balance will be disturbed (as windings has zero resistance). The result is that the primary draws a current i_1 from the source so as to create m.m.f. $F_1 = i_1 N_1$; which at all time cancels out the load caused m.m.f. $i_2 N_2$ so that ϕ is maintained constant independent of the load current flow.

Thus $$i_1 N_1 = i_2 N_2$$

$$\frac{i_1}{i_2} = \frac{N_2}{N_1} = \frac{1}{K} \qquad ...(5)$$

Obviously i_1 and i_2 are the phase for positive current directions marked on the diagram (primary current in at the dotted terminal and secondary current out of the dotted terminal) Since flux f is independent load, so that e_2 and v_2 must always equal e_2 as the secondary resistance is less. Therefore from equations (3) and (5)

$$\frac{i_1}{i_2} = \frac{N_2}{N_1} = \frac{v_2}{v_1} \qquad ...(6)$$

or $$v_1 i_1 = v_2 i_2 \qquad ...(7)$$

which means that the instantaneous power into primary equals the instantaneous power of the secondary; a direct consequence of the assumption (1) which means a loss less transformer.

In terms of r.m.s. value equation (6) will be written as

$$\frac{I_1}{I_2} = \frac{N_2}{N_1} = \frac{1}{K}$$

Which implies that current in an ideal transformer transform in inverse ratio of windings turns. Equation (7) in terms of r.m.s. value will read $V_1 I_1 = V_2 I_2$ *i.e.* the VA output balanced by VA input.

Figure 10.5 shows the schematic of ideal transformer of Fig. 10.6 with dot marks identifying similar polarity ends. It was already seen above that V_1 and V_2 are in phase and so are I_1 and I_2. Now

$$\frac{V_1}{V_2} = \frac{N_1}{N_2} \text{ and } \frac{I_1}{I_2} = \frac{N_2}{N_1}$$

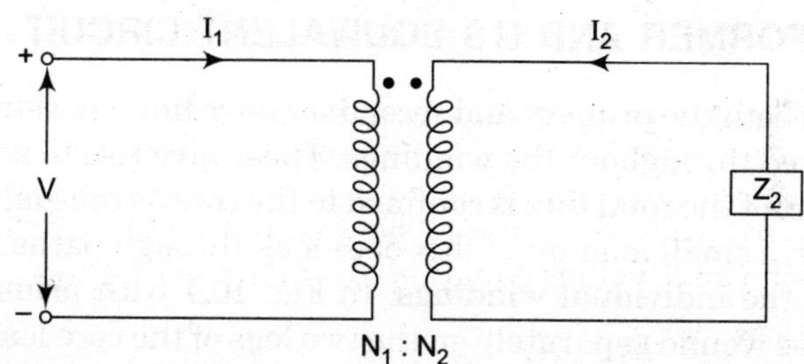

Fig. 10.5 *Schematic diagram of ideal transformer.*

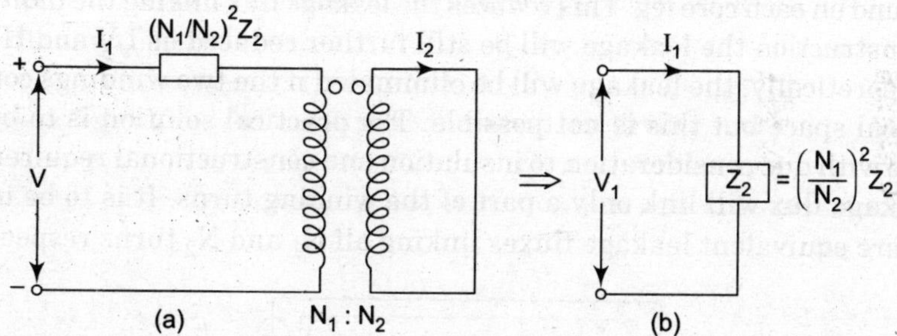

Fig. 10.6 *Ideal transformer referring impedance from secondary to primary.*

Dividing out above equations.

$$\frac{V_1}{V_2} = \frac{N_1}{N_2}$$

$$\frac{I_1}{I_2} = \frac{N_2}{N_1}$$

$$\frac{V_1}{I_1} = \left(\frac{N_1}{N_2}\right)^2 \frac{V_2}{I_2} = \left(\frac{N_1}{N_2}\right)^2 Z_2$$

$$Z_1 = \left(\frac{N_1}{N_2}\right)^2 Z_2 = K^2 Z_2 = Z_2'$$

Therefore, we can say the impedance in secondary side when seen on the primary side in transformed in the direct ratio of the square of turns Fig. 10.6 (*a* and *b*).

Similarly impedance Z_1 from the primary side can be referred to the secondary as

$$Z_1 = \left(\frac{N_2}{N_1}\right)^2 Z_1 = \left(\frac{1}{K}\right)^2 Z_1$$

Transferring impedance from one side to other side of transformer is known as referring the impedance to the other side.

It may be concluded that is an ideal transformer voltages are transformed in the direct ratio of turns; currents in the inverse ratio and impedances in the direct ratio squared; while power and VA remain unaltered.

10.3 REAL TRANSFORMER AND ITS EQUIVALENT CIRCUIT

In the real transformer; both the primary and secondary have finite resistances R_1 and R_2; which are uniformly distributed throughout the windings. These give rise to associated copper losses (I^2R). While a major part of the total flux is confined to the core as mutual flux ϕ linking both the primary and secondary, a small amount of flux does leak through paths which lie mostly in air and link separately in the individual windings. In Fig. 10.7 with primary and secondary for simplicity assumed to be wound separately on the two legs of the core leakage flux ϕ_{l_1} caused by primary m.m.f. I_1N_1 links primary windings itself and ϕ_{l_2} caused by I_2N_2 links the secondary windings. Thereby causing self-leakages of the two windings. Half the primary and half the secondary is wound on each core leg. This reduces the leakage flux linking the individual windings. In shell type construction the leakage will be still further reduced as LV and HV pancakes are interleaved. Theoretically, the leakage will be eliminated if the two windings could be placed in the same physical space but this is not possible. The practical solution is to bring the two as close as possible with due consideration to insulation and constructional requirements. Actually some of the leakage flux will link only a part of the winding turns. It is to be understood here that ϕ_{l_1} and ϕ_{l_2} are equivalent leakage fluxes linking all N_1 and N_2 turns respectively.

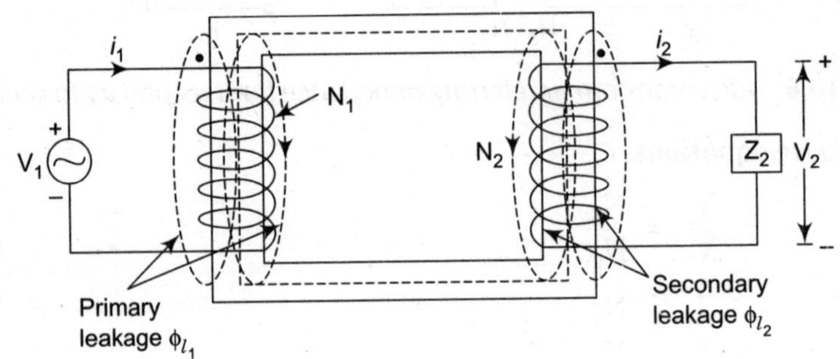

Fig. 10.7 *Real transformer.*

As the leakage flux paths in air for considerable parts of their path lengths, winding m.m.f. and its self-linkage caused by leakage flux are linearly related in each winding therefore contributing constant leakage inductances of both primary and secondary windings. These leakages are distributed throughout the winding quite uniformly.

Both resistances and leakage reactance's of the transformer winding series effect and for low operating frequencies at which the transformer is commonly used (50 Hz or 60 Hz). These can be regarded as lumped parameters. The real transformer is Fig. 10.7 can now be represented as a semi-ideal transformer having lumped resistance R_1 and R_2 and leakage reactance symbolized as X_{l1} and X_{l2} in series with corresponding windings as shown in Fig. 10.8.

The semi-ideal transformer draws magnetizing current to set up the mutual flux ϕ and to provide the power loss in the core; it however has no winding resistances and devoid of any leakage. The induced e.m.f. of the semi-ideal transformer is E_1 and E_2 which differ respectively from primary and secondary terminal voltages V_1 and V_2 by small voltage drop in winding resistance and leakage reactance (R_1, X_{l1} for primary and R_2, X_{l2} for secondary). The ratio of transformation is

$$K = \frac{N_1}{N_2} = \frac{E_1}{E_2} = \frac{V_1}{V_2} = \frac{I_2}{I_1}$$

Because the resistance and leakage reactance of the primary and secondary are so small in a transformer that $E_1 \approx V_1$ and $E_2 \approx V_2$.

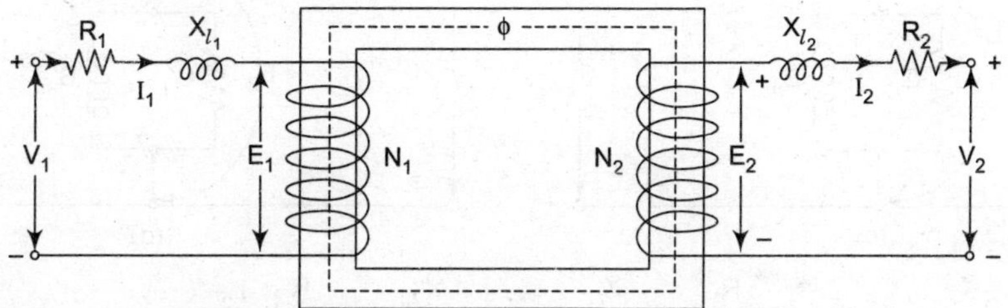

Fig. 10.8 *Circuit model of semi-ideal transformer.*

10.3.1 Equivalent Circuit

The current I_1 flowing in the primary of semi-ideal transformer can be visualized to comprise two components as below:

(*i*) Exciting current I_o whose magnetizing component I_m creates mutual flux ϕ and whose core loss component I_c provide the loss associated with alternation of flux.

(*ii*) A load component I'_2 which counter balances the secondary m.m.f. $I_2 N_2$ so that the mutual flux remain constant independent of load, determined only by E_1.

Thus $$I_1 = I_o + I'_2$$

where $$\frac{I'_2}{I_2} = \frac{N_2}{N_1}$$

The exiting current I_o can be represented by the circuit model of Fig. 10.7 so that semi-ideal transformer of Fig. 10.8 is now reduced to the true ideal transformer. The corresponding circuit (Equivalent circuit) modelling the behaviour of a real transformer is drawn in Fig. 10.9(*a*) where for ease of drawing the core not shown for the ideal transformer.

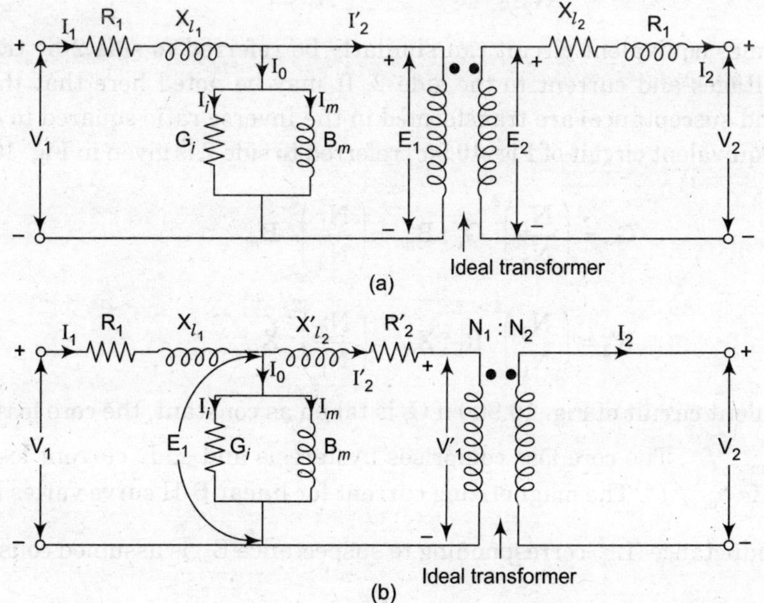

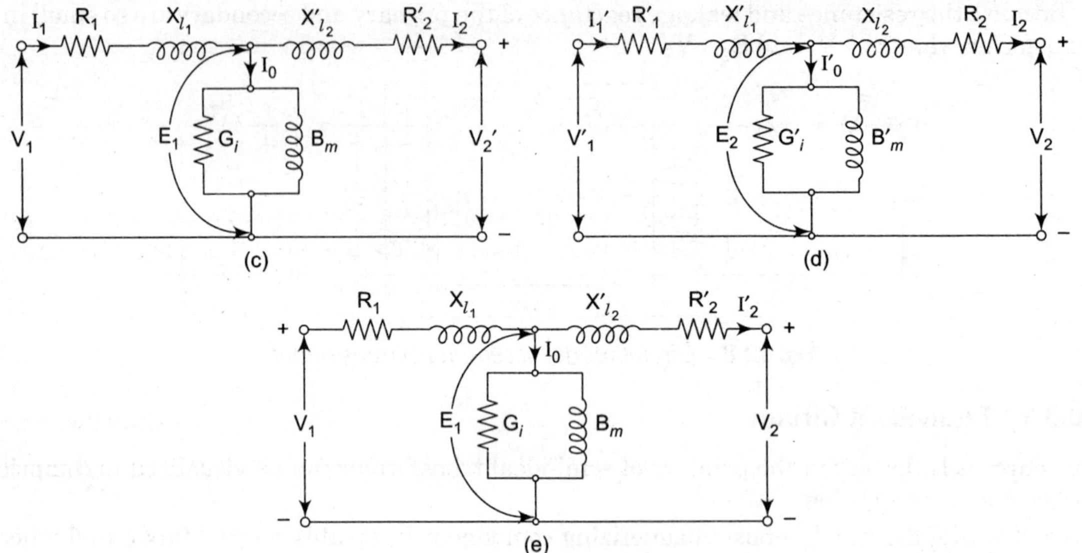

Fig. 10.9 *Development of transformer equivalent circuit.*

The impedance ($R_2 + jX_{l2}$) on the secondary side of ideal transformer can now referred to its primary side resulting in the equivalent circuit of Fig. 10.9(*b*).

Wherein $$X'_{l2} = \left(\frac{N_1}{N_2}\right)^2 X_{l2} \text{ and } R'_2 \left(\frac{N_1}{N_2}\right)^2 R_2$$

The load voltage and current referred to the primary side are

$$V'_2 = \left(\frac{N_1}{N_2}\right) V_2 \text{ and } I'_2 = \left(\frac{N_2}{N_1}\right) I_2$$

The transformer equivalent circuit can similarly be referred to side 2 by transforming all impedances, voltages and current to the side 2. It may be noted here that the admittances (conductance and susceptance) are transformed in the inverse ratio squared in contrast to impedances. The equivalent circuit of Fig. 10.9(*c*) referred to side 2 is given in Fig. 10.9(*d*) where in

$$G'_i = \left(\frac{N_1}{N_2}\right)^2 G_i \; ; B'_m = \left(\frac{N_1}{N_2}\right)^2 B_m$$

$$R'_1 = \left(\frac{N_2}{N_1}\right)^2 R_1 \; ; X'_{l1} = \left(\frac{N_2}{N_1}\right)^2 X_{l1}$$

In the equivalent circuit of Fig. 10.9(*e*) if G_i is taken as constant, the core loss is assumed to vary as $E_1^2 \propto \phi_{max}{}^2 f^2$. The core loss comprises hysteresis and eddy current loss expresses as ($K_h\, \phi_{max\,1.6}\, \phi + Ke\, \phi_{max}{}^2 f^2$. The magnetizing current for linear B-H curve varies proportional to $\phi_{max} \propto \frac{E_1}{f}$. If inductance ($L_m$) corresponding to suspectance B_m is assumed constant,

$$I_m = \frac{E_1}{2\pi f\, L_m}.$$

10.4 APPROXIMATE EQUIVALENT CIRCUIT

In constant freq. (50 Hz) power transformer approximate form of equivalent circuit are commonly used. It was observed that winding resistance and leakage reactance's are very small, $V_1 \approx E_1$ even under conditions of load. Therefore the exciting current drawn by the magnetizing branch (G_i ll B_m) would not be affected significantly by shifting it to the input terminals *i.e.* it is now excited by V_1 instead of E_1 as shown in Fig. 10.9 (*a*). It may also be observed that with this approximation, the current through R_1X_{l1} is now I_2 rather than $\bar{I}_1 = \bar{I}_0 = \bar{I}_2$. Since I_o is very small (less than 5% of full load current), this approximation changes the voltage drop insignificantly. Thus it is basically a good approximation. The winding resistance and reactance being in series can now be combined into equivalent resistance and reactance of the transformer as seen from the approximate side. Remembering that all qualities in equivalent circuit are referred either to the primary or to secondary. The equivalent resistance referred either side can be dropped so the impedance becomes equal to equivalent reactance as shown in Fig. 10.10(*d*).

Here

R_{eq} (equivalent resistance) = $R_1 + R_2$

X_{eq} (equivalent reactance) = $X_{l1} + X_{l2}$

Z_{eq} (equivalent impedance) = $R_{eq} + j\,X_{eq}$

In computing voltages from approximate equivalent circuit the parallel magnetizing branch has no role to play and can therefore be ignored.

The approximate equivalent circuit offers excellent computational ease without any significant loss in the accuracy of results. Further, the equivalent resistance and reactance as used in the approximate equivalent circuit offer an added advantage in that these can be readily measured experimentally.

10.5 PHASOR DIAGRAM

For the approximate equivalent circuit of Fig. 10.10(b).

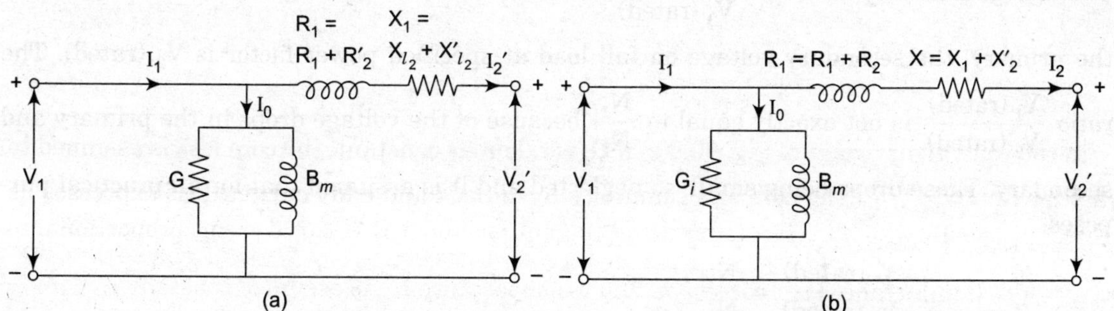

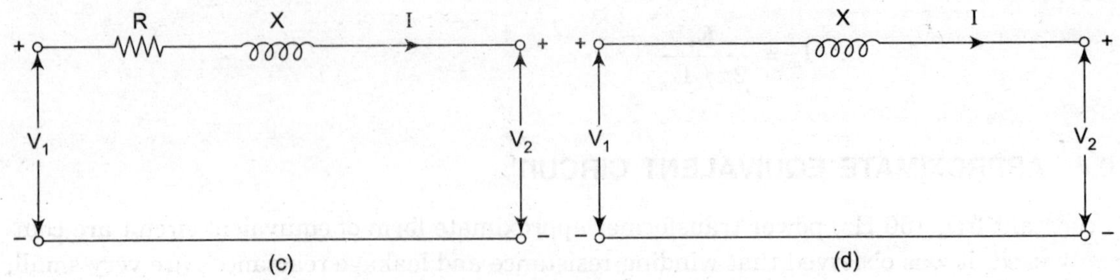

Fig. 10.10 *Approximate equivalent circuit of a transformer.*

$$\bar{V}_2 = \bar{V}_1 - \bar{I}Z$$

$$\bar{V}_1 = V_2 + I(R + jX)$$

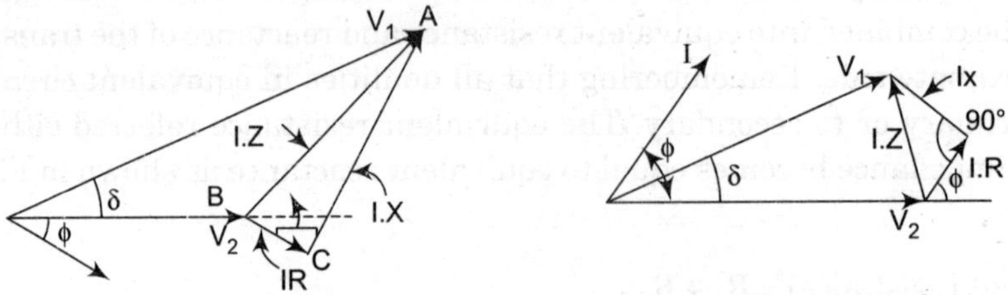

(a) Lagging power factor

(b) Leading power factor

Fig. 10.11 *Phasor diagram of transformer for the approximate equivalent circuit.*

It is observed from these phasor diagrams that for the phase angle indicated

$V_2 < V_1$ for lagging power factor

and $V_2 > V_1$ for leading power factor

In the phasor diagrams the angle δ is such that V_2 leads V_2. This is an indication of fact that power flows from side 1 to side 2 of the transformer this angle is quite small and is related to the value of the equivalent reactance, resistance of the transformer being negligible.

10.6 NAME PLATE RATING

The voltage ratio is specified as $\frac{V_1 \text{ (rated)}}{V_2 \text{ (rated)}}$. It means that when voltage V_1 (rated) is applied to the primary, the secondary voltage on full load at specified power factor is V_2 (rated). The ratio $\frac{V_1 \text{ (rated)}}{V_2 \text{ (rated)}}$ is not exactly equal to $\frac{N_1}{N_2}$ because of the voltage drops in the primary and secondary. These drops being small; so neglected and it is assumed that for all practical purposes.

$$\frac{V_1 \text{ (rated)}}{V_2 \text{ (rated)}} = \frac{N_1}{N_2}$$

The rating of transformer is specified in units of VA or KVA or MVA depending on its size.

$$KVA\,(rated) = V\,(rated) \times I\,(full\,load)/1000$$

where V and I are referred to one particular side.

The transformer name plate also specifies the equivalent impedance, but not actual ohm. It is expressed as the percentage voltage drop as:

$$I\,(full\,load) \times Z \times 100/V\,(rated)$$

Where all quantities must be referred to anyone side.

10.7 TRANSFORMERS LOSSES

The transformer has no moving parts so that its efficiency is much higher than of rotating machines. Following losses takes place in transformer:

1. *Core loss:* It consists of hysteresis and eddy current losses resulting from alternations of magnetic flux in the core. The core loss is constant for a transformer operated at constant voltage and frequency; as for all power transformer frequency is kept constant.
2. *Copper loss:* This loss occurs in windings resistance when the transformers carries its load current, varies as the square of the loading expressed as a ratio of the full load.
3. *Load (stray) loss:* It is largely results from leakage fields inducing eddy-current its tank wall and conductors.
4. *Dielectric-loss:* The loss takes place in insulating material (dielectric material) particularly oil and solid insulation.

Power losses in the transformer is basically due to core loss or iron loss (constant) and copper loss (variable).

The above losses and parameters of equivalent circuit can be easily determined by two simple tests without actually load it.

10.8 TRANSFORMER TESTING

Due to two major difficulties it is not feasible to test the large transformer on direct load are (*a*) the large amount of energy has to be wasted in such a test (*b*) it is very difficult to arrange testing load for large transformer for direct testing. Thus the performance of a transformer should be calculated from its equivalent circuit parameters; which are determined by conducting indirect testing or non-loading test. In these test the power consumption is very less. The two indirect tests are the open circuit test (OC test) and short circuit test (SC test).

1. *Open circuit or no-load test:* In this test one winding is open circuited and a voltage usually rated voltage at rated frequency is applied to the other winding. The voltage, current and power at the terminals of these windings are measured. It is normally convenient to apply the test voltage to the winding; that has a voltage rating equal to that of the available power source (Usually from LV side), while the HV side is kept open circuited.

   The Fig. 10.12 gives the circuit diagram, the wattmeter gives the no load power loss (P_o), ammeter gives no load current or excitation current (I_o) and the voltmeter gives test voltage V. The no load current or excitation current I_o is very small (2-6% of rated current) and R_1 and X_1 are also small, so E_1 can be regarded equal to V_1 by neglecting the series Impedance or Ohmic loss or copper losses as are negligible. Therefore power input on no load equals the core loss.

i.e. $P_o = P_i$ (Iron loss)

$$I_2^2 R_1 = 0$$

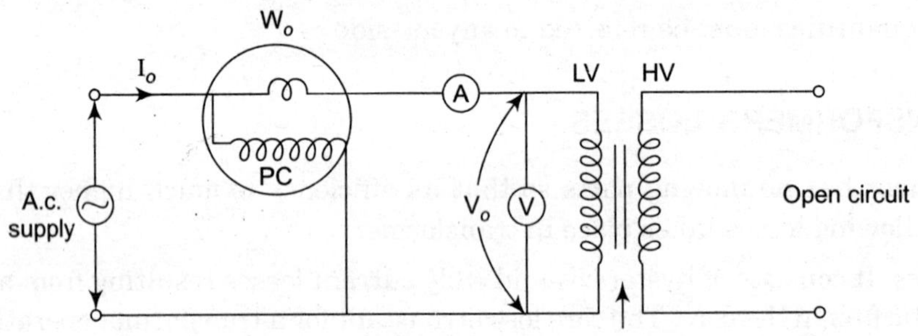

Fig. 10.12 *Circuit diagram for open circuit test.*

So
$$P_i = P_o$$
$$Y_o = G_i - jB_m$$
$$Y_o = \frac{I_o}{V_1}$$
$$V_1^2 G_i = P_o$$
$$G_i = \frac{P_o}{V_1^2}$$

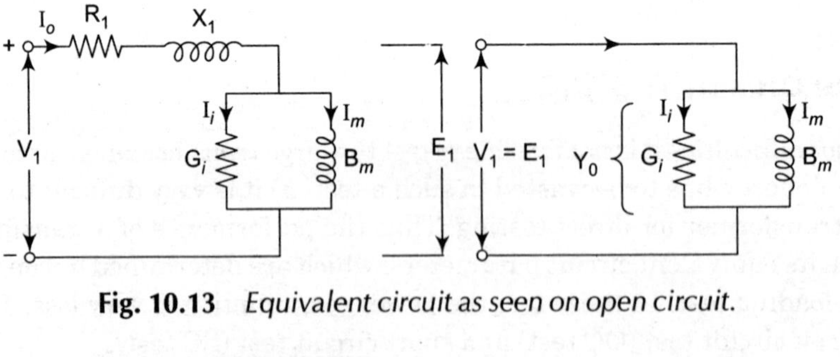

Fig. 10.13 *Equivalent circuit as seen on open circuit.*

So
$$B_m = \sqrt{(Y_o^2 - G_i^2)}$$

where θ_o = No load power factor

$$\theta_o = \cos^{-1}\left(\frac{P_o}{V_o I_o}\right)$$

2. *Short circuit test:* In this test, one windings is short circuited across its terminals. For convenience of supply arrangement voltage and current to be handled the test is usually

conducted from the HV side of the transformer while the LV side is short circuited as shown in Fig. 10.14(*a*). Since the transformer resistances and leakage resistance are very small so the voltage V_{sc} needed to circulate the short circuited current or full load current is very low *i.e.* 5-8% of the rated voltage. Though the choice of the winding to be short-circuited is usually determined by the measuring equipments available for the test.

Therefore, the exciting current under short circuit condition is only about 0.1 to 0.5% of the full load current (I_o at the rated voltage is 2-6% of the full load current) so the shunt branch of the equivalent circuit can altogether be neglected giving the equivalent circuit of Fig. 10.14(*b*).

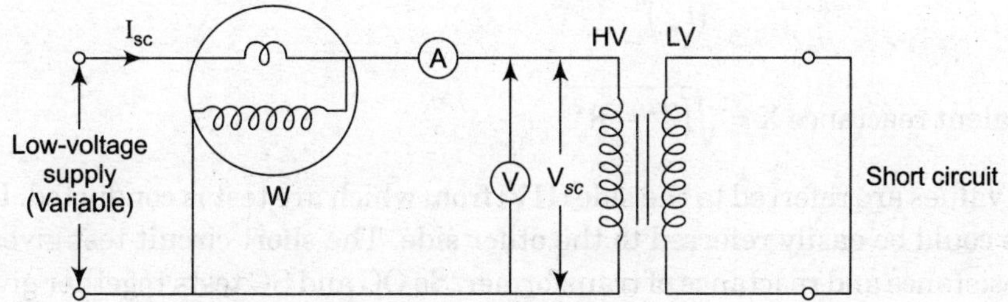

Fig. 10.14 *(a) Circuit diagram for short circuit test on transformer.*

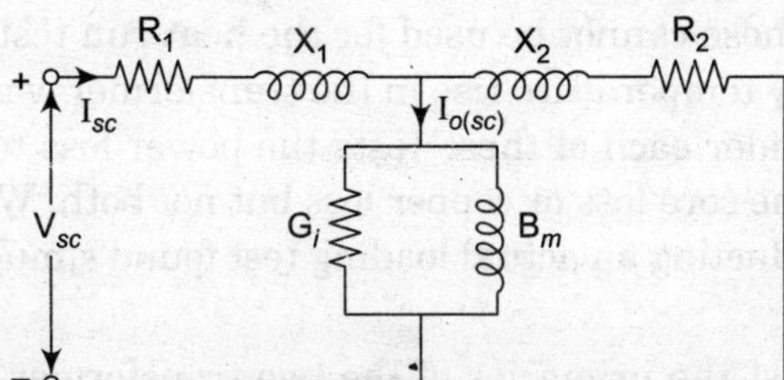

Fig. 10.14 *(b) Equivalent circuit under short circuit condition as seen from the HV side.*

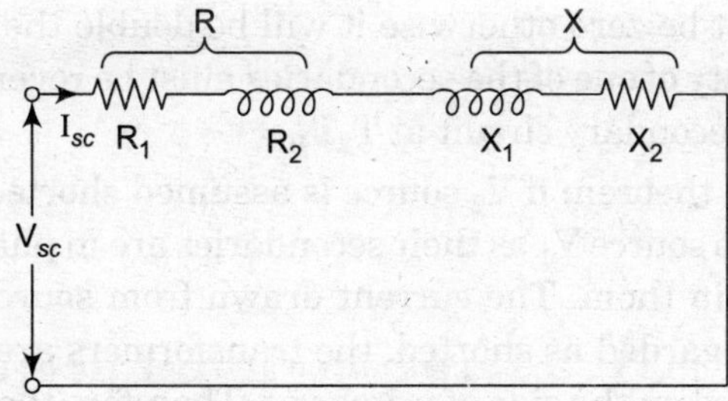

Fig. 10.15 *Equivalent circuit under short circuit condition.*

While conducting the SC test, the supply voltage is gradually raised from zero till the transformer draws full-load current (usually by Auto Transformer). The meter readings under these conditions are:

$$\text{Voltage} = V_{sc}$$

$$\text{Current} = I_{sc}$$

$$\text{Power input} = P_{sc}$$

Since the transformer is excited at very low voltage, the iron loss is neglected, the power input corresponds only to copper loss *i.e.*

$$P_{sc} = P_c = \text{Copper loss}$$

From equivalent circuit for Fig. 10.15, the circuit parameters are computed as below:

$$Z = \frac{V_{sc}}{I_{sc}} = \sqrt{(R^2 + X^2)}$$

$$\text{Equivalent resistance } R = \frac{P_{sc}}{(I_{sc})^2}$$

$$\text{Equivalent reactance } X = \sqrt{(Z^2 - R^2)}$$

These values are referred to the side (HV) from which are test is conducted. If desired, the values could be easily referred to the other side. The short circuit test gives the equivalent resistance and reactance of transformer. So OC and SC tests together give the parameter of the approximate equivalent circuit.

3. *Sumpner's (back to back) test:* The OC and SC tests on a transformer yield its equivalent circuit parameters these cannot be used for the heat run test wherein the purpose is to determine the steady temperature rise in the transformer was fully loaded continuously this is so because under each of these tests the power loss to which the transformer is subjected is either the core loss or copper loss but not both. While by the Sumpner's test method without conducting an actual loading test found simultaneously on two identical transformers.

   In the Sumpner's test the primaries of the two transformers are connected in parallel across the rated voltage supply (V_1), while the two secondaries are connected in phase opposition as shown in Fig. 10.16. For the secondaries to be in phase opposition the voltage across T_2T_4 must be zero otherwise it will be double the rated secondary voltage in which case the polarity of one of the secondaries must be reversed. Current at low voltage (V_2) is injected into secondary circuit at T_2T_4.

   As per superposition theorem if V_2 source is assumed shorted, the two transformers appear in open circuit to source V_1 as their secondaries are in phase opposition and therefore no current can flow in them. The current drawn from source V_1 is thus $2I_o$ and power is $2P_o$. When V_1 is regarded as shorted, the transformers are series connected across V_2 and are short circuited on the side of primaries. Therefore the impedance seen at V_2 is 2Z and when V_2 is adjusted to circulate full load current (I_{fl}). The power fed is 2Pc (twice the full load copper loss of each transformer). Thus in the Sumpner's test; while the transformers are not supplying any load full iron loss occurs in their cores and full. Copper loss occurs in their windings. Net power input to the transformers being $(2P_o + 2P_c)$ .The heat run test could therefore be conducted on the two transformers while only losses are supplied.

   In Fig. 10.16, the auxiliary voltage source is included in the circuit of secondaries, the test could also be conducted by including the auxiliary source in the circuit of primaries.

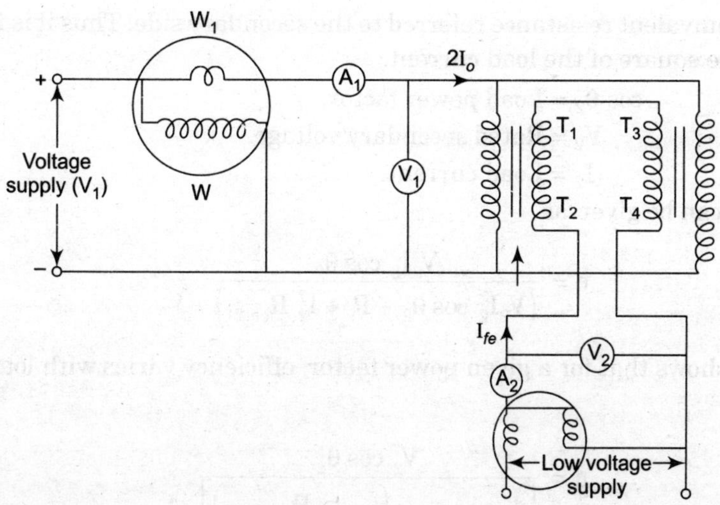

Fig. 10.16 *Sumpner's test on two identical single-phase transformers.*

10.9 TRANSFORMER EFFICIENCY

Power and distribution transformers are designed to operate under condition of constant r.m.s. voltage and frequency. The total losses occurring in transformer under loaded condition can be divided with two groups:

1. Iron losses that occur in the magnetic core. These include hysteresis and eddy current losses.
2. Copper losses occurring in the windings of transformer.

Iron losses are assumed as constant losses because the core flux is practically independent of load. Since the transformer is a static device, there are no rotational losses such as windage and frictional losses in a rotating machine.

The efficiency of a transformer is in the range of 96–99%. The efficiency of a transformer is defined as the ratio of the useful power output to the input power thus

$$\eta = \text{Output/input}$$
$$= 1 - \text{losses/(output + losses)}$$
$$= \text{Output/(output + losses)}$$

The accurate method of determining efficiency would be to find the losses from open circuit test and short circuit test method.

$$\eta = \frac{V_2 I_2 \cos\theta_2}{(V_2 I_2 \cos\theta_2 + P_i + P_c)}$$

where P_i = Iron loss

$$P_c = \text{Copper loss} = I_1^2 R_1 + I_2^2 R_2$$
$$= I_2^2 R_{eq(2)}$$

where $R_{eq(2)}$ = Equivalent resistance referred to the secondary side. Thus it is found that copper losses vary as the square of the load current.

$\cos\theta_2$ = Load power factor.

V_2 = Rated secondary voltage.

I_2 = Load current.

So, efficiency can be given as

$$\eta = \frac{V_2 I_2 \cos\theta_2}{\left(V_2 I_2 \cos\theta_2 + P_i + I_2^2 R_{eq(2)}\right)}$$

above equation shows that for a given power factor, efficiency varies with load current. Therefore

$$\eta = \frac{V_2 \cos\theta_2}{\left[V_2 \cos\theta_2 + \left(\frac{P_i}{I_2 + I_2 R_{eq(2)}}\right)\right]}$$

For maximum value of efficiency 'η' for given $\cos\theta_2$, the denominator of above equation must have the least value. The condition for maximum η obtained by differentiating the denominator and equating it to zero is

$$\frac{d}{dI_2}\left\{\frac{V_2 \cos\theta_2 + P_i}{I_2 + I_2 R_{eq(2)}}\right\} = 0$$

$$-\frac{P_i}{I_2^2} + R_{eq(2)} = 0$$

$$I_2^2 R_{eq(2)} = Pi$$

Hence, total copper losses (variable) = Core loss (constant).

Hence the efficiency of the transformer is maximum at a load that makes the total copper losses equal to the core loss (constant losses).

'All-day' efficiency: The all day efficiency of a transformer is the ratio of the total energy output (kWh) in a 24-h a day to the total energy input in the same time. The distribution are normally operating on a varying load cycle, while the core losses are constant independent of the load.

Hence the efficiency of such transformers should be measured on the energy basis.

$$\text{All day efficiency} = \frac{\text{Output in kilowatt-hours}}{\text{Input in kilowatt-hours}}$$

It is an important figure of merit for distribution transformers, which feed daily load cycle varying over a wide load range. Higher energy efficiencies are achieved by designing distribution transformers to yield maximum efficiency at less than full load. This is achieved by restricting, the core flux to lower values by using a relatively larger core cross-section.

10.10 VOLTAGE REGULATION

Generally all the loads are designed to operate practically at constant voltage. It is, therefore necessary that the output voltage of a transformer must stay within a normal limits as the load and its power factor vary. The load terminal voltage changes because of the voltage drop in the leakage reactance of the transformer. Therefore, the voltage regulation is used to determine the voltage drop/change characteristic, of the transformer.

Voltage regulation is defined as the change in magnitude of the secondary voltage, when full load of specified power factor supplied at rated voltage in thrown off *i.e.* reduced to no load with primary voltage and frequency held constant, as percentage of the rated load terminal voltage.

In term of symbols

$$\% \text{ Voltage regulation} = \left(\frac{V_{20} - V_{2.fl}}{V_{2.fl}}\right) \times 100$$

where V_{2fl} = Rated secondary voltage while supply full load at specified power factor.

V_{20} = Secondary voltage when loads is thrown off.

Figure 10.17 shows the transformer equivalent circuit referred to the secondary side and Fig. 10.18 gives its phasor diagram. The voltage drops IR and IX are very small in a well designed transformer. As a result the angle δ between V_1 and V_2 is of negligible order, so that

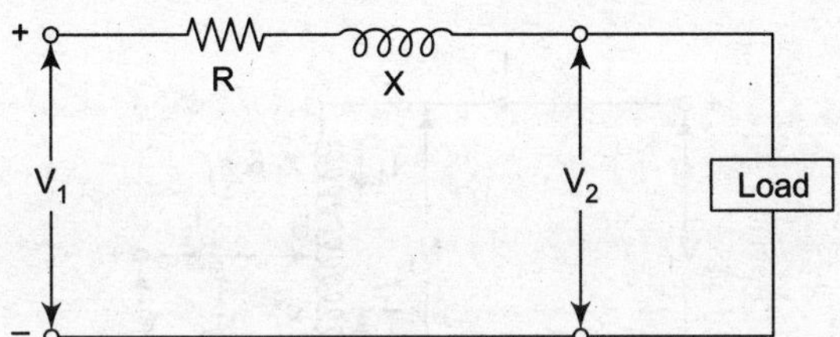

Fig. 10.17 *Approximate equivalent circuit referred to secondary.*

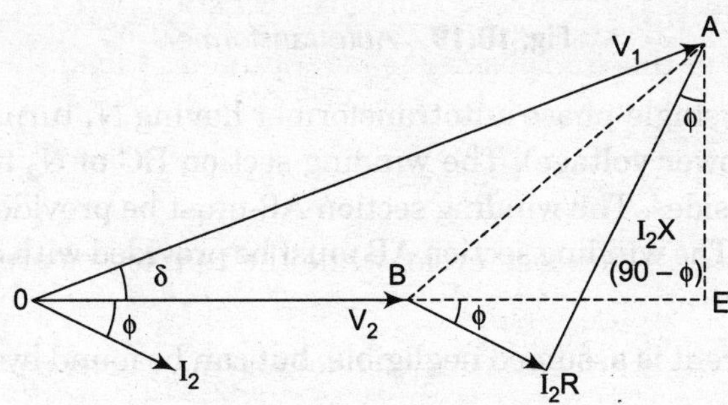

Fig 10.18 *Phasor diagram.*

$$V_1 = V_2 + I\,\angle\theta\,(R + jX)$$

$$V_1 = OE$$

$$V_1 - V_2 = I\,(R \cos\theta + X \sin\theta);\ \theta \text{ lagging}$$

$$= I\,(R \cos\theta - X \sin\theta\, R);\ \theta \text{ leading}$$

When the load is thrown off

$$V_{20} = V_1$$

Therefore $\quad V_{20} - V_2 = I\,(R \cos\theta \pm X \sin\theta)$

where $I = I_2$ is full load secondary current and V_2, the full load secondary voltage [Equal to the value of V_2 (rated)].

Thus,

$$\%\ \text{Voltage regulation} = \left(\frac{(V_{20} - V_2)}{V_2}\right) \times 100 = I\left[\frac{(R\cos\theta \pm X\sin\theta)}{V_2}\right] \times 100.$$

10.11 AUTOTRANSFORMERS

In the two-winding transformers the windings are electrically isolated when the primary and secondary windings are electrically connected so that a part of primary and secondary windings is common, as shown in Fig. 10.19, the transformer is known as an autotransformer. The autotransformer is specially economical where the voltage ratio is less than 2, because of electrical isolation is note necessary. The autotransformer has lower reactance, smaller exciting current, lower losses and better voltage regulation compared to two-winding transformer. The major applications are variable voltage power supplies, starting of induction motor, inter connection of HV systems at voltage levels with transformation ratio less than 2.

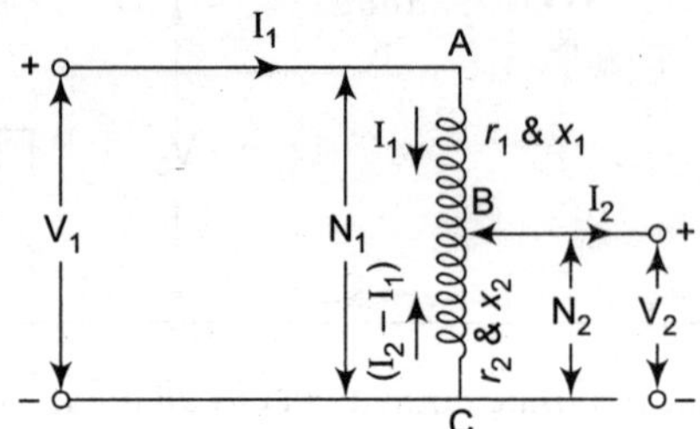

Fig. 10.19 *Autotransformer.*

Figure 10.19 shows a single-phase autotransformer having N_1 turns primary with N_2 turns tapped for secondary (lower voltage). The winding section BC of N_2 turns is common to both primary and secondary sides. The winding section AB must be provided with extra insulation, being at higher voltage. The winding section AB must be provided with extra insulation being at higher voltage.

The magnetizing current is assumed negligible, but can be found by no-load test and may be used.

From Fig. 10.19, the two-winding voltage and turn ratio is

$$K = \frac{N_1 - N_2}{N_2} = \frac{V_{AB}}{V_{BC}} = \frac{N_1}{N_2} - 1 \qquad ...(1)$$

Two autotransformer voltage and turn ratio is

$$K' = \frac{V_{AB} + V_{BC}}{V_{BC}} = \frac{V_{AB}}{V_{BC}} = \frac{N_1}{N_2} \quad ...(2)$$

from above two equation, it is established that

$$K' = 1 + K \quad ...(3)$$

Equation (3) clearly shows that the transformation ratio of two windings is more when connected as on autotransformer than when used a two-winding transformer further when K is small or K′ is close to unity.

The VA rating when used as a two-winding transformer

$$(VA)_{TW} = (V_1 - V_2)\, I_1 = V_2\,(I_2 - I_1) \quad ...(4)$$

While, when used as an autotransformer its rating is

$$(VA)_{auto} = V_1 I_1 = V_2 I_2$$

From equation (4)

$$\frac{(V_1 - V_2)}{V_2} = \frac{(I_2 - I_1)}{I_1} = K$$

Which gives

$$V_1 - V_2 = KV_2$$

or

$$(I_2 - I_1) = KI_1$$

$$I_2 = (K + 1)\, I_1$$

or

$$I_1 = \frac{I_2}{(K + 1)}$$

Putting value of $(V_1 - V_2)$ & I_1 in equation (4)

$$(VA)_{TW} = KV_2 \cdot \frac{I_2}{(K+1)} = \frac{K}{(K+1)} \cdot V_2 I_2$$

or

$$(VA)_{TW} = \frac{K}{K+1} \cdot (VA)_{auto}$$

or

$$(VA)_{auto} = \left(1 + \frac{1}{K}\right)(VA)_{TW}$$

$$= \frac{N_1}{(N_1 - N_2)}(VA)_{TW} \quad ...(5)$$

Therefore

$$(VA)_{auto} > (VA)_{TW} \quad ...(6)$$

From equation (5) it is clear that two-windings connected as autotransformer will have a greater VA rating than when connected as an two-winding transformer, because of VA transfers through not only induction but also through conduction.

The material used in autotransformer and two-winding transformer of same volt-ampere rating is compared as:

$$W_{auto} = \frac{I_1(N_1 - N_2) + (I_2 - I_1)N_2}{I_1 N_1 + I_2 N_2}$$

$$= 1 - \frac{2I_1N_2}{2I_1N_1} \qquad (\because \; I_1N_1 = N_2I_2)$$

$$= 1 - \frac{N_2}{N_1} = 1 - \frac{V_2}{V_1} \qquad ...(7)$$

where, G stands for weight of winding material.

While constant current density in winding conductor is assumed.

From equation (7) saving in conductor material can be found

$$G_{TW} - G_{auto} = \frac{1}{K'} G_{TW}$$

= Saving of conductor material is using autotransformer

i.e., $\frac{1}{K'}$ = 0.1, saving is 10%, while $\frac{1}{K'}$ = 0.9 saving is 90%. Therefore use of auto is close to unity transformer is more economical when truncation.

SOLVED NUMERICAL PROBLEMS

Example 1. *A 3000/230 V, 50 Hz, single phase transformer is built in a core having an effective cross-sectional area of 120 cm² and 60 turns in the low-voltage winding. Calculate :*

(a) the value of the maximum flux density.

(b) the number of turns on the high voltage winding.

Solution: Given

$$E_1 = 3000 \text{ V}, E_2 = 230 \text{ V and } f = 50 \text{ Hz}$$
$$A_i = 120 \text{ cm}^2 = 120 \times 10^{-4} \text{ m}^2$$
$$E_2 = \sqrt{2\pi}\, \pi\, \phi_m f N_2 = 4.44\, B_m\, A_i f N_2$$
$$B_m = \frac{E_2}{4.44\, A_i\, f\, N_2} = \frac{230}{4.44 \times 120 \times 10^{-4} \times 50 \times 60}$$
$$= \frac{230 \times 10000}{1598400} = 1.438 \text{ teslans (T)}$$

We have relation,

$$\frac{E_1}{E_2} = \frac{N_1}{N_2}$$

or

$$N_1 = \frac{E_1 \times N_2}{E_2} = \frac{3000 \times 60}{230} = 782.6$$

The number of turns should be even *i.e.*, N_1 = 782 or 784.

Example 2. *The no-load voltage ratio in a single phase, 50 Hz, core type transformer is 3300/250. Calculate the number of turns in each winding, if the flux is to be 0.06 Wb.*

Solution: Given,

$$\text{No-load voltage ratio} = \frac{300}{250}$$

No-load voltage of low voltage winding = 250 V

$$\text{flux } \phi = 0.06 \text{ Wb}$$

$$\text{frequency } f = 50 \text{ Hz}$$

Induced e.m.f. in low voltage winding (secondary) of the transformer is given by

$$E_2 = 4.44 f\phi N_2$$

or

$$250 = 4.44 \times 50 \times 0.06 \times N_2$$

$$N_2 = \frac{250}{4.44 \times 50 \times 0.06} = \frac{250}{13.32} = 18.76$$

Number of poles should be even *i.e.*, 18 or 20

$$N_2 = 18 \text{ or } 20$$

Number of turns in high voltage winding

$$N_1 = N_2 \times \frac{V_1}{V_2} = 18 \times \frac{3300}{250} = 237.6$$

$$N_1 = 236 \text{ or } 238.$$

Example 3. *A 240/110 V single phase transformer takes an input of 280 VA at no-load and at rated voltage. The core loss is 130 W.*

Calculate:

(*i*) *The iron-loss component of no-load current;*

(*ii*) *The magnetizing component of no-load current; and*

(*iii*) *No-load power factor.*

Solution: We know that

$$V_1 . I_o = \text{Input voltage}$$

$$V_1 . I_o = 280$$

$$I_o = \frac{280}{110} = 2.545 \text{ A.}$$

(*i*) No-load p.f. $\cos \phi_0$

$$\text{Core loss} = V_1 I_o . \cos \phi_0$$

$$110 = 280 \cos \phi_0$$

$$\cos \phi_0 = \frac{110}{280} = 0.392$$

(*ii*) Core loss component of no-load current

$$I_W = I_o \cos \phi_0 = \frac{\text{Core loss}}{V_1} = \frac{130}{240} = 0.541 \text{ A}$$

(*iii*) Magnetizing component of no-load current

$$I_M = \sqrt{I_0^2 - I_W^2} = \sqrt{(2.54)^2 - (0.541)^2}$$

$$= \sqrt{6.451 - 0.292} = \sqrt{6.159} = 2.481 \text{ A.}$$

Example 4. *20 kVA, 3300/210 V, 50 Hz single phase transformer has the following resistance and leakage reactances:*

$$r_1 = 0.6\ \Omega \qquad x_1 = 2.5\ \Omega$$
$$r_2 = 0.002\ \Omega \qquad x_2 = 0.03\ \Omega$$

Calculate the equivalent resistance and reactance referred to secondary side.

Solution: Resistance of primary

$$r_1 = 0.6\ \Omega$$

Resistance of primary referred to secondary,

$$r_1' = 0.6\left(\frac{210}{3300}\right) = 0.038\ \Omega$$

Resistance of secondary

$$r_2 = 0.002\ \Omega$$

Equivalent resistance of the transformer referred to secondary side,

$$\bar{R}_2 = r_2 + r_1' = 0.002 + 0.038$$
$$= 0.040\ \Omega$$

Leakage reactance of primary

$$x_1 = 2.5\ \Omega.$$

Leakage reactance of primary referred to secondary

$$x_1' = 2.5\left(\frac{210}{3300}\right)^2$$
$$= 2.5 \times (0.063)^2 = 0.010\ \Omega$$

Leakage reactance of secondary

$$x_2 = 0.03\ \Omega.$$

Equivalent leakage reactance of transformer referred to secondary size,

$$\bar{X}_2 = x_2 + x_1'$$
$$= 0.03 + 0.010 = 0.040\ \Omega.$$

Example 5. *Calculate the regulation of a transformer in which ohmic loss is 2% of the output and reactance drop is 4% of the voltage when the power factor is (a) 0.8 lagging; (b) 0.8 leading; and (c) unity.*

Solution:

(*a*) Regulation at 0.8 p.f. lagging

$$= R_{e_2} \cos \phi_2 + X_{e_2} \sin \phi_2$$
$$= 2 \times 0.8 + 4 \times 0.6$$
$$= 1.6 + 2.4 = 4\%$$

(*b*) Regulation at 0.8 P.f. leading

$$= R_{e_2} \cos \phi - X_{e_2} \sin \phi = 1.6 - 2.4 = -0.8\%.$$

(*c*) Regulation at unity power factor

$$= R_{e_2} = 1\%.$$

Example 6. *A single-phase, 90 kVA, $\frac{3000}{300}$ V, 50 Hz transformer has an impedance drop of 10% and resistance drop of 8%. Calculate the following:*

(i) Regulation at full load 0.9 power factor lagging;

(ii) The value of the power factor at which regulation is zero.

Solution:
$$\frac{I_2 Z_{e_2}}{V_2} \times 100 = 10$$

$$I_2 Z_{e_2} = \frac{10\,V}{100} = \frac{10 \times 300}{100} = 30 \text{ V}$$

$$\frac{I_2 R_{e_2}}{V_2} \times 100 = 8$$

$$I_2 R_{e_2} = \frac{8\,V_2}{100} = \frac{8 \times 300}{100} = 24 \text{ V}$$

$$I_2 X_{e_2} = \sqrt{(I_2 Z_{e_2})^2 - (I_2 R_{e_2})^2} = \sqrt{(30)^2 - (24)^2}$$

$$= \sqrt{900 - 576} = \sqrt{324} = 18 \text{ V}$$

(*a*) Approximate voltage regulation at lagging power factor

$$= \frac{I_2 R_{e_2} \cos \phi_2 + I_2 X_{e_2} \sin \phi_2}{V_2}$$

$$= \frac{10 \times 0.9 + 18 \times 0.6}{300}$$

$$= 9 + 10.8 = 19.8 \text{ PU}$$

(*b*) For zero regulation, the power factor must be leading

$$\therefore \quad \frac{I_2 R_{e_2} \cos \phi_2 - I_2 X_{e_2} \sin \phi_2}{V_2} = 0$$

$$\tan \phi_2 = \frac{I_2 R_{e_2}}{I_2 X_{e_2}} = \frac{10}{18} = 0.56$$

$\therefore$ Power factor $\cos \phi_2 = \cos (\tan^{-1} 0.56)$

$$= \cos 29.05$$

$$\cos \phi_2 = 0.87 \text{ (leading).}$$

Example 7. *A 3400 V, single-phase, 50 Hz transformer has a no-load current of 3 A at 25 V and no-load, losses are 700 W. The resistance of primary winding is 0.7 ohm. Two-third of the core losses are hysteresis losses. If the primary voltage and frequency are 430 V and 80 Hz, calculate the new value of core losses.*

Solution: Copper loss in primary winding at no load

$$= (3)^2 (0.7)$$

$$= 9 \times 0.7 = 6.3 \text{ W}$$

Core loss $= 700 \times 6.3 = 693.7$ W

Hysteresis loss $= 693.7 \times \frac{2}{3} = 462.4 \text{ W}$

Eddy current loss $= 693.7 \times \frac{1}{3} = 231.2 \text{ W}$

As both primary voltage and frequency are increased in same ratio, flux density will be constant.

New value of hysteresis loss $= (462.4)\left(\frac{80}{50}\right)$

$= 739.84 \text{ W}$

New value of eddy current loss

$$= (231.2) \times \left(\frac{80}{50}\right)^2$$

$$= 231.2 \times 2.56 = 591.872 \text{ W}$$

Total core (Iron loss) at 430 V and 80 Hz

$$= 739.84 + 591.87$$

$$= 1331.71 \text{ W.}$$

Example 8. *At 340 V and 50 Hz, the total core loss of a transformer was found to be 1800 W. When the transformer is supplied at 180 V and 26 Hz, the core loss is 600 W. Calculate the hysteresis and eddy current loss at 340 V and 50 Hz.*

Solution: $\frac{V_1}{f_1} = \frac{340}{50} = 6.8$

and $\frac{V_2}{f_2} = \frac{180}{26} = 6.8$

Since $\frac{V_1}{f_1} = \frac{V_2}{f_2} = 6.8$

So, flux density remain constant. Hence,

$$\frac{P_i}{f} = x + yf$$

$\therefore$ $\frac{1800}{50} = x + 50y$ ...(*i*)

and $\frac{600}{26} = x + 25y$ ...(*ii*)

On solving equation (*i*) and equation (*ii*)

$$x + 50y = 36.00$$

$$\underline{x} + \underline{26}y = \underline{30.76}$$

$$50y - 26y = 36.00 - 30.76$$

$$24y = 5.24$$

$$y = \frac{5.24}{24}$$

$$y = 0.218$$

and putting in (1) for x,

$$x + 50 \times 0.218 = 36$$

$$x = 36 - 10.916 = 25.08.$$

Therefore, at 50 Hz

$\therefore$ P_h (hysteresis loss) $= xf = 25.08 \times 50 = 1254.2$ W and P_e (eddy current loss) $= yf^2$
$= 0.218 \times (50)^2 = 545$ W.

Example 9. *A single-phase, 330/540 V transformer give the following test results:*

Open circuit test: 330 V, 1 A, 60 watts on I.v. side

Short circuit test : 30 V, 8 A, 90 watts on h.v. side

Calculate the circuit constants and draw the equivalent circuit with the help of obtained parameters:

Solution: Open circuit test: Given,

$$V_1 = 330 \text{ V}$$

$$I_o = 1 \text{ A}$$

$$P_i = 60 \text{ watts}$$

$$P_i = V_1 I_o \cos \phi_0$$

$$\cos \phi_0 = \frac{P_i}{V_i \cdot I_0} = \frac{60}{330 \times 1} = 0.182$$

$$I_W = I_o \cos \phi_0 = 1 \times 0.182 = 0.182 \text{ A}$$

$$I_m = I_o \sin \phi_0 = \sqrt{I_0^2 - I_W^2}$$

$$= \sqrt{(1)^2 - (0.182)^2} = \sqrt{1 - 0.033}$$

$$I_m = \sqrt{0.967} = 0.983 \text{ A}$$

Now
$$R_0 = \frac{V_1}{i_W} = \frac{330}{0.182} = 1813.18 \ \Omega$$

and
$$X_0 = \frac{V_1}{I_\mu} = \frac{330}{0.983} = 335.70 \ \Omega$$

Short Circuit test:

Given,

$$V_{2\,(sc)} = 30 \text{ V}$$

$$P_{i\,(sc)} = 90 \text{ watts}$$

$$I_{2\,(sc)} = 8 \text{ A}$$

Also,
$$\frac{N_1}{N_2} = \frac{V_1}{V_2} = \frac{330}{540} = 0.62$$

Voltage in low voltage side

$$V_{1\,(sc)} = V_{2\,(sc)} \times \frac{N_1}{N_2} = 30 \times \frac{330}{540} = 18.4 \text{ V}$$

Primary full load current (*l.v.* side)

$$\Rightarrow \qquad I_{1\,sc} = I_{2\,(ac)} \times \frac{N_2}{N_1} = 12 \times \frac{540}{330} = 19.64 \text{ A}$$

Now, power during short-circuited condition

$$P_{i\,(sc)} = I^2_{1(sc)}\, R_{e_1}$$

$$R_{e_1} = \frac{P_{i\,(sc)}}{I^2_{1\,(sc)}} = \frac{90}{(19.64)^2} = \frac{90}{385.540} = 0.234\ \Omega$$

$$Z_{e_1} = \frac{V_{1\,(sc)}}{I_{1\,(sc)}} = \frac{18.40}{19.64} = 0.9368\ \Omega.$$

Therefore,

$$X_{e_1} = \sqrt{Z^2_{e_1} - R^2_{e_1}} = \sqrt{(0.936)^2 - (0.234)^2}$$

$$= \sqrt{0.8776 - 0.0547}$$

$$= \sqrt{0.8229} = 0.907\ \Omega.$$

From above obtained parameters, the equivalent circuit can be drawn as

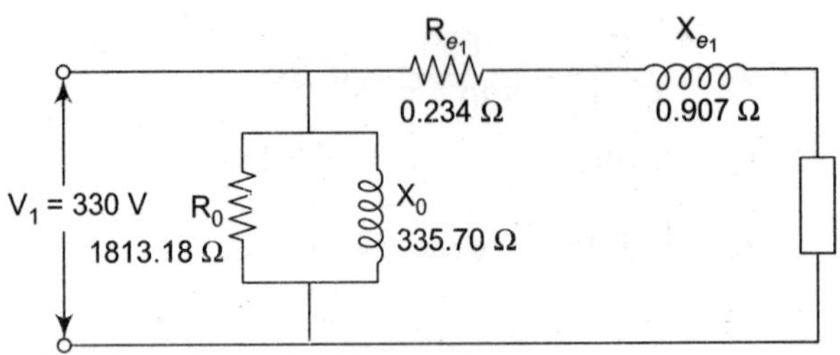

Fig. 10.20

Example 10. *A transformer is rated at 120 kVA. At full load operated condition its copper loss is 1000 watts and its core loss is 940 watts. Determine the following :*

(*i*) *The efficiency at full load and unit power factor*

(*ii*) *The efficiency at half load, 0.8 power factor*

(*iii*) *The efficiency at 80% full load and 0.7 power factor*

(*iv*) *The load kVA at which maximum efficiency will occur*

(*v*) *The maximum efficiency at 0.85 power factor.*

Solution:

$$P = 120 \text{ kVA} = 120 \times 10^3 \text{ VA}$$

$$P_{ifl} = 1000 \text{ W}, P_i = 940 \text{ W}.$$

So, efficiency is given by

$$\eta = \frac{mP\cos\phi_2}{mP\cos\phi_2 + P_i + m^2 P_{ifl}}$$

where,

$$m = \frac{\text{Load (given)}}{\text{Full load}}$$

(*i*) At full load condition $m = 1$ and $\cos \phi_2 = 1'$

$$\eta = \frac{1 \times 120 \times 10^3 \times 1}{1 \times 120 \times 10^3 \times 1 + 940 + (1)^2 \times 1000}$$

$$= \frac{120 \times 10^3}{120 \times 10^3 + 940 + 1000}$$

$$= \frac{120 \times 10^3}{120 \times 10^3 + 1940} = \frac{120 \times 1000}{121940}$$

$$= 0.984 = 98.4\%$$

(*ii*) At half loaded condition,

$$m = \frac{1}{2} \cdot \cos \phi_2 = 0.8$$

Efficiency at half load

$$\eta = \frac{\frac{1}{2} \times 120 \times 10^3 \times 0.8}{\frac{1}{2} \times 120 \times 10^3 \times 0.8 + 940 + \left(\frac{1}{2}\right)^2 \times 1000}$$

$$= \frac{60 \times 1000 \times 0.8}{60 \times 10^3 \times 0.8 + 940 + 250} = \frac{60 \times 1000 \times 0.8}{60 \times 1000 \times 0.8 + 1190}$$

$$= \frac{48 \times 1000}{48 \times 1000 + 1190} = \frac{48000}{49190} = 0.975$$

$$= 97.5\%.$$

(*iii*) Given, $\cos \phi_2 = 0.7$

Now, at 80% of full load

$$m = \frac{80}{100} = 0.8$$

Efficiency at that time

$$\eta = \frac{0.8 \times 120 \times 10^3 \times 0.7}{0.8 \times 120 \times 10^3 \times 0.7 + 940 + (0.8)^2 \times 1000}$$

$$= \frac{0.56 \times 120 \times 10^3}{0.56 \times 120 \times 10^3 + 940 + 640}$$

$$= \frac{0.56 \times 120 \times 10^3}{0.56 \times 120 \times 10^3 + 1580}$$

$$= \frac{67200}{67200 \times 1580} = \frac{67200}{68780} = 0.978 = 97.8\%$$

(*iv*)

$$P_{max} = P_{fl} \sqrt{\frac{P_i}{P_{ifl}}} = 120 \sqrt{\frac{940}{1000}}$$

$$= 120 \times 0.969 = 116.34 \text{ kVA}$$

(*v*) Maximum efficiency

$$\eta_{max} = \frac{P_{max} \cos \phi_2}{P_{max} \cos \phi_2 + 2P_i}$$

$$= \frac{116.34 \times 10^3 \times 0.85}{116.34 \times 10^3 \times 0.85 + 2 \times 940}$$

$$= \frac{98889}{98889 + 1880} = \frac{98889}{100769}$$

$$= 0.981 = 98.1\%.$$

Example 11. *Open-circvit and short circuit test m a 10 kVA, $\frac{240}{330}$, 50 Hz, 1–ϕ transformer gave the following test results:*

o.c. test 240 V, 2 A, 90 watts (l.v. side)

s.c. test 50 V, 10.5 A, 160 watts (h.v. side)

Determine the efficiency and regulation of the transformer at full load 0.8 power factor lagging.

Solution: From o.c. test, core loss P_i = 90 W.

From s.c. test in h.v. side, I_{2sc} = 10.5 A

Copper loss at short-circuit current

$$R_{2\,sc} = I_{2sc}^2 R_{e_2} = 160 \text{ W}$$

$$R_{e_2} = \frac{P_{2sc}}{I_{2sc}^2} = \frac{160}{(10.5)^2} = \frac{160}{110.25} = 1.451\ \Omega.$$

Now,

$$\text{kVA} = \frac{V_2 \cdot I_{2fl}}{1000}$$

$$10 = \frac{330 \times I_{2fl}}{1000}$$

$$I_{2fl} = \frac{10 \times 1000}{330} = 30.30 \text{ A}$$

Copper loss at full load,

$$P_{cfl} = I_{2fl}^2 R_{e_2} = (30.30)^2 \times 1.451$$

$$= 1332.281 \text{ W}$$

Efficiency at full load

$$= \frac{\text{Output (full load)}}{\text{Output (full load)} \times \text{Copper loss (full load)} + \text{Core loss}}$$

$$= \frac{V_2 \cdot I_2 \cos \phi}{V_2 \cdot I_2 \cos \phi + P_{cfl} + P_i} = \frac{10 \times 1000 \times 0.8}{10 \times 1000 \times 0.8 + 1332.2 + 90}$$

Efficiency

$$= \frac{8000}{8000 + 1332.2 + 90} = \frac{8000}{9422.2}$$

$$= 0.849 = 84.9\%.$$

From short circuit list, $V_{2sc} = 50$ V

$$V_{2sc} = I_{2sc} \times Z_{e_2}$$

$$Z_{e_2} = \frac{V_{2sc}}{I_{2sc}} = \frac{50}{10.5} = 4.7619\ \Omega.$$

$$R_{e_2}^2 + X_{e_2}^2 = Z_{e_2}^2$$

$$X_{e_2} = \sqrt{Z_{e_2}^2 - R_{e_2}^2} = \sqrt{(4.7619)^2 - (1.451)^2}$$

$$= \sqrt{22.675 - 2.1054} = 4.535\ \Omega.$$

Hence, $\cos\phi_2 = 0.8,$

$$\phi_2 = \cos^{-1} 0.8$$

$$\phi_2 = 36.86$$

$$\sin\phi_2 = \sin(36.86) = 0.6$$

Voltage regulation $= \frac{I_2}{V_2}\left[R_{e_2}\cos\phi + X_{e_2}\sin\phi\right]$

$$= \frac{10.5}{330}[1.451 \times 0.8 + 4.535 \times 0.6]$$

or $= 0.03\ [1.1608 + 2.721] = 0.1164$ Pu.

$= 1.164\%.$

EXERCISES

1. Write the working principle of a single phase transformer.
2. Draw and explain the phasor diagram of transformer on load.
3. Write the properties of ideal transformer.
4. Explain why the tr·nsformer cannot work in DC system.
5. Draw and explain the equivalent circuit diagram of transformer (i) referred to primary (ii) referred to secondary.
6. Discuss the various types of losses occurring in single phase transformer.
7. Explain open circuit test and short circuit test of single phase transform. Also explain why on open circuit test hv winding is kept open while on short circuit test LV kept shorted?
8. Explain, why transformers are rated in VA or KVA or MVA?
9. Prove that is single phase transformer maximum efficiency is obtained when copper losses are equal to core losses.
10. What is voltage regulation in the transformer and how it affects its efficiency.
11. Explain in the single phase transformer frequency does not changes while transforming the voltage level.
12. Prove that on transforming voltage is transformer the power remains constant.
13. Explain the working principle of auto transformer and its application.
14. Prove that as compared to single phase two winding transformer there is copper saving.
15. Prove that transformation ratio in auto transformer is greater that two winding transformer.

11

Three-Phase Induction Machine

11.1 INTRODUCTION

The most commonly and widely used motor in any industry is induction motor. The induction motor is a.c. operated machine and very much similar to transformer in its operation and construction. Induction motors are morl rugged, require less maintenance and are less expensive than d.c. machine of equal kilowatt and speed ratings. It is widely used for industrial drive due to low cost, efficient and reliable. It has good speed regulation and high starting torque. It has a reasonable overload capacity. Like other electrical rotating machine, an induction machine can also be used as an induction generator, if the rotor is rotated by a primemover. However, induction generators have restricted applications as a source of power supply. Induction motors are called generalised transformer because both depends upon Faraday's law of electromagnetic induction for its operation and both have same equivalent circuit and phaser diagram. The only difference is that induction motor has rotating part but transformer is static device.

11.2 CONSTRUCTION OF INDUCTION MACHINE

Three-phase induction motor consists of two main parts namely stator which is stationary part and the rotor which is rotating part of induction motor. The rotor is placed inside the stator and is supported on both the side by two end-shields which house the bearings. The stator is built up of high grade alloy steel lamination to reduce the eddy current losses. The laminations are slotted on the inner periphery and are insulated from each other. Energy is supplied to the stator. Energy is transferred to the rotor windings through electromagnetic induction and hence such machines are called induction motors.

Rotor is one of the important part of induction motor. It is also built up of thin laminations of the same material as stator. The rotor core is a laminated steel cylinder having slots, in which aluminium conductors are die-cast or copper conductors are wound parallel to the shaft. The stator windings are made from three separate windings and are insulated from each other and displaced in space by 120° electrical.

Stator frame and yoke are also very important part of induction motor as they support and give proper stand to the motor. Stator frame supports the stator and the yoke give proper stand to motor.

Types of rotor: As per construction point of view, the rotor of three-phase induction motors are classified in two category.

(*i*) Cage rotor and

(*ii*) Slip ring or wound rotor.

Cage rotor: The rotor aluminium conductors in a squirrel cage induction motor does not need any insulation from the rotor-core, as the current will flow through the parts of minimum resistance that is through the rotor conductors. The rotor bars are shorted at both the rotor ends by end-rings. The slots of rotor are not made parallel to the rotor shaft but are skewed at a certain angle with the shaft to reduce the magnetic noise during motor operation, to produce a more uniform torque and prevent the possible magnetic locking. The magnetic locking is also termed as cogging. During cogging the rotor and stator teeth attract each other due to magnetic effect, so there is no rotation of rotor.

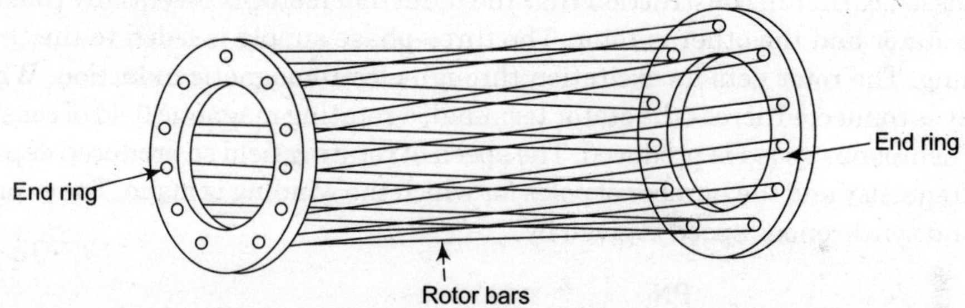

Fig. 11.1 *Cage rotor (Slightly skewed).*

Slip ring or wound rotor: The rotor winding is made on the insulated rotor slots with copper conduct similar to the stator winding. The stator windings are connected in star. The three free ends are brought to the three slip rings mounted on the shaft. Slip rings are also insulated from the rotor shaft. The external resistance can be connected with the rotor winding through brush and slip-ring connection. The external resistance provides the facility to increase the starting torque and decrease the starting current. It also helps in controlling the speed of three-phase induction motor.

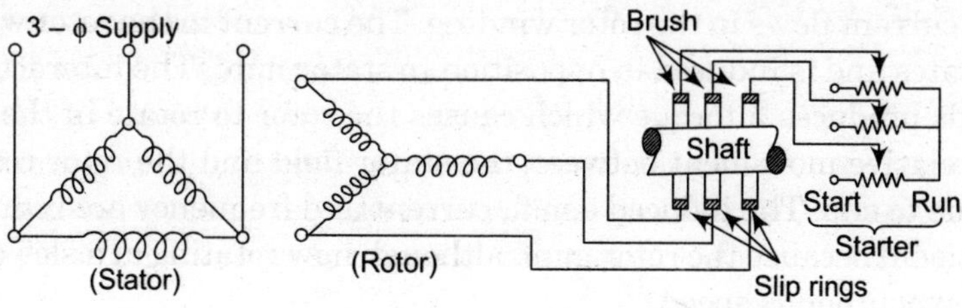

Fig. 11.2 *Wound rotor induction motor.*

Comparison between Cage Rotor and Slip-ring Induction Motor

Cage rotor	*Slip-ring or Wound rotor*
1. Robust in construction and low cost	1. Bulky due to presence of brush and high cost.
2. Requires less maintenance	2. Regular maintenance is needed.
3. Absence of brush eliminate the chances of sparking	3. Sparking takes place due to brush.
4. Highly efficient and high power factor	4. Less efficient and low power factor than cage rotor.
5. Low starting torque and higher starting current than wound rotor	5. High starting torque and low starting current.
6. External resistance cannot be added. So, fail to control the speed.	6. External resistance can be added to rotor circuit in order to control the speed.

11.3 WORKING PRINCIPLE OF A THREE-PHASE INDUCTION MOTOR

As it is discussed earlier in construction that the induction motor is essentially consists of two parts one is stator and the other is rotor. The three-phase supply is feded to the three-phase stator winding. The rotor gets its excitation through electromagnetic induction. When three-phase supply is connected across the stator terminal, a rotating magnetic field of constant magnitude at synchronous speed is produced. The speed of rotating field so produced depends upon the supply frequency and the number of poles for which the winding is made. The expression for frequency and synchronous speed is given by

$$f = \frac{PN_s}{120}$$

$$N_s = \frac{120\,f}{P}$$

where $N_s \rightarrow$ Synchronous speed

$P \rightarrow$ Number of poles

$f \rightarrow$ Supply frequency.

The rotor winding cuts the rotating field and an e.m.f. is induced in the rotor winding, when the rotor is at rest, the frequency of this e.m.f. is same as the supply frequency. If the rotor circuit is closed, a current flows in the rotor winding. The current in the rotor winding produces an mmf which rotates and is induced in opposition to stator mmf. The interaction of the stator and the rotor fields produces a torque which causes the rotor to rotate in the direction of the stator field. The relative movement between the stator field and the rotor conductors is ultimately reduced due to slip. The induced e.m.f., current and frequency are reduced. The torque remains unidirectional because the rotor mmf, although now rotating at a slower speed relative to the stator viz., synchronous speed.

11.4 PRODUCTION OF ROTATING FIELD

The three-phase winding is displaced in space by 120° and the current supplied is displaced in time by 120°. When this three-phase current is supplied to three-phase stator winding, a magnetic flux is produced which rotates in space.

Let the instantaneous fluxes be given by

$$\phi_1 = \phi_m \sin \omega t$$
$$\phi_2 = \phi_m \sin (\omega t - 120°)$$
$$\phi_3 = \phi_m \sin (\omega t - 120°)$$

Consider the figure shown below:

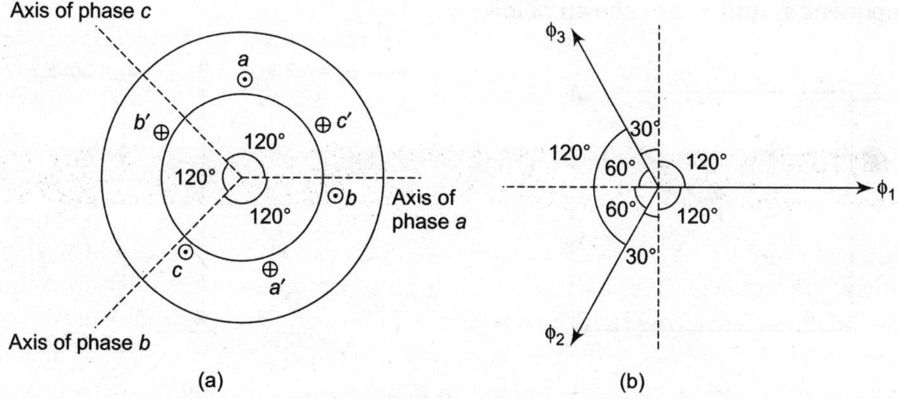

Fig. 11.3

The resultant flux produced by this system may be determined by resolving the horizontal and vertical components with respect to the physical axes as shown in Fig. 11.3(*b*).

The resultant vertical component of flux is given by

$$\phi_1 = 0 - \phi_2 \cos 30° + \phi_3 \cos 30°$$
$$= \cos 30° \, [-\phi_m \sin (\omega t - 120°) + \phi_m \sin (\omega t + 120°)]$$
$$= \frac{\sqrt{3}}{2} \phi_m \, [-(\sin \omega t \cos 120° - \cos \omega t \sin 120°) + (\sin \omega t \cos 120° + \cos \omega t \sin 120°)]$$
$$= \frac{\sqrt{3}}{2} \phi_m \, (2 \cos \omega t \sin 120°) = \frac{\sqrt{3}}{2} \phi_m \times 2 \cos \omega t \times \frac{\sqrt{3}}{2}$$
$$\phi_v = \frac{3}{2} \phi_m \cos \omega t \qquad ...(1)$$

Now, resolving the resultant horizontal component of flux is given by

$$\phi_h = \phi_1 - \phi_2 \cos 60° - \phi_3 \cos 60°$$
$$= \phi_1 - (\phi_2 + \phi_3) \cos 60°$$
$$= \phi_1 - \frac{1}{2} (\phi_2 + \phi_3)$$

$$= \phi_m \sin \omega t - \frac{1}{2} [\phi_m \sin (\omega t - 120°) + \phi_m \sin (\omega t + 120°)]$$

$$= \phi_m \sin \omega t - \frac{\phi_m}{2} (\sin \omega t \cos 120° - \cos \omega t \sin 120° + \sin \omega t \cos 120° + \cos \omega t \sin 120°)$$

$$= \phi_m \sin \omega t - \frac{\phi_m}{2} \times \left(2 \sin \omega t \left(-\frac{1}{2}\right)\right)$$

or $$\phi_h = \frac{3}{2} \phi_m \sin \omega t \qquad ...(2)$$

The component ϕ_v and ϕ_h are shown below:

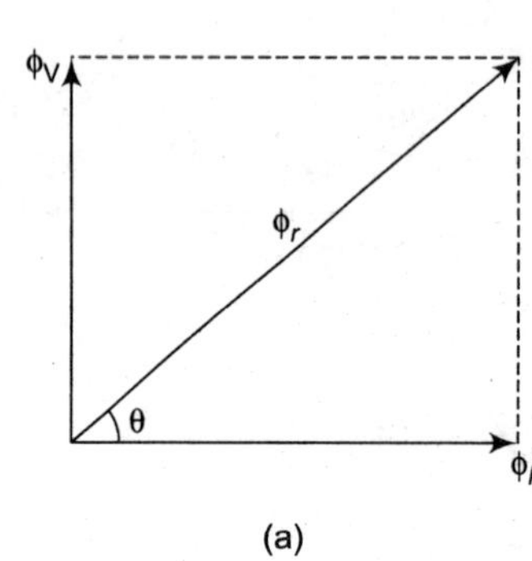

(a)

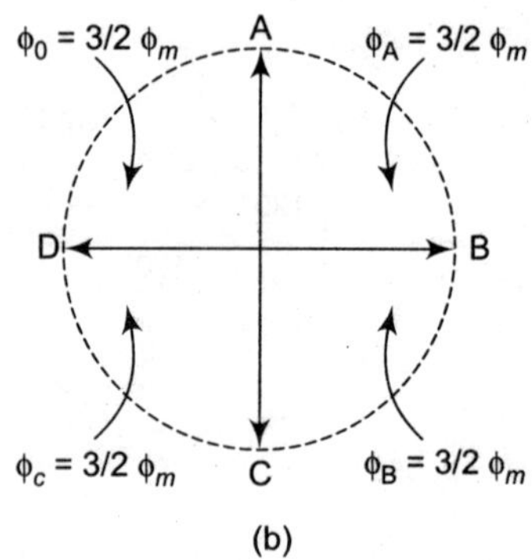

(b)

Fig. 11.4

Now, the total resultant flux

$$\phi_r = \sqrt{\phi_v^2 + \phi_h^2} = \sqrt{\left(\frac{3}{2} \phi_m \sin \omega t\right)^2 + \left(\frac{3}{2} \phi_m \cos \omega t\right)^2}$$

$$= \frac{3}{2} \phi_m \sqrt{\sin^2 \omega t + \cos^2 \omega t}$$

or $$\phi_r = \frac{3}{2} \phi_m \qquad ...(3)$$

Also, $$\tan \theta = \frac{\phi_v}{\phi_h} = \frac{\frac{3}{2} \phi_m \cos \omega t}{\frac{3}{2} \phi_m \sin \omega t}$$

$$= \cos \omega t = \tan \left(\frac{\pi}{2} - \omega t\right) \qquad ...(4)$$

∴ $$\theta = \frac{\pi}{2} - \omega t$$

Therefore from equation (4) it can be concluded that

(*i*) when $\omega t = 0°$ at that time $\theta = 90°$ and position A is achieved as shown in Fig. 11.4(*b*)

(*ii*) when $\omega t = 90°$, $\theta = 0$ and position B is achieved, as shown in Fig. 11.4(*b*)

(*iii*) when $\omega t = 180°$, $\theta = -90$ and position C is achieved, as shown in Fig. 11.4(*b*)

(*iv*) when $\omega t = 270°$, $\theta = -180°$ and position D is achieved, as shown in Fig. 11.11(*b*)

Slip : The speed of the rotor relative to the stator rotating field is called slip. In other words, the difference between synchronous speed and the rotor speed is also known as slip. It is usually expressed as a fraction of the synchronous speed. In usual practice slip is expressed either as per unit or as a percentage of synchronous speed.

$$\text{Per unit slip} = \frac{\text{Synchronous speed } (N_s) - \text{Rotor speed } (N_r)}{\text{Synchronous speed } (N_s)}$$

and percentage slip %
$$S = \frac{N_s - N_r}{N_s} \times 100$$

The value of slip at full-load varies from about 6 % for small motors to about 2 % for all large motors.

At stand still the rotor speed is zero and so slip will be

$$S = \frac{N_s - N_r}{N_s}$$

as
$$N_r = 0$$
$$S = 1$$

The rotor cannot run at synchronous speed because then there will be no rotor e.m.f. and no rotor current and torque. If rotor is required to run at synchronous speed then some external driving torque is compulsory. If the rotor speed becomes greater than synchronous speed *i.e.*, $N_r > N_s$, the slip becomes negative, the rotor torque opposes the external driving torque and the machine starts acting as induction generator.

11.5 E.M.F. AND CURRENT IN INDUCTION MOTOR

The voltage applied to each phase is constant and purely sinusoidal and alternating in nature at frequency f_1, the flux per pole ϕ is sinusoidally distributed in space around the air gap and is rotating at synchronous speed and the stator and rotor should posses same number of poles and phases.

At stand still condition, the e.m.f. induced in each phase of stator and rotor is given by

$$E_1 = 4.44\, K_{b1}\, K_{p1}\, f_1 N_1 \phi \quad ...(i)$$
$$E_2 = 4.44\, K_{b2}\, K_{p2}\, f_2 N_2 \phi \quad ...(ii)$$

where, E_1 = e.m.f. induced per phase in stator volts

E_2 = e.m.f. induced per phase in rotor volts

ϕ = Flux per pole in weber

f_1 = Supply frequency in Hz

N_1 and N_2 = Number of turns in series per phase of the stator and rotor respectively.

K_{b1}, K_{b2} = Breadth factor of stator and rotor winding respectively.

K_{p1}, K_{p2} = Pitch factor of stator and rotor winding respectively.

At standstill condition the frequency of e.m.f. induced in the rotor is same as supply frequency.

The rotor current can be obtained at two different cases depending upon the condition of slip.

(*i*) At standstill conditions

Let E_{20} = e.m.f. induced for phase of rotor at standstill.

R_2 = Resistance per phase of the rotor.

X_{20} = Reactance per phase of the rotor at standstill.

$= 2\pi f_1 L_2$

At standstill condition the rotor frequency is equal to the supply frequency.

Z_{20} = Rotor impedance per phase at standstill

I_{20} = Rotor current per phase at standstill

So, $Z_{20} = R_2 + jX_{20}$

$$I_{20} = \frac{E_{20}}{Z_{20}}$$

Power factor at standstill condition is

$$\cos \phi_2 = \frac{R_2}{Z_{20}} = \frac{R_2}{\sqrt{R_2^2 + X_{20}^2}}$$

(*ii*) Current in rotor at some slip S :

The induced e.m.f. for phase in the rotor winding at slip *s* is

$$E_{2S} = E_{20}$$

Rotor winding resistance per phase = R_2

Rotor winding reactance per phase at slip *s* is

$$X_{2S} = 2\pi f_2 L = 2\pi (sf_1) L = SX_{20}$$

Rotor winding impedance for phase at slip *s* is

$$Z_{2S} = R_2 + jX_{2S} = R_2 + jsX_{20}$$

Rotor current at slips *s* is

$$I_{2S} = \frac{E_{2S}}{Z_{2S}}$$

Power factor at slip is

$$\cos \phi_{2S} = \frac{R_S}{Z_{2S}}$$

where the subscript 2, 0 and *s* represents rotor standstill condition and at some slip *s*.

11.6 FREQUENCY OF ROTOR VOLTAGE AND CURRENT

The frequency of rotor winding is of variable nature but the frequency of stator winding is same as that of the supply frequency and it is given by

$$f = \frac{PN_s}{120} \qquad ...(i)$$

The rotor frequency depends upon the slip. At the standstill condition the rotor frequency is same as that of the supply frequency. The rotor frequency depends upon the difference between the synchronous speed and the rotor speed and the slip at that instant and is given by:

$$f_r = \frac{P(N_s - N_r)}{120} \qquad ...(ii)$$

Dividing equation (*ii*) by equation (*i*), we get

$$\frac{f_r}{f} = \frac{P(N_s - N_r)/120}{PN_s/120} = \frac{N_s - N_r}{N_s}$$

$$\frac{f_r}{f} = s$$

$$f_r = sf \qquad ...(iii)$$

Therefore, rotor current frequency = Per unit slip × Supply frequency when the rotor is at standstill

$$N_r = 0$$

$$s = \frac{N_s - N_r}{N_s} \text{ and } f_r = f$$

Induction Motor: A Generalised Transformer

The induction motors are also known as generalised transformer due to the following facts: The e.m.f. equation of transformer is given by

$$E = 4.44 \,.\, f \,.\, N \,.\, \phi$$

and e.m.f. equation of induction motor is

$$E_1 = 4.44 \,.\, K_{b1}\, K_{pl}\, f_1\, N_1\, \phi$$

and $$E_2 = 4.44 \,.\, K_{b2} \,.\, K_{p2}\, f_1\, N_2\, \phi$$

So, the e.m.f. equation of transformer and induction motors are similar, the only difference is that due to rotating part in induction motor, the extra term pitch factor and breadth factor is involved. The phasor diagram and the equivalent circuit of transformer and induction motor are similar to each other. The operating principle in both the cases are same *i.e.*, Faraday's law of electromagnetic induction and both are a.c. operated machine.

The important differences between transformer and induction motors are:

(*i*) The losses in an induction motor are higher due to presence of rotating part. So, efficiency is lower than transformer.

(*ii*) In an induction motor the magnetic leakages and leakage reactances of rotor and stator are higher than that in a transformer.

(*iii*) The magnetic circuit of an induction motor has an air gap and this creates the per unit value of magnetising current much higher than that of a transformer.

11.7 EQUIVALENT CIRCUIT OF INDUCTION MOTOR

The equivalent circuit of induction motor enables us to evaluate the performance characteristics at steady-state conditions by ordinary network solution method.

Rotor Circuit Diagram

Rotor power input = Rotor copper loss + Rotor power developed.

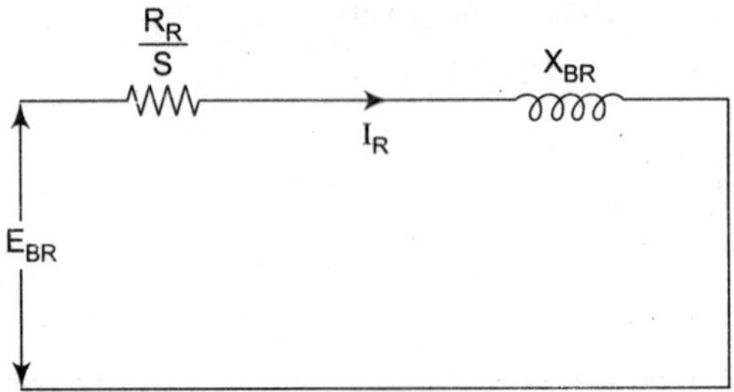

Fig. 11.5

The rotor resistance can be expressed as

$$\frac{R_R}{S} = R_R + R_R\left(\frac{1-S}{S}\right)$$

where, R_R = Rotor resistance

E_{BR} = Blocked rotor-induced voltage per phase

X_{BR} = Rotor reactance of standstill condition

S = Slip

Therefore, power developed by motor is given by

$$P_d = \omega_R T$$

as T_d = Torque developed

ω_R = Angular velocity of rotor

Hence, total torque developed by motor

$$T_d = \frac{\text{Rotor power developed}}{\text{Angular velocity of rotor}} \text{ in Newton-Metre.}$$

where $\omega_R = \frac{2\pi N_R}{60}$ rad/sec

N_R = Rotor speed in RPM at which power is developed.

11.8 STATOR CIRCUIT DIAGRAM

Let R_S = Stator phase winding resistance in ohm

X_S = Leakage reactance of stator phase winding in ohm

The equivalent circuit of stator is given below in Fig. 11.6.

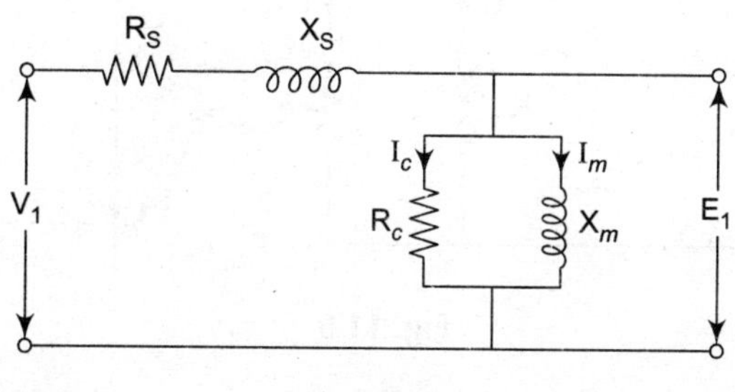

Fig. 11.6

The stator equivalent diagram of induction motor on per phase bars is given below in Fig. 11.7.

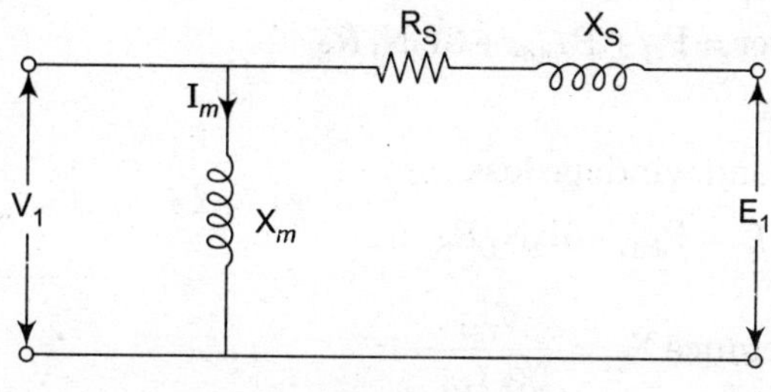

Fig. 11.7

Equivalent circuit diagram of induction machine on a per phase basis referred to stator is given below in Fig. 11.8.

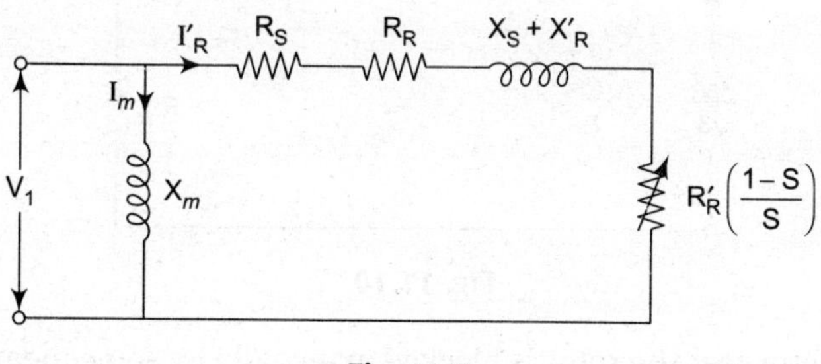

Fig. 11.8

At light loads (small slip), the resistor representing the mechanical power $R_R \dfrac{(1-S)}{S}$ is large. Hence I_R is small as compared to magnetising current I_m. Hence the circuit behaves purely inductive as X_m is the dominating element in the circuit and therefore power factor is small.

11.9 DETERMINATION OF EQUIVALENT CIRCUIT PARAMETERS

(*i*) **No Load Test:** The motor is run without load and with rated voltage supplied to the stator which is shown below in Fig. 11.9.

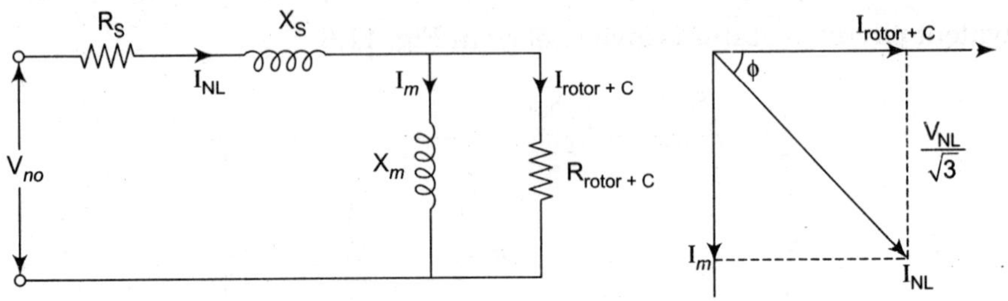

Fig. 11.9

V_{no} = Line to line stator voltage at no load, volts

I_{no} = Line current at no-load, amps

P_{no} = Three-phase power input at no load, watts

So, Input-power = $P_C + P_{rotor} + 3I_2N_LR_S$

where P_C = Core loss

P_{rotor} = Friction and windage loss

Hence, $P_{rotor} + C = P_{NL} - 3I_2N_LR_S$

∴ Magnetising reactance $X_m = \dfrac{V_{LN}}{\sqrt{3}\, I_{NL}}$

(*ii*) **Blocked Rotor Test:**

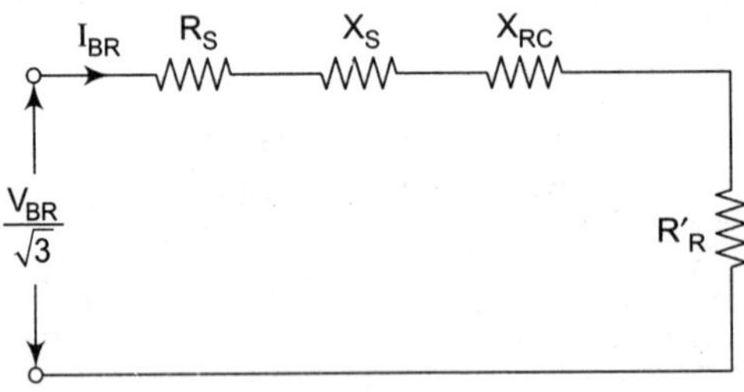

Fig. 11.10

In this blocked rotor test the rotor is blocked externally by some means and prevent rotor from turning.

Hence, $N_R = 0$ and ultimately at standstill $s = 1$

Here, V_{BR} = Line to line stator voltage at blocked rotor condition

I_{BR} = Line current at blocked rotor condition

P_{BR} = Three-phase power input

In this case the magnetising reactance is neglected and

$$R_R^1 = \frac{R_R^1}{S}$$

In terms of stator:

Equivalent motor impedence

$$Z_e = \frac{V_{BR}}{\sqrt{3}\, I_{BR}}$$

$$= \left(R_S = R_R^1\right) + J\,(X_S + X_{R'})$$

Equivalent motor resistance

$$R_e = \frac{R_{BR}}{3I_{BR}^2}\, R_S + R_{R'}$$

Equivalent motor reactance

$$X_e = \sqrt{Z_e^2 = R_e^2} = X_S + X_{R'}$$

Rotor resistance in terms of stator

$$R_R' = R_e - R_s \quad \therefore \quad X_S = X_R'$$

Turn ratio, $\alpha = \dfrac{N_S}{N_R} = \dfrac{V_L L}{V_S}$

where N_s = Synchronous speed rpm

N_r = Rotor speed in rpm

Actual rotor resistance and reactance will be

$$X_R = \frac{X_R'}{X_L} \text{ and } R_R = \frac{X_R'}{X_L}$$

11.10 PHASOR DIAGRAM OF INDUCTION MOTOR

The phasor diagram of three-phase induction motor is very much similar to the phasor diagram of a transformer. The phaser diagram of three-phase induction motor is shown below:

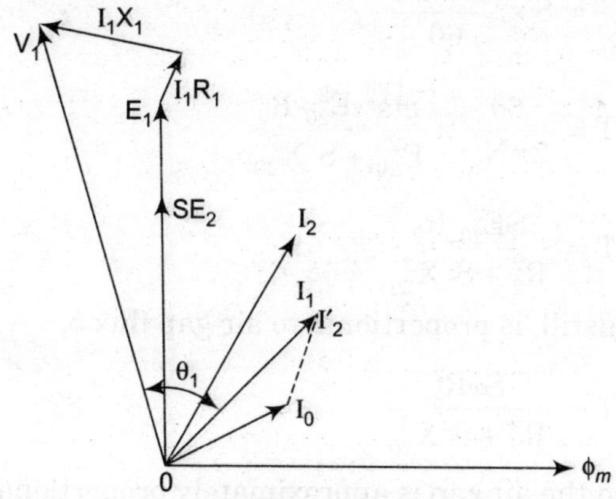

Fig. 11.11

In between stator and rotor there is air gap in which electromagnetic field is produced. The flux so generated in air gap is common to both stator and rotor so, taken as reference line. ϕ_m is mutual flux and induces an e.m.f. E_1 in the stator winding. This induced e.m.f. can be taken as a voltage rise considering E_1 as a voltage drop in the stator winding E_1 loads ϕ_m by 90°. SE_2 is the voltage induced in rotor per phase. I_0 is the no load current of the motor. I_2^1 is the component of stator current which balance rotor current. The rotor stator current I_1 is the phasor sum of I_0 and I_2^1. The stator impressed voltage is the resultant of the component consisting of E_1, voltage drop I_1R_1 which is in phase with I_1 and voltage I_1X_1 in quadrature with leading I_1.

In order to overcome the leakage reactance drop in the stator winding is

$$V_1 = E_1 + I_1 (R_1 + jX_1)$$

and θ_1 is phasor diagram is the power factor of the stator power. All the referred quantites in phasor diagram is per phase quantity.

11.11 THREE-PHASE MOTOR CHARACTERISTICS

Developed torque T_d: The total electrical power developed in the rotor is equal to $mE_2I_2 \cos \phi_2$ where m is the number of phases. This electrical power is lost as $I^2{}_R$ (loss in the rotor circuit)

$$\text{Rotor } I^2{}_R \text{ loss} = mE_2I_2 \cos \phi_2$$

$$= \frac{ms\, E_{20}\, SE_{20}}{\sqrt{R_2^2 + (SX_{20})^2}} = \frac{R^2}{\sqrt{R_2^2 + (SX_{20})^2}}$$

$$= \frac{ms^2\, E_{20}^2\, R_2}{R_2^2 + s^2X_{20}^2} \quad \text{...(i)}$$

Now, we have equation of power transfer from stator to rotor

$$\text{Rotor input} = \frac{2\pi N_S T}{60}$$

Also it follows that

$$\text{Roter} \quad I^2{}_R - \text{loss} = S \times \text{Rotor input}$$

or

$$\frac{ms^2\, E_{20}^2\, R_2}{R_2^2 + S^2X_{20}^2} = S \times \frac{2\pi N_S T}{60}$$

or

$$T = \frac{60}{2\pi N_S} \times \frac{ms^2\, E_{20}^2\, R_2}{R_{20}^2 + S^2X_{20}^2} \quad \text{...(ii)}$$

Therefore,

$$T \propto \frac{SE_{20}^2\, R_2}{R_2^2 + S^2X_{20}^2} \quad \text{...(iii)}$$

Since rotor e.m.f. at standstill, is proportional to air-gap flux ϕ,

$$T \propto \frac{S\phi R_2^2}{R_2^2 + S^2X_{20}^2} \quad \text{...(iv)}$$

Since the flux produced in the air gap is approximately proportional to the supply voltage to the stator and consequently T is proportional to the square of the stator applied voltage, V_1.

Therefore, $T \propto V_1^2$

By this obtained result equation (*iii*) can be written as

$$T \propto \frac{SE^2R_2}{R_2^2 + S^2X_{20}^2}$$

or

$$T = \frac{K\,SE^2R_2}{R_2^2 + S^2X_{20}^2} \quad ...(v)$$

Torque- slip characteristics: The torque equation is given as

$$T = \frac{K\,SE^2R_2}{R_2^2 + (SX_2)^2}$$

where R_2 and X_{20} are rotor resistance and reactance per phase respectively and k_1 is proportionately constant,

when $s = 0$, $T = 0$ and hence the torque-slip curve starts from origin *i.e.* zero.

At normal speeds, close to the synchronous speed SX_{20} will be small and is negligible with respect to rotor resistance R_2.

Therefore, $T \propto S$, wiiere R_2 and the supply voltage is constant.

Since $S \propto T$, for low value of slip, torque-slip curve is approximately a straight line. As slip increases the torque also increases and becomes maximum when $S = \frac{R_2}{X_{20}}$ when the slip further increases with increases in motor load, then rotor resistance R_2 becomes negligible compared to SX_{20}. Therefore, larger values of slip.

So, we get $T \propto \frac{S}{(SX_{20})^2} \times \frac{1}{S}$

Hence, the curve obtain following above condition will be a rectangular hyperbola.

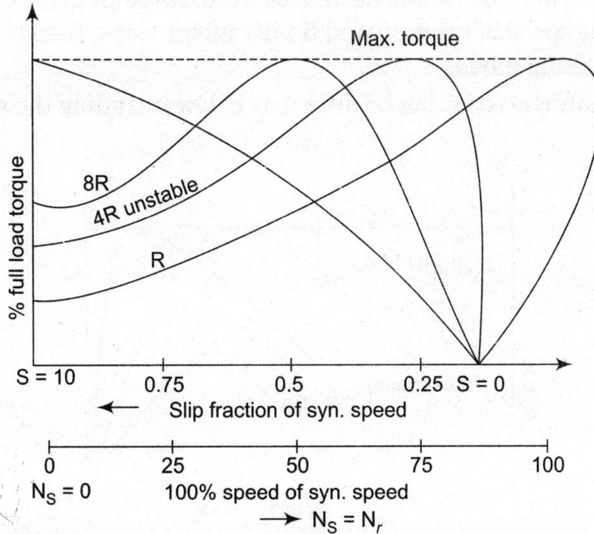

Fig. 11.12

The torque-slip characteristics of three-phase induction motor works three different modes *i.e.*

(*i*) **Motering mode :** At synchronous speed $s = 0$, this results torque to be zero. When the speed is near to synchronous speed, the slip in very small and the term $(SX_{20})^2$ in torque equation becomes negligible as compared to the rotor resistance R_2.
Therefore,

$$T = \frac{K_1 S}{R_2}$$

when R_2 is constant,

$$T = K_2 S \qquad ...(i)$$

when $$K_2 = \frac{K_1}{R_2}$$

above equation conclude that the torque is proportional to the slip. Hence, when the slip is small, the torque-slip curve will be a straight line. In this condition the motor will be in motoring mode.

(*ii*) **Generating mode:** In this mode of operation, the slip gradually increases with the increase in load with the increase in the term $(SX_{20})^2$, so that R_2^2 term in torque equation become negligible and ultimately neglected.

Therefore, $$T = \frac{K_3 R_2}{(SX_{20})^2}$$

This, torque becomes inversely proportional to the slip towards the stand still conditions. In this mode the torque-slip curve is represented by a rectangular hyperbola. This mode of operation is generating mode.

(*iii*) **Braking mode:** When the torque decreases from the point of maximum torque resulting in decrease in the speed of rotation and finally motor slops, then it can be said that motor is working in Braking mode.

Torque speed-characteristics can be drawn as below including the above mentioned mode of operation is

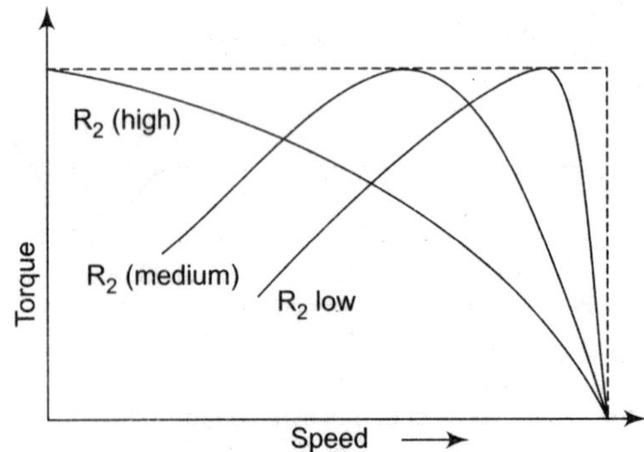

Fig. 11.13

11.12 STARTING OF THREE-PHASE INDUCTION MOTOR

During the operation of 3 – ϕ induction motor, the stator is fed with 3 – ϕ a.c. supply, a rotating magnetic field is produced and the rotor starts rotating. Thus, a three-phase induction motor is self-starting. At the time of starting the rotor speed is zero, ultimately results slip as unity and draws very large starting current. The purpose of employing the starter in three-phase induction motor is to protect the motor form overload and no-load condition and reduce the heavy starting current.

Starting of Squirrel Cage Motors

Three types of starter are commonly used for starting the cage motors:

(*i*) Star-delta starting

(*ii*) Direct-on-line starting

(*iii*) Autotransformer starting.

(*i*) **Star-delta starting :** This method of starting is very common type of starting and extensively used. A star-delta starter is used for a cage motor, designed to run normally on delta-connected stator winding. Figure 11.14(*a*) shows the connection diagram of a three-phase induction motor with a star-delta starter.

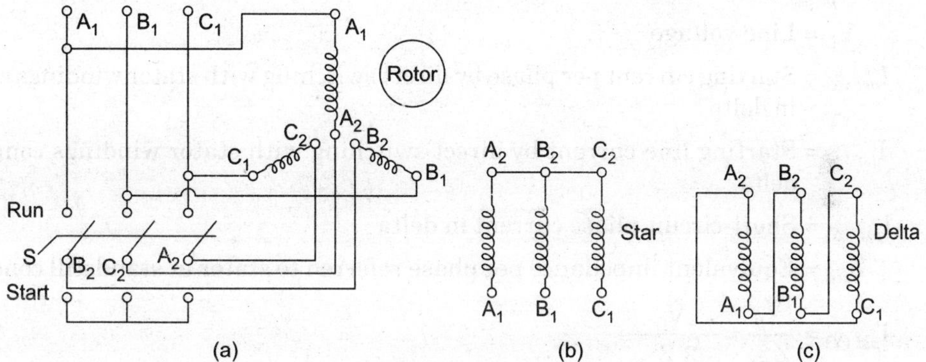

Fig. 11.14

When switch s is in the start position, the stator windings are connected in star as shown in Fig. 11.14(*b*). As soon as the motor picks up speed *i.e.*, 80 % of its rated value, the change over switch *s* is thrown quickly to the run position which connects stator winding in delta Fig. 11.14(*c*). So, by connecting the stator winding, first in star and then in delta, the line current drawn by the motor at starting is reduced to one-third as compared to the starting current during delta connection.

During starting, when the stator windings are star connected, each stator phase gets a voltage $\frac{V_L}{\sqrt{3}}$ as V_L is line voltage. Since the torque developed by an induction motor is directly proportional to the square of applied voltage *i.e.*,

$$T \propto \frac{SE^2R_2}{R_2^2 + (SX_{20})^2}$$

Star-delta starting helps in reducing the starting torque to one-third obtainable by direct-on-line starting.

This type of starting is generally known as reduced voltage starting.

Theoratical concept of star-delta starting: At the time of starting, the stator windings are connected in star and therefore voltage across each phase winding is equal to $\frac{1}{\sqrt{3}}$ times the line voltage.

Let V_L = Line voltage

$I_{pst\,(Y)}$ = Starting current per phase with starter windings connected in star.

$I_{lst\,(Y)}$ = Starting line current with stator winding in star during star connection, line current = phase current

$$I_L = I_{ph}$$

i.e., $I_{lst\,(Y)} = I_{pst\,(Y)}$

If V_p = Phase voltage

V_L = Line voltage

$I_{pst\,(\Delta)}$ = Starting current per phase by direct switching with stator windings connected in delta

$I_{lst\,(\Delta)}$ = Starting line current by direct switching with stator windings connected in delta

$I_{pse\,(\Delta)}$ = Short-circuit phase current in delta

Z_{eid} = Equivalent impedance per phase referred to stator at standstill condition.

$$I_{pst\,(Y)} = \frac{V_p}{Z_{eio}} = \frac{V_L}{\sqrt{3}\,Z_{e10}}$$

$$I_{pt\,(\Delta)} = \frac{V_p}{Z_{e10}}$$

Now, for delta connection,

Line current $= \sqrt{3}$ phase current

$$\therefore \qquad I_{lst\,(\Delta)} = \sqrt{3}\,I_{pst(\Delta)} = \frac{\sqrt{3}\,V_L}{Z_{e10}}$$

$$\frac{\text{Starting line current with star-delta starting}}{\text{Starting line current with direct switching in delta}} = \frac{I_{pst(Y)}}{I_{lst(\Delta)}} = \frac{\left(\dfrac{V_L}{\sqrt{3}\,Z_{e10}}\right)}{\sqrt{3}\left(\dfrac{V_L}{Z_{e10}}\right)} \qquad \ldots (i)$$

Thus, in star-delta starter, the starting current from the supply is one-third of that with direct switching in delta.

Also,

$$\frac{\text{Starting torque with star-delta starting}}{\text{Starting torque with direct switching in delta}} = \frac{\left(\dfrac{V_L}{\sqrt{3}}\right)^3}{V_L^2} \qquad ...(ii)$$

Here, the starting torque with star-delta starter is reduced to one-third with direct switching in delta.

Now,

$$\frac{\text{Starting torque with star-delta starting}}{\text{Full-load torque with stator winding in delta}} = \left(\frac{\left(Ipst(Y)^2 \,.\, \dfrac{R_2}{I}\right)}{2\pi n_s}\right) + \left[\frac{I_{pft(\Delta)}}{2\pi n_s} \cdot \frac{R_2}{S_{ft}}\right]$$

$$= \left[\frac{I_{pst(\Delta)}}{I_{pft(\Delta)}}\right] S_{fl} \,.$$

where, $I_{pfl(\Delta)}$ = Full-load phase current with winding in delta.

But $$I_{pst(Y)} = \frac{\dfrac{V_L}{\sqrt{3}}}{Z_{e10}}$$

$$I_{pst(\Delta)} = \frac{V_L}{Z_{e10}}$$

$\therefore$ $$I_{pst(Y)} = \frac{1}{\sqrt{3}} I_{pst(\Delta)}$$

$$I^2_{pst(Y)} = \frac{1}{3} I^2_{pst(\Delta)}$$

$$\frac{\text{Starting torque with star-delta starting}}{\text{Full-load torque with stator winding in delta}} = \left(\frac{I_{pst(Y)}}{I_{pft(\Delta)}}\right) S_{fl} = \frac{1}{3}\left[\frac{I_{pst(\Delta)}}{I_{psl(\Delta)}}\right] S_{fl}$$

(*ii*) **Direct on-line starting:** In this method of starting of squirrel cage motor, the full voltage is connected across the stator terminals, large current is drawn by the windings. This is due to the reason that, at starting the induction motor behaves as a short circuited transformer with its secondary, *i.e.*, rotor separated from the primary *i.e.*, the stator by a small air-gap.

The connection diagram for direct-in-line starting is shown in Fig. 11.15(*a*).

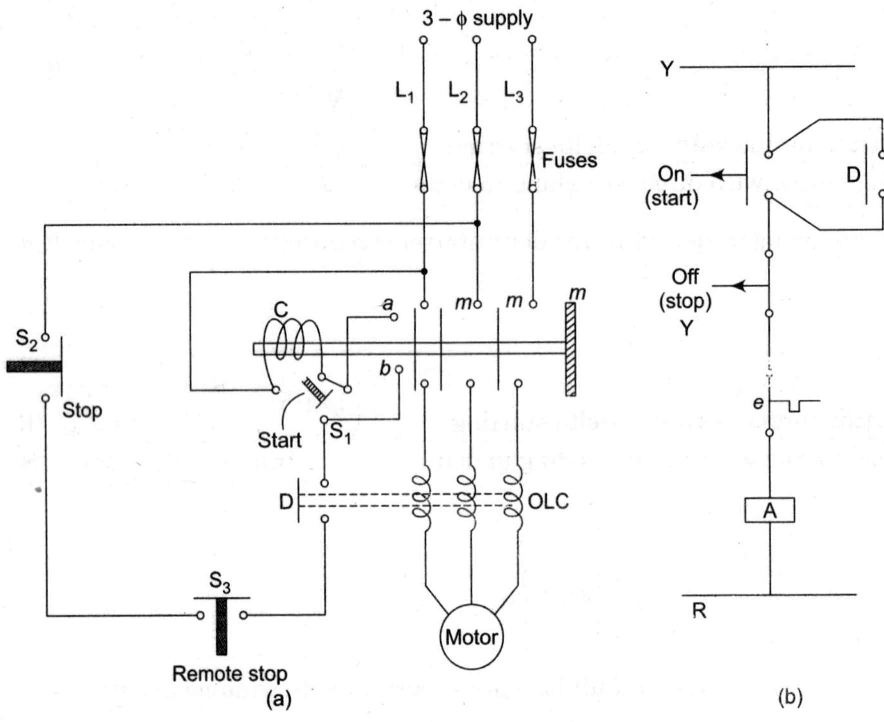

Fig. 11.15 *Schematic diagram of circuit.*

(*iii*) **Autotransformer Starting:** An autotransformer with different tappings at a voltage lower than the rated voltage is used for starting a three induction motor.

During starting a low voltage is applied through the least tap. As the motor picks up speed, the tap position is switched over to higher voltage tap position.

Theory of autotransformer starting: An autotransformer starter consists of an transformer and a switch as shown in Fig. 11.16.

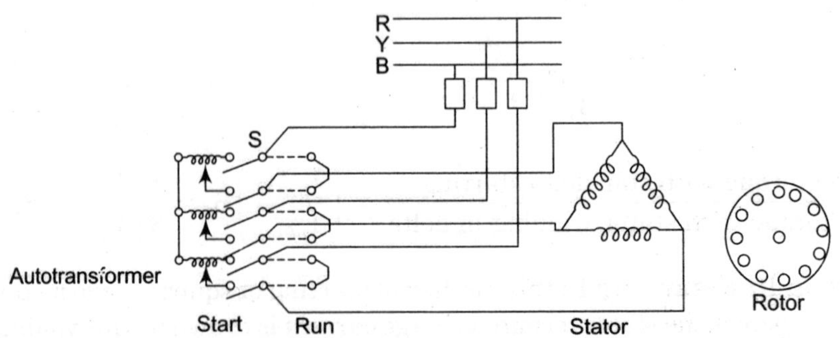

Fig. 11.16 *A manual Autotransformer starter.*

When switch s is put on start position, a reduced voltage is applied across the motor terminals, when the motor picks up speed nearly 80 % of its normal speed, the switch is put to RUN position. Then the autotransformer is cutout of the circuit and full rated voltage gets applied across the motor terminal. Normally two or three voltage steps are used during starting and then the autotransformer is completely cutout from the circuit.

Let I_s be the starting current with direct starting. If the full load current and n be the transformation ratio of autotransformer. As the voltage applied to each phase is reduced to $\frac{1}{n}$ of the line voltage.

So, current in each phase of motor $= \frac{I_s}{n}$

$$\text{Line current} = \frac{\text{Motor current}}{\text{Transformation ratio}}$$

i.e., line current $= \frac{1}{n^2} \cdot I_s$

Therefore, starting torque

$$= \text{Full load torque} \left(\frac{\text{Motor starting current}}{\text{Full load current}}\right)^2$$

$$= \text{Full load torque} \left(\frac{\frac{I_s}{n}}{I_f}\right)^2$$

$$= \frac{1}{n^2} \times \text{Starting torque (direct starting).}$$

11.13 SPEED CONTROL OF THREE-PHASE INDUCTION MOTOR

The speed of rotor of $3-\phi$ induction motor is given by

$$N_r = (1 - S)\, N_s \qquad \ldots(i)$$

and

$$N_s = \frac{120 f}{P} \qquad \ldots(ii)$$

Therefore, putting the value of equation (*ii*) in equation (*i*), the rotor speed becomes

$$N_r = (1 - S)\, \frac{120 f}{P} \qquad \ldots(iii)$$

where, N_r = Speed of rotor in rpm

S = Slip of induction motor

P = No. of poles of induction motor

f = Supply frequency

From equation (*iii*) it is seen that the speed of 3 – ϕ induction motor can be controlled by controlling or changing the supply frequency, no. of poles and slip.

The methods which are randomly used for the speed control of 3 – ϕ induction motor are as follows:

1. *By changing the applied frequency*: The synchronous speed of three-phase induction motor is given by

$$N_s = \frac{120f}{P}$$

where, ϕ = Supply frequency

P = No. of poles.

So, by changing the frequency the speed of rotor can be controlled.

2. *By changing the number of stator poles:* Change of number of poles is achieved by having two or more independent stator windings in the same slots. Each winding gives a different number of poles and this results different synchronous speed.
3. *Rotor rheostat control:* In this method of speed control an additional resistance is connected in the rotor circuit. It is only applicable to the slip ring induction motor. So, by this way the speed can be reduced by inserting a resistance in the rotor circuit.
4. *Slip energy recovery:* The slip power is feded an auxiliary induction motor mechanically coupled to the main or primary motor as shown in Fig. 11.17.

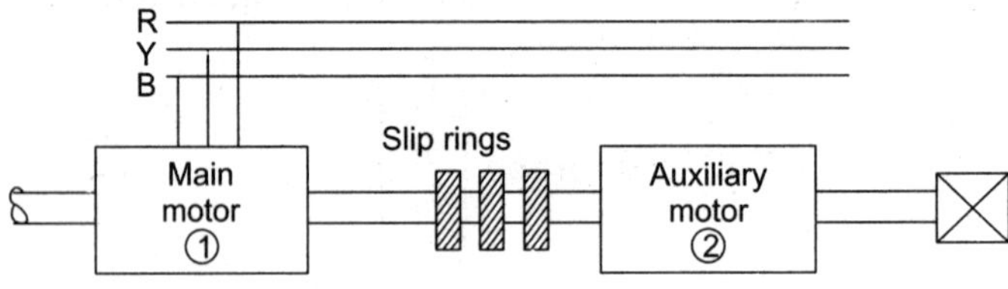

Fig. 11.17

This 3 – ϕ supply is given to the stator of main motor. The auxiliary motor is fed from the rotor circuit of the main motor.

Let f be the supply frequency and P_1 and P_2 be the no. of poles of main motor and auxiliary motor respectively. Then the speeds so obtained is

$$\text{Main motor alone} = \frac{f}{p_1}$$

$$\text{Aux. motor alone} = \frac{f}{p_2}$$

$$\text{Differential cascade} = \frac{f}{(p_1 - p_2)}$$

$$\text{Cummulative cascade} = \frac{f}{(p_1 + p_2)}$$

So, by this way, the speed can be varied or controlled.

Induction generator: If the rotor speed is increased above the synchronous speed, the induced e.m.f. will be in phase opposition to its sub-synchronous position because the rotor conductors are moving faster than stator rotating field and the line linkages of the rotor current will start flowing in phase opposition to applied voltage. This mechanism results in an induction generator.

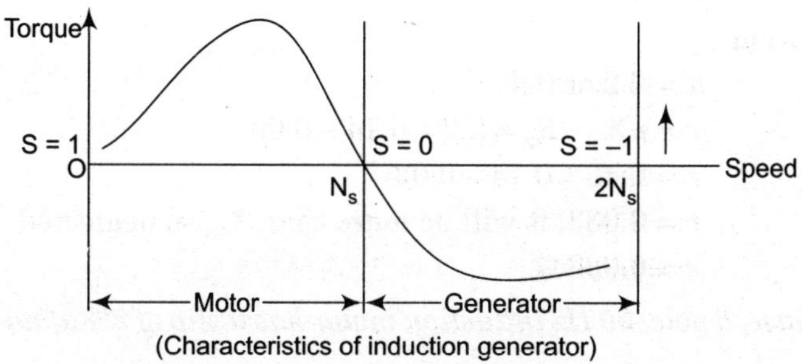

(Characteristics of induction generator)

Fig. 11.18

SOLVED NUMERICAL PROBLEMS

Example 1. *A 6-pole, 3-phase, 50 Hz induction motor has rotor resistance 0.03 ohm and reactance 0.14 ohm respectively per phase. Calculate speed at maximum torque. What resistance is to be added to give* $\frac{3}{4}$ *maximum torque at starting?*

Solution: Number of poles = 6 poles

Supply frequency = 50 Hz

$$\therefore \quad N_s = \frac{120 f}{P} = \frac{120 \times 50}{60} = 100 \text{ rpm.}$$

$$R_2 = 0.03\ \Omega\ ;\ X_2 = 0.14\ \Omega$$

The condition for maximum torque is

$$R_2 = S_m X_2$$

$$\therefore \quad S_m = \frac{R_2}{X_2} = \frac{0.03}{0.14} = 0.21$$

Speed at maximum torque

$$N = (1 - S_m)\, N_1$$

$$= (1 - 0.2) \times 1000 = 800 \text{ rpm.}$$

If $T_{st} = \frac{3}{4}$ maximum torque, then the relation between T_{st} and T_{max} is

$$\frac{T_{st}}{T_{max}} = \frac{2n}{1 + n^2}$$

where, $n = \frac{R_2 + r}{T_{max}}$ and r = Resistance to be added.

$$\therefore \quad \frac{2n}{1+n^2} = \frac{3}{4}$$

or $$n^2 - 8n + 3 = 0$$

Solving above equation

$$n = 2.2 \text{ or } 0.4$$

$$\therefore \quad r = nX_2 - R_2 = 2.2 \times 0.14 - 0.03$$

$$r = 0\,45 \times 0.14 - 0.03$$

$r = 0.033$, it will be more than X_2, so neglected

$$r = 0.033\ \Omega.$$

Example 2. *A three-phase, 6 pole, 50 Hz induction motor has a slip of 2% at no-load and 4% at full load. Calculate*

(*a*) *synchronous speed*

(*b*) *no-load speed*

(*c*) *full-load speed*

(*d*) *frequency of rotor current at stand still*

(*e*) *frequency of rotor current at full load.*

Solution:

(*a*) $N_s = \frac{120 f}{P} = \frac{120 \times 50}{6} = 1000$ rpm.

(*b*) $s = \frac{N_s - N_r}{N_s}$

$$N_r = N_s(1-s) = 1000\left(1 - \frac{2}{100}\right)$$

$$= 1000(1 - 0.02)$$

$= 980$ rpm at no-load.

(*c*) $N_r = N_s(1-s) = 1000\left(1 - \frac{4}{100}\right)$

$$= 1000(1 - 0.04)$$

$= 960$ rpm at full load.

(*d*) When rotor is stand still

$$S = \frac{N_s - N_r}{N_s}$$

$$s = 1, f_2 = sf_1$$

$$= 1 \times 50 \text{ Hz} = 50 \text{ Hz}.$$

(*e*) At full load, rotor current frequency

$$f_2 = sf_1 = s \times 50$$

$$= \frac{4}{100} \times 50 = 2 \text{ Hz}$$

Example 3. *A 3-phase, 50 Hz induction motor has a full-load speed of 1480 rpm. For this motor, calculate the following:*

(*a*) *Number of poles*

(*b*) *Full-load slip and rotor frequency*

(*c*) *Speed of stator field with respect to*

(*i*) *stator structure and*

(*ii*) *rotor structure*

(*d*) *Speed of rotor field with respect to*

(*i*) *stator structure and*

(*ii*) *rotor structure*

Solution:

(*a*) Using formulae

$$N_s = \frac{120\,f}{P}$$

$$1480 = \frac{120 \times 50}{P}$$

$$P = \frac{120 \times 50}{1480} = 4.05$$

Number of poles must be even so, number of poles will be either 4 or 6.

(*b*) Synchronous speed

$$N_s = \frac{120\,f}{P} = \frac{120 \times 50}{4} = 1500 \text{ rpm}$$

$\therefore$ Slip $\quad s = \frac{N_s - N_r}{N_s} = \frac{1500 - 1480}{1500} = 0.014$

Rotor frequency, $\quad f_2 = sf_1$

$$= 0.014 \times 50 = 0.7 \text{ Hz.}$$

(c) (*i*) Speed of stator field w.r.t. stator structure

$$= N_s = 1500 \text{ rpm.}$$

(*ii*) Speed of stator field w.r.t. rotor structure

$$1500 - 1480 = 20 \text{ rpm}$$

(*d*) (*i*) Speed of rotor field w.r.t. stator structure

= (Mechanical speed of rotor) + (Speed of rotor field w.r.t. rotor structure)

$$= 1480 + 21 = 1501 \text{ rpm.}$$

(*ii*) Speed of rotor field w.r.t. rotor structure

$$= \frac{120 \times (\text{Rotor frequency})}{\text{Poles}} = \frac{120 \times 0.7}{4} = 21 \text{ rpm}$$

Example 4. *A 8 pole induction motor is fed from 50 Hz supply. If the rotor emf frequency at full load is 4 Hz. Calculate the full-load slip and speed.*

Solution: $f_1 = 50$ Hz and $f_2 = 4$ Hz

Rotor frequency is

$$f_2 = sf_1$$
$$4 = s \times 50$$
$$s = \frac{4}{50} = 0.08$$

and also
$$N_s = \frac{120 f}{P} = \frac{120 \times 50}{8} = 750 \text{ rpm}$$
$$N_r = N_s (1 - s)$$
$$= 750 (1 - 0.08) = 690 \text{ rpm.}$$

Example 5. *A 6-pole, 3-phase, 420 V, 50 h.p., 50 Hz star connected induction motor has the following parameter per phase:*

$r_1 = 1.2\ \Omega$ $\qquad x_1 = 0.25\ \Omega$
$r'_2 = 0.14\ \Omega$ $\qquad x'_2 = 0.30\ \Omega$

The stator core losses equals 1200 W, friction and windage losses are equal to 850 W. The no-load current is 16 amperes at p.f. of 0.08 lagging, when the motor is operating at a slip of 0.028, calculate

(*a*) *Input line current and power factor*
(*b*) *Torque developed by motor*
(*c*) *Horse power output*
(*d*) *Efficiency.*

Solution:

(*a*) $V_1 = \frac{420}{\sqrt{3}} = 242.49 \angle OV$

$$I'_2 = \frac{V_1}{r_1 + \frac{r'_2}{s} + j(x_1 + x'_2)} = \frac{242.4\angle 0}{\left(0.12 + \frac{0.14}{0.028}\right) + j(0.25 + 0.30)}$$

$$I'_2 = \frac{242.4 \angle 0}{5.12 + j(0.55)}$$

$$I_0 = 16 \angle -85$$
$$= I_1 = I_2' + I_0$$

(*b*) Rotor copper loss

$$= 3(I_2')^2 r_2' = 3.$$

Example 6. *A 6-pole, three-phase, 230 V induction motor develops a torque of 18.37N-m when the stator current is 60 A. Find the stator current and torque if the stator voltage is reduced to 130 V.*

Solution: We obtain earlier that torque is directly proportional to V_1^2

Therefore, new value of torque

$$= 18.37\left(\frac{130}{230}\right)^2 = 5.87 \text{ N-m}$$

The stator current $= 60 \times \dfrac{130}{230}$

$= 33.9$ amperes.

Example 7. *A slip ring induction motor runs at 280 rpm on full load when connected from 50 Hz supply. Calculate*

(*a*) *number of poles*

(*b*) *slip*

(*c*) *slip for full-load torque if rotor resistance is doubled*

(*d*) *rotor copper losses with added rotor resistance if the original value of rotor copper loss is 230 W.*

Solution:

(*a*) $N_s = 280$ rpm

$$P = \frac{120f}{N_s} = \frac{120 \times 50}{280} = 21.42$$

Number of poles must be even so, P will be either 20 or 22.

(*b*) $N_s = \dfrac{120 \times 50}{20} = 300$ rpm

Slip $s = \dfrac{300 - 280}{300} = 0.06$

(*c*) It is seen earlier that for same torque, *s* is directly proportional to r_2'. Since $r_2' \propto r_2$, so the slip *s* is directly proportional to r_2. Since the rotor resistance is doubled. Therefore,

$s = 0.12$

(*d*) The rotor current is same as before viz., the full load current. As the rotor resistance is doubled so, rotor copper loss will be double.

Rotor copper loss $= 2 \times 230$

$= 460$ watts.

Example 8. *A 3-phase induction motor takes 4 times full-load current on direct starting and develops 2.5 times full load torque. What must be applied voltage to obtain full load torque at starting and what will be the line current?*

Solution: $\dfrac{\text{Full load torque}}{\text{Starting torque with direct}} = \dfrac{1}{2.5} = \dfrac{1}{K^2}$

Starting

or $$K^2 = 2.5$$

$$K = \sqrt{2.5} = 1.58$$

Applied voltage $= \dfrac{1}{1.58} = 0.632$ times the line voltage

Starting line current $= \dfrac{1}{K^2} \times 4 \times$ Full load current

$= \dfrac{4}{2.5} \times$ Full load current

$= 1.6 \times$ Full load current.

Example 9. *A 40 hp, 430 volt, 3-phase, 50 Hz induction motor with star connected stator winding gave the following test result:*

No load test: Applied line voltage 430 V, line current 20 A, watt meter reading 5040 and –3050 watts.

Block rotor test: Applied line voltage 35.6 volts, line current 55 A, watt meter readings 2050 and 750 watts.

Find the parameters of the equivalent circuit.

Solution: No load test:

Applied voltage per phase $= \dfrac{430}{\sqrt{3}} = 248.02$ volts

$$W_0 = 5040 - 3050 = 1990$$

$$\cos \theta_0 = \frac{1990}{3 \times 248.2 \times 20} = \frac{1990}{14892} = 0.134$$

$$R_0 = \frac{248.2}{20 \times 0.134} = \frac{248.2}{2.68} = 92.61\ \Omega$$

$$X_0 = \frac{248.2}{20 \sin \theta_0} = 9.32\ \Omega$$

Block rotor test:

Applied voltage per phase $= \dfrac{35.6}{\sqrt{3}} = 20.56$ V

$$W_{sc} = 2050 + 750$$
$$= 2800 \text{ watts}$$

$$\cos \theta_{sc}, = \frac{2800}{3 \times 20.56 \times 55} = \frac{2800}{3 \times 20.56 \times 55} = 0.825$$

$$r_1\, r_2' = \frac{20.56}{55} \times 0.825 = 0.3084$$

and $$x_1 + x_2' = \frac{20.56}{55} \sin \theta_{sc} = 0.20 \text{ ohm.}$$

Example 10. *The input to a 3-phase 50 Hz, 6 pole induction motor is 200 kW. Stator losses are 10 kW. Mechanical losses are 5 kW and full load slip is 0.06. Calculate:*

(*a*) *frequency of rotor e.m.f. at standstill*

(*b*) *frequency of rotor e.m.f. at full load*

(*c*) *rotor copper loss*

(*d*) *efficieny of the motor.*

Solution:

(*a*) Frequency of rotor e.m.f.

$$f_2 = sf_1$$

at stand still

$$s = 1$$

$$f_2 = 1 \times f_1 = 50 \text{ Hz}$$

(*b*) At full load, slip $s = 0.06$

frequency of rotor e.m.f.

$$f_2 = 0.06 \times 50 = 3 \text{ Hz}$$

(*c*) Rotor input = 200 – 1.0 = 190 kW

$$\text{Rotor copper loss} = s \times \text{Rotor input}$$

$$= 0.06 \times 190 = 11.4 \text{ kW}$$

(*d*) Motor output = 190 – 11.4 – 5

$$= 190 - 16.4 = 173.6 \text{ kW}$$

$$\text{Efficiency of motor} = \frac{\text{Motor output}}{\text{Motor input}} = \frac{173.6}{200}$$

$$= 0.868 = 86.8\%.$$

Example 11. *A 3-phase, 50 Hz, 420 V, 4-pole star connected induction motor has following equivalent circuit parameters and stator losses is 3 kW.*

$$R_0 = 110 \text{ ohm}, \quad X_0 = 22 \text{ ohm}$$

$$r_1 = 0.2 \text{ ohm}, \quad x_1 = 0.5 \text{ ohm}$$

$$r_2' = 0.3 \text{ ohm}, \quad x_2 = 0.5 \text{ ohm}.$$

The full-load slip is 0.04. Calculate:

(*a*) *motor speed*

(*b*) *stator current at no-load*

(*c*) *stator current at full-load*

(*d*) *motor input at full-load*

(*e*) *rotor copper losses*

(*f*) *motor output and efficiency.*

Solution:

(*a*) $N_r = N_s (1 - s)$

and

$$N_s = \frac{120 f}{P}$$

$$N_r = \frac{120 f}{P} (1 - s) = \frac{120 \times 50}{4} (1 - 0.04)$$

$$N_r = 1500 \times 0.96 = 1440 \text{ rpm}$$

(*b*) Stator voltage per phase = $\frac{420}{\sqrt{3}}$ = 230.94 V

No load current

$$I_0 = \frac{V_1}{R_0 + jX_0} = \frac{230.94}{110 + j22}$$

$$= 1.86 \angle -10.3° \text{ A}$$

(c) $$I_2' = \frac{V_1}{r_1 + jx_1 + jx_2' + \frac{r_2'}{s}} = \frac{230.94}{0.2 + j0.5 + j0.5 + \frac{0.3}{0.04}}$$

$$I_2' = 42.73 \angle -10.09° \text{ A}$$

Stator current at full load

$$= I_0 + I_2'$$

Motor power factor

$$= \cos(-10.09) = 0.89 \text{ lagging.}$$

(d) Motor input = $\sqrt{3} \times 420 \times 42.73 \times 0.89$

= 27664.3 watts

= 27.66 kW.

(e) Stator losses = 3 kW

Rotor input = 27.66 – 3

= 25.66 kW.

Rotor copper loss = 0.04 × 25.66 = 1.026 kW.

(*f*) Motor output = 25.66 – 1.026 – 3

= 25.66 – 4.02 = 21.64 kW

Motor efficiency $= \frac{21.64}{25.66} = 0.843 = 84.3\%$.

EXERCISES

1. Describe the construction and working of a 3-φ induction motor.
2. Why 3-φ induction motors are called as 'Generalised Transformer'?
3. Compare cage and wound 3-phase induction motor.
4. What do you understand by 'slip' in an induction motor?
5. Derive an expression for the frequency of rotor current in an induction motor.
6. Explain the reason why an induction motor does not run at synchronous speed.
7. Why the power factor of a three-phase induction motor is lagging?
8. What is a star-delta stator?
9. What are the various method of starting of three-phase induction motor? Explain in brief.
10. Why the rotor core loss is negligible in 3-φ induction motor?
11. Explain torque-slip characteristics of 3-φ induction motor.

12. Draw and explain phasor diagram of a three-phase motor.

13. Derive equation for torque developed in an induction motor.

14. Explain various methods of testing of 3-ϕ induction motor. How are the parameters of equivalent circuit determined from the test results?

15. What do you understand by 'slip energy recovery'?

16. Draw the equivalent circuit of 3-ϕ induction motor also specify the various circuit parameters.

17. Why starters are necessary for induction motor starting? Name different starting methods for 3-phase induction motors.

18. Discuss briefly the various methods of speed control of three-phase induction motors.

19. Explain in brief:

(*i*) Motoring mode

(*ii*) Generating mode

(*iii*) Braking mode.

With respect to torque-slip characteristics.

20. What is an induction generator?

21. If an 8-pole induction motor running from a supply of 50 H has an induced e.m.f. in the rotor frequency 1.3 Hz, determine the slip and speed of the motor.

22. Calculate the speed of a 4-pole induction motor having a slip of 5% at full load with a supply frequency of 50 Hz. What will be the speed of a 2-pole alternator supplying the motor?

23. A 4-pole, 50 Hz induction motor runs with a slip of 0.01 P.u. in full load. Calculate the frequency of the rotor current:

(*a*) at stand still and

(*b*) On full load.

24. A 4-pole induction motor runs at 1450 rpm at a frequency of 50 Hz supply. Find the percentage slip and the frequency of rotor current.

25. A 3-phase, 200 V induction motor has 4-poles. Find the rotor speed if

(*a*) frequency is 60 Hz and slip is 0.05

(*b*) frequency is 50 Hz and slip is 0.04.

26. A 6-pole, 50 Hz, 3-phase induction motor has a rotor resistance of 0.02 Q per phase and the stand still reactance of 0.5 Ω per phase. Determine the speed at which the maximum torque is developed.

27. A 3-phase, 6 pole, 400 V, 50 Hz induction motor has a speed of 930 r.p.m. in full load. Calculate the slip. How many complete alternations will the rotor voltage make per minute?

28. A 3-phase, 6-pole, 40 hp, 380 V star connected induction motor has the following parameters :

$$r_1 = 0.2\ \Omega\ ;\ x_1 = 0.25\ \Omega.$$
$$r_2' = 0.12\ \Omega\ ;\ x_2' = 0.3\ \Omega.$$

At no-load the motor. The stator core losses are 1100 W and rotational losses are 850 W. If slip is 2.3%, find (*a*) Input line current (*b*) Power factor

(*c*) Developed torque (d) Efficiency

29. A 3-phase, 60 Hz, 6-pole induction motor has a full load slip of 0.04. Find frequency of rotor current at the instant of starting and at full load.

30. A 4-pole, 3-phase, 240 V, 50 Hz induction motor has the following parameters of its circuit model referred to stator side are

$$r_1 = 1.3\ \Omega\ ;\ x_1 = 1.2\ \Omega.$$
$$r_2' = 0.4\ \Omega\ ; x_2' = 1.2\ \Omega.$$
$$X_m = 25\ \Omega$$

Rotational losses are 800 W.

(*a*) For a speed of 1440 rpm, calculate the input current, power factor, torque and efficiency.

(*b*) Calculate maximum torque and the slip at which it occurs.

31. The following test results were obtained on a 6.5 kW, 380 V, 4-pole, 50 Hz, delta-connected induction motor with a stator resistance of 2.3 Ω/phase :

No-load	380 V,	4.5 A,	410 Ω
Blocked rotor	130 V,	20 A,	1440 Ω

Calculate the braking torque developed when the motor, running with a slip of 0.04 has two of its supply terminals suddenly interchanged.

32. A 3-phase, 400 V, 50 Hz induction motor takes a power input of 35 kW at its full-load speed of 890 r.p.m. The total stator losses are 2 kW and the friction and windage losses are 1.5 kW. Calculate :

(*a*) slip (*b*) rotor ohmic losses

(*c*) shaft torque (*d*) efficiency

33. The stator of a 3-phase slip-ring induction motor is connected to 50 Hz supply. At the rotor terminals, a frequency of 40 Hz is required. Find the possible speeds at which the rotor must be driven. What is the ratio of slip-ring e.m.fs. at these speeds at no-load condition?

34. A 3-phase induction motor is designed to operate at rated voltage and rated frequency f. In case both source voltage and frequency are respectively changed to

(a) $\frac{V}{2}, \frac{f}{2}$ (b) $V, \frac{f}{2}$ (c) $\frac{V}{2}, f$ (d) $V, 2f$

Find the maximum and starting torques in terms of their rated value neglecting all stator losses.

35. A 380 V, 50 Hz, 3-phase, delta connected, 6-pole induction motor has the following stand still impedances per phase referred to stator :

$$Z_1 = 1.3 + j2.3\ \Omega$$
$$Z_2 = 1.42 + j2.2\ \Omega$$

Neglecting magnetizing impedance, determine

(*a*) goss output power at a slip of 0.04

(*b*) the change in line current when the stator terminals are switched from delta to star at a slip of 0.04.

(*c*) the resistance to be added in each rotor phase to obtain a starting torque of 330 Nm.

12

Single-Phase Induction Motor

12.1 INTRODUCTION

The single-phase motors are manufactured in fractional kilowatt range to be operated on single-phase supply and for use in several application like fans, washing machines, refrigerators, blowers, centrifugal pumps etc. Again single-phase induction motors very small in sizes $\left(\frac{1}{300} \text{ to } \frac{1}{20} \text{ hp}\right)$ are used in toys, hair dryers, vending machines etc. The a.c. series motor, also known as universal motor is very widely used in portable tools, vacuum cleaners and various kitchen equipments. Single-phase motors are less satisfactory compare to three-phase induction motor and it operates at low power factor.

12.2 SINGLE-PHASE INDUCTION MOTOR

A single-phase induction motor is similar to that of three-phase induction motor in physical appearance. The single-phase induction motor has a distributed single-phase winding while in case of 3-ϕ induction motor, the stator has distributed three phase windings. The three phase supply is given to 3-ϕ induction motor. If one phase of 3-ϕ induction motor is opened during the running condition with moderate load, it is observed that the motor continues to run at slightly lower speed.

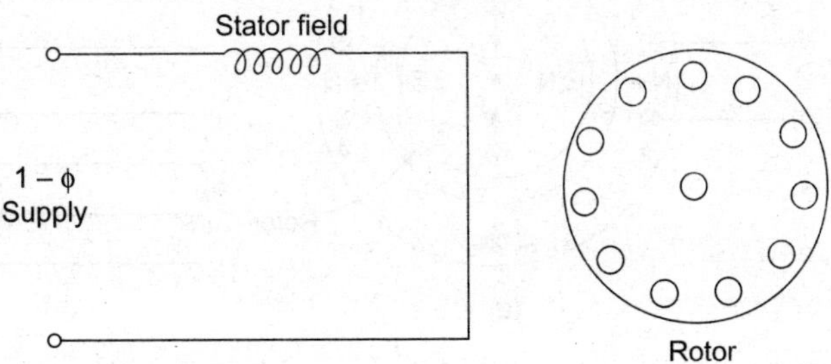

Fig. 12.1 *Single-phase induction motor.*

By opening one of the phase from three-phase, the three-phase motor becomes single-phase motor.

(Conditions in 1-ϕ induction motor during starting) the rotor has short circuited cage winding and the stator and rotor is magnetically coupled having uniform air gap between stator and rotor.

12.3 CONSTRUCTION OF SINGLE-PHASE INDUCTION MOTOR

A single-phase induction motor is similar to the three-phase induction motor except its stator is fed with single-phase supply. The rotor construction is identical to that of a three-phase squirrel cage induction motor. In other words, only three-phase rotor can be interchanged with that of a single-phase rotor.

The stator and rotor of single-phase induction motor are not electrically connected but are magnetically coupled and there is uniform air-gap between the stator and rotor. The stator slots are distributed uniformly, so a single-phase double-layer winding is employed. A simple single phase winding would produce no rotating magnetic field and no starting torque. It is therefore necessary to split the stator winding into two parts, each displaced in space on the stator to make the motor self-starting.

12.4 WORKING PRINCIPLE OF SINGLE-PHASE INDUCTION MOTOR

A single-phase induction motor consists of a stator on which a single-phase winding is provided and a rotor having cage winding. When single-phase supply is given to the stator winding, a pulsating magnetic field is produced. The pulsating magnetic field means the field which is build up in one direction, falls to zero, and then again build up in opposite direction. The axis of this field is stationary in the horizontal direction as shown in Fig. 12.2. The pulsating field will induce an e.m.f. in the rotor conductor by transformer action. Since the rotor has a closed circuit, current will flow through the rotor conductors.

The direction of induced e.m.f. and current in the rotor conductor is shown in Fig. 12.2.

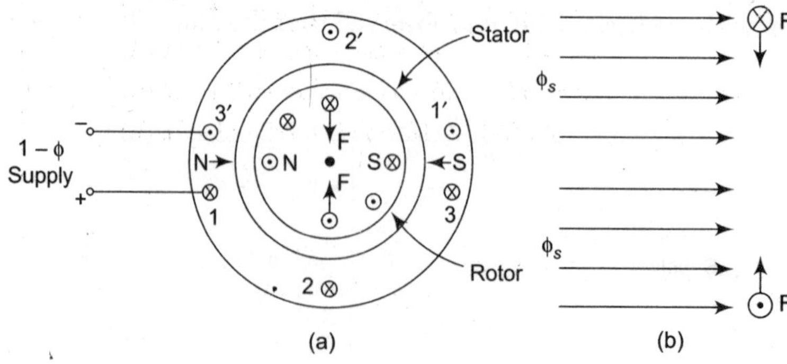

Fig. 12.2 *(a) When rotor is stationary the rotor field produced by transformer e.m.f. aligned with stator field; (b) Direction of force.*

The two sets of forces shown in figure will cancel each other and the rotor will experiences no torque. The torque is zero as the stator and rotor magnetic fields are aligned. Therefore, when a single-phase supply is given to the stator winding, the rotor does not rotate. A starting torque is therefore provided to the rotor which enables the rotor to pick up speed in any direction.

The rotor of single-phase induction motor develops torque when it is given an initial rotation can be explained with the help of two theories namely

(*i*) Cross-field theory

(*ii*) Double revolving field theory.

Both the theories are fairly complicated and has no advantage over each other in numerical calculations. Almost similar results are achieved with both the theories. Both the theories explains how a torque is produced while rotor rotate.

12.5 DOUBLE REVOLVING FIELD THEORY OF SINGLE-PHASE INDUCTION MOTOR

This theory is based on the fact that the pulsating magnetic field produced by the stator winding can be expressed as the sum of two oppositely rotating fields of same magnitude and strength. The magnitude of each of these fields will be equal to one-half of the maximum field strength of the stator pulsating field.

If two sinusoidally distributed mmfs having constant magnitude $\frac{F_{max}}{2}$ are rotating in opposite directions, their combined effect is equivalent to one pulsating field F_{max} cos ωt varying between + F_{max} and – F_{max} as shown in Fig. 12.3.

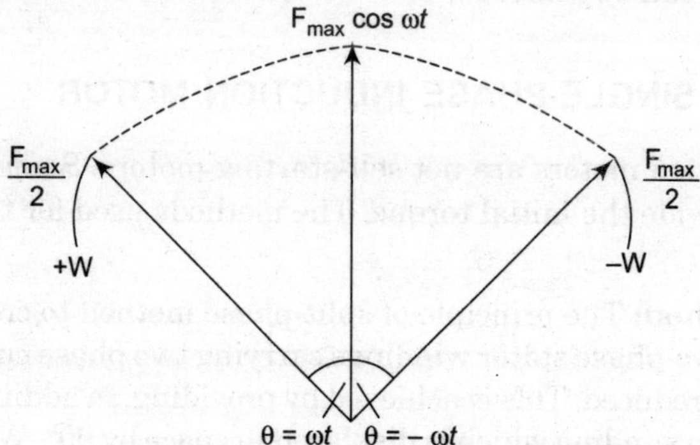

Fig. 12.3 *Double revolving field*

Therefore, a pulsating fields can be resolved into two rotating fields of half amplitude rotating in opposite direction at synchrous speed.

The stator mmf F_s can be written as

$$F_s = F_i \cos \theta \qquad ...(i)$$

where, $\theta \rightarrow$ electrical space angle from stator coil axis

$F_i \rightarrow$ instantaneous value of mmf at the coil axis.

The instantaneous value of mmf at the coil axis is directly proportional to the instantaneous stator current. Therefore,

$$F_i = F_{max} \cos \omega t \quad ...(ii)$$

Putting the value of equation (*ii*) in equation (*i*) we get

$$F_s = F_{max} \cos \omega t \cos \theta \quad ...(iii)$$

The obtained equation (*iii*) can be rewritten as

$$F_s = \frac{1}{2} F_{max} \cos (\theta - \omega t) + \frac{1}{2} F_{max} \cos (\theta + \omega t) \quad ...(iv)$$

From equation (*iv*), it is observed that the two waves each has a peak value of half the amplitude of the pulsating wave travels in opposite direction. The wave having argument $(\theta - \omega t)$ travels in forward direction while wave having argument $(\theta + \omega t)$ travel in reversed direction of θ.

It is therefore, concluded that the stationary pulsating magnetic field can be resolved into two rotating fields of equal magnitude and moving at synchronous speed in opposite direction at the same frequency. The theory based in such a resolution of pulsating field into two rotating fields is called the double revolving field theory.

When the rotor is standstill, the induced voltages are equal and opposite resulting in two equal and opposite torques, which cancel each other. So net torque when rotor is at standstill is zero. If rotor is provided initial rotation by auxiliary means in either direction the torque due to the rotating field acting in the direction of initial rotation will be more than the torque due to other rotating field. Therefore, the motor will develop a net positive torque in the same direction at the initial rotation. Once the motor starts rotating the auxiliary arrangement provided to initialize the rotation can be removed.

12.6 STARTING OF SINGLE-PHASE INDUCTION MOTOR

The single-phase induction motors are not self-starting motors. Some special means must be provided in order to provide the initial torque. The methods used for this purpose is described below:

(*a*) **Split-phase Method:** The principle of split-phase method to create at starting a condition similar to a two-phase stator windings carrying two phase currents so that a rotating magnetic field is produced. This is achieved by providing, in addition to main single-phase winding, a starting winding which is displaced in space by 90°. With the main winding on the stator slots. The resistance and reactance of the two winding circuits are made such that, when connected across a single-phase supply the current flowing through the winding will have a considerable time phase difference. The two currents I_a and I_m flows in auxiliary and main winding and because of the different resistance to reactance ratio in the two windings, the current I_m and I_a are out of phase by about 45° as shown in Fig. 12.4. The motor is equivalent to an unbalanced two-phase motor.

By this two-phase arrangement, a revolving field a created and motor develops the starting torque.

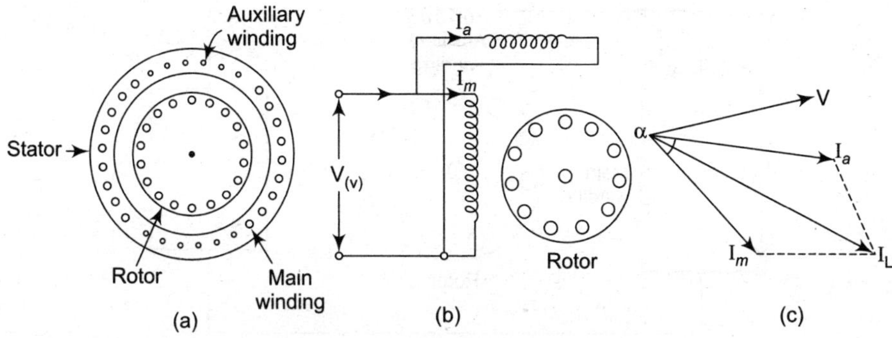

Fig. 12.4 *Split-phase induction motor.*

(*b*) **Capacitor Start Motor:** It has also two windings on the stator that is main winding and auxiliary winding which are space displaced by 90°. The time displacement between the main winding and auxiliary winding current is obtained by inserting an additional capacitor in series with the auxiliary winding. By the use of capacitor of appropriate value the power factor of the motor is improved and also the auxiliary winding current I_a in auxiliary winding at standstill is made to lead I_m current of main winding by about 90°. Thus due to the use of capacitor the current I_a and I_m displaced by 90° resulting much larger starting torque than in resistance start split phase motor. The figure for capacitor start motor is drawn in Fig. 12.5:

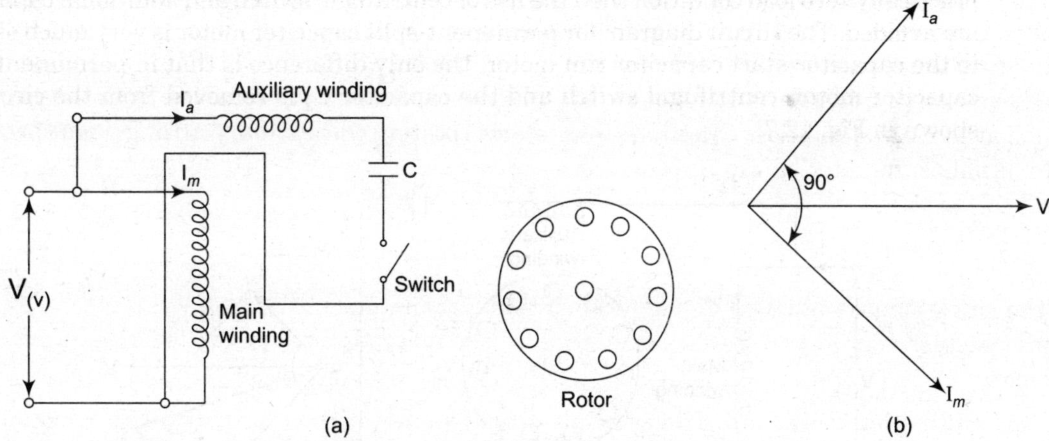

Fig. 12.5 *Capacitor start single phase IM.*

The capacitor and the auxiliary windings are made for the short time duration and the switch used in circuit disconnects them when the motor achieves about 75% speed.

(*c*) **Capacitor-start Capacitor-Run Motor:** In this type of motor the auxiliary winding contain two capacitors which are parallel to each other with a centrifugal switch as shown in Fig. 12.6.

The two capacitors C_1 and C_2 are considered in circuit during starting. After the motor has picked up speed, the centrifugal switch opens thus disconnecting capacitor C_2 from the circuit and the capacitor C_1 and the auxiliary winding remains in the circuit during running condition and the motor work as a balance two-phase motor. The two-phase operation

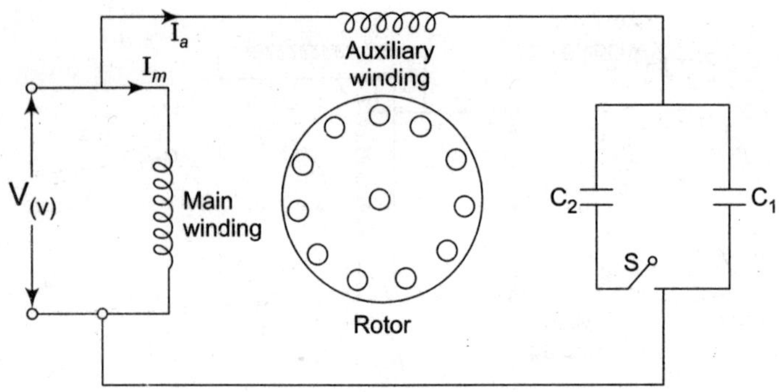

Fig.12.6 *Capacitor-start capacitor 1-phase IM.*

results better efficiency and due to presence of auxiliary winding and capacitor C_1 in the circuit during running condition improves the power factor. The cost of designing this type of motor is higher than capacitor start motor because of the fact that the auxiliary winding and capacitor C_1 is considered in circuit for continuous duty rating. Two capacitors C_1 and C_2 are required due to the reason that the size of capacitor needed for fair starting torque is very much different from the size of capacitor required for best operating condition. Capacitor C_1 has a value of about a few μF while C_2 is much larger in the range of 50 to a few 100 μF. Capacitor C_2 carries current only at starting then disconnect from the circuit.

(*d*) **Permanent-split Capacitor Motor:** In several applications where motor starts under practically zero load condition then the use of centrifugal, switch and additional capacitors are avoided. The circuit diagram for permanent-split capacitor motor is very much similar to the capacitor-start capacitor run motor, the only difference is that in permanent-split capacitor motor, centrifugal switch and the capacitor C_2 is removed from the circuit as shown in Fig. 12.7:

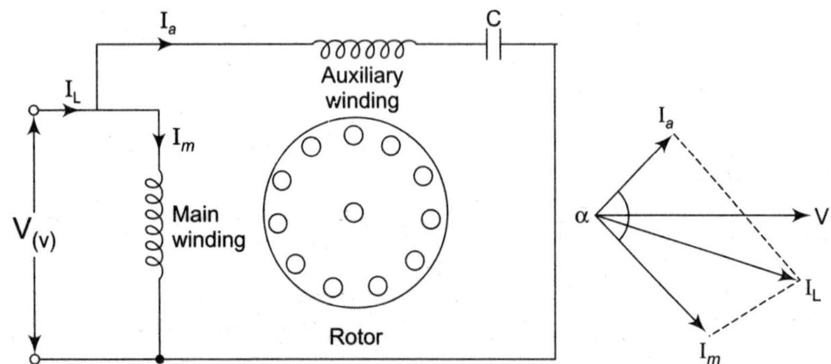

Fig. 12.7 *Parmanent-split capacitor induction motor.*

Both the windings as well as capacitor remain in the circuit during starting and running condition. The current in the auxiliary winding is very small at starting due to the lower value of capacitor. Therefore the starting torque is very small in this motor. This motor is used in fans, blowers and other loads which required small starting torque.

(*e*) **Shaded-pole Motor:** Single-phase induction motors using shaded-poles called shaded pole motors are manufactured for very small ratings and are used in applications where low starting torques are required. Some of the applications of shaded-pole type induction

motors are in fans, blowers, fans used in air heater, slide and film projector, advertisement display device etc. It is less expensive and have low cost. A shaded pole motor has a salient pole construction as shown in Fig. 12.8. Low resistance copper bands called shading bands are placed so as to surround approximately half of each pole face.

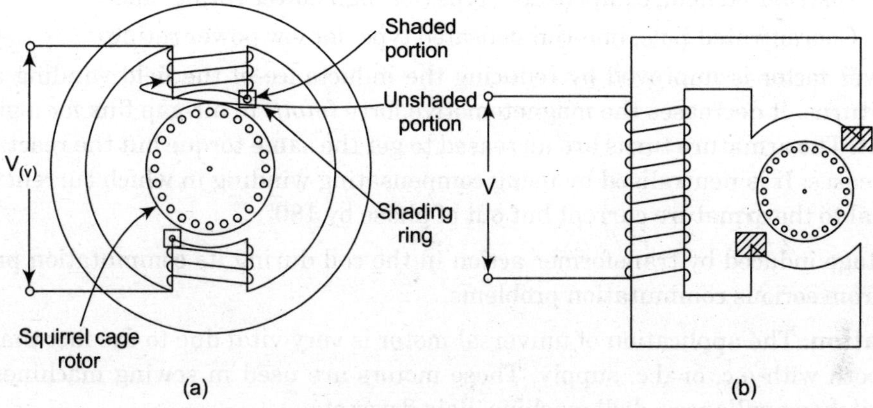

Fig. 12.8 *Shaded pole type induction motor.*

The e.m.f. induced in the shading band causes a current to flow in the band. This current opposes the change of flux creating a time lag between the flux in shaded and unshaded portions of the pole. The flux in shaded portion of pole will lag the flux in other portion. This phenomenon results in rotating magnetic field with rotates from direction unshaded to shaded portion of the pole creating a starting torque.

(*f*) **Reluctance Motor: Reluctance motor works on the principle that when a magnetic material placed in the magnetic field, a force of magnetic pull exerts in the magnetic material tending to bring the piece of magnetic material in the densest portion of the magnetic field. Thus magnetic material align itself at the place where the reluctance is minimum. The diagram for reluctance motor is shown in Fig. 12.9.**

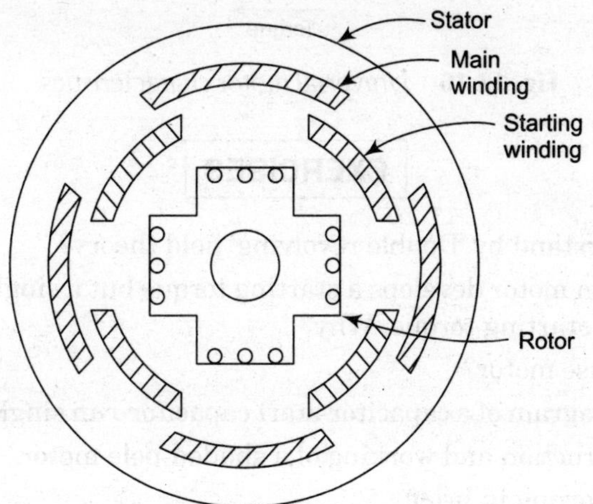

Fig. 12.9 *Reluctance motor.*

(g) **Universal Motor:** The universal motor is that type of motor which operates both with a.c. and d.c. supply. The speed and the output is same either operated with a.c. or d.c. The armature of the universal motor is same as that of ordinary series motor. These are of two types:

(i) Distributed field, compensated type (for high power rating) and

(ii) Concentrated-pole, non-compensated type (for low power rating)

The power factor is improved by reducing the inductance of the field winding and hence number of turns. It decreases the magnetomotive force (*mmf*) in air-gap flux for a given specified current. The armature turns are increased to get the same torque but the reactance of the motor increases. It is neutralised by using compensating winding in which current is directly proportional to the armature current but out of phase by 180°.

The voltage induced by transformer action in the coil during its commutation process and protect it from serious commutation problems.

Application: The application of universal motor is very vital due to the fact that it can be operated both with a.c. or d.c. supply. These motors are used in sewing machines, vacuum cleaners, kitchen appliances, drill machine, hair-dryer etc.

The speed-torque characteristics for d.c. and a.c. input supply is shown in Fig. 12.10:

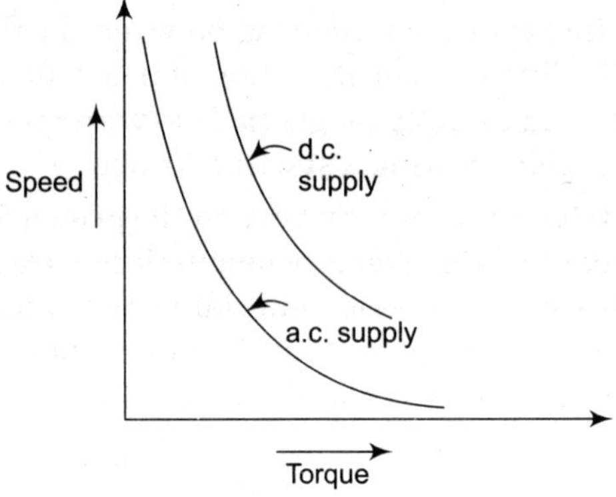

Fig. 12.10 *Universal motor characteristics.*

EXERCISES

1. What do you understand by 'Double revolving' field theory?
2. A 3-phase induction motor develops a starting torque but a single-phase induction motor does not develop a starting torque. Why?
3. What is a split-phase motor?
4. Draw the circuit diagram of a capacitor-start capacitor-run single-phase induction motor.
5. Describe the construction and working of a shaded-pole motor.
6. Write notes on following in brief:

   (a) Starting of single-phase induction motor

   (b) Capacitor-start motor

(*c*) Shaded-pole motor

(*d*) Reluctance motor

(*e*) Universal motor.

7. In what way is a capacitor-start capacitor-run motor different from a permanent split capacitor motor?

8. What is compensating winding? In which motor it is used?

9. Explain the principle and working of shaded-pole motor.

10. Why universal motor is called as universal?

11. List some of the applications of single-phase motors.

12. With reference to double revolving field theory explain why single-phase induction motor is not self-starting.

13. Explain principle and working of following motors:

(*a*) phase-split

(*b*) capacitor-start single-phase motor

(*c*) capacitor-start and capacitor-run single phase induction motor.

14. Explain why universal motor can operate from both d.c. and a.c. supply.

15. Why should the auxiliary winding in a capacitor start motor be disconnected after the motor has picked up speed.

16. Differentiate between reluctance motor and universal motor.

13

Direct Current Machine (Generators)

13.1 INTRODUCTION

d.c. machine is generally an alternating current machine in which the current vary sinusoidally and is furnished with a special device named 'Commutator'. The commutation is the process by which the a.c. is converted into d.c. and vise-versa. The commutation process also helps in keeping the armature mmf field stationary in space.

The d.c. generator is the device, which converts the mechanical energy into electrical energy. It is based on the principal that when the conductor is rotated in a magnetic field, an induced voltage would generate in the conductor. d.c. generators are used only for special applications and local d.c. power generation. It is due to the reason that commutation is required to rectify the internal generated a.c. voltage and thus making of higher rating d.c. power generators is not feasible.

13.2 CONSTRUCTION OF D.C. MACHINE

A d.c. generator mainly consists of four main parts they are: (1) Field windings or field system (2) Armature windings or armature of d.c. machine, (3) Commutator and (4) Brushes shown in Fig. 13.1.

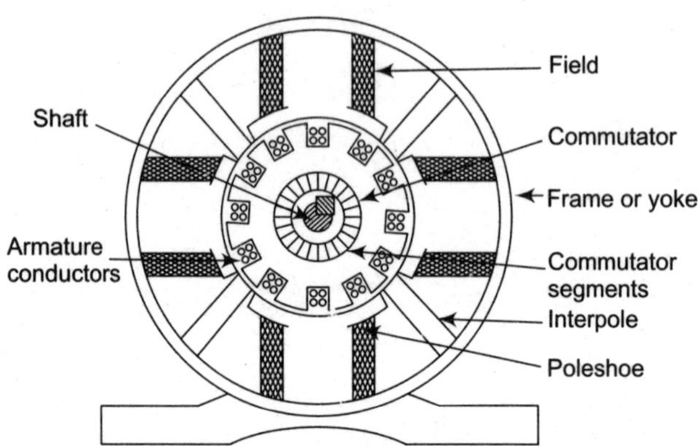

Fig. 13.1 *Main parts of a four-pole d.c. machine.*

Field system

The magnetic field system is stationary part of the d.c. machine. The main purpose of the field system is to produce a uniform magnetic field in the air gap, within which the armature rotates. Electromagnets are preferred in comparison with permanent magnet due to its greater magnetic effect.

The outer frame or yoke is a hollow cylinder of cast steel. Even numbers of poles are bolted to the yoke. The yoke is one of the essential parts of d.c. machine. It is necessary due to the following reason:

(*a*) It provides a path for the magnetic circuit

(*b*) It supports the pole core and acts protective shielding to the machine.

As the poles project inwards they are called as salient poles. Each pole core has a pole shoe having a curved surface. The pole shoe performs the following two requirements: (*a*) It supports the field coils, (*b*) It reduces the reluctance by increasing the cross sectional area of the magnetic circuit.

Armature

The armature is the rotating part of the d.c. machine. The armature of the d.c. machine consists of a shaft upon which a laminated cylindrical core is mounted called armature core. The armature core has slots in its outer surface. The laminations are insulated from each other and tightly clamped together. The purposes of using these laminations is to minimize the eddy circuit loss. The connected arrangement of slots and conductor is called as armature winding. Basically two types of windings are used in d.c. machine, they are lap and wave winding.

Lap Winding

In lap winding, the finish end of one coil is connected to a commutator segment and to the start end of the adjacent. The armature coil is connected to the commutator by the lap winding form. The ends of each armature coil segments connected to the commutator so as to bring the equalization between total number of parallel path and the total number of poles of the machine. Coil under the same pole and similarly all remaining coils are connected.

Let A = Number of parallel paths

P = Number of poles of machine

So, according to the above statement in lap winding

$$A = P$$

It can be remembered easily by the spelling of lap winding. In LAP both A and P is present so, it can be remembered here as $A = P$.

Wave Winding

In wave winding pattern the ends of the armature coils are connected to the segment at same distance apart so as to obtain only two parallel paths. These two parallel paths are obtained between the positive and negative brushes.

For wave winding A = 2
where A = Number of parallel paths.

Commutator

The commutator is one of very important part of d.c. machine which rotates with the armature. The main function of commutator is to convert d.c. to a.c. and vice-versa, the commutator also helps to keep the magnetic flux stationary in space. Generally alternating voltage is produced in the coil, which is rotating in a magnetic field, but direct current is required in the external circuit. For this purpose commutator is needed. Each commutator segment is connected to the ends of the armature coils. The commutator receives the current from the brushes, which are also placed on the rotating armature.

Brushes

Generally the carbon brushes are used in the d.c. machine. The brushes are generally needed to collect from the armature winding. Two or more carbon brushes are placed in the commutator for uniform distribution of current. It distributes the current in both the cycle. Each brush is supported in a metal box called brush box. The current so produced in the armature winding passes through the commutator and then to the external circuit by these brushes.

13.3 BASIC MAGNETIC CIRCUIT OF D.C. GENERATOR

The magnetic circuit of a 4-pole d.c. generator is shown in Fig. 13.2. The dotted lines shows the main flux path. The air gap is the space between the armature and the pole face. The flux so produced by the field winding of a generator links the pole cores, air gap, armature core and the yoke.

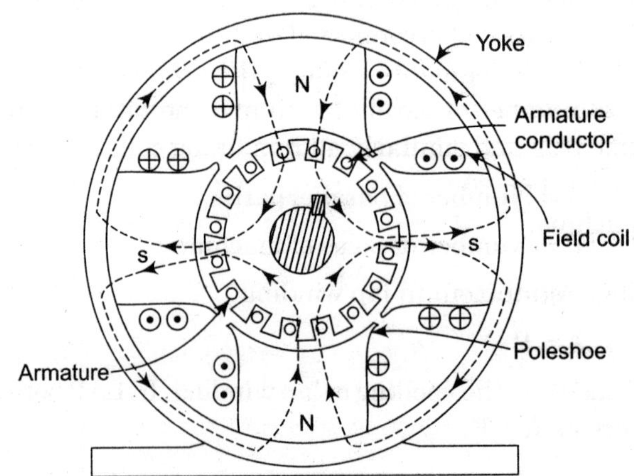

Fig. 13.2 *Magnetic circuit of d.c. generator.*

Generator action: d.c. generators are used only for special applications and the local d.c. power generation. It is due to the reason that Commutator is required to rectify the internal a.c.

voltage and thus making of higher rating d.c. power generators is not feasible. Also d.c. motors are versatile machine, since these provide high starting and decelerating torques.

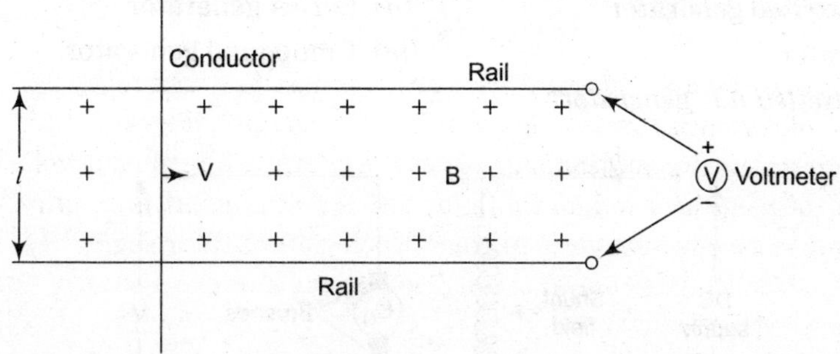

Fig. 13.3 *Generator action.*

Consider a conductor of length '*l*' perpendicular to a uniform magnetic field of flux density 'B' moves with a relative velocity '*v*' with respect to the field, shown in Fig. 13.3.

Then,

B = Magnetic flux density in Tesla (T)

l = Length of conductor in meter (m)

v = Velocity of conductor in the field in m/s

When the conductor motion is not perpendicular to the field, the equation is given by,

$$e = Blv \sin \theta \text{ volts.}$$

13.4 EXCITATION OF D.C. MACHINE

In d.c. machine the field coils or field winding is excited by the current in order to produce the magnetic flux. The production of the useful magnetic flux with the application of electric current to the field windings is generally known as excitation. Generally two types of excitation are possible in a d.c. machine, they are

(*i*) separately-excited d.c. machine.

(*ii*) self-excited d.c. machine.

Separately-Excited D.C. Machine

In this method of excitation the field winding is energized or excited by a separate d.c. source. With the application of separate d.c. source the field coils are energized and produces the magnetic flux. This type of excitation is known as separate excitation.

Self-Excited Machine

In this method of excitation the field coils or field winding is excited or energized by the machine itself. So, the excitation of field winding by machine itself is to produce magnetic flux is known as self-excitation.

The d.c. machines are distinguished according to the connection of field and armature winding. The main types of d.c. machines are categorized below:

(*i*) Separately-excited generator (*ii*) Series generator

(*iii*) Shunt generator (*iv*) Compound generator.

(*i*) *Separately-excited d.c. generator:*

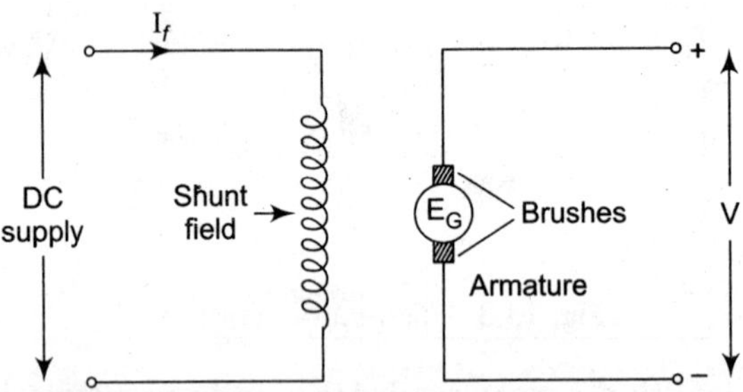

Fig. 13.4 *Separately-excited d.c. generator.*

As we have discussed earlier that in separate-excitation the field winding is energized by a separate d.c. source is produced magnetic flux. The connection diagram for separately-excited d.c. generators is drawn in Fig. 13.4.

In the diagram,

I_f = Field current

V_t = Terminal voltage (armature)

(*ii*) *Series generator:* In series generator, the principal of operation is base on self-excitation method. The circuit or connection Fig. 13.5, for the series generators is given in Fig. 13.5:

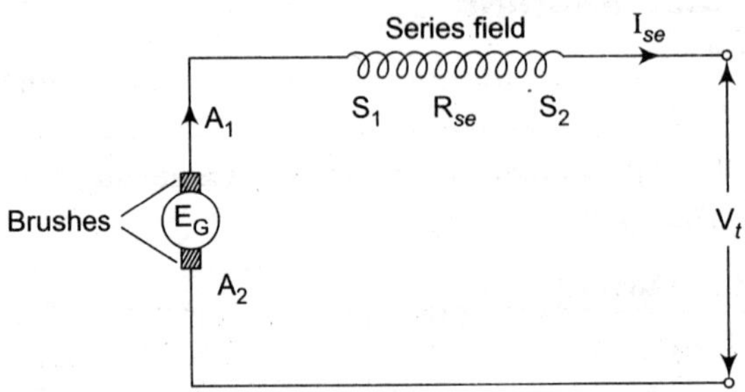

Fig. 13.5 *Series generators.*

where, I_a = Armature current

I_{se} = Series field current

R_{se} = Series field winding

V_t = Terminal voltage

In series generators, the field winding or field coils are in series with the armature windings. The operations of series a generator is based on the armature current, which flows through the series field winding and then magnetic flux is generated. Since the armature current is large, the series field winding consists of few turn of wire of large cross sectional area.

(*iii*) *Shunt generator:* The shunt generator is also self-excited d.c. machine. In this type of machine, the field winding or coils are in parallel to the armature winding. The connection diagram for the shunt generators is drawn in Fig. 13.6.

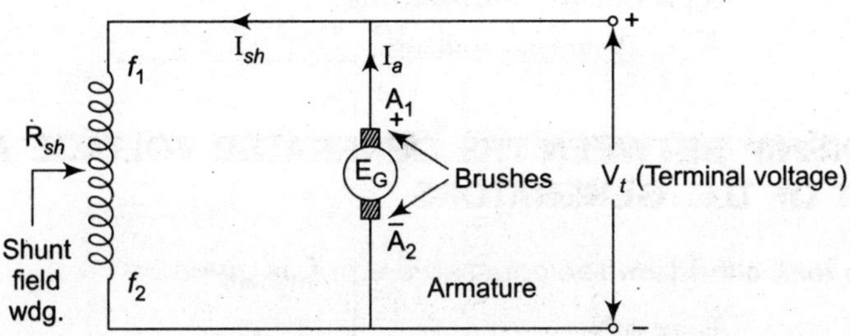

Fig. 13.6 *D.c. shunt generators.*

The shunt field is made of large number of turns of fine wire as it receive the full output voltage and small field winding current as shown in Fig. 13.6.

where R_{sh} = Resistance of shunt field winding

I_{sh} = Shunt field current

V_t = Terminal voltage

I_a = Armature current.

(*iv*) *Compound generator:* This is also self-excited d.c. machine in which, the machine itself excites the field winding. In compound generation both the series and parallel field windings are connected across the armature. The machine having series and field winding both are known as compound generation. The two types of connection are possible in compound generator. They are

(*a*) short shunt compound generation

(*b*) long shunt compound.

The connection diagrams for both the types of connection are shown in Fig. 13.7:

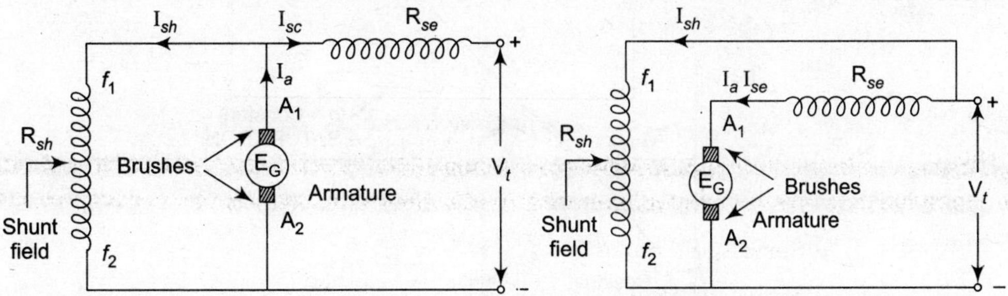

Fig. 13.7 *Short shunt and long shunt compound generators.*

In figure the connection diagram of short shunt compound generator is shown. In short-shunt, the shunt field is connected in parallel with the armature. Whereas in the long-shunt compound generator both series and shunt fields are parallel with the armature.

I_a = Armature current

I_{sh} = Shunt field current

I_{se} = Series field current

R_{sh} = Shunt field winding

R_{se} = Series field winding

V_t = Terminal voltage.

13.5 RELATIONSHIP BETWEEN THE GENERATED VOLTAGE AND EXCITATION CURRENT OF D.C. GENERATORS

At no load: At no load condition the generated e.m.f. is given by

$$E_G = \phi.N$$

or

$$E_G = K\phi N$$

where,

K = Constant of proportionality

ϕ = Magnetic flux

N = Speed of rotation in RPM.

E_G depends directly in the flux which is produced by the ampere-turns of the field coils when the speed is constant. Since the number of turns is constant, the flux depends on only the field current.

When $I_f = 0$ is with the field circuit open circuited a small voltage E_S is appeared due to effect of residual magnetism. When the field current increases the generated e.m.f. also increases linearly up to the knee of the magnetization curves. After this knee point is achieved, increases in I_f causes saturation of magnetic field.

From the above discussion we can conclude that a larger increase in field current is required to obtain desired increase in voltage above the knee point or in saturation region. The magnetization curve for separately excited d.c. machine is drawn in Fig. 13.8.

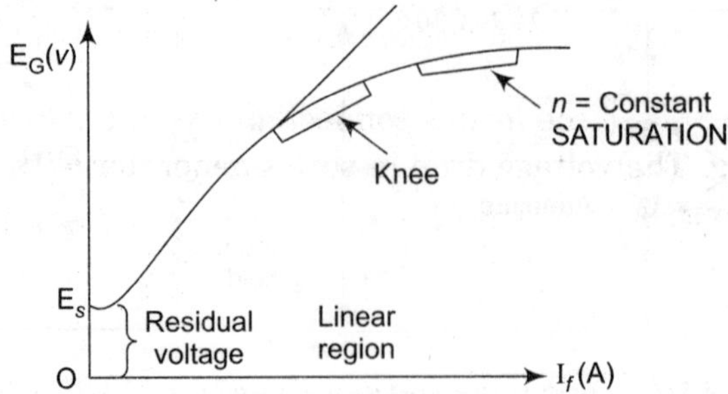

Fig. 13.8 *Magnetization (no load) characteristics of a separately excited d.c. machine.*

In above curves, the curves is not obtained from the origin but shifted a little above on the ordinate, due to the residual magnetism.

13.6 APPLICATION OF LOAD

(*a*) **Separately excited d.c. generator:** Generally the terminal voltage decreases when the armature winding resistance increases and the expression is given as

$$V_t = E_G - I_a.R_a$$

where, V_t = Terminal voltage

I_a = Armature current (load current)

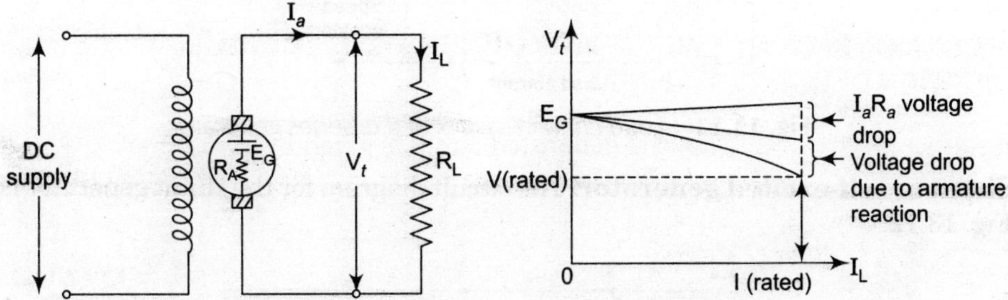

Fig. 13.9 *Load characteristics of a separately excited d.c. generator.*

Another factor for decrease in terminal voltage is the armature reaction. This armature reaction is discussed later on. Due to this the field flux decreases and distorted, which ultimately results in decreases of terminal voltage shown in Fig. 13.9.

(*b*) **Series generator (with load):** The circuit diagram for series generator is as shown in Fig. 13.10:

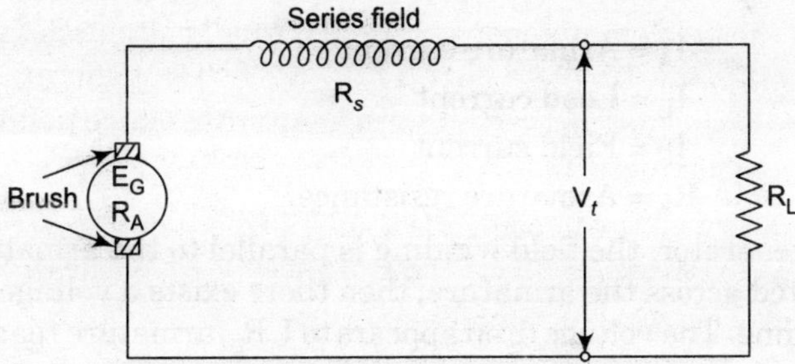

Fig. 13.10 *Circuit diagram for series generator with load.*

In series generator when the load is connected then there is a voltage drop in field and armature winding. The voltage drop in series generator with the application of load is given by expression:

$$V_t = E_G - I_a (R_A + R_S)$$

where, V_t = Terminal voltage

I_a = Armature current

R_A = Armature resistance

R_S = Series field resistance

The load characteristics of d.c. series generator can be drown as shown in Fig. 13.11.

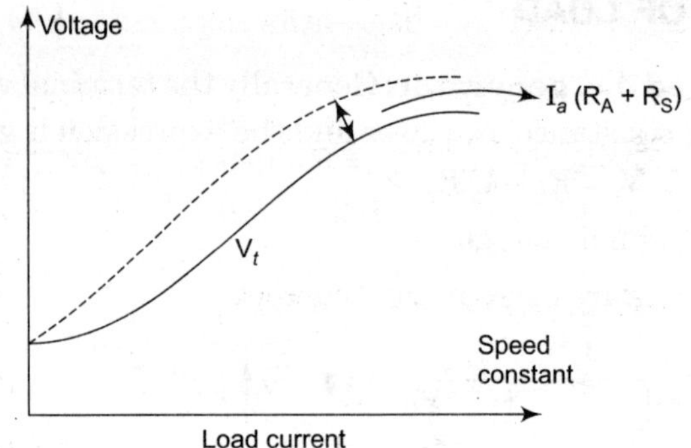

Fig. 13.11 *Load characteristics of d.c. series generator.*

(*c*) **Shunt or self-excited generator:** The circuit diagram for the shunt generator is shown in Fig. 13.12.

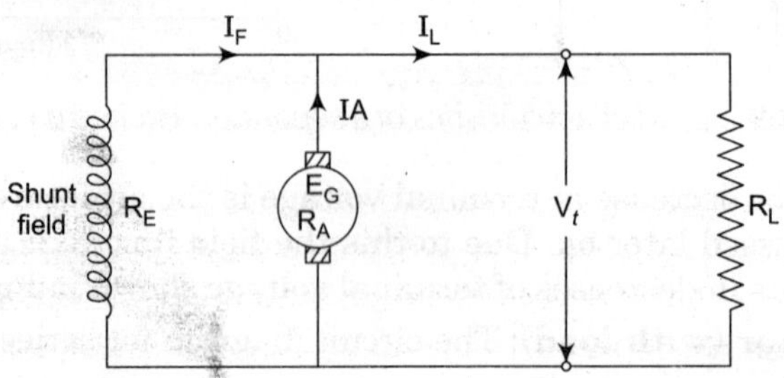

Fig. 13.12 *Circuit diagram for shunt generator.*

where I_A = Armature current

I_L = Load current

I_F = Field current

R_A = Armature resistance.

In the shunt generator, the field winding is parallel to the armature winding. When the load is connected across the armature, then there exists a voltage drop in field as well as armature winding. The voltage drop appears to I_aR_a, armature reaction and weak end flux. The load of characteristics d.c. shunt generator is as shown in Fig. 13.13.

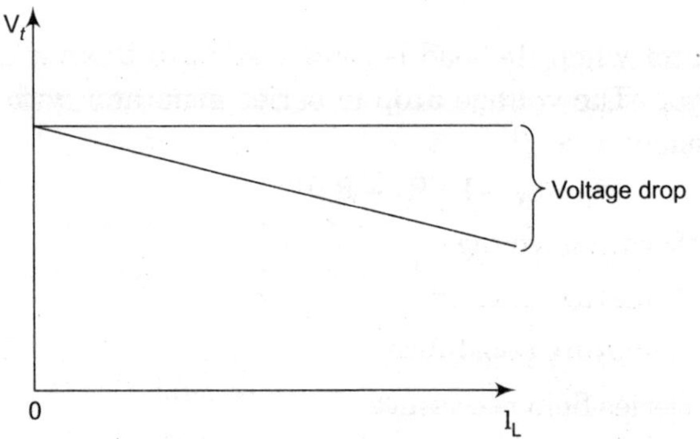

Fig. 13.13 *Load characteristics of d.c. shunt generator.*

(*d*) **Compound Generator:** The circuit diagram for compound generator is shown in Fig. 13.14.

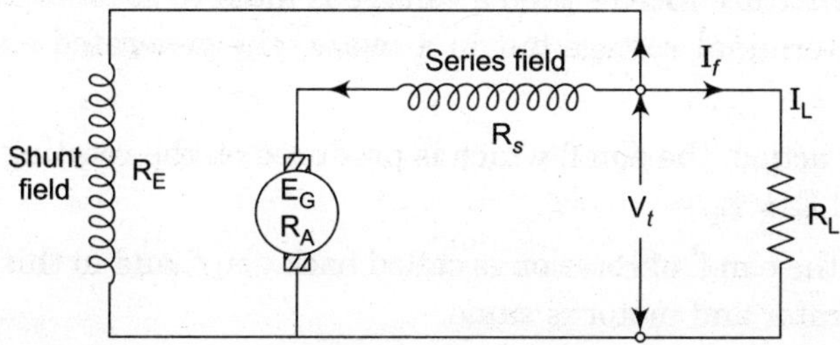

Fig. 13.14 *Circuit diagram for compound generator.*

In d.c. compound generator the shunt and series field is connected in parallel with the armature. When the load is connected across the armature, the voltage drop appears in series, shunt and armature winding (I_AR_A). This voltage drop appears due to weak end flux and distortion of flux path from the original.

The load characteristic of dc compound generation is shown in Fig. 13.15.

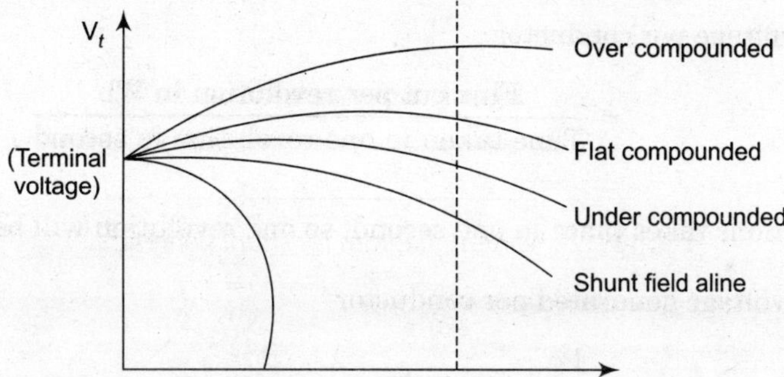

Fig. 13.15 *Load characteristics of d.c compound generator.*

13.7 E.M.F. EQUATION OF D.C. MACHINE

The elementary form of a.c. and d.c. generation is shown in Fig. 13.16.

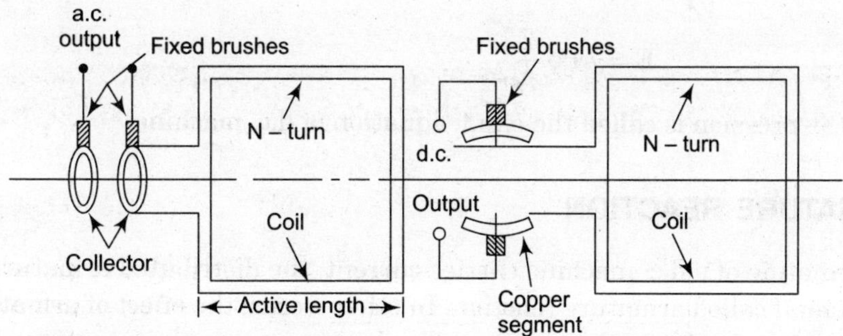

Fig. 13.16 *Elementary form of (a) a.c. generator (b) d.c. generator.*

When the field coils are energized then magnetic flux ϕ generated due to which the armature rotates. When the armature rotates then a'voltage is induced in its coils. In a generator, e.m.f. is greater than the terminal voltage but in a motor, the generated e.m.f. is less than of the terminal voltage.

During generator action, the e.m.f. which is produced on the rotating armature is called the generated e.m.f. and $E_r = E_g$.

In case of motion the e.m.f. of rotation is called back e.m.f. and in this case $E_r = E_b$. The final result for both generator and motor is same.

Let ϕ = Flux in Weber (Wb)

A = Number of parallel paths

Z = Total number of conductors

P = Total number of paths

n = Speed of rotation in revolution per second (rps)

Therefore, $\frac{Z}{A}$ = Number of armature conductors.

So, generated voltage per conductor

$$= \frac{\text{Flux cut per revolution in Wb}}{\text{Time taken in one revolution in second}}$$

Since, n revolutions takes place in one second, so one revolution will be made in $\frac{1}{n}$ second.

Thus, average voltage generated per conductor

$$= \frac{P\phi}{\frac{1}{n}} = P\phi n \text{ volts}$$

Therefore total voltage generated is given by

E = (Average voltage per conductors). (Number of conductors per path)

$$E = P\phi n \cdot \frac{Z}{A}$$

or

$$E = nP\phi \frac{Z}{A}$$

The above expression is called the e.m.f. equation of d.c. machine.

13.8 ARMATURE REACTION

When the armatuie of a d.c. machine carries current, the distributed armature windings produce its own mmf called armature reaction. In other words, the effect of armature mmf on the main field flux distribution in the air gap is also known as armature reaction.

The armature mmf produces two undesirable effects on the main field flux and they are (*i*) Reduction in the main field flux

(*ii*) Distortion of the main field flux wave.

(*iii*) According to the above obtain expression of generated e.m.f.

$$E \propto \phi$$

When the ϕ decreases, the generated voltage and torque also decreases. The figure illustrating the space distribution is shown in Fig. 13.17.

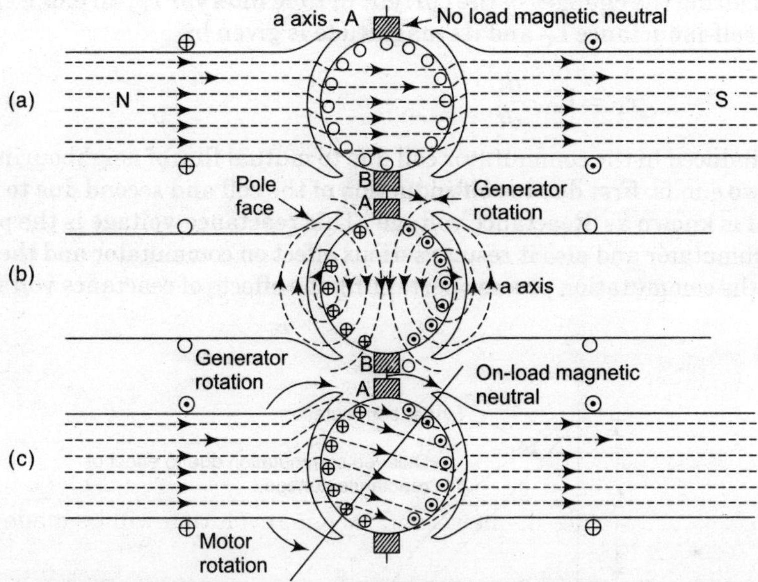

Fig. 13.17 *(a) Main pole flux, (b) Armature flux and (c) Resultant of main pole flux and armature flux.*

13.9 COMMUTATION PROCESS

When the armature coils reaches the brush, it carries current $\left(=\frac{I_a}{a}\right)$ in one direction but soon after the armature coil traversed the brush width, the current in the coil gets reversed to $-\frac{I_a}{a}$. This reversal of current in the armature coil is known as commutation process.

The two main function of commutation is:

(*i*) It converts the unidirectional or direct current into the alternating current in the armature coil and vice-versa.

(*ii*) It also helps in keeping the armature mmf stationary in space.

Generally the current generated in armature of d.c. generator is of alternating nature. When armature conductors are under North Pole, these current flows in the direction and when it faces the South Pole, the flow of current takes place in opposite direction.

When the armature coil passes the brush width the current in the armature coil is reversed. When a brush spans two commutator segments, the winding element connected to those segments is short circuited. This period of short circuit of armature coil by a brush creates the reversal of current and also it brings its maximum value in reverse direction.

The time during short-circuit condition is known as the period of commutation. It is denoted by T_C.

In the commutated coil, the current changes from $\frac{I_a}{a}$ (*i.e.*, I_c) to $-\frac{I_a}{a}$ (*i.e.* $-I_C$) in commutated period T_C. During the changes of the current in time interval T_C, an e.m.f. e_C is induced in the coil due to it self-inductance L_C and its magnitude is given by

$$e_C = L_C \cdot \frac{di}{dt}$$

Also, e.m.f. is induced in the commutator coil due to mutual flux of neighbouring coils. Therefore the sum of two e.m.fs. first due to self-induction of the coil and second due to mutual flux of neighbouring coil is known as 'Reactance voltage'. This reactance-voltage is the principal cause of sparking in commutator and also it results serious effect on commutator and the carbon brush. The diagram for the commutation process pertaining the effects of reactance voltage is shown in Fig. 13.18.

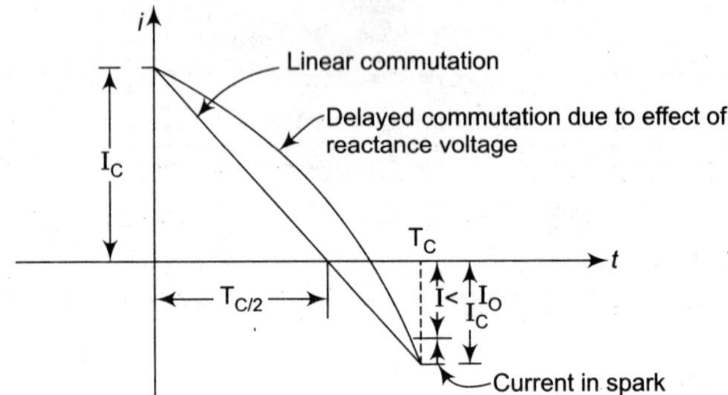

Fig. 13.18 *Delayed commutation (due to reactance voltage).*

13.10 METHOD FOR DECREASING THE EFFECT OF COMMUTATION

The effect of reactance voltage 'e_C' due to which sparking occur can be reduced though not be completely eliminated by the following methods:

(*i*) **By Using Brushes:** The type of brush material affects the commutation process considerably. The various type of brushes are carbon, electro-graphite, copper-graphite etc. Carbon brushes are used for small d.c. machines, while in general electro-graphite is used frequently. Due to the use of the brushes the reactance voltage responsible for sparking is reduced.

(*ii*) **By Using Interpoles:** The interpoles are the auxiliary poles, which are placed between the main poles. The interpoles are fitted to the yoke and are known as commutating pole or compoles. The use of the interpoles also helps in reducing the reactance voltage, which results in reduction of sparking at the commutator. The figure of interpole for d.c. machine is shown in Fig. 13.19.

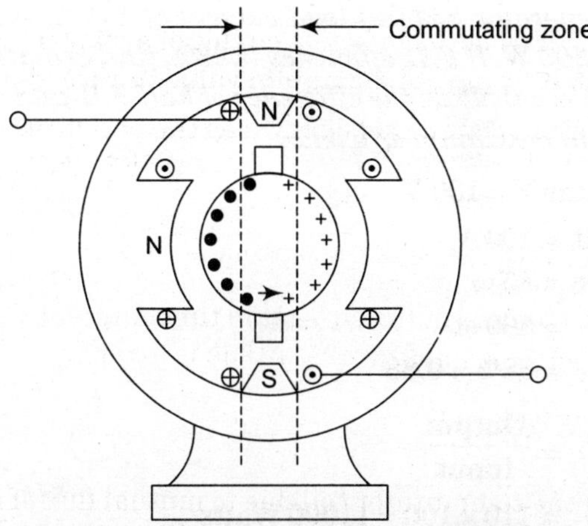

Fig. 13.19 *Inter poles for d.c. machine.*

SOLVED NUMERICAL PROBLEMS

Example 1. *A 10 pole d.c. generator has an armature with the following specifications slots: 120 with six conductor per slot, each capable of taking 100 amp winding lap. If the speed is 900 rmp find the flux per pole to produce a voltage of 250 volts on no-load. If the terminal voltage on full load is 240 volts what is the rated output of the machine?*

Solution: Number of poles = 10

Slots = 120

Conductors/slot = 6

Capacity of conductor = 100 A

Speed = 900 rpm

Voltage = 250 V

∴ Total number of conductors, Z = 6 × 120 = 720

We know e.m.f. $E_g = \dfrac{\phi ZN}{60} \times \dfrac{P}{A}$

Since it is lap wound $E_g = \dfrac{\phi ZN}{60}$

∴ $\phi = \dfrac{E_g \times 60}{ZN} = \dfrac{250 \times 60}{720 \times 900} = 2.31 \times 10^{-2}$ Wb

Terminal voltage = 240 V

Since the number of poles = 10 = Number of parallel path

Fixed current = Current carrying capacity of conductor × 10

= 100 × 10 = 1000A

∴ Rated power = 1000 × 240 = 240 kW. **Ans.**

Example 2. *A 110 V shunt generator has full load current of 100 A, shunt field resistance of 55 ohms and constant losses of 500 W. If F.L. efficiency is 88%, find armature resistance. Assuming voltage to be constant at 100 V, calculate the efficiency of half F.L. and at 505 overload. Find the load current corresponding to maximum efficiency.*

Solution: Voltage of machine V = 100 V

I_L – Load current = 100 A

R_f – Shunt field resistance = 55 ω

Stray losses = 500 ω

F.L. efficiency = 88% = 0.88

$$\text{Efficiency} = \frac{\text{Output}}{\text{Input}}$$

$$\text{Output} = 110 \times 100 = 11000 \text{ Watts}$$

$$\therefore \text{ Input} = \frac{11000}{0.88} = 12500 \text{ Watts}$$

$$\therefore \text{ Losses} = \text{Input} - \text{Output} = 12500 - 11000 = 1500 \text{ W}$$

Losses = Stray losses + Shunt field Cu losses + Arm Cu losses

$$\text{Shunt field Cu loss} = V \times I_{sh} = 110 \times \frac{110}{55}$$

$$= 220\text{W}$$

$$\therefore \text{ Cu losses} = 1500 - 500 - 220 = 780 \text{ Watts}$$

Let us assume shunt field current is negligible compared with armature current.

$$\therefore \text{ Cu losses} = 780 \text{ W} = I^2 r$$

$$\therefore \text{ Armature resistance} = \frac{\text{Cu losses}}{I^2} = \frac{780}{(100)^2} = 0.078 \text{ W}$$

If the load is half,
$$I = \frac{1}{2} = \frac{100}{2} = 50 \text{ A} = I_a$$

$$\therefore \text{ Cu losses} = Ia^2 R_a = (50)^2 \times 0.078 = 195 \text{ W}$$

$$\text{Losses} = 195 + 500 + 200 = 915 \text{ W}$$

$$\text{Output} = 50 \times 110 = 5500 \text{ W}$$

$$\therefore \ \% \quad n = \frac{\text{Output}}{\text{Output} + \text{Losses}} = \frac{5500}{5500 + 915} \times 85.73\%$$

If the machine is 50% overloaded

then
$$I = 1.5\, I = 150 \text{ A}$$

$$\therefore \text{ Cu losses} = I^2 a\, R_a = (150)^2 \times 0.078 = 1755 \text{ W}$$

$$\therefore \text{ Losses} = 1755 + 500 + 220 = 2475 \text{ W}$$

$$\therefore \ \% \quad n = \frac{\text{Output}}{\text{Output} + \text{Losses}} = \frac{110 \times 150}{(110 \times 150) + 2475} = 86.95\%$$

Condition for maximum efficiency is Cu losses = Constant losses

Constant losses = Stray losses + Shunt Cu losses

= 500 + 220 = 720

$$I = \sqrt{\frac{\text{Losses}}{R}} = 96.1\text{A. } \textbf{Ans.}$$

Example 3. *A pole, wave wound, 220 V d.c. shunt generator has a full load current of 50 A and a shunt field current of 2 A. Find the demagnetising and cross-magnetising ampere turns per pole if the brushes are advanced through 3 commutator segments at full-load. There are 125 segments in the commutator of the machine. Assume the machine to have single turns coils.*

Solution: Number of poles = 6 pole wave wound

Terminal voltage $V = 220\text{ V}$

Load current $I_a = 50\text{ A}$

Field current $I_f = 2\text{ A}$

Number of commutator segments = 125

The machine has single turn coil.

∴ Advance of brushes = 3 commutator segment

$$\therefore \quad \theta_m = \frac{360° \times 3°}{125} = 8.69°$$

2 × Number of commutator segments = Number of conductors.

Since every commutator segment will have two conductors

$$Z = 2 \times 125 = 250$$

Armature current $I_a = I + I_f = 50 + 2 = 52\text{ A}$

$$\therefore \quad \text{Number of AT/pole} = Z \times I \times \frac{\theta_m}{360}$$

where I = Armature current/parallel path $= \frac{52}{2}$

Since it is wave winding,

$$\therefore \quad \text{AT/Pole} = 250 \times \frac{52}{2} \times \frac{8.64}{360} = 156 \text{ AT/Pole.}$$

$$\text{Similarly, AT/Pole (cross-magnetising)} = ZI\left(\frac{1}{2P} - \frac{\theta}{360}\right)$$

$$= 250 \times \frac{52}{2}\left(\frac{1}{2 \times 6} - \frac{8.64}{360}\right) = 385 \text{ AT/Pole. } \textbf{Ans.}$$

Example 4. *A 10 pole lay wound be d.c. generator has 700 armature conductors. The total armature current is 1600 A. Find the number of compensating winding conductors in each pole face to give full armature reaction compensation if each pole covers 70% of the pole pitch.*

Solution: Number of total armature conductors/pole $= \frac{700}{10} = 70$

Number of active armature conductor/pole = 0.7 × 70 = 49

Number of parallel = 10

Number of compensating winding conductors = $\frac{49}{10} = 4.9$

≈ 5 Cadin/pole. **Ans.**

Example 5. *Two generators (shunt) are together supplying 100 a to a load. Generator 1 gives 400 V on no load and 300 V when supplying 100 A. Generator second gives 400 V on no load 320 V when supplying 100 A. Find the current supplied by each and the voltage across load.*

Solution: Let I_1 be the current supplied by first generator. Then the current supplied by second generator is $100 - I_1$. The terminal voltage V is given by

$$V = 400 - \frac{I_1}{100}(400 - 300)$$

Also $$V = 400 - \frac{100 - I_1}{100}(400 - 300)$$

Solve above two eqns.

$$100\,V - 40000 + I_1(100) = 0$$

$$100\,V + 100\,I_1 - 40000 = 0 \quad ...(i)$$

and second

$$100\,V - 40000 + (100 - I_1)(80) = 0$$

$$100\,V - 40000 + 80000 - 80\,I_1 = 0$$

$$100\,V - 80\,I_1 - 32000 = 0 \quad ...(ii)$$

Subtract eqns. (*i*) and (*ii*)

$$100\,I_1 + 80I_1 - 40000 + 32000 = 0$$

$$180\,I_1 = 8000$$

$$I_1 = \frac{8000}{180} = 44.44$$

$$V = 355.56$$

Current supplied by second generator = $100 - I_1 = 65.66$. **Ans.**

Example 6. *A series generator having a total resistance (i.e., armature and series field winding) of 0.5 ohm is running at 1000 rpm and is delivering 5 kW at a voltage of 100 V if the speed is rauised to 2000 rpm and load adjusted to 10 kW find the new armature current and terminal voltage.*

Solution: Since the armature and field currents are the same let this current be denoted by I.

Initial value of $I = \frac{500}{100} = 50A$

Initial generated e.m.f = $100 + 50 \times 0.5 = 125$ V

$$\text{e.m.f.} \times (N)\,I = kNI$$

$$125 = R \times 1000 \times 50$$

$$k = \frac{1}{400}$$

New condition $I = \frac{10000}{V}$

Generated e.m.f. $= V + \frac{10000}{V} \times 0.5$

This must be equal to $k \times 2000 \times \frac{10000}{V}$

$$\frac{1}{400} \times 2000 \times \frac{10000}{V} = V + \frac{4000}{V}$$

hence $V = 214.4$

$$I = \frac{10000}{214.4} \; 45.53. \textbf{ Ans.}$$

EXERCISES

1. Explain the construction and working of a d.c. machine and also list their function.

2. Describe the construction of armature, field magnet and commutator of a d.c. machine.

3. Derive the equation for e.m.f. induced in a d.c. machine.

4. What is 'commutation process'? Explain the various function of commutator in a d.c. machine.

5. (*a*) What is the function of interpole in a machine?

(*b*) What happens if the interpoles winding is disconnected?

(*c*) What can be the effect if the interpoles produce very strong commutating field?

6. Explain the phenomenon armature reaction with suitable diagram.

7. What is compensating winding? What is its function? Where is it housed? How is it connected?

8. Discuss the effect of armature reaction is d.c. generator.

9. Explain the difference between demagnetising and cross-magnetising armature mmf in d.c. machine.

10. Describle the process self-excitation in a d.c. machine. What condition must be fulfilled for the machine to self-exite?

11. Draw and label the connection diagram of separately excited, shunt, series and compound generator.

12. What is difference between long shunt and short shunt compound generator?

13. Distinguish between the internal and internal characteristics be derived from the external characteristics of d.c. generator.

14. Explain the resistance commutation in brief.

15. Draw the circuit model of d.c. generator and obtain the expressing for induced e.m.f. in some.

16. What is the basic difference between lap and wave winding?

17. What are the various methods which improves the commutation process?

18. A shunt generator produces 450 A at 230 V. The resistance of the shunt field and armature are 50 W and 0.025 ohm respectively. Find out the armature voltage drop.

19. A 2-pole generator with wave wound armature has 51 slots each having 24 conductors. The flux per pole is 0.01 Wb. Find speed at which the generator should run.

20. In a 10-pole, lap wound d.c. generator, the number of active armature conductors per pole is 50. Obtain the number of compensating conductor per pole required.

21. A 100 kW, 250 rpm, d.c. generator has 6 poles. The armature has 83 slots each holding 8 conductors, and the winding is wave wound. Find out the no-load flux per pole at which and e.m.f. of 500 volts is induced.

22. A 4-pole, 1200 rpm generator with a lap wound armature has 65 slots and 12 conductors per slot. The flux per pole is 0.02 Wb. Calculate the e.m.f. induced in the armature.

23. A 4-Pole, 1200 rpm, lap wound d.c. generator has 760 conductors. If the flux per pole is 0.02 Wb. Calculate the e.m.f. generated.

24. In a d.c. machine if p is the number of poles, N is the armature speed is rpm then obtain the frequency of magnetic reversals.

25. In a d.c. generator running at 1600 rpm gives 240 V d.c. If the speed is propped to 1400 rpm without change in flux. Calculate the new e.m.f.

26. If the no-load voltage of a certain generator is 220 V and the rated load voltage is 200 V. Calculate the voltage regulation.

27. A shunt generator running at 600 rpm has an induced e.m.f. of 200 Volts. If the speed increases to 750 rpm. Calculate the induced e.m.f.

28. A 4 pole d.c. generator has 490 armature conductors and an armature current of 140 A. The brushes have been given a load of 10° actual. Find out the demagnetising AT/Pole if the machine (a) lap wound and (b) wave wound.

14

D.C. Machine (Motor)

14.1 INTRODUCTION

A motor is a type of machine in which the electrical energy is converted into mechanical energy. In both the cases that is generator and motor and induced e.m.f. is generated in armature coil. The e.m.f. generated in case of generator is greater than that of terminal voltage V_t while e.m.f. generated in motor (back e.m.f.) is less than that of terminal voltage. From the constructional point of view, both generator and motor is very much similar.

14.2 COUNTER E.M.F. OR BACK E.M.F. IN D.C. MOTORS

When a voltage is applied to the d.c. motors the current will start to flow into the positive brush, through the commutator into the armature winding. Like generators, the motor has also armature winding. The armature windings of generators and motor are identical to each other. Therefore conductors under the North Pole field carries current in one direction while all conductors under the South Pole field carries current in opposite direction.

When the armature carries a current it will produce a magnetic field of its own, which will interact with the main field. As a consequence a force developed on all the conductors, which tends to turn the armature. The figure for the armature current direction for clockwise rotation and resulting direction of counter e.m.f. in a d.c. motor is shown in Fig. 14.1.

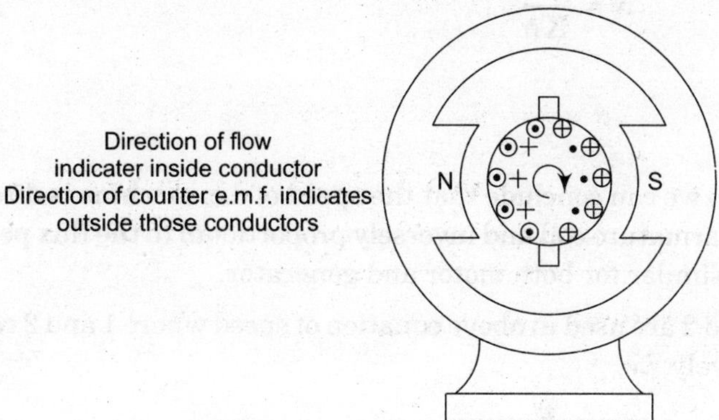

Fig. 14.1 *Back e.m.f. direction.*

Due to motor action the armature rotates and the armature conductors continuously cut through this resultant field. Therefore voltage generated in very same conductors which experiences force.

If no load is connected to the motor then the back e.m.f. or counter e.m.f. will be nearly equal to the applied voltage. The power developed by the armature I, in this case is just that power needed to overcome the rotational losses. All this mean that the armature current I_A is controlled and limited by the counter e.m.f. E_C, therefore

$$I_A = \frac{(V_L - E_C)}{R_A}$$

where V_L is applied voltage across the armature winding and RA is the resistance of armature winding.

14.3 SPEED OF D.C. MACHINE

The speed of d.c. machine is expressed in rpm or rps (revolution per minute/second) the generated e.m.f. equation of a d.c. machine is given

$$E = \frac{PN\phi Z}{60\ Z} \qquad ...(i)$$

where E = Generated e.m.f. in volts

P = Total member of poles

Æ = Flux in Weber

Z = Total number of conductor and

A = Number of parallel path

Now from above equation of e.m.f., we get

$$N = \left(\frac{60\ A}{PZ}\right), \frac{E}{\phi} \qquad ...(ii)$$

or

$$N = \frac{E}{K\phi} \qquad ...(iii)$$

where $K = \frac{PZ}{60A}$

From equation (*iii*) we can conclude that the speed of d.c. machine is directly proportional to the e.m.f. induced in armature coil and inversely proportional to the flux per pole in Weber. The equation of e.m.f. is similar for both motor and generator.

If the suffixes 1 and 2 are used in above equation of speed where 1 and 2 represents initial and final values respectively, *i.e*.

$$N_1 = \frac{E_1}{K\phi_1} \qquad ...(i)$$

$$N_2 = \frac{E_2}{K\phi_2} \quad ...(ii)$$

Dividing equation (*i*) by equation (*ii*), we get

$$\frac{N_1}{N_2} = \left(\frac{E_1}{E_2}\right)\left(\frac{\phi_1}{\phi_2}\right).$$

14.4 TORQUE PRODUCTION IN D.C. MACHINE

The electromagnetic torque produced in d.c. machines is expressed in terms of interaction between main field flux ϕ_f and armature mmf F_a. The torque is the very important parameter of the d.c. machine and can be derived as under. The voltage equation of a d.c. motor is

$$V_t = E + I_a R_A \quad ...(i)$$

Multiplying the above equation with armature current I_a, we get

$$V_t I_A = E\,I_a + I_a^2 R_A$$

$$V_t I_a = \text{Electrical power input to the armature + Ohmic losses}$$

where $I_a^2 R_A$ = Ohmic loss in the armature

Let T_e be the electromagnetic torque developed in the armature circuit.

Mechanical power developed by armature,

$$P_m = \omega T_e = 2\pi n T_e \quad ...(ii)$$

Also we know that,

$$P_m = EI_a = \omega T_e = 2\pi n T_e$$

where, P_m = Mechanical power

Since, $E = \dfrac{nP\phi Z}{A}$, putting this value in equation (*ii*), we get

$$\left(\frac{nP\phi Z}{A}\right) I_a = 2\pi n T_e$$

or

$$T_e = \left(\frac{nP\phi Z}{2\pi n A}\right) I_a = \left(\frac{P\phi z}{2\pi A}\right) I_a$$

or

$$T_e = \left(\frac{PZ}{2\pi A}\right) \phi I_a \quad ...(iii)$$

The above equation is electromagnetic torque equation of d.c. machine.

$$T_e = K\phi I_a \quad ...(iv)$$

where, $K = \dfrac{PZ}{2\pi A}$

Hence from the above equation we can conclude that the electromagnetic torque T_e developed in d.c. machine is directly proportional to the product of the flux and the current flowing in armature *i.e.*, $T_e \propto \phi I_a$.

In both d.c. generator and d.c. motor the torque equation is same.

14.5 CLASSIFICATION OF D.C. MOTORS

D.c. motors are broadly classified into three categories. They are,

(*i*) d.c. series motor

(*ii*) d.c. shunt motor

(*iii*) d.c. compound motor

(*i*) **d.c. Series Motor:** In d.c. series motor the field winding is connected in series with the armature winding which is shown in Fig. 14.2.

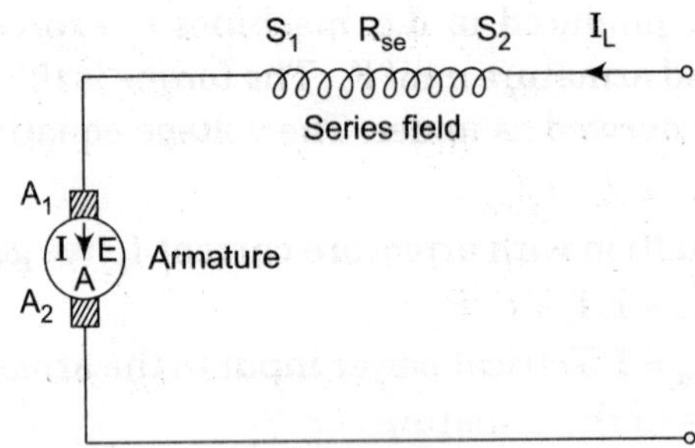

Fig. 14.2 *Circuit diagram of d.c. series motor.*

In series motor the load current is equal to the armature current *i.e.*,

$$I_L = I_A$$

By Kirchhoff's current law, for the above circuit

$$I = I_L = I_A \quad ...(i)$$

where I_f = Series field current

I_A = Armature current and

I = Total resultant current in the motor.

The terminal voltage V_t can be written as

$$V_t = E + I\,(R_A + R_f) \quad ...(ii)$$

Multiplying (*i*) and (*ii*) we obtain the power expression of the d.c. series motor and can be given by

$$V.I = EI + I^2(R_A + R_f) \quad ...(iii)$$

So, the power input = Mechanical power developed + Losses in the armature winding + Losses in field winding.

or

$$V.I = P_M + I^2 R_A + I^2 R_f \quad ...(iv)$$

From equations (*iii*) and (*iv*), it is clear that the

Power input in a d.c. series motor is equal to the mechanical power developed if the losses are neglected.

For d.c. series motor torque is directly proportional to the product of flux and the armature current and is given by

$$T \propto \phi I_A$$

where T = Torque of d.c. motor, N-m

ϕ = Flux produced, Wb

I_A = Armature current, amp

Therefore $$\frac{N_2}{N_2} = \left(\frac{E_{b2}}{E_{b1}}\right) \cdot \left(\frac{\phi_1}{\phi_2}\right)$$ ...(*iv*)

Prior to the saturation the flux produced is directly proportional to the armature current.

$$\phi = I_A$$ ...(*v*)

Therefore equation (*v*) can be rewritten as

$$\frac{N_2}{N_2} = \left(\frac{E_{b2}}{E_{b1}}\right) \cdot \left(\frac{I_{A1}}{I_{A2}}\right)$$ ...(*vi*)

So consequently from (*v*) and (*vi*), we get the torque is directly proportional to the square of armature current *i.e.*

$$T \propto I_a^2$$

where N = Speed of rotation in rpm

E_b = Back e.m.f. produced in armature in volts.

(*ii*) **d.c. Shunt Motor:** In d.c. shunt motor the field winding is connected is parallel with the armature winding. The circuit diagram is as shown in Fig. 14.3.

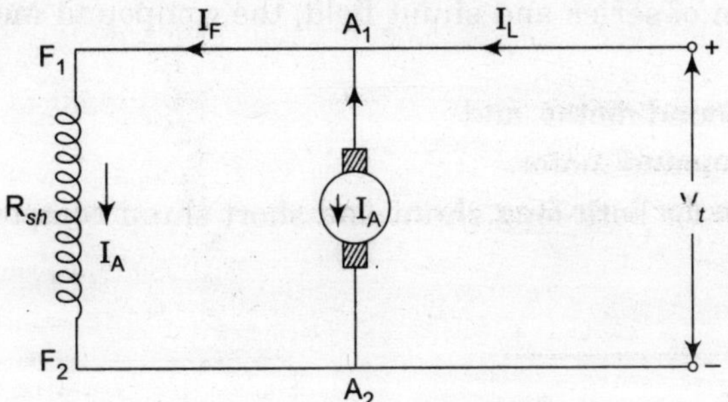

Fig. 14.3 *Circuit diagram of d.c. shunt motor.*

By KCL at node *n*, we get

$$I_L - I_{sh} - I_A = 0$$

or $$I_L = I_{sh} + I_A$$ ...(*i*)

where I_L = Load current in Amp

I_{sh} = Shunt field current in Amp

I_A = Armature current in Amp

and the voltage equation can be given by the following equation

$V_t = I_{sh} R_{sh}$. For, shunt field winding,

and $V_t = E + I_A R_A$. For, armature winding circuit

Now power expression can be given by

Power input = Mechanical power developed + Losses in the armature + Losses in the field

$$V_t I = P_m + I_A^2 R_A + I_{sh}^2 R_{sh}$$
$$= P_m + I_A^2 R_A + V_t I_{sh}$$

Since $V_t = I_{sh} R_{sh}$

or $P_m = V_t I - V_t I_{sh} - I_A^2 R_A = V_t(I - I_{sh}) - I_A^2 R_A$

$$= V_t I_A - I_A^2 R_A = (V_t - I_A R_A) I_A.$$

Therefore, $P_m = EI_A$

So, mechanical power developed is equal to the product of e.m.f. induced in motor and the current flowing in the armature (all losses are neglected).

For d.c. shunt motor, the flux $\phi_1 = \phi_2$

Therefore, $$N_2 N_1 = \frac{E_{b2}}{E_{b1}} \quad ...(i)$$

From above expression, we obtain that the torque produced in motor is directly proportional to that of armature current.

$$T_a \propto I_a$$

or $$T_a = KI_a \quad ...(ii)$$

(*iii*) **d.c. Compound Motor:** In d.c. compound motor both series and shunt field are used. According to connection of series and shunt field, the compound motors are classified in two category:

(*a*) Long shunt compound motor, and

(*b*) Short shunts compound motor.

The circuit diagrams for both long shunt and short shunt compound motors are shown in Fig. 14.4.

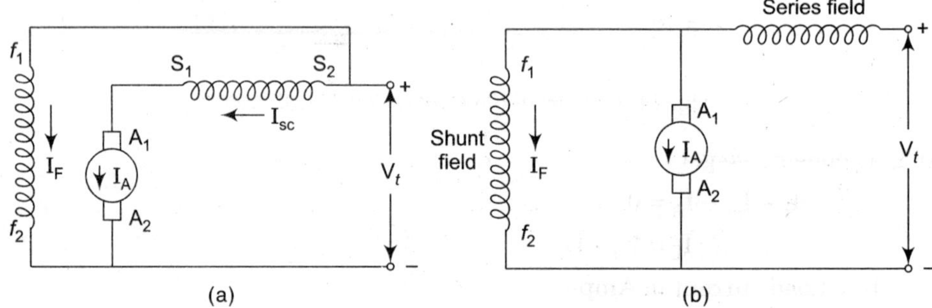

Fig. 14.4 *d.c. compound motor (a) Long shunt and (b) Short shunt.*

The long shunt compound motor is similar to the long shunt compound generator. Hence in case of motor,

$I_{se} = I_A$ *i.e.* the series field current is equal to the armature current.

and $I_L = I_a + I_{sh}$

where I_L = Load current

I_A = Armature current

I_{sh} = Shunt field winding current.

The d.c. compound motor may be cumulatively compounded or differentially compounded.

14.6 STARTING OF D.C. MOTOR

Why we need starter in d.c. motor?

In d.c. motor we know that,

$$V_t = E_b + I_A R_A$$

where V_t = Terminal or supply voltage in volts

E_b = Back e.m.f. of motor in volts

I_A = Current flowing in armature conductor in ampere

R_a = Armature resistance in ohm.

When the motor is at rest the back e.m.f. E_b is equal to zero. With the application of voltage to the stationary armature, the armature of the d.c. motor draws a very large current because a very low resistance of the armature. Hence during starting we apply voltage from a low value, which is increased gradually to rated voltage. Also for overcoming this starting problem some resistance is inserted in series with the armature circuit to offset for the absence of back e.m.f.

The starter is used in d.c. motor for the reason listed below:

(*i*) Causes heavy sparking at the brushes and may destroy the commutator and brush gear.

(*ii*) Large torque is developed causing mechanical shock to the shaft.

(*iii*) Such heavy current cannot be drawn from the supply.

Starter for shunt motor: A starter used for a shunt connected d.c. motor is shown in Fig. 14.5. In the Fig. 14.5, L, F and A denote line, field and armature respectively.

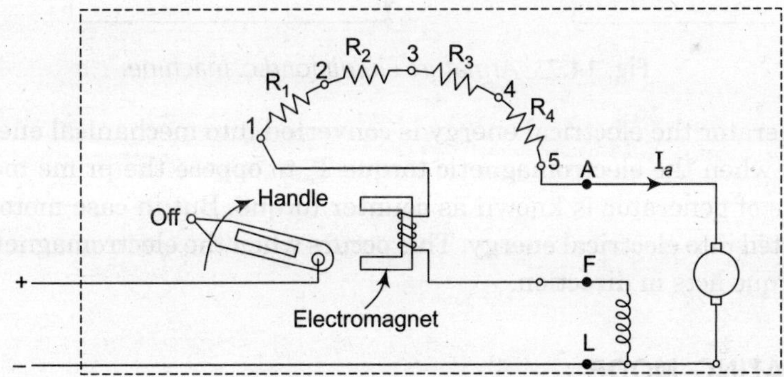

Fig. 14.5 *d.c. Three-point starter.*

In order to start the motor the switch S, is closed and starter arm is moved to position 1 as indicated 1, 2, 3, 4 and 5 are the several starting resisters. A holding coil is used which holds the switch and keeps in it required position. The holding coil is also connected to the shunt field in series. The starting resistor bank is connected in series with the armature winding in-order to avoid high current passage from the armature.

After the armature has accelerated sufficiently on first contact, the starter arm is slowly-moved to following contacts until the iron keeper on the arm is held by the holding coil electro-magnet. The diagram for the variation in armature current during starting period is shown in Fig. 14.6.

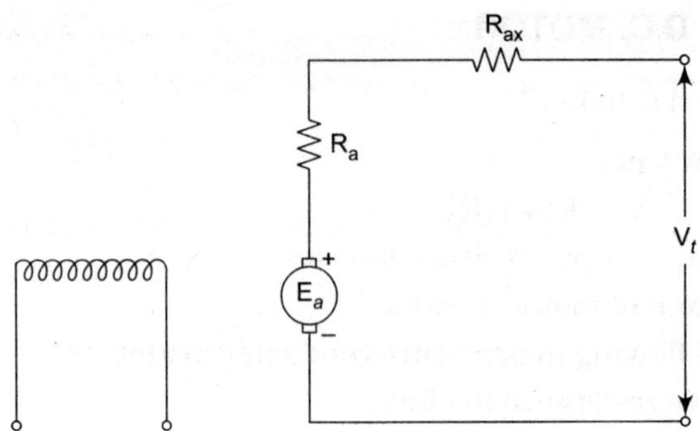

Fig. 14.6 *Armature current at start-up period.*

Circuit model of d.c. machine: The circuit diagram for the armature of d.c. generator as well as d.c. motor is shown in Fig. 14.7:

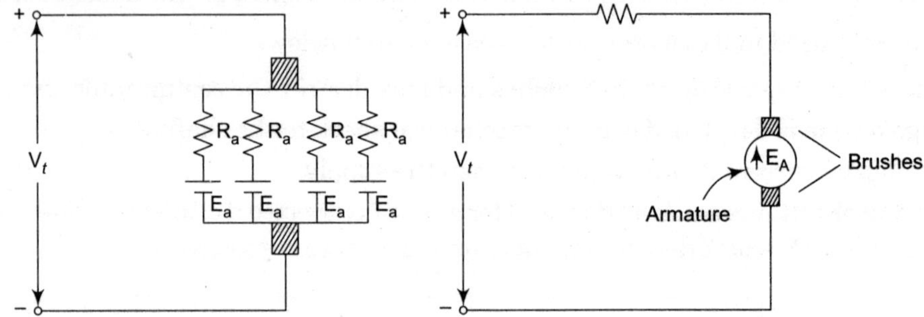

Fig. 14.7 *Armature circuit for d.c. machine.*

In case of generator the electrical energy is converted into mechanical energy. This conversion occurs only when the electromagnetic torque T_e to oppose the prime moves torque. This torque T_e in case of generator is known as counter torque. But in case motor the mechanical energy is converted into electrical energy. This occurs when the electromagnetic torque and the prime mover torque acts in direction.

14.7 GENERATING MODE

The circuit for generating mode is shown in Fig. 14.8.

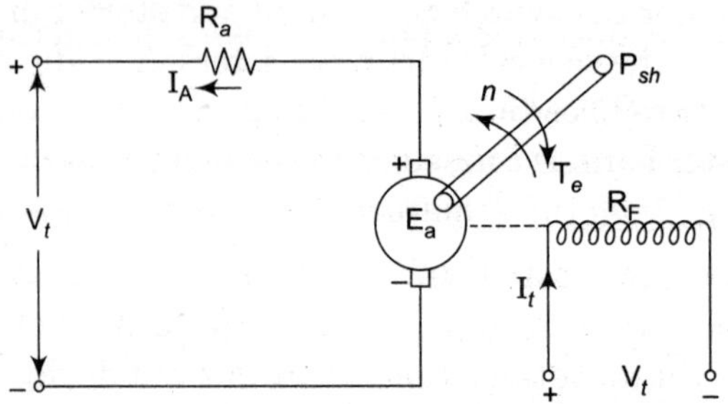

Fig. 14.8 *Generating mode.*

The d.c. machine operates in generating mode when the armature current I_A and the generated e.m.f. E_A are in the same direction but in generating mode the speed and electromagnetic torque in opposite direction.

For d.c. generators;

Mechanical power input = Electrical power output + (Losses in armature winding + Losses in field winding)

and $$V_t = E - I_A R_A$$

where, V_t = Terminal voltage in volts

I_A = Armature current in amperes

R_A = Armature resistance in ohm.

Also, drop across the brush is usually taken as 1 or 2 V.

Therefore the above equation becomes

$$V_t = E - I_A R_A - 2$$

Multiplying above equation by I_A, we get

$$V_t I_4 = EI_A - I_A R_A - 2I_A$$

or Electrical power output = Electromagnetic power – Ohmic loss – Brush contact loss.

Motoring mode: The circuit diagram for the motoring mode of d.c. machine is shown in Fig. 14.9.

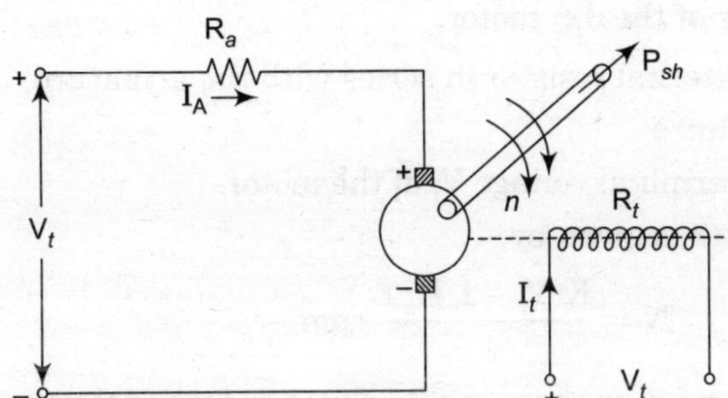

Fig. 14.9 *Motoring mode.*

In motoring mode, the direction of armature current I_a and the armature e.m.f. is in opposite direction having different polarity. The speed and the electromagnetic torque T_e is in the same direction during the motion operation.

For motion:

Electrical power outputs = Mechanical power – Losses in armature – Losses in field

and, $$V_t = E + I_A R_A$$

where all the terminology V_t, E, I_A and R_a are same as in generator.

When brush contact resistance is considered the equities becomes.

$$V_t = E + I_A R_A + 2$$

Multiplying above equation by I_a, we get

$$V_t I_A = EI_A + I_A^2 R_A + 2I_A$$

Electrical power input = Electromagnetic power + Losses + Brush contact.

14.8 SPEED CONTROL OF D.C. MOTOR

Before describing the various methods of speed control of d.c. motor, some terminology which are frequently used are as follows:

Base speed: This is the speed at which the motor runs at rated armature voltage and rated field current. The base speed is generally equal to the name plate speed of motor.

Speed regulation: If the speed changes no load to full load $\Delta\omega_m$, then speed regulation is defined as the ratio of $\Delta\omega_m$ to rated speed ω_m or full load speed.

$$\text{Speed regulation} = \frac{(\text{No load} - \text{Full load speed})}{\text{Full load speed}} = \frac{(n_{SL} - n_{FL})}{n_{FL}} = \frac{\Delta\omega_m}{\omega_m}$$

$$\text{Percentage speed regulation} = \left(\frac{\Delta\omega_m}{\omega_m}\right) \cdot 100$$

Speed range: The speed range of d.c. machine is defined as the ratio of the maximum speed to the minimum speed of the motor.

Constant power drive: If the motor shaft power remains constant over a given speed range, then the system on which it is investigated is called constant power drive.

Speed control of d.c. motor: The speed control of d.c. motor can be performed by varying the following parameter of the d.c. motor:

1. By inserting an external resistor in series with the armature
2. By changing the flux ϕ
3. By changing the terminal voltage V_t of the motor.

The speed of d.c. motor is given by

$$N = \frac{K(V_t - I_a R_a)}{\phi} \text{ rpm}$$

Hence speed can be varies by changing R_a, flux ϕ or terminal voltage V_t.

1. *By inserting an external resistor in series with the armature:* When an external resistor is connected in series with armature, the current through the armature is reduced, resulting in reduction of speed of the motor Fig. 14.10.

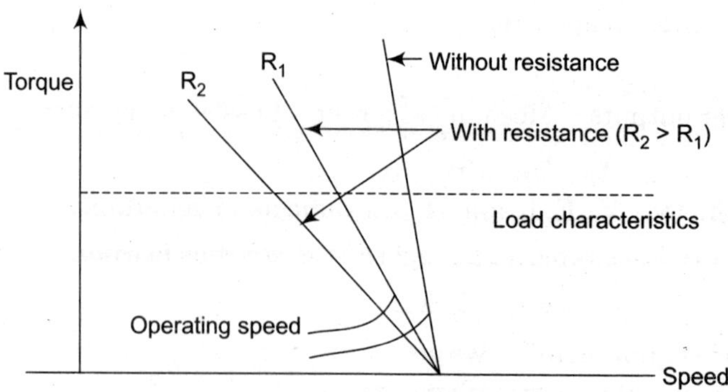

Fig. 14.10 *Resistor control.*

This method of speed control is very much simple and less expensive but due to the following disadvantage is not in practice:

(*i*) By adding resistance in the armature circuit the speed of the motor, compared to that without converting resistance is always lower.

(*ii*) At no-load condition this method is not applicable.

(*iii*) A considerable part of power is lost due to insertion of external resistance in armature as I_a^2 R losses occur.

(*iv*) It also results in loss of constant speed characteristics and the speed control is limited to 50% of rated speed.

The figure for armature resistance speed control of d.c. machine is shown in Fig. 14.10.

2. **By changing the flux φ:** It is one of the important methods of speed control of d.c. motor. In this method of speed control, the working flux φ is changed. In order to change the working flux φ, a resistance is connected in series with the field winding. Normally the resistor used is of variable nature and called as a field rheostat. Due to the insertion of resistance in field winding, the field current decreases and the speed is increased with a reduction in φ. From the expression of speed *i.e.*

$$N = \frac{K(V - I_a R_a)}{\phi} \text{ rpm}$$

It is clear that speed of the d.c. motor is inversely proportional to the working flux φ. So when φ decreases ultimately the speed increases. The figure for the field current control shown in Fig. 14.11:

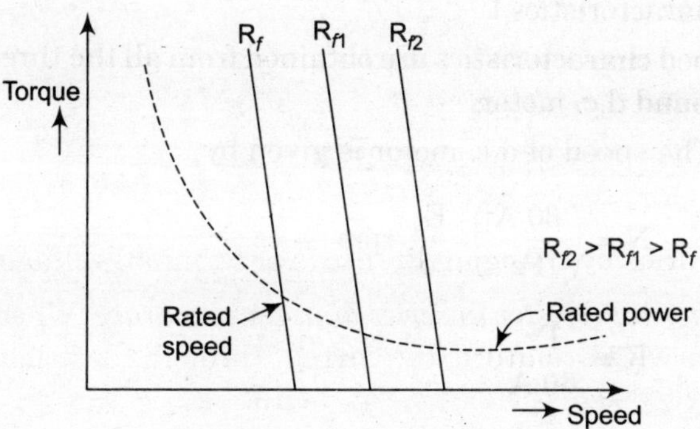

Fig. 14.11 *Flux control.*

The disadvantage of method: (*i*) The speed can be raised only at which the motor normally runs and the load is fixed, (*ii*) The speed is increased without any reduction in shaft load.

3. *By changing the terminal voltage V_t of the motor:* This method is similar to the armature control method. The voltage control method lowers or decreases the speed. From the expression of speed, we see that the speed is directly proportional to terminal voltage:

$$N = \frac{K(V_t - I_a R_a)}{\phi} \text{ rpm}$$

Therefore by decreasing terminal voltage the speed of d.c. motor can be decreased. This method of speed control is similar to the armature control method, however it does not have any drawback and not much additional power wastage. The Fig. 14.12 shows the armature voltage speed control of d.c. motor.

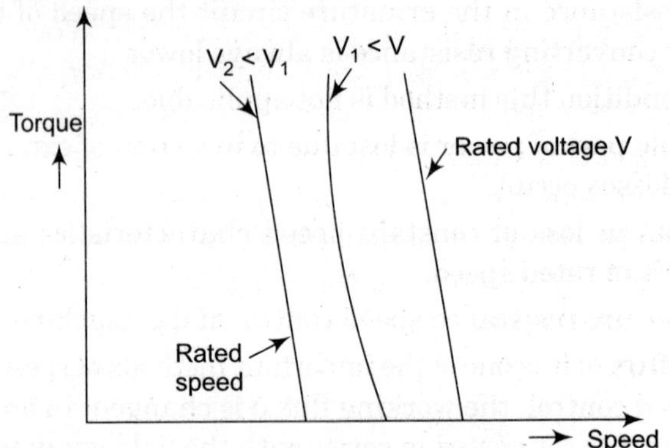

Fig. 14.12 *Armature voltage speed control.*

14.9 CHARACTERISTICS OF D.C. MOTORS

In a d.c. motor the e.m.f. E_a generated in the armature is called the back or counter e.m.f. In d.c. motor the supply voltage is generally constant. It has several characteristics

(*i*) Speed – Armature current characteristics

(*ii*) Torque – Armature current characteristics

(*iii*) Speed – Torque characteristics

All the above mentioned characteristics are obtained from all the three types of d.c. motor *i.e.* series, shunt and compound d.c. motor.

d.c. Shunt Motor: The speed of d.c. motor is given by

$$N = \left(\frac{60\ A}{Pz}\right) \cdot \frac{E_a}{\phi}\ \text{rpm}$$

and
$$K = \frac{Pz}{60\ A}$$

Therefore,

$$N = \frac{E_a}{K\phi} \quad ...(i)$$

But in case of motor the terminal voltage is given by the expression

$$V_t = E_a + I_a R_a$$

$$E_a = V_t - I_a R_a \quad ...(ii)$$

Putting the value of E_a from equations (*ii*) to (*i*), we get

$$N = \frac{(V_t - I_a R_a)}{K\phi} \quad ...(iii)$$

14.10 SPEED ARMATURE CURRENT CHARACTERISTICS

During constant supply voltage V_t and constant field current I_f the motor speed is effected by the I_aR_a drop and the demagnetizing effect of armature reaction. When the armature current I_a is increased, then the demagnetization effect of armature reaction also increases which result in reduction of field flux ϕ. Therefore, the speed of motor increases. But with the increases in I_a, the voltage drop I_aR_a increases and the numeration $(I_t - I_aR_a)$ decreases as a result the speed of motor decreases. In above equation (*iii*), when the armature current is increased the numeration decrement is more than that of denominator. So, the speed of d.c. motor drops slightly from its no-load speed N_{nl}. At no-load condition the armature current is negligible. So the speed of shunt motor at no-load is given by

$$N_{nl} = \frac{V_t}{K\phi} \qquad \ldots(iv)$$

If the effect of armature reaction is neglected, then the denominator will became constant. The figure, which illustrates the speed-current characteristic, is shown in Fig. 14.13.

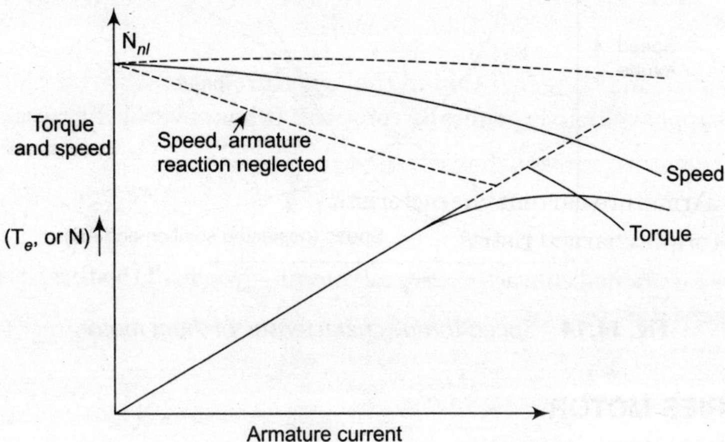

Fig. 14.13 *Speed-current and torque-current characteristic.*

14.11 SPEED-TORQUE CHARACTERISTICS (SHUNT MOTOR)

Under the steady-state condition, the expression for speed and torque is given below:

Speed, $$N = \frac{V_t - I_a R_a}{K\phi}$$

and Torque, $$T_e = K\phi I_a$$

or $$T_a = \frac{I_e}{K\phi}$$

Putting the value of I_a in above equation, we get

$$N = \frac{V_t}{K\phi} - \left(\frac{T_e}{K\phi}\right)\frac{R_a}{K\phi}$$

$$N = \frac{1}{K\phi}\left[\frac{V_t - T_e R_a}{K\phi}\right]$$

$$N = \frac{V_t}{K\phi} - \frac{R_a T_e}{K^2\phi^2}$$

$$N = \frac{N_{nl} - R_a T_e}{K^2\phi^2}$$

From above equation if the electromagnetic torque t_e increases the speed of d.c. shunt motor decreases. It is also necessary that for larger T_e, I_a must be large. The air gap flux ϕ reduces due to saturation and armature reaction.

The speed torque characteristics for d.c. shunt motor is shown In Fig. 14.14:

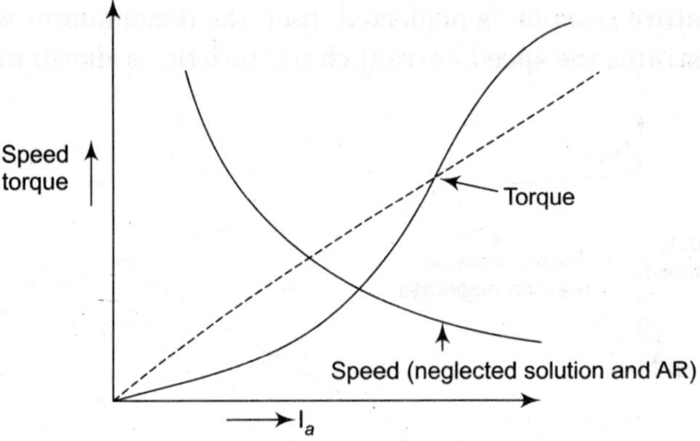

Fig. 14.14 *Speed-torque characteristic of shunt motor.*

14.12 D.C. SERIES MOTOR

In d.c. series motor as explained earlier the armature current I_a is equal to the field current I_f.

Speed current characteristics: The terminal voltage in case of d.c. series motor is expressed as

$$V_t = E_a + I_a(R_a + R_{se}) \quad ...(i)$$

and

$$E_a = K\phi N \quad ...(ii)$$

$$E_a = V_t - I_a(R_a + R_{se}) \quad ...(iii)$$

By the help of equation (*ii*), we get

$$K\phi N = V_t - I_a(R_a + R_{se})$$

or

$$N = \frac{V_t}{K\phi} - \frac{I_a(R_a + R_{se})}{K\phi}$$

Let $\phi = DI_a$, where D is any constant

$$N = \frac{V_t}{KDI_a} - \frac{I_a(R_a + R_{se})}{KDI_a} \quad ...(iv)$$

If saturation and armature reaction is neglected, then the main flux ϕ is directly proportional to the armature current. At no-load, the armature current is very small. Thus the voltage drop, $I_a (R_a + R_{se})$ at no load is almost nearly zero as compared to terminal voltage V_t. From above equation the no-load speed N_{nl} is

$$N_{nl} = \frac{V_t}{K\phi} = \frac{V_t}{KDI_a} \quad ...(v)$$

Speed is directly proportional to the terminal voltage and inversely proportional to I_a. The no-load speed of d.c. series motion becomes abnormally high due to small I_a. This is the reason that the series motor must always start and operate under load condition. The speed current characteristic of d.c. series motor is shown in Fig. 14.15:

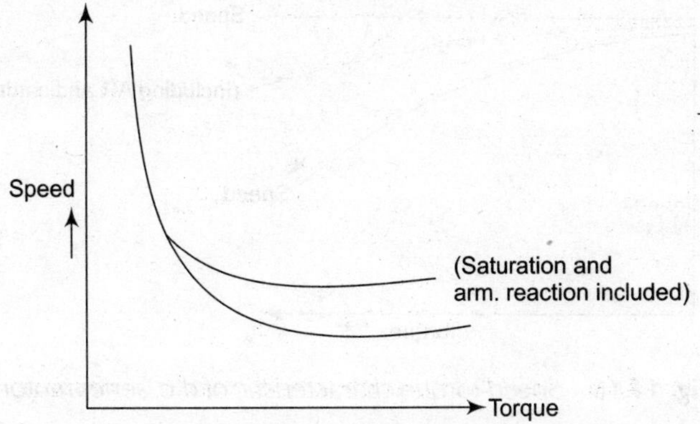

Fig. 14.15 *Speed current characteristic of d.c. series motor.*

Torque-Current Characteristics: Assuming armature reaction and saturation negligible,

$$\phi = DI_a$$

Therefore, $$T_e = K\phi I_a$$

or $$= KDI_a^2 = D_l I_a^2 \quad ...(i)$$

From the above expression, it is clear that the torque is directly proportional to the square of armature current and therefore the torque-current characteristic is parabolic, in Fig. 14.13.

Speed-torque Characteristics: The expressions for torque is mentioned above in torque-current characteristics that is

$$T_e = K\phi I_a$$

or $$T_e = K\phi I_a^2$$

or $$I_a^2 = \frac{T_e}{KD}$$

or $$I_a = \sqrt{\left(\frac{T_e}{KD}\right)}$$

Putting the value of I_a in following equation

$$N = \frac{V_t}{KDI_a} - \frac{(R_a + R_{se})}{KD}$$

Therefore
$$N = \frac{V_t}{\sqrt{KDT_e}} - \frac{R_a\ R_{se}}{KD}$$

The saturation and armature reaction is being negligible. The speed-torque characteristics is a hyperbola as shown in Fig. 14.16.

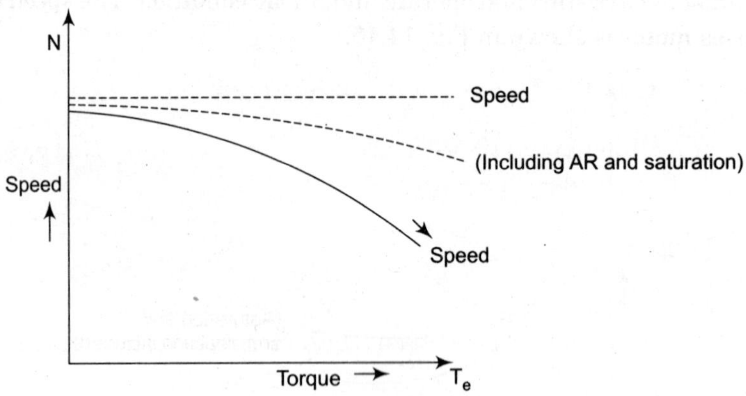

Fig. 14.16 *Speed-torque characteristic of d.c. series motor.*

14.13 D.C. COMPOUND MOTOR

The characteristic of cumulatively compound motor is explained below, while differently compound motor is rarely used.

Speed-current Characteristics: We obtained earlier the expression for long shunt compound motor in which the series and shunt fields are in parallel with the armature. For long shunt compound motor.

$$V_t = E_a + I_a (R_a + R_s)$$

and
$$E_a = K\phi N$$
$$= K(\phi_{sh} + \phi_{se})N \qquad ...(i)$$

where, ϕ_{sh} and ϕ_{se} are the shunt and series flux respectively.

$$N = \left\{\frac{1}{K(\phi_{sh} + \phi_{se})}\right\}(V_t - I_a(R_a + R_s)) \qquad ...(ii)$$

From above two expressions it is clear that when the armature current I_a increases then, ϕ_{sh} also increases as a result the denominator of equation-2 increases but numerator decreases. Therefore we can conclude that the speed of the d.c. compound motor decreases as the armature current I_a increases. The speed-current characteristic of d.c. compound motor is shown in Fig. 14.17.

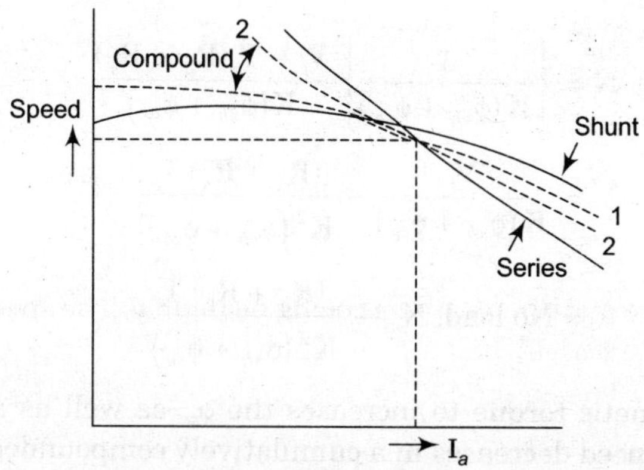

Fig. 14.17 *Speed-current characteristic of compound motor.*

Torque-current Characteristics: The electromagnetic torque T_e is given by

$$T_e = K\phi I_a$$

$$T_e = K(\phi_{sh} + \phi_{se})\, I_a \text{ for d.c. compound motor.}$$

If the shunt field is stronger than series field then the torque current characteristics is denoted by curve 1 and when the series field is stronger than shunt field; denoted by curve 2. The Fig. 14.18 illustrating the torque-current characteristics is shown in Fig. 14.18.

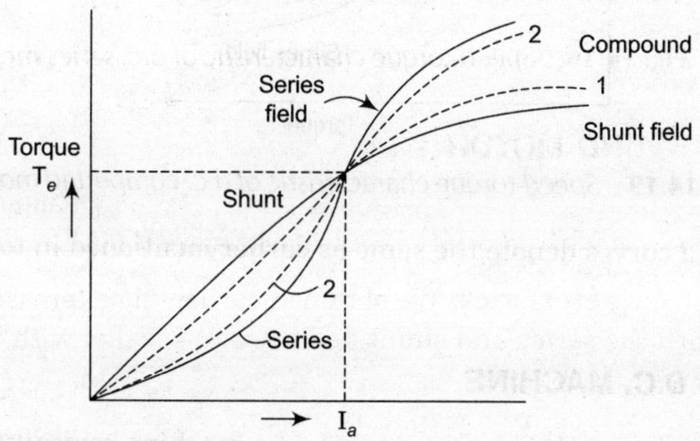

Fig. 14.18 *Torque-current characteristic of compound motor.*

Speed-torque Characteristics: The equation of torque for d.c. compound motor is given by

$$T_e = K(\phi_{sh} + \phi_{se})I_a \qquad \ldots(i)$$

or

$$I_a = \frac{T_e}{K(\phi_{sh} + \phi_{se})} \qquad \ldots(ii)$$

Putting the value of I_a in given equation(speed equation)

$$N = \frac{1}{K(\phi_{sh} + \phi_{se})}[V_t - I_a(R_a + R_s)]$$

We get,
$$N = \left\{\frac{1}{K(\phi_{sh} + \phi_{se})}\right\} \frac{V_t - T_e(R_a + R_s)}{K(\phi_{sh} + \phi_{se})}$$
$$= \frac{V_t}{K(\phi_{sh} + \phi_{se})} - \frac{(R_a + R_s)\,T_e}{K^2(\phi_{sh} + \phi_{se})^2}$$
$$= \text{No load, } N_{nl} - \frac{(R_a + R_s)\,T_e}{K^2(\phi_{sh} + \phi_{se})^2} \quad \ldots(iii)$$

When the electromagnetic torque to increases the ϕ_{se} as well as armature current I_a increases. As a result the speed decreases in a cumulatively compounded motor due to the addition of flux ϕ_{sh} and ϕ_{se}. In other words, the flux $(\phi_{sh} + \phi_{se})$ becomes more due to which the speed drops. Speed of d.c. motor is inversely proportional to the working flux. The figure speed-torque characteristic is shown in Fig. 14.19.

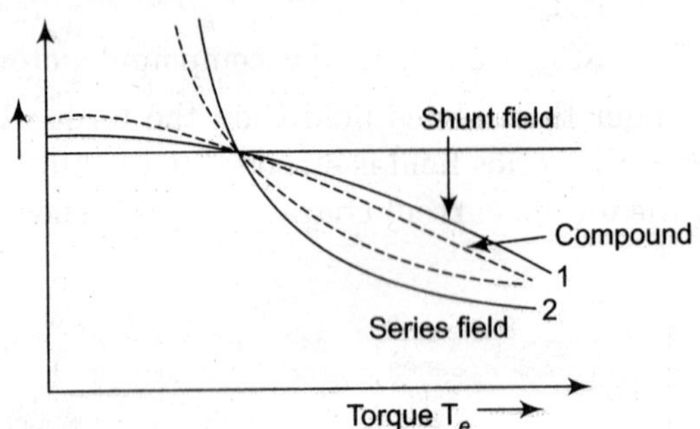

Fig. 14.19 *Speed-torque characteristic of d.c. compound motor.*

The number 1 and 2 curves denote the same as earlier mentioned in torque-current characteristics.

14.14 LOSSES IN D.C. MACHINE

In d.c. machine like all the machine losses occur. In d.c. machine basically three of losses occur.

(*i*) Cooper loss
(*ii*) Iron loss
(*iii*) Rotational loss.

Copper loss: The copper losses occur in d.c. machine due to the resistance of the windings. The current flowing through these windings creates ohmic loss (*i.e.* I^2R losses). In all the windings *i.e.* field winding, interpoles, armature winding and compensating winding the ohmic loss takes place due to resistance in all the windings.

Iron loss: Due to the rotation of armature in magnetic field the iron parts as well as the conductors cut the magnetic field flux. An e.m.f. is induced in the iron part, which causes current to flow through these parts. These are the eddy currents. Losses due to eddy current are called eddy current loss. This can be reduced by varnished lamination of iron core, another loss occurring in iron part is due to Hysteresis. The hysteresis loss occur as the armature continuously moves through the alternating stationary magnetic field poles. Both the losses *i.e.* eddy current loss and hysteresis loss combinedly called iron loss.

Rotational loss: The rotational loss consists of bearing friction loss, friction of the brushes riding in the commutator and the windage loss. This windage loss is small as compared to the other losses.

14.15 EFFICIENCY OF D.C. MACHINE

The efficiency of any machine is defined as the ratio of output to the input. In general the efficiency is given by,

$$\eta = 1 - \text{Losses/input}$$

For motor,

$$\eta_m = 1 - \frac{\text{Losses}}{(V_t I_L + \text{Losses})}$$

or

$$\eta_m = 1 - \frac{\left(I_a^2 R_a + V_f I_f + W_0\right)}{V_t I_L}$$

For generators,

$$\eta_g = 1 - \frac{\text{Losses}}{(V_t I_L + \text{Losses})}$$

or

$$\eta_g = 1 - \frac{\left(I_a^2 R_a + V_f I_f + W_0\right)}{\left(V_t I_L \; I_a^2 R_a + V_f I_f + W_0\right)}$$

where V_t and I_L are the output voltage and output current respectively for generator and for motor V_t and I_L are the input quantities and W_0 windage loss.

SOLVED NUMERICAL PROBLEMS

Example 1. *A 400 V pole shunt motor has an armature current 60 A the flux per pole 0.08 Wb. The armature resistance is 0.3 ohm and brush contact drop is 1 volt per brush. If the machine has 780 armature conductors and a wave winding, find the full load speed of the machine.*

Solution:

$$E_b = 400 \times 60 \times 0.3 \times 2 = 380 \text{ V}$$

$$E_b = \frac{P\phi ZN}{60 \text{ A}}$$

$$N = \frac{60 \; E_b \; A}{P\phi Z} = \frac{60 \times 380 \times 2}{4 \times 0.08 \times 780} = 192 \text{ rpm. } \textbf{Ans.}$$

Example 2. *A 140 V d.c. shunt motor has an armature resistance of 0.2 ohm and a field resistance 70 ohm. The full load line current is 40 A and the full load speed is 1800 rpm. If the brush contact drop is 3 V. Find the speed of the motor at half load.*

Solution: $\text{If} = \dfrac{140}{70} = 2\text{A}$

At full load, $I_a = I_L - I_f = 40 - 2 = 38 \text{ A}$

$$E_b = V - I_a R_a - \text{brush drop}$$
$$= 140 - 38 \times 0.2 - 3 = 129.4 \text{ V}.$$

Since flux is constant, $E_b \times N = kN$

$$129.4 = k(1800)$$

or $$k = \frac{129.4}{1800}$$

when the load is reduced to half, the armature current will also be halved and will be 19 A.

$$E_b' = 140 - 19 \times 0.2 - 3 = 133.2 \text{ V}$$

The new speed N_1 is given by

$$133.2 = kN_1$$

$$N_1 = \frac{1}{K}\, 1133.2 = \frac{133.2 \times 1800}{129.4}$$
$$= 1952.8 \text{ rpm. } \textbf{Ans.}$$

Example 3. *A250 V shunt motor on no-load runs at 1000 rev-per min. and takes 5 A. The total armature and shunt field resistances are respectivily 0.2 ohm and 210 ohm. Calculate the speed when loaded and taking a current of 50 A, if armature reaction weakens the field by 3%.*

Solution: V-Terminal voltage = 250 V

N_1 = Speed = 1000 rpm

R_a = Armature resistance = 0.2 Ω

R_f = Field resistance = 250 Ω

I_{L1} = Load current after armature reaction = 50 A

I_{L2} = Load current before armature reaction = 5A

Weakening of flux = 3%

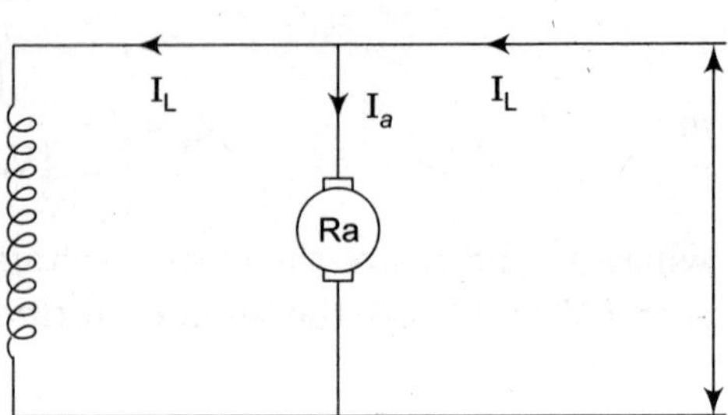

$\therefore$ $$\phi_2 = 0.97 f_1$$

$\therefore$ $$I_f = \frac{250}{250} = 1\text{A}$$

$\therefore$ $$I_{a1} = I_{L1} - I_f = 5 - 1 = 4$$
$$E_{b1} = V - I_a R_a = 250 - 4 \times 0.2 = 249.2$$
$$E_{b2} = V - I_a R_a = 250 - 49 \times 0.2 = 240.2$$

We know for shunt machine

$$\frac{E_{b1}}{E_{b2}} = \frac{\phi_1 N_1}{\phi_1 N_1}$$

$$N_2 = \frac{\phi_1 N_1}{\phi_2} \times \frac{E_{b2}}{E_{b1}}$$

$$= \frac{1}{0.97} \times 1000 \times \frac{240.2}{249.2} = 994 \text{ rev/min. } \textbf{Ans.}$$

Example 4. *A 220 V shunt motor has an armature resistance of 0.5 ohm and takes a current of 40 A on full load. By has much the main flux be reduced to raise the speed by 50% if the developed torque is constant?*

Solution: For a d.c. machine $\dfrac{E_{b1}}{E_{b2}} = \dfrac{\phi_1 N_1}{\phi_2 N_2}$

or $$\frac{N_2}{N_1} = \frac{E_{b2}}{E_{b1}} \times \frac{\phi_1}{\phi_2}$$

Since torque is constant

$$\phi_1 I_{b1} = \phi_2 I_{a2}$$

$$I_{a2} = I_{a1} \times \frac{\phi_1}{\phi_2} \; 40x$$

where $$x = \frac{\phi_1}{\phi_2}$$

$$E_{b1} = 220 \times (40 \times 0.5) = 200 \text{ V}$$

$$E_{b2} = 220 - (40x \times 0.5) = (220 - 20x) \text{ V}$$

$$\frac{N_2}{N_1} = \frac{3}{2}$$

$$\frac{3}{2} = \frac{(220 - 20x)}{200} \times x$$

$$x_2 - 11x + 15 = 0$$

$$x = \frac{11 \pm \sqrt{121 - 60}}{2} = \frac{11 \pm 7.81}{2}$$

$$x = 9.4 \text{ or } 1.6$$

Value of 0.4 is rejected as it does not give the necessary increase in speed.

$$\frac{\phi_1}{\phi_2} = 1.6; \; \frac{\phi_2}{\phi_1} = \frac{\phi_2}{\phi_1} = \frac{1}{1.6}$$

$$\frac{\phi_1 - \phi_2}{\phi_1} = \frac{1.6 - 1}{1.6} = \frac{3}{8}$$

$\therefore$ Percentage change in flux $= \dfrac{3}{8} \times 100 = 37.5\%$. **Ans.**

Example 5. *A 4 pole series-wound fan motor runs normally at 600 rev/min on a 250 V supply, taking 20 A. The filed coilsare connected all in series. Estimate the speed and the current taken by the motor of the coils are reconnected in two parallel groups of two in series. The load torque is increases as the source of the speed. Assume that the flux is directly proportional to the current and ignore losses.*

Solution: When the coils are connected in two parallel groups current through each becomes $\dfrac{I_{a2}}{2}$, where I_{a2} is the new armature current.

$$\therefore \qquad \phi_2 \propto \frac{I_{a2}}{2}$$

$$T_a \propto \phi I_a$$

$$\propto N_2 - \text{(given)}$$

$$\phi_1 I_{a1} \propto N_1^2 \text{ and } \phi_2 I_{a2} \propto N_1^2$$

$$\left(\frac{N_2}{N_1}\right)^2 = \frac{\phi_1\, I_{a2}}{\phi_1\, I_{a1}}$$

Since losses are negligible, field coil resistance, as well as armature are negligible. It means that negligible. Hence back e.m.f. in each case equals the supply voltage.

$$\frac{N_2}{N_1} = \frac{E_{b2}}{E_{b1}} \times \frac{\phi_1}{\phi_2} = \frac{\phi_1}{\phi_2}$$

Putting this value in above, we get

$$\left(\frac{\phi_1}{\phi_2}\right)^2 = \frac{\phi_2 I_{a1}}{\phi_1 I_{a1}} \quad \text{or} \quad \frac{I_{a2}}{I_{a1}} = \left(\frac{\phi_1}{\phi_2}\right)^2$$

Now $\phi_1 \propto 20$ and $\phi_2 \times \frac{I_{a1}}{2}$

$$\frac{I_{a2}}{20} = \left(\frac{20}{\frac{I_{a1}}{2}}\right)^3 \quad \text{or} \quad I_{a1} = [20 \times (40)^3]^{1/4}$$

$$= 33.64 \text{ A}$$

Form $\quad \frac{N_2}{N_1} = \frac{2I_{a1}}{I_{a2}}$, $N_2 = 714$ rpm. **Ans.**

Example 6. *A series motor takes a line current so A and develops a torque 100 Nm. Calculate the torque when (a) the line current is 35 A assuming that the magnetic circuit is unsaturated and (b) the line current is 60 A. Assuming that an increase in line current from 30 A to 60 A causses 60% increases in field flux.*

Solution:

(*a*) $T_e \propto \phi I_a \propto I_a^2$

$$T_e = 100\left(\frac{35}{30}\right)^2 = \frac{9.225}{9} = 136.11 \text{ N-m}$$

(*b*) $T_e \propto \phi I_a$

Now value of torque $= 100 \times 1.6 \times \frac{60}{30} = 320$ N-m. **Ans.**

EXERCISES

1. What are the various methods of speed control of d.c. motor? Explain in brief.
2. Which method of speed control speeds lower than rated speed? Explain reasons.
3. What is the function of a starter in a d.c. motor?
4. Why in the speed of a d.c. shunt motor is nearly constant, whereas the speed of a d.c. series motor is variable?
5. What do you understand by 'back e.m.f. in d.c. motor'? Explain the principle of torque production in a d.c. motor.
6. Derive the torque equation of a d.c. motor.
7. Sketch and explain the speed-load characteristics of following d.c. machines:
   (*i*) Series motor (*ii*) Shunt motor
   (*iii*) Cumulatively compounded motor. (*iv*) Differentially compounded motor.
8. Why the starting current is very high in a d.c. motor and how the used of starter reduces the starting current to a safe value?
9. Why starter is necessary for the d.c. motors? Explain the various types of starter in d.c. motor with their operating principle.
10. What is the function of 'no volt release' and 'over volt release' in a starter? Discuss the operation of these two features in a shunt motor starter.
11. Explain the torque-slip characteristic of d.c. motor.
12. Starting from first principle, derive an expression for the electromagnetic torque of a d.c. motor.
13. Differentiate between the generator action and motor action of a d.c. machine.
14. Explain the phenomena between generator action and motor action of a d.c. machine.
15. Explain various differences between long shunt and short shunt compound motor with suitable diagram.
16. Explain why a small change in motor speed and back e.m.f. will produce longer changes in armature current of a d.c. motor.
17. Explain the function of commutator in a d.c. motor. Also explain how it is different from commutation in generator.
18. A series motor has armature resistance of 0.8 ohm and field resistance of 0.4 ohm. It takes a current of 14 A from a 220 V supply and runs at 800 rpm. Find the speed at which it will run when connected in series with a 4 ohm resistance and taking the same current at the same supply voltage.
19. A d.c. motor develops a torque of 140 Nm at 1400 rpm. If the motor now runs at 1000-rpm, then obtain the new developed torque.
20. A d.c. series motor develops a torque of 20 Nm at 34 of load current. If the current is increased to 6 A, the calculate the new torque developed.
21. The armature of a shunt motor contains 0.8 ohm resistance. The motor is to run on a 120 V circuit. If the motor is suddenly thrown on the circuit while the armature is standing still (back e.m.f. in normal running being 110 V). Obtain the current drawn by the motor.

22. A 230 V d.c. motor takes a no-load current of 2 A and runs at 1200 rpm. In case the full load current is 40 A then calculate the speed at full load.
23. A d.c. motor develops a torque of the 110 Nm at 400 rpm. At 300 rpm obtain the torque developed.
24. A 4-pole wave wound dc armature has 294 conductors. Find out
    (a) flux per pole to generate 230 V at 1400 rpm.
    (b) electromagnetic torque at this flux when the armature current is 120 A.
25. The maximum current during starting for a 300 V shunt motor is to be limited to 125 A. The resistance of armature is 0.25 ohm. Find the resistance elements for a 12 element starter.
26. A shunt generator delivers 50 kW at 240 V and 380 rpm. The armature and field resistance are 0.04 and 40 ohm respectively. Calculate the speed of the machine when working as a motor taking an input of 40 kW at 240 V. Assume 1V per brush for contact drop.
27. A 220 V series motor is which the total armature and the field resistance is 0.1 W is working with unsature field, taking 100 A and running at 800 rpm. Find out the speed at which the motor will run developing half the torque.

15

Synchronous Machines

Alternating voltage is generated when armature is being rotated in a uniform magnetic field with stationary field poles. Alternating voltage can also be generated in stationary armature conductors when field poles are rotated. In alternators or synchronous generator, armature is kept stationary, while field system of a synchronous machine is excited by direct current and armature is rotated by a prime mover *i.e.,* by steam turbines, hydraulic turbines or diesel engines. The advantages of rotating field system and stationary armature are:

1. The voltage level of the field winding is comparatively low, and hence could be excitated with the help of slip rings.
2. In this an armature is a usually required to generate high voltages. In the stationary armature winding, high voltage terminals of the armature winding can be connected directly to the terminals.
3. In a stationary armature system, winding and insulation are less exposed to mechanical stress and vibration.

Synchronous machines are called "Synchronous" as their speed, is directly related to the supply frequency and running at the synchronous speed.

$$n_s = \frac{120\,f}{\text{P}} \text{ rpm}$$

where n_s is the synchronous speed in rpm,

P is the number of magnetic poles

f = Supply frequency in Hz (hertz).

15.1 CONSTRUCTION

Similar to other rotating machines, a synchronous machine consists of two basic parts *i.e.,* stator and rotor. The stator is stationary part *i.e.,* armature winding and rotor is rotating part called rotor *i.e.,* field winding the rotor is centred within the stator so that the rotor axis is concentric with that of the stator. The space between the outside of the rotor and inside of the stator is called the air gap.

The rotor is slotted to accommodate the field winding. Rectangular slots with tapered teeth are milled out in the rotor, as a result rectangular conductors can be used for the field winding.

The rotor is mounted on a shaft. The shaft is supported in bearings so that the rotor is free to rotate. The rotor is solidly fastened to the shaft so that the rotor and the shaft can rotate at the same speed. Therefore, the terms 'rotor speed', 'shaft speed' or machine speed all are the same thing and can be used interchangeably. The rotor and stator each have three basic parts, the core, the windings and the insulation. The purpose of the rotor and stator cores is to conduct the magnetic field through the coils of the windings.

The winding conductors are either of copper or aluminium and that could be in the form of coils of wire or heavy bars depending upon the current carrying capacity required. Normally each winding consists of several coils in series or series-parallel combination depending upon the voltage and current requirements of the machine. The ends of certain windings are brought out to terminals to allow easy contraction of supply.

Since the speed of rotor is high so insulation system requires should also have good mechanical properties therefore normally mica, asbesto and hard composition insulating materials are used.

In some machines, the windings of either the rotor or the stator may be placed around projecting magnetic pole pieces, called salient poles. When a core has salient poles of the coils of the winding are wound around the waists of the pole pieces. These narrower parts of the salient poles are called the pole cores. The shaped ends of the poles are called the pole shoes. Their purpose is to provide the correct flux density distribution in the air gap.

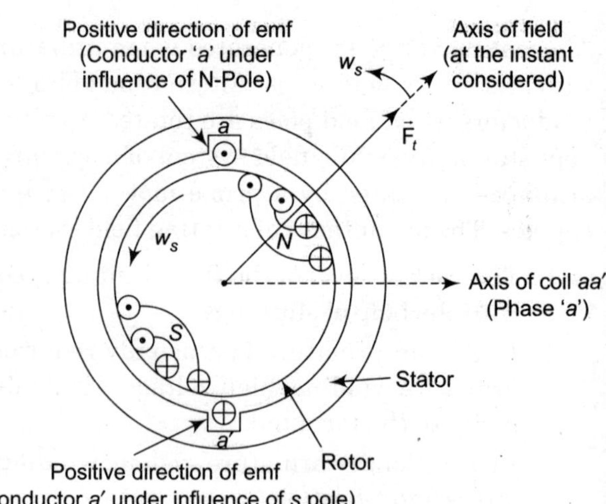

Fig. 15.1 *(Construction of Synchronous Machine)*

15.2 PRINCIPLE OF OPERATION OF ALTERNATORS

Alternators operate on the fundamental principles of electromagnetic induction. In the alternator, the stator accommodating a three-phase winding is stationary, while the field system rotates. Figure 15.2 shows the schematic diagram of a three-phase alternator with stationary armature and rotating field system. When the field system is rotated by a prime mover, the stator conductors are cut by the magnetic flux and hence an e.m.f. is induced in the stator conductors which is given by,

$$e = Blv \text{ volts}$$

where B = Instantaneous value of flux density, in tesla

l = Length of the stator conductor, in metre

v = Speed of the conductor, in m/s.

As the l and v are constant, the induced e.m.f. e is directly proportional to flux density. Hence the e.m.f. wave for the stator phase will be identical to the flux density wave. If the flux density produced by the field winding is sinusoidal in its distribution around the air gap, the voltage induced in the phase coils will be sinusoidal as shown in Fig. 15.2. In salient pole synchronous machines, a sinusoidal wave of flux density is closely obtained by shaping the pole shoe in such a way that the air gap gradually increases from centre to the pole tips. However, in non-salient pole synchronous machines the air gap is more or less uniform. In this type of synchronous machines a sinusoidal flux density is obtained by distributing the field winding in several slots. A further improvement on the output voltage waveform is made by making a coil span for the stator winding less than the pole pitch.

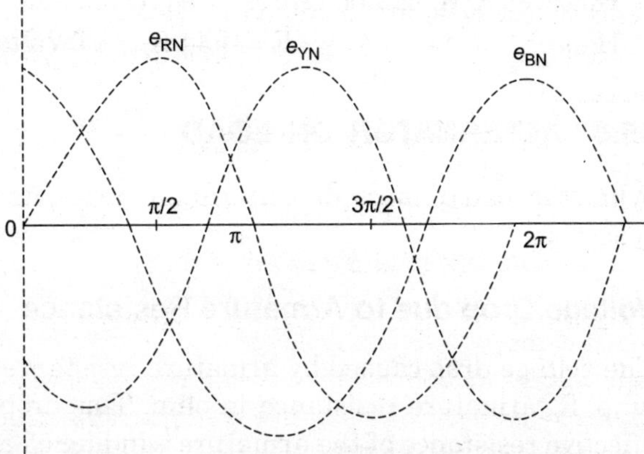

Fig. 15.2 *Voltage wave shapes of 3-phase alternator.*

The frequency of generated e.m.f. depends upon the number of poles P and the speed at which the alternator is being run by the prime mover is given by,

$$f = \frac{PN}{120} \text{ Hz}.$$

15.3 E.M.F. EQUATION

Let P be the number of poles, ϕ the flux per pole in Weber, N speed of the alternator in rpm; T the total number of turns per phase, K_d the distribution factor K_c the coil span or pitch factor, and f the frequency of the generated e.m.f. in Hz, then

Flux cut by each conductor per revolution = $P\phi$

Flux cut by each conductor per second = $P\phi \times \frac{N}{60}$

Thus average e.m.f. generated in each conductor = $\frac{P\phi N}{60}$ Volt

Since one turn consists of two active conductors,

The average e.m.f. generated per turn = $2\,\frac{P\phi N}{60}$ Volts

Hence average e.m.f. generated per phase = $2\,\frac{P\phi N}{60} \times T$ Volts

But $\quad f = \frac{PN}{120}$ Hz

So, average e.m.f. per phase = $4f\phi T$ Volts

Rms value of e.m.f. generated per phase

$$E = 1.11 \times 4f\phi T = 4.44 f\phi T \text{ Volts}$$

Considering distribution factor and chording factor of winding; the rms value of the e.m.f. generated per phase is given by

$$E = 4.44\, K_d K_c f \phi T$$

However K_dK_c can be represented by a single term K_w, known as the winding factor.

Hence $$E = 4.44\ K_w f \phi T \text{ Volts.}$$

15.4 ALTERNATOR ON LOAD

When the load connected to alternator varies, its terminal voltage changes due to the following reasons:

Voltage Drop due to Armature Resistance

The voltage drop caused by armature resistance per phase is IR_a where I is phase current in amp, R_a armature resistance in ohm. This drop, however, can be neglected practically. The effective resistance of the armature winding of generator is greater than the conductor resistance as measured by direct current. Due to alternating current, the additional energy loss are as following :

1. Eddy current loss
2. Hysteris loss
3. Loss due to unequal distribution of current.

Armature Leakage Reactance

When the load current flows through the armature winding, it builds up the local flux which cuts the winding and counter e.m.f. is generated. This effect produces armature reactance that is equal to $2\pi f_L$, where L is in H. This armature reactance is called leakage reactance X_L, this leakage flux is proportional to the armature current, because the magnetic path it follow is not normally saturated. X_L varies some what with the position of the armature and field poles. It is usually assumed to be constant.

Armature Reaction

When the load current flows in the stator conductors it produces a magnetic field which has a cross-magnetizing, demagnetizing or magnetizing effect upon the main flux due to the field windings. Such an effect of the armature currents upon the main flux is termed armature reaction. The armature reaction depends on the power factor of the load.

Synchronous Reactance

The combination of leakage reactance X_L along with the armature reactance is called synchronous reactance.

$$X_s = X_L + X_a \text{ ohm}$$

15.5 SYNCHRONOUS IMPEDANCE

It is given as follows:

$$Z_s = R_a + j(X_L + X_a) = R_a + jX_s$$

$$Z_s = \sqrt{R_a^2 + X_s^2}\,.$$

15.6 EQUIVALENT CIRCUIT MODEL AND ITS PHASOR DIAGRAM

Figure 15.3 shows the equivalent circuit model of a synchronous generator alternator.

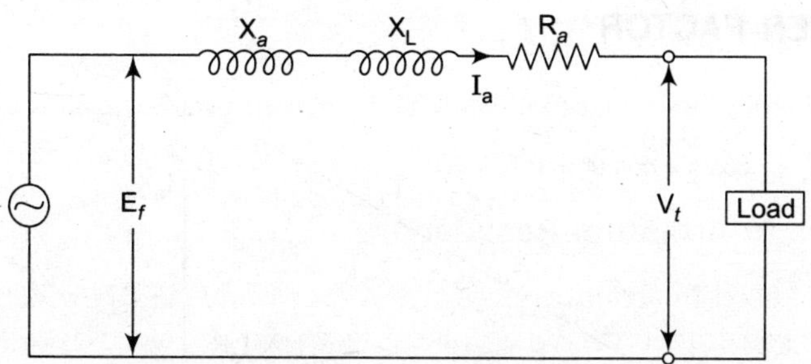

Fig. 15.3 *Equivalent circuit model an alternator.*

15.7 LAGGING POWER FACTOR

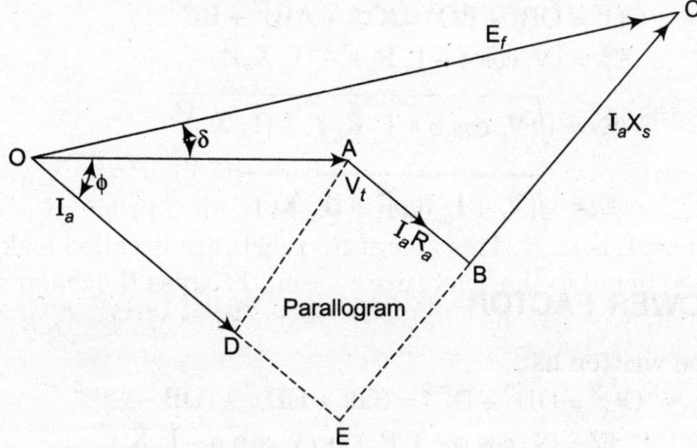

Fig. 15.4 *Phasor diagram of alternator at lagging power factor*

$OC = E_f$ = No load e.m.f.

$OA = V_t$ = Terminal voltage, in volt

I_a = Armature current, amp

R_a = Effective armature resistance, in ohm

X_s = Synchronous reactance

$\cos \phi$ = Power factor

δ = Power angle or torque angle

$AB = I_aR_a$

$BC = I_aX_s$

From Fig. 15.4 it can be written as

$$OC^2 = OE^2 + EC^2 = (OD + DE)^2 + (EB + BC)^2$$

$$E_f^2 = (V_t \cos \phi + I_a R_a)^2 + (V_t \sin \phi + I_a X_s)^2$$

or $$E_f = \sqrt{(V_t \cos\phi + I_a R_a)^2 + (V_t \sin\phi + I_a X_s)^2}$$

15.8 UNITY POWER FACTOR

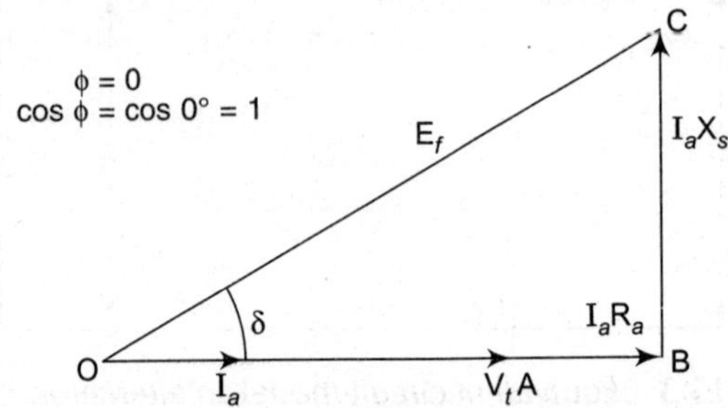

Fig. 15.5 *Phasor diagram of alternator at unity power factor*

From Fig. 15.5, it can be written as,

$$OC^2 = OB^2 + BC^2 = (OA + AB)^2 + BC^2$$

$$E_f^2 = (V_t \cos\phi + I_a R_a)^2 + (I_a X_s)^2$$

$$E_f = \sqrt{(V_t \cos\phi + I_a R_a)^2 + (I_a X_s)^2}$$

$$E_f = \sqrt{(V_t + I_a R_a)^2 + (I_a X_s)^2}$$

[$\phi = 0°$, $\cos\phi = \cos 0° = 1$]

15.9 LEADING POWER FACTOR

For Fig. 15.6, it can be written as

$$OC^2 = OD^2 + DC^2 = (OE + ED)^2 + (DB - CB)^2$$

$$E_f^2 = (V_t \cos\phi + I_a R_a)^2 + (V_t \sin\phi - I_a X_s)^2$$

or $$E_f = \sqrt{(V_t \cos\phi + I_a R_a)^2 + (V_t \sin\phi - I_a X_s)^2}$$

The angle between E_f and V is called the power angle or torque angle of the alternator.

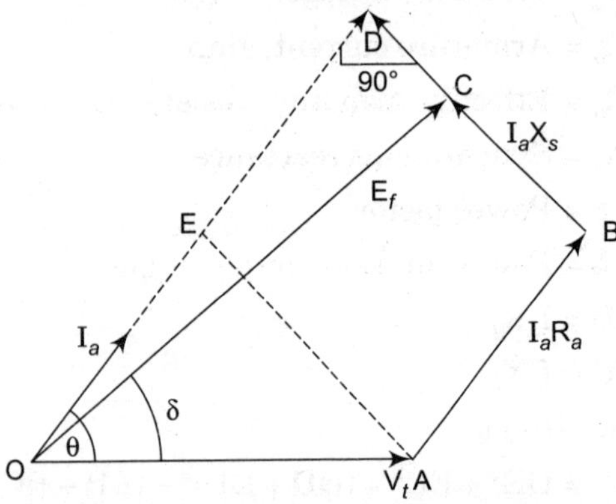

Fig. 15.6 *Phasor diagram of an alternator at leading power factor.*

Measurement of synchronous impedance: To measure the synchronous impedance the following tests are performed:

(*i*) Open circuit characteristic test.

(*ii*) Short circuit characteristic test.

Open Circuit Characteristic (OCC) Test

The alternator is run at the rated synchronous speed the armature terminals are kept open *i.e.*, all loads are disconnected, as shown in Fig. 15.7.

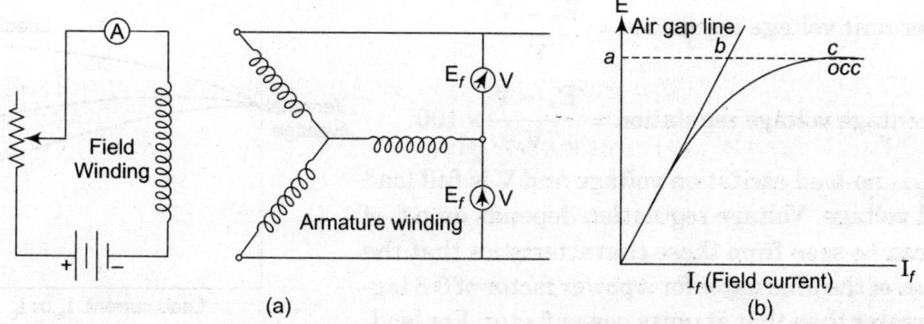

Fig. 15.7 *Circuit diagram and characteristic curve on OCC test of an alternator.*

Initially the field current is set to zero, then gradually the field current is gradually increased so that the final value of E_f should be 25 to 30% higher than the rated voltage. Figure 15.7 shows the OCC with field current along abscissa and e.m.f. along ordinate. The OCC will not be straight line because of saturation in the iron part of the magnetic circuit.

Short Circuit Characteristic (SCC) Test

To find the short circuit characteristic (SCC), the speed of the alternator be rated speed. The armature terminals are short circuited by an ammeter as shown in Fig. 15.8(*a*). The field current is gradually increased from zero until the short circuit armature current has reached its the full load value. The reading should be taken in a short time to avoid armature overheating. Fig. 15.8(*b*) shows the SCC of an alternator.

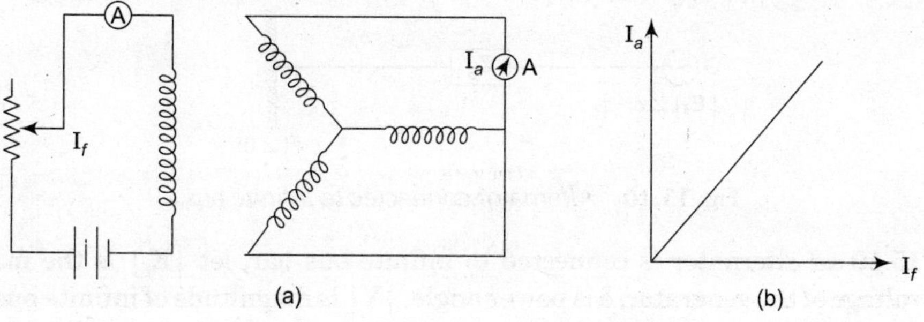

Fig. 15.8 *Circuit diagram and characteristic curve on SCC test of an alternator.*

Synchronous impedance Z_s can be defined with the above two test results as follows:

$$Z_s = \frac{\text{Open circuit terminal voltage for a certain field current}}{\text{Short circuit current for the same field current}}$$

15.10 VOLTAGE REGULATION

Voltage regulation of a synchronous machine is defined as the change in terminal voltage expressed as a percentage of rated voltage when load at a given power factor is removed. The speed and field current remains constant.

$$\therefore \quad \text{Per unit voltage regulation} = \frac{E_f - V}{V}$$

$$\text{and percentage voltage regulation} = \frac{E_f - V}{V} \times 100$$

where E_f is no-load excitation voltage and V is full load terminal voltage. Voltage regulation depends on p.f. of load. It can be seen from these characteristics that the regulation of the alternator for a power factor of 0.8 lagging is greater than that at unity power factor. For leading power factor, voltage regulation may be positive or negative due to decrease of E_f is shown in Fig. 15.9.

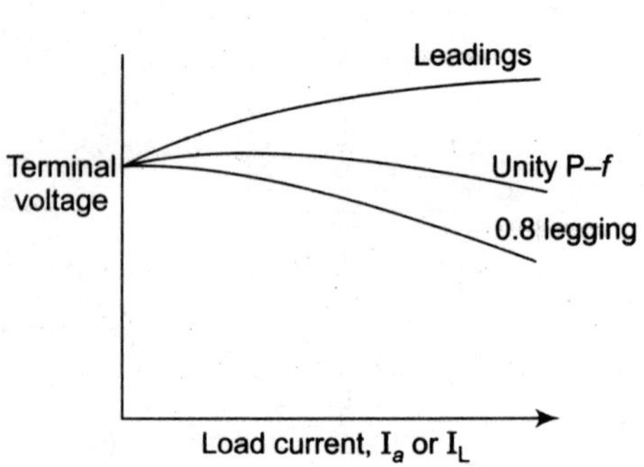

Fig. 15.9 *Load characteristics of alternator*

15.11 IMPORTANCE OF VOLTAGE REGULATION IS AS FOLLOWS

(*i*) Voltage rise must be known during throwing all the load because the winding insulation must be able to with stand the voltage.

(*ii*) The usage of automatic voltage control is determined by voltage regulation.

(*iii*) Voltage regulation affects the steady-state short-circuit conditions and stability.

(*iv*) Voltage regulation also affects the parallel operation.

15.12 POWER DELIVERED BY ALTERNATOR

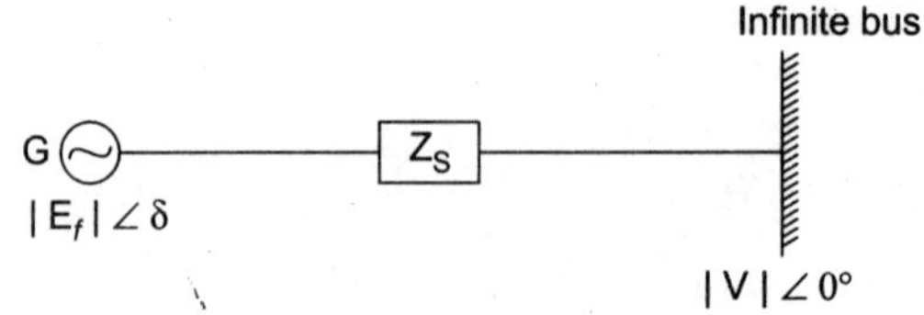

Fig. 15.10 *Alternator connected to infinite bus.*

In Fig. 15.10 an alternator is connected to infinite bus bar, let $|E_f|$ is the magntiude of excitation voltage of the generator, δ is power angle, $|V|$ is magnitude of infinite bus voltage, I_a is armature current, Z_s is synchronous impedance of the alternator, R_a is effective resistance, and X_s is synchronous reactance.

Then, $Z_s = |Z_s| \angle \phi$

$$Z_s = (R_a + jX_s) = \sqrt{R_a^2 + R_s^2} \angle \tan^{-1}\left(\frac{X_s}{R}\right)$$

Therefore current, $I_a = \dfrac{|E_f| \angle \delta° - |V| \angle 0°}{Z_s \angle \phi°}$ and $I_a{}^* = \dfrac{|E_f| \angle - \delta° - |V|}{|Z_s| \angle - \phi°}$

Complex power delivered by alternator per phase to infinite bus is given by,

$$S = (P + jQ) = |V| I_a^*$$

$$= |V| \left[\frac{|E_f| \angle - \delta - |V|}{|Z_s| \angle - \phi}\right] = \left[\frac{|E_f||V|}{|Z_s|} \angle \phi - \delta - \frac{|V|^2}{|Z_s|} \angle \phi\right]$$

∴ Real power, $P = \dfrac{|E_f||V|}{|Z_s|} \cos(\phi - \delta) - \dfrac{|V|^2}{|Z_s|} \cos \phi$

and Relative power, $Q = \dfrac{|E_f||V|}{|Z_s|} \sin(\phi - \delta) - \dfrac{|V|^2}{|Z_s|} \sin \phi$

Practically R_a is very small and it can be neglected

$$\phi = \tan^{-1}\left(\frac{X}{R_a}\right) = 90°$$

or $|Z_s| = |X_s|$

So, the real power, $P = \dfrac{|E_f||V|}{|X_s|} \cos(90° - \delta)$

$$P = \frac{|E_f||V|}{|X_s|} \sin \delta$$

where $P_m = \dfrac{|E_f||V|}{|X_s|}$ is maximum power limit.

While, $Q = \dfrac{|E_f||V|}{|X_s|} \sin(90 - \delta) - \dfrac{|V|^2}{X_s}$

$$Q = \frac{|V|}{|X_s|}\left(|E_f| \cos \delta - |V|\right)$$

Following three cases arises:

(*i*) If $|E_f| \cos \delta > V$, alternator is over excited and it delivers reactive power to the infinite bus. Under this condition, the alternator is operating at lagging power factor.

(*ii*) If $|E_f| \cos \delta = V$, alternator is normally excited and it neither delivers nor absorbs reactive power from the infinite bus. Under this condition, the alternator is operating at unity power.

(*iii*) If $|E_f| \cos \delta < V$, alternator is under excited and it absorbs reactive power from the infinite bus. Under this condition, the alternator is operating at leading power factor.

15.13 SYNCHRONOUS MOTORS WORKING PRINCIPLE

A three-phase synchronous motor is quite similar to a three-phase alternator in basic construction. As such it consists of a three-phase winding on the stator and a field winding on the rotor. Figure 15.11 shows the schematic diagram of a three-phase synchronous motor, illustrating its principle of operation. A three-phase balanced supply is fed to the three-phase stator winding, as a result, a magnetic field of constant magnitude rotating at synchronous speed is produced in the stator.

Let the stator have two poles N and S. These poles will be rotating at synchronous speed. Let the direction of rotation be clockwise as shown in Figure 15.11. Suppose the rotor of the synchronous motor is stationary in the position shown in figure and the stator poles N and S occupy the positions *a* and *b* respectively at a certain instant. Under this condition, a repulsive force (like poles) will act upon the rotor and it has a tendency to rotate in the anticlockwise direction. The positions of the stator poles N and S will be interchanged alter half a period, *i.e.*, N is at *b* and S at *a*. Hence, a force of attraction will act upon the rotor and it has a tendency to rotate in a clockwise direction. However, the rotor will not be in a position to respond to this alternating torque and such it would remain at rest. Hence, a three-phase synchronous motor does not have a self-starting torque. Now suppose the rotor is rotated by an external means in a clockwise direction and if the rotation of the rotor is to be maintained, then two essential conditions must be satisfied, viz., (*i*) stator poles S and N must be at the position *a* and *b* respectively as in Fig. 15.11 and (*ii*) the rotor and stator poles must continued to occupy the same relative positions. The second condition can be fulfilled only when the rotor keeps in step with the revolving field of the stator. Hence, the rotor of three-phase synchronous motor must run at a synchronous speed fixed by the number of poles and the frequency.

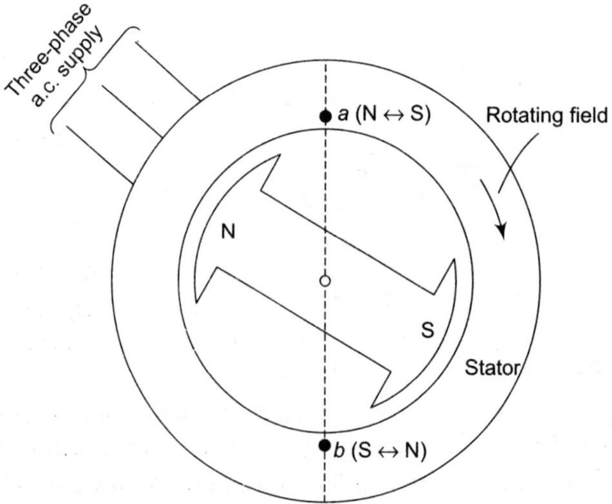

Fig. 15.11 *Principle of operation of synchronous motor.*

In order to start the three-phase synchronous motor, it is brought up to the rated speed by some external means. Then, it is synchronized to the supply under this condition the motor is floating in the bus bar. Now the prime mover is disconnected and the machine will start behaving as synchronous motor drawing power from the supply.

The equivalent and phasor diagram of synchronous motor is shown in Fig. 15.12.

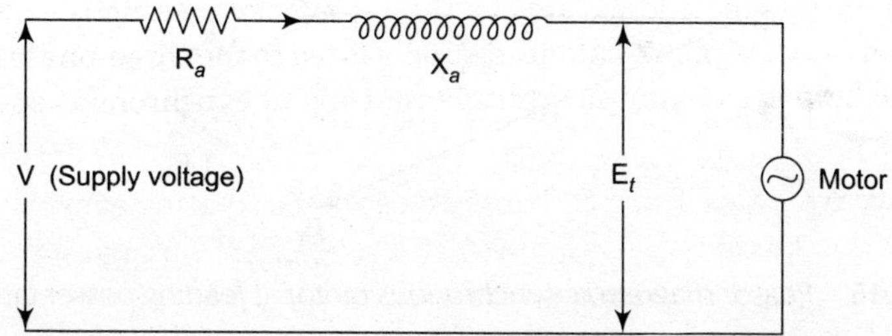

Fig. 15.12 *Equivalent circuit diagram of synchronous motor.*

At lagging power factor,

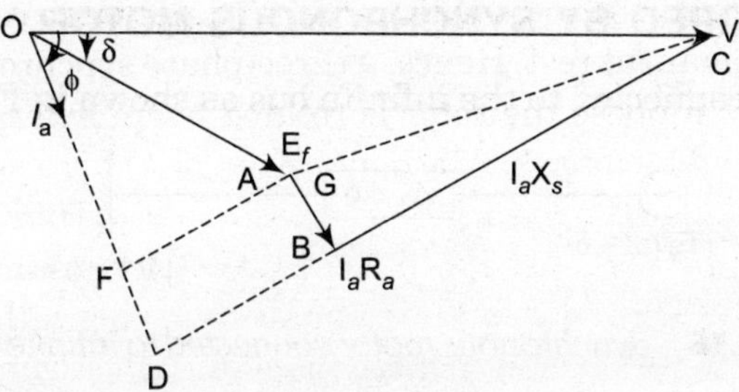

Fig. 15.13 *Phasor diagram of synchronous motor at lagging power factor.*

$$E_f = OG = \sqrt{(OD - FD)^2 + (CD - CB)^2}$$

$$E_f = \sqrt{\left[(V\cos\phi - I_aR_a)^2 + (V\sin\phi - I_aX_a)^2\right]}$$

At unity power factor,

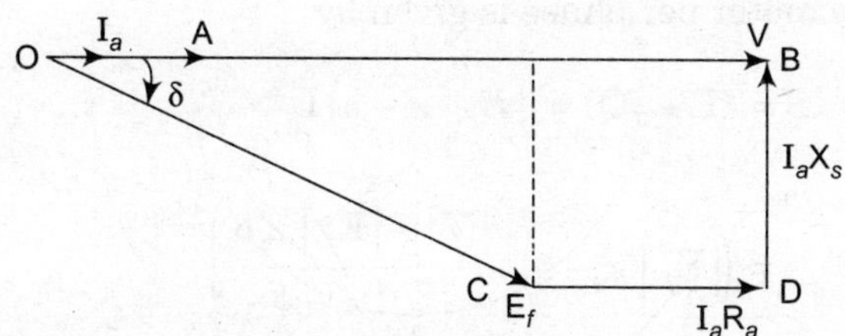

Fig. 15.14 *Phasor diagram of synchronous motor at unity power factor.*

$$E_f = OC = \sqrt{(OB - CD)^2 + BD^2}$$

$$E_f = \sqrt{(V - I_a R_a)^2 + (I_a X_s)^2}$$

At leading power factor,

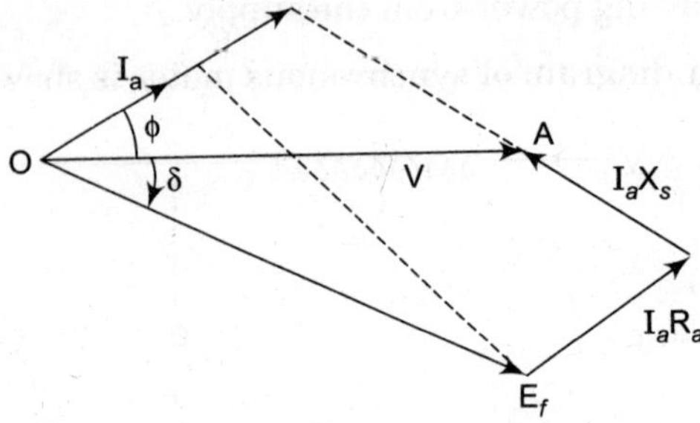

Fig 15.15 *Phasor diagram of synchronous motor at leading power factor.*

$$E_f = \sqrt{\left[(V \cos \phi - I_a R_a)^2 + (V \sin \phi + I_a X_s)^2\right]}$$

15.14 POWER DEVELOPED BY SYNCHRONOUS MOTOR

The synchronous motor is connected to the infinite bus as shown in Fig. 15.16.

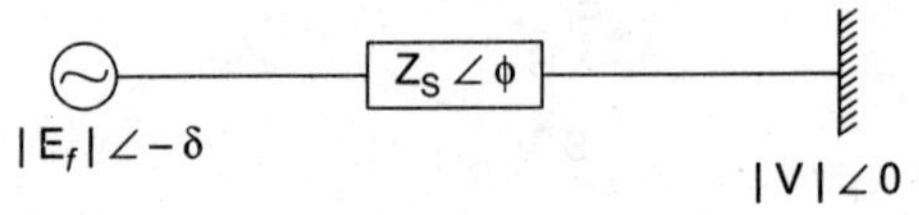

$Z_S \angle \phi$

$|E_f| \angle -\delta$ $\qquad$ $|V| \angle 0$

Fig. 15.16 *Synchronous motor connected to Infinite bus.*

The current drawn $\qquad I_a = \dfrac{|V| - |E_f| \angle -\delta}{Z_s \angle \phi}$

Congugate of current

$\therefore \qquad I_a^* = \dfrac{|V| - |E_f| \angle \delta}{Z_s \angle -\phi}$

Power developed by the motor per phase is given by

$$S = (P + jQ) = \left(|E_f| \angle -\delta\right) I_a^*$$

$$= \left(|E_f| \angle -\delta\right) \left(\frac{|V| - |E_f| \angle \delta}{|Z_s| \angle -\phi} \right)$$

$$= \frac{|E_f||V|}{|Z_s|} \angle \phi - \delta - \frac{|E_f|^2}{|Z_s|} \angle \phi$$

$\therefore$ Real power, $$P = \frac{|E_f||V|}{|Z_s|}\cos(\phi - \delta) - \frac{|E_f|^2}{|Z_s|}\cos\phi$$

Reactive power $$Q = \frac{|E_f||V|}{|Z_s|}\sin(\phi - \delta) - \frac{|E_f|^2}{|Z_s|}\sin\phi$$

Since armature resistance is very small so it can be neglected:

$$P = \frac{|E_f||V|}{X_s}\sin\delta \ (\phi = 90°)$$

$$\therefore \quad P = P_m \sin\delta$$

where $P_m = \dfrac{|E_f||V|}{X_s}$ is the maximum power developed by the motor,

Gross torque developed by the motor is

$$T = \frac{P_m}{w_s} = \frac{P_m}{2\pi N_s}$$

where N_s is synchronous speed of motor in r.p.s.

Since power at the shaft $P_{sh} = P_m$ – Rotational losses

$\therefore$ Gross torque developed by the motor is given as:

$$T = \frac{P_{sh}}{w_s} = \frac{P_{sh}}{2\pi N_s}$$

Power input to the motor is given by

$$S_i = P_i + jQi$$

$$= V I_a^*$$

$$= V\left[\frac{|V| - |E_f|\angle\delta}{|Z_s|\angle -\phi}\right] = \frac{|V|^2}{|Z_s|}\angle\phi - \frac{|E_f||V|}{|Z_s|}\angle(\delta + \phi)$$

$$P_i = \frac{|V|^2}{|Z_s|}\cos\phi - \frac{|E_f||V|}{|Z_s|}\cos(\delta + \phi)$$

$$P = |V|^2 \text{ since } R_a \text{ is very small so } \phi = 90°$$

$$P_i = -\frac{|E_f||V|}{Z_s}\cos(\delta - 90°)$$

$$P_i = -\frac{|E_f||V|\sin\delta}{Z_s}$$

and $$Q_i = \frac{|V|^2}{|Z_s|}\sin\phi - \frac{|E_f||V|}{|Z_s|}\sin(\delta + \phi)$$

$$Q_i = \frac{|V|^2}{|X_s|} - \frac{|E_f||V|}{|X_s|}\cos\delta$$

Neglecting R_a being very small

so $$Q_i = \frac{|V|}{|X_s|}\left[|V| - E_f \cos\delta\right]$$

The field excitation is increased still further, so that e.m.f. induced is E_3 and the stator current I_3, leading to applied voltage by an angle δ_3 as shown in Fig. 15.15. Therefore, the following be concluded:

1. Synchronous motor draws a lagging current, when it is under excited.
2. At normal excitation, it draws minimum armature current.

SOLVED NUMERICAL PROBLEMS

Example 1. *A three-phase, 50 Hz, 20 poles salient pole alternator with star connected stator winding has 180 slots on the stator. Each slot consists of 8 conductors. The flux per pole is 25 mWb and is sinusoidally distributed. The coils are full pitch. Calculate (i) the speed; (ii) the generated e.m.f. per phase and (iii) the line e.m.f. where $K_d = 0.96$ and $K_c = 1.0$.*

Solution:

(*i*) Frequency $f = 50$ Hz

Number of poles, $P = 20$

Speed at which the alternator must be run, $N = \frac{f \times 120}{N}$

$$= \frac{50 \times 120}{20} = 300 \text{ rpm}$$

Since $K_d = 0.96$ and $K_c = 1.0$

Winding factor $K_d = K_d \cdot K_c = 0.96 \times 1.0 = 0.96$

Total number of conductors, $Z = 180 \times 8 = 1440$

Conductors per phase $= \frac{1440}{3} = 480$

Turns per phase $= \frac{480}{2} = 240$

Flux per pole, $\phi = 25 \times 10^{-3}$ Wb

Generated e.m.f. per phase, $E_{Ph} = 4.44 \text{ kW} f \phi T$ Volts

$= 4.44 \times 0.96 \times 50 \times 25 \times 10^{-3} \times 240$

$= 1279$ Volts.

(*ii*) The stator is star connected, as such

the line e.m.f. $E_L = \sqrt{3}\, E_{Ph}$

$= \sqrt{3} \times 1279 = 2215$ Volts. **Ans.**

Example 2. *A 3.3 kV, three-phase star connected alternator has a full load current of 100 A. Under short circuit condition, it takes 5 A field current to produce full-load short circuit current. The e.m.f. of open circuit for the same excitation is 700 volt (line to line). The armature resistance is 1.0 Ω/phase. Determine synchronous reactance per phase and regulation for (i) 0.8 Pf lagging and (ii) 0.9 Pf leading.*

Solution:

Terminal voltage/phase during O.C.C. = $\frac{700}{\sqrt{3}}$ = 404.15 Volt

Synchronous impedance is expressed as

$$Z_s = \left.\frac{\text{O.C. voltage}}{\text{S.C. voltage}}\right|_{\text{for same field excitation}}$$

$$= \frac{404.15}{100} = 4.042\ \Omega$$

Synchronous reactance $X_s = \sqrt{Z_s^2 - R_a^2}$

$$= \sqrt{(4.042)^2 - (1)^2} = 3.92\ \Omega$$

(*i*) Percentage regulation for 0.8 p.f. lagging

Rated phase voltage, $V = \frac{3.3 \times 1000}{\sqrt{3}}$ = 1905.3 Volts.

$$E_0 = \sqrt{(V \cos\phi - I R_a)^2 + (V \sin\phi + I X_s)^2}$$

$$= \sqrt{(1905.3 \times 0.8 + 100 \times 1)^2 + (1905.3 \times 0.6 + 100 \times 3.92)^2}$$

$$= \sqrt{2638155.6 + 1545496.5} \qquad \left\{\begin{array}{l}\cos\phi = 0.8\\ \text{so, } \phi = 36.86\\ \sin\phi = 0.6\end{array}\right.$$

= 2045.39 Volt

∴ Percentage regulation = $\frac{E_0 - V}{V} \times 100 = \frac{2045.39 - 1905.3}{1905.3} \times 100 = 7.35\%$

(*ii*) Percentage regulation for 0.8 p.f. leading :

$$E_0 = \sqrt{(V \cos\phi - I R_a)^2 + (V \sin\phi - I X_s)^2} \qquad \left\{\begin{array}{l}\cos\phi = 0.9\\ \phi = 25.84°\\ \sin\phi = 0.44\end{array}\right.$$

$$= \sqrt{(1905.3 \times 0.9 + 100 \times 1)^2 + (1905.3 \times 0.44 - 100 \times 3.92)^2}$$

$$= \sqrt{(1814.77)^2 + (446.33)^2} = 1868.85 \text{ Volts}$$

$$\therefore \text{ Percentage regulation} = \frac{E_0 - V}{V} \times 100$$

$$= \frac{1868.85 - 1905.30}{1205.30} \times 100 = -1.91\%. \textbf{ Ans.}$$

Example 3. *A 440 V, single-phase synchronous motor gives a net output mechanical power of 7.5 kW and operates at 0.9 p. f. lagging. If effective resistance is 0.8 Ω. If the iron and mechanical losses are 550 W and excitation losses are 750 W. Calculate (i) armature current and (ii) efficiency.*

Solution:

Given supply voltage = 440 V

Net output mechanical power = 7.5 Ω

Power factor = 0.9 lagging

Effective resistance = 0.8 Ω

Iron and mechanical losses = 550 W

Excitation losses = 750 W

(*i*) *Armature Current:*

Let I be the input armature current

Therefore,

Power input = VI cos φ

= 440 × 1 × 0.9 = 396 IW

Power output = Power input – Mechanical losses – Copper losses

$$7500 = 396\,I - 550 - I^2 \times 0.8$$

$$0.8\,I^2 + 550 + 7500 - 396\,I = 0$$

$$0.8\,I^2 - 396\,I + 8050 = 0$$

$$I^2 - 4951 + 10062.5 = 0$$

$$I_{1,2} = \frac{-b + \sqrt{b^2 - 4ac}}{2a}$$

$$I_{1,2} = \frac{495 \pm \sqrt{(-495)^2 - 4 \times 1 \times 10062.5}}{2 \times 1}$$

$$I_{1,2} = \frac{495 \pm \sqrt{204781}}{2}$$

$$I_{1,2} = \frac{495 \pm 452.53}{2}$$

$$I_1 = \frac{495 + 452.53}{2} = 473 \text{ amp (Practically not possible)}$$

$$I_2 = \frac{495 - 452.53}{2} = 21.24 \text{ amp}$$

So consider, armature current = 21.24 amp.

(*ii*) *Efficiency:*

Power input including excitation losses

$$= (440 \times 21.24 \times 0.9 + 750) = 9161.04$$

$$= 9161.04 \text{ W or } 9.16 \text{ kW}$$

Efficiency $\eta = \frac{\text{Output}}{\text{Input}} \times 100 = \frac{7500}{9161.04} = 81.86\%.$ **Ans.**

Example 4. *A three-phase star-connected, 1000 KVA, 11000 volt alternator has rated current of 50.0 A. The armature resistance of winding per phase is 0.50 Ω. The test results are given below:*

O.C. test: Field current = 12.5 A, Voltage between lines = 422 V

S.C. test: Field current = 12.5 A, Line current = 5010 Amp.

Determine the full-load voltage regulation of the alternator for 0.8 p.f. lagging.

Solution:

The rating = 1000 KVA, 11000 V, I = 50.0 Amp, $R_a = 0.50\ \Omega$

Since alternator is star connected, therefore

$$I_{Ph} = I_L = 50.0 \text{ Amp, O.C. line voltage} = 422 \text{ volt}$$

Then, synchronous impedance $Z_s = \frac{\text{O.C. phase voltage}}{\text{Rated current per phase}}$

$$= \frac{\frac{422}{\sqrt{3}}}{50.0} = 4.87 \text{ ohm.}$$

Synchronous reactance $X_s = \sqrt{Z_s^2 - R_a^2}$

$$= \sqrt{(4.87)^2 - (0.50)^2} = 4.85\ \Omega$$

$$E_{Ph} = V = \frac{11000}{\sqrt{3}} = 6350.85 \text{ V} = 6351 \text{ volt}$$

(*i*) Voltage regulation at 0.8 p.f. lagging

$$\cos\phi = 0.8,\ \phi = 36.86 \text{ so, } \sin\phi = 0.6$$

Therefore, $K_0 = \sqrt{(V\cos\phi + IR_a)^2 + (V\sin\phi + IR_s)^2}$

$$= \sqrt{(6351 \times 0.8 + 50 \times 0.5)^2 + (6351 \times 0.6 + 50 \times 4.85)^2}$$

$$= \sqrt{(5105.8)^2 + (4053.1)^2} = 6518.96 \text{ volt}$$

$$\therefore \quad \text{Percentage regulation} = \frac{E_0 - V}{V} \times 100$$

$$= \frac{6518.96 - 6351}{6351} \times 100 = 2.64\%. \textbf{ Ans.}$$

Example 5. *A three-phase star-connected alternator is delivering 20 MW and 8 MVAR to an infinite bus at 11 kV. The alternator has synchronous impedance of (0 + j5) Ω. Determine the load angle and the excitation e.m.f of the alternator.*

Solution: Since $\overline{E_0} = \overline{V} + \overline{I}\,\overline{Z_s}$

Taking V as reference, $VI \cos\phi = 20 \text{ MW}$

$$= 20 \times 10^6 \text{ M}$$

$$VI \sin\phi = 8 \text{ MVAR} = 8 \times 10^6 \text{ VAR}$$

$$\therefore \quad \tan\phi = \frac{8}{20} = 0.4 \quad \text{or} \quad \phi = \tan^{-1}(0.4)$$

$$\phi = 21.8°$$

Load current,
$$I = \frac{20 \times 10^6}{\sqrt{3} \times 11000 \times \cos(21.8)} = 1130.6 \text{ amp.}$$

Excitation e.m.f. of the alternator (per phase) is expressed as:

$$\overline{E_f} = \left(\frac{11 \times 1000}{\sqrt{3}} \angle 0°\right) + (1130.6 \angle -21.8°) \times 5 \angle 90°$$

$$6350.8 + 5653 \angle 68.2° = 6350.8 + 2099.3 + j5248.7$$

$$= 8450.1 + j5248.7$$

$$= 9947.51 \angle 31.84°$$

$$P = \frac{E_0 \cdot V}{X_s} \sin\delta$$

$$\sin\delta = \frac{PX_s}{E_0 V} = \frac{\frac{20}{3} \times 10^6 \times 5}{9947.51 \times 63508}$$

$$\sin\delta = 0.53$$

$$\delta = \sin^{-1}[0.53] = 31.84°$$

Load angle $\delta = 31.84°$. **Ans.**

EXERCISES

1. A 500 V, 33 KVA, three-phase alternator, has an effective resistance of 0.3 Ω. A field current of 11 amp produces an armature current of 200 A on short circuit and an e.m.f. of 400 volt on open circuit. Calculate the full load regulation at Pf 0.8 lagging.
2. A three-phase 11 KVA, 440 V, 50 Hz, star-connected alternate supplies the rated load at 0.8 P_f lagging. If the armature resistance is 0.5 Ω and synchronous reactance is 8 Ω, find the power angle and voltage regulation.
3. A 500 V, three-phase synchronous motor gives a net output mechanical power of 7.5 kW and operates at 0.8 p.f. lagging. Its effective resistance is 0.75 Ω. If the iron and friction losses are 500 W and excitation losses are 800 W, estimate the armature current. Calculate the efficiency.
4. A three-phase, 50 Hz 600 KVA, 6000 V star-connected alternator has an effective resistance of 0.2 Ω. A field current of 10 A produces 480 V on open circuit and a field current of 5 A gives armature current of 105 amp. Calculate the voltage regulation of this alternator at 0.8 power factor lagging.
5. A three-phase, 3.3 kV, star connected synchronous motor has a synchronous reactance of 5 Ω per phase. The input to the motor is 1000 kW at normal voltage and the induced line e.m.f. is 4 kV. Calculate the line current, power factor, and mechanical power output. Take stray losses as 20 kW. Neglect the armature resistance.
6. Calculate the voltage induced per phase in a three-phase, 50 Hz, alternator having a flux per pole of 0.1515 Wb. The number of conductors are 360. Assume full pitch coil with a distribution factor of 0.96.
7. A three-phase 10-pole star connected alternator runs at 720 rpm. It has 120 stator slots with 10 inductors per slot. If the flux per pole is 0.056 Wb, determine the phase and line induced e.m.f.
8. Derive an expression for the voltage induced in an alternator, phase consisting of a number of full pitch coils joined in series. Assume the air gap flux to have sinusoidal distribution.
9. Explain what is meant by the synchronous reactance of an alternator and voltage regulation.
10. Why a 3-phase synchronous motor does not have any self-starting torque? Also discuss with a suitable phasor diagram the behaviour of three-phase synchronous motor at no load.
11. Explain the effect of change of load on the performance of three-phase synchronous motor.
12. In case of a three-phase synchronous motor, explain how the power factor at which the motor operates can be controlled by varying the field excitation.

16

Electrical Measurements and Measuring Instruments

16.1 INTRODUCTION

This is the process in which an unknown quantity is compared with standard quantity and the result so obtained is followed by some numeric value and dimension.

According to Lord Kelvin, "I often say that when you can measure what you are speaking about and can express it in numbers you know something about it; when you cannot express it in numbers your **knowledge is of a meagre and unsatisfactory kind.**"

16.2 CLASSIFICATION OF MEASURING INSTRUMENTS

Measuring instruments are of two types:

1. Absolute instrument
2. Secondary instrument.

Absolute Instrument

In this type of instrument the unknown quantity is compared in terms of the physical constant of instrument and the measurement is done by deflection of the pointer. This type of instrument called absolute instrument. For example, Tangent Galvanometer.

Secondary Instrument

These type of instruments are made by calibration of absolute instrument and the measurement of unknown parameters are determined by the deflection of pointer only. Secondary instruments work in two modes:

1. Analog mode
2. Digital mode

1. **Analog mode:** In analog mode of operation, the signals, which are used, vary in a continuous fashion and takes on an infinite number of values in any given range as shown in Fig. 16.1. The analog signals are sinusoidal in nature.

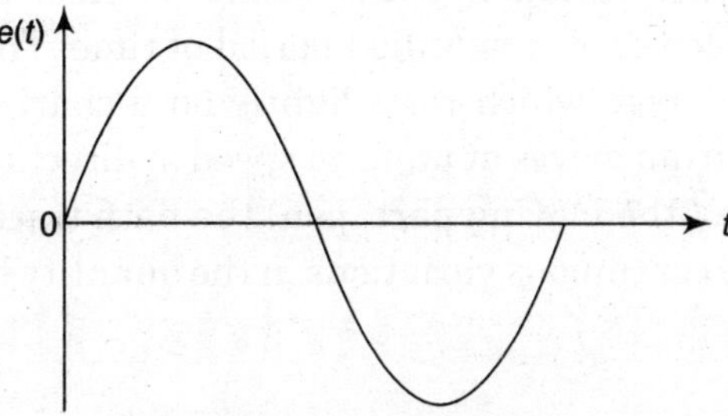

Fig. 16.1 *Continuous signal.*

The instrumental working on analog mode of operation is known as analog instruments. The analog instruments are those type of instruments in which the quantity under measurement is expressed by the deflection of pointer in the instrument.

2. **Digital mode:** The signals, which are used in the digital mode of operation, are discrete in nature *i.e.* discontinuous in the form of pulses. These signals vary in discrete steps and thus take up only finite different values in a given range as shown in Fig. 16.2.

The instruments working in digital mode of operation is known as digital instruments are those instruments in which the quantity under measurement is expressed in decimal number.

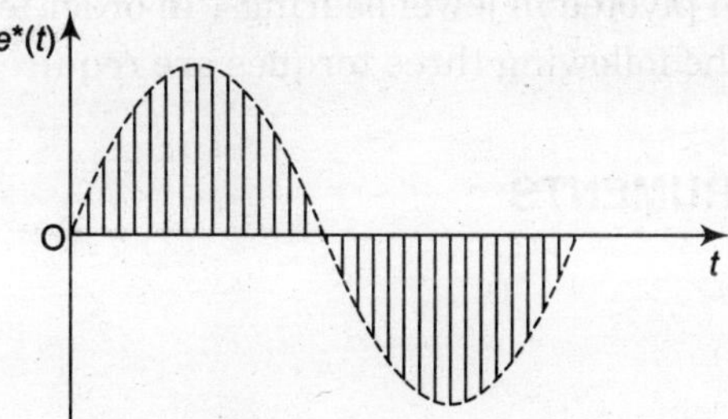

Fig. 16.2 *Discrete signal.*

The analog instruments are three types:

(*i*) Indicating instrument

(*ii*) Integrating instrument

(*iii*) Recording instruments.

(*i*) **Indicating instruments:** Indicating instruments are the types of instrument in which the values of quantity under measurement are measured by the deflection of the pointer on a graduated scale. For example, ammeter, voltmeter, wattmeter, etc.

(*ii*) **Integrating instruments:** Integrating instruments are those instrument which gives the values of quantity under measurement in a specified duration of time, amount of electrical energy (kWh) supplied to a circuit in a specified duration of time. Energy meter, Ampere hour meter is the best example of integrating instrument it gives.

(*iii*) **Recording instruments:** Recording instruments are those instruments, which gives a continuous record of the variation of the quantity under measurement. For example, current, voltage and power over specified period of time. The moving part of the instrument carries an inked pen, which rests lightly on a chart or graph sheet carried by a revolving drum. The drum moves at uniform speed in direction perpendicular to the direction of the deflection of the moving part (pen) the path traced out on the chart or graph sheet by pen gives the continuous variations in the quantity being recorded. For example, X-Y plotter.

16.3 CLASSIFICATION OF INSTRUMENTS

Indicating instruments are of the following types:

(*i*) Moving iron instrument
(*ii*) Moving coil instrument
(*iii*) Dynamometer type instrument
(*iv*) Hot wire type instrument
(*v*) Electrostatic type instrument
(*vi*) Induction type instrument

All such instrument essentially consists of a pointer moving over a calibrated scale and attached to the moving system pivoted in jewel bearings. In order to ensure the proper operation of indicating instruments, the following three torques are required:

16.4 TORQUE IN INSTRUMENTS

1. Deflecting torque. (T_d)
2. Controlling torque (T_c)
3. Damping torque (D)

1. **Deflecting torque:** One important requirement in an indicating instrument is the arrangement for producing deflecting or operating torque when the instrument is connected in the electrical circuit to measure the given electrical quantity. The deflecting torque causes the moving system to move from zero position when the instrument is connected to the circuit to measure the given electrical quantity.

2. **Controlling torque:** Due to the action of deflecting torque the pointer will continue to move indefinitely and shall be independent of the value of electrical quantity to be measured. This necessitates that the controlling torque must be provided. This controlling torque should oppose the deflecting torque and should with increase with the deflection of the moving system so that the pointer is brought to rest at a position when the opposing torques are equal. The controlling torque performs the following two functions:

(*i*) It opposes the deflecting torque and increases with the deflection of the moving system.

(*ii*) It brings the pointer back to zero position after the removal of deflecting torque.

The controlling torque is provided by the following methods:

(*a*) Spring control (*b*) Gravity control

(*a*) *Spring control:* The spring control is provided in most of the indicating instruments. Basically the controlling torque is obtained either by twisted suspension wire or two spirally wound springs. In spring control the two hairsprings are made from non-magnetic alloy. Such as phosphor-bronze. The two springs are wound in opposite direction on the spindle S of the moving system. The two springs are labelled as S_1 and S_2 shown in Fig. 16.3.

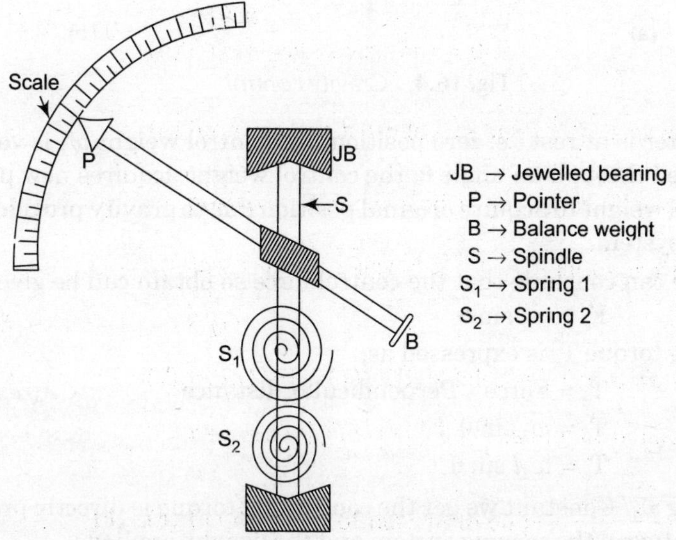

Fig. 16.3 *Spring control.*

The spring so used should not be subjected to fatigue and should have very low specific resistance and temperature coefficient. The two springs are wound in opposite direction as when the deflecting torque is applied to moving system; one spring closes while other opens. Resulting in reduction in effect of temperatures change of two springs. The control torque is due to the combined torsion of spring S_1 and S_2.

The control torque so, produced is directly proportional to the angular deflection θ of the pointer.

Let, T_c = Controlling torque

θ = Angular deflection

Therefore

$$T_c \propto \theta$$

or, $$T_c = k\,\theta$$

where k is the spring constant.

(*b*) *Gravity control:* In gravity controlling, a small adjustable weight w_1 is attached to the moving system. Another weight w_2 is also attached to the moving system in order to counter balance the weight of moving system as shown in Fig. 16.4.

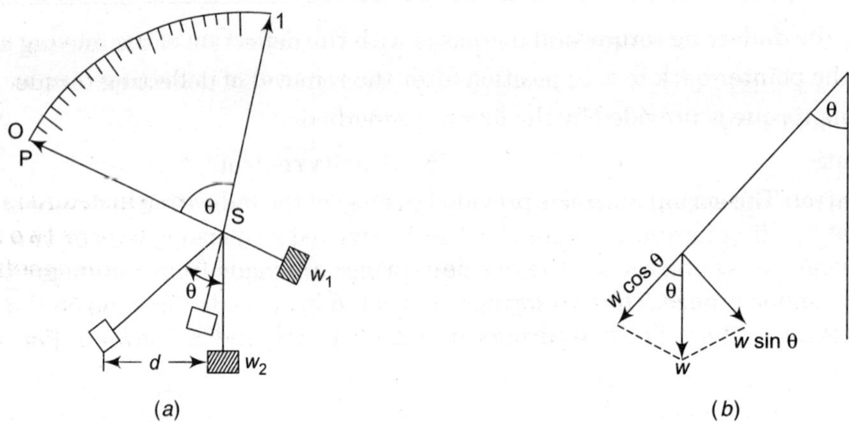

Fig. 16.4 *Gravity control.*

When the pointer is at rest *i.e.* zero position, the control weight w_1 is vertical. When the pointer deflected through an angle θ, the control weight acquires new position. The tendency of control weight to acquire original position due to gravity provides the controlling to the moving system.

From above we can conclude that the control force so obtain can be given by,

$$F_c = w_1 \sin \theta$$

And controlling torque T_c is expressed as,

$$T_c = \text{Force} \times \text{Perpendicular distance}$$

$$T_e = w_1 \sin \theta . l$$

$$T_c = w_1 l \sin \theta$$

Here, assuming $w_1 l$ Constant we get the controlling torque is directly proportional to the sine of angle between the moving system and the weight-applied w_1.

3. **Damping torque:** If the moving system is acted upon by deflecting torque alone then the pointer due to its inertia will oscillate about final position before coming to rest. These oscillations are undesirable and must be prevented. In order to avoid oscillation of the pointer and to bring it quickly to its deflected position, damping torque is provided, which opposes the movement in backward or forward of the pointer and operates only when the system is moving.

It is worthwhile to mention here that damping torque acts only when the pointer is in motion. When the pointer is in a particular final position though deflecting and controlling torques are acting on the moving system but the damping torque is zero because the pointer is steady and there is no movement of the moving system. In fact damping torque act like a brake on the moving system.

The degree of damping decides the behaviour of the moving system. If the instrument is under-damped the pointer oscillate about the final position and take some time to come to rest. On the other hand, if the instrument is over-damped, the pointer will become slow and lethargic. However, if the degree of damping is adjusted to such a value that the pointer moves and quickly come to its final position, the instrument is said to be deadbeat. Figure 16.5 shows the graph for under-damping, over-damping and critical-damping (deadbeat).

Method commonly used to provide damping are:

(*i*) Air friction damping

(*ii*) Eddy current damping

(*iii*) Fluid damping.

Classification of moving iron instruments

MI instruments are of two types:

1. Attraction type
2. Repulsion type.

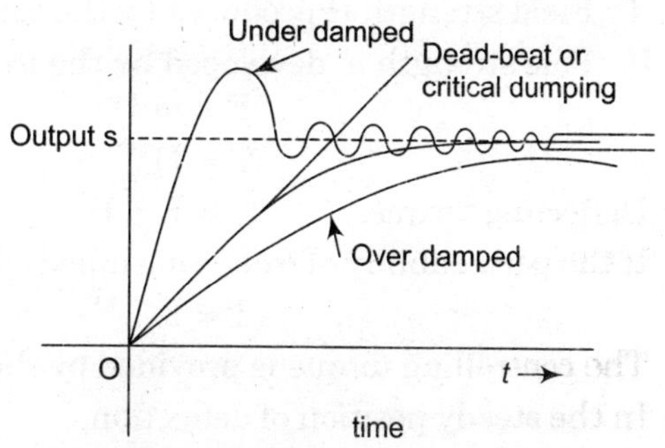

Fig. 16.5 *Graphes for damping*

1. Attraction type moving instruments

Principle: These instruments are based on the principle that when an unmagnetised soft iron piece is placed in the magnetic field, then the piece is attracted towards the coil. The moving system of the instrument is attached to the soft iron piece and the operating current is passed through a coil placed near it. The operating current sets up magnetic field, which attracts the iron piece and thus moves the pointer, which is attached to moving part over the scale.

Construction: It consists of hollow cylindrical coil or solenoid, which is kept fixed as shown in Fig. 16.6. An oval shaped soft iron piece is attached to the spindle such a way that it can move in or out of the coil. The pointer is attached to the spindle so that it is deflected with the motion of the soft iron piece. The controlling torque on the moving system is provided by spring control method while damping is provided by air friction.

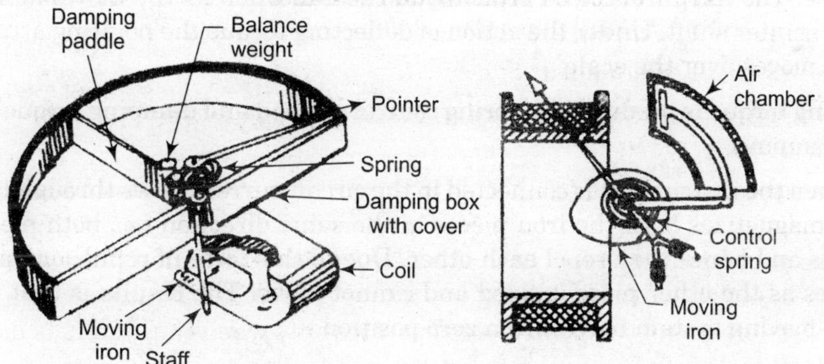

Fig. 16.6 *Moving iron attraction type constructional feature.*

Working: When the instrument is connected in the circuit the operating current flows through the coil the current sets up magnetic field in the coil. In other words, the coil behaves like a magnet and therefore it attract the soft iron piece towards it. The result is that the pointer attached to the moving system moves from zero position.

If current in the coil is reversed the direction of magnetic field also reverses and so does the magnetism produced in the soft iron piece. Therefore, the direction of deflecting torque remains unchanged. It follows, therefore that such instruments can be used for both d.c. as well as a.c. work.

Deflecting torque: The force F pulling the soft iron piece towards the coil depends upon:

I. Field strength H produced by the coil

II. Pole strength m developed by the iron piece

$$F \propto m\,H$$

$$F \propto H^2$$

Deflecting torque, $T_d \propto F \propto H^2$

If the permeability of iron is assumed constant, $H \propto I$

$$T \propto I^2$$

The controlling torque is provided by the springs, $T_e \propto \theta$ (deflection)

In the steady position of deflection,

$$T_d = T_c$$

$$\theta \propto I^2 \qquad \text{(for d.c.)}$$

$$\propto I_{rms}^2 \qquad \text{(for a.c.)}$$

Since deflection $\theta \propto I^2$ therefore scale of such instruments is non-uniform being crowded in the beginning. In order to make the scale of such instruments uniform, suitably shaped iron piece is used.

2. Repulsion type moving instruments

Principle: These instruments are based on the principle of repulsion between the two iron pieces similarly magnetized.

Construction: It consists of a fixed cylindrical hollow coil which carries the operating current (see Fig. 16.6). Inside the coil there are two soft iron pieces or vanes one of which is fixed and other is movable. The fixed iron piece attached to the coil whereas the movable iron piece is attached to the pointer shaft. Under the action of deflecting torque the pointing attached to the moving system moves over the scale.

The controlling torque is produced by spring control method and damping torque is provided by air friction damping.

Working: When the instrument is connected in the circuit current flows through the coil. The magnetic field magnetizes both the iron pieces in the same direction *i.e.*, both pieces become similar magnets and hence they repel each other. Due to this force of repulsion only movable iron piece moves as the other piece is fixed and cannot move. The results is that the pointer attached to the moving system moves from zero position.

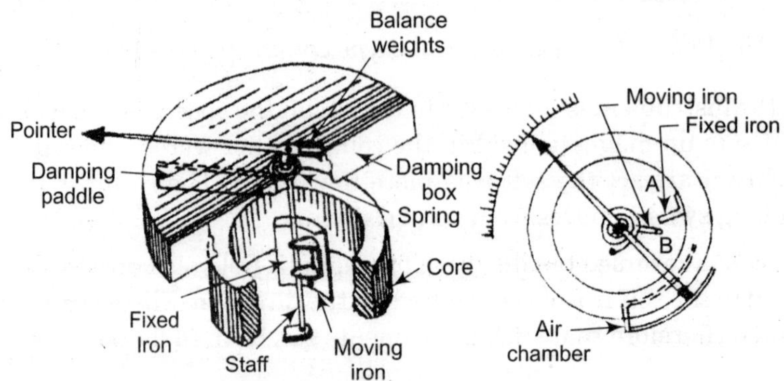

Fig. 16.7 *Moving iron repulsion type constructional feature.*

Deflecting torque: The deflecting torque result due to the repulsion between the similarly changed iron pieces. If the two pieces develop pole strength m_1 and m_2 respectively, then,

Instantaneous deflecting torque $\propto$ repulsive force

$$\propto m_1 m_2$$

Since pole strengths developed are proportional to H,

Instantaneous deflecting torque,

$$T_d \propto H^2$$

Assuming constant permeability H α current I through coil

Deflecting torque $\quad T_d \propto I^2$

Controlling torque provided by springs $T_c \propto \theta$

In the steady position of deflection,

$$T_d = T_c$$

$$\theta \propto I^2$$

$$\propto I^2 \quad \text{(for d.c.)}$$

$$\propto I_{rms} \quad \text{(for a.c.)}$$

Since deflection θ is proportional to square of current through the coil therefore scale of such instruments is non-uniform; being crowded in the beginning. However, scale of such instruments can be made uniform by using torque shaped iron pieces.

16.5 MERITS AND DEMERITS OF MOVING IRON INSTRUMENTS

Merits: The moving iron instruments have the following advantages:

1. These are cheap, robust and simple in construction.
2. The instruments can be used for both a.c. as well as d.c. circuits.
3. These instruments have high operating torque.
4. These instruments are reasonably accurate.

Demerits: The following are the disadvantages of moving iron instruments:

1. Such instruments have non-uniform scale.
2. These instruments are not very sensitive.
3. Errors are introduced due to change in frequency in case of a.c. measurements.
4. Higher power consumption.

16.6 MOVING COIL INSTRUMENTS

Moving coil instruments are of two types namely:

1. Permanent magnet type (for d.c. work only)
2. Dynamo-meter type (for both a.c. and d.c. work)

Permanent Magnet Type Moving Coil Instruments

These instruments are employed either as ammeters or voltmeters and can be used for d.c. work only.

Principle: This type of instrument is based on the principle that when a current carrying conductor is placed in a magnetic field mechanical force acts on the conductor. The coil placed field and carrying operating current is attached to the moving system. With the movement of the coil the pointer moves over the scale.

Construction: It consists of a powerful permanent magnet with soft iron pieces and light rectangular coil of many turns of fine wire wound on aluminium former inside which is an iron core shown as in Fig. 16.7. The purpose of coil is to make the field uniform. The coil is mounted on the spindle and acts as the moving element. The current is led into and out of the coil by means of the two control hair springs one above and the other below the coil. The springs also provide the controlling torque. Eddy current damping is provided by aluminium former.

Working: When the instrument is connected in the circuit, operating current flows through the coil. This current carrying coil is placed in the magnetic field produced by the permanent magnet and therefore mechanical force act in the coil. As the coil is attached to the moving system the pointer moves over the scale.

It may be noted here that if current direction is reversed the torque will also be reversed since the direction of the field of the permanent magnet is same. Hence the pointer will move in the opposite direction, *i.e.*, it will go on the wrong side of zero (below zero). In other words, these instruments work only when current in the circuit is passed in a definite direction *i.e.*, for d.c. circuit only.

It is worthwhile to mention here that such instruments are called permanent magnet moving coil instruments because a coil moves in the field of permanent magnet.

Deflecting Torque

When the current is passed through the coil a deflection torque is produced due to the reaction between permanent magnet field and the magnetic field of the coil as shown in Fig. 16.8.

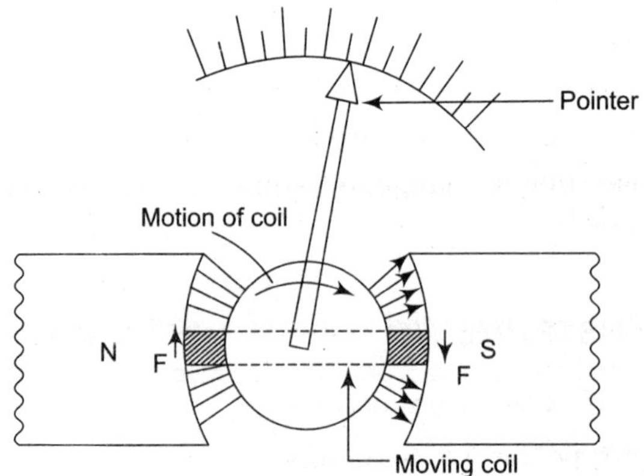

Fig. 16.8 *Constructional feature of PMMC instrument.*

Let B = Flux density in $\frac{Wb}{m^2}$ in the air gap between the magnetic poles and iron core.

l = Active length of the coil side in meters.

N = Numbers of turns of coil

If a current of I amperes flows in direction shown, then force F acting on each coil side is given by F = BIlN Newton's

Deflecting torque $\quad T_w$ = Force × Perpendicular distance

$= F \cdot 2r$ Newton meters

Where, r is the distance of coil side from the axis of the rotation in meters (see Fig. 16.7)

or $\quad T_d = N\,B\,I\,l.2\,r$ Newton meters

If Am^2 (= l.2r) is the face area of the coil then;

$T_d = N\,B\,I\,A$ Newton meters

i.e., Deflection torque = Ampere turns on coil (NI). Area of coil (A). Flux density B

Since N, B and A are constant therefore,

$$T_d \propto I \qquad ...(i)$$

Since the control by springs therefore controlling torque is proportional to the angle of deflection *i.e.*,

$$T_c \propto I \qquad ...(ii)$$

The pointer will come to rest at a position when $T_d = T_c$

From equations (*i*) and (*ii*), we get

$$\theta \propto I$$

Thus the deflection θ is directly proportional to the operating current. Therefore such instruments have uniform scale.

Advantages:

(*i*) Uniform scale

(*ii*) Very effective eddy current damping

(*iii*) Power consumption is low

(*iv*) No Hysteresis loss

(*v*) As working field is very strong therefore such instruments are not affected by stray fields.

(*vi*) Such instruments require small operating current

(*vii*) Very accurate and reliable.

Disadvantages:

(*i*) Such instruments cannot be used for a.c. measurements.

(*ii*) Some errors are caused due to the ageing of control springs and the permanent magnet.

(*iii*) Costlier as compared to moving iron instruments.

Extension of Range of PMMC Instruments

The range of PMMC instruments can be extended by using shunt and multipliers:

Extension of range of ammeter by using shunt: The shunts are basically used to bypass the high current in the circuit. The shunt is very low resistor, which is connected in parallel with the ammeter. The basic movement of meter of a d.c. ammeter in a PMMC is *d'*Arsonoval galvanometer. The coil winding of a basic movement meter is small and light and can carry very small currents. Figure 16.9 shows the basic movement and its shunt in ammeter circuit.

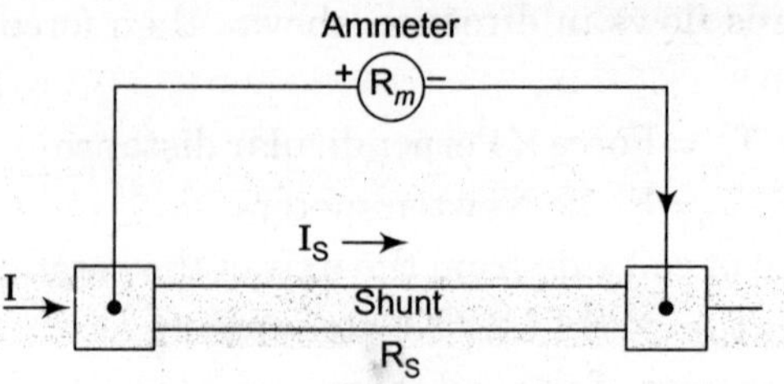

Fig. 16.9 *Ammeter shunt.*

By KVL, the shunt resistance can be calculated.

Where in circuit,

R_{sh} = Shunt resistance

I_{sh} = Shunt current

R_m = Resistance of movement in coil.

I_m = Deflection current of movement

I = Current to be measured.

In above circuit shunt is connected in parallel with the meter movement. Therefore, the voltage drop across shunt and movement will be same. And can be given as

$$I_{sh} R_{sh} = I_m R_m$$

or
$$R_{sh} = \frac{I_m R_m}{I_{sh}}$$

Also
$$I_{sh} = (I - I_m)$$

Therefore we can write

$$R_{sh} = \frac{I_m R_m}{(I - I_m)}$$

or
$$\left(\frac{I}{I_m}\right) - 1 = \frac{R_m}{R_{sh}}$$

or
$$I - I_m = 1 + \frac{R_m}{R_{sh}}$$

This ratio of total current to the current in the movement is called as multiplying power of shunt.

Therefore, multiplying power

$$M = \frac{I}{I_m} = 1 + \frac{R_m}{R_{sh}}$$

or shunt resistance,
$$R_{sh} = \frac{R_m}{(m - 1)}$$

Requirement for shunt:

(*i*) The temperature coefficient for shunt and instrument should be low.

(*ii*) Shunt resistance should be time invariant.

(*iii*) Should have low thermal electromotive force with copper.

(*iv*) Voltmeter multiplier.

Dynamometer Type Instruments

These instruments are the modified form of permanent magnet type. Here operating field is produced not by a permanent magnet but by another fixed coil. The moving system and the control system are similar to those of permanent magnet type. Such instruments can be used for both a.c. and d.c. circuits. They can be used as ammeters and voltmeters but are generally used as wattmeters.

Principle: These instruments are based on the principle that mechanical force exists between the current carrying conductors.

Construction: A dynamometer type instruments essentially consists of a fixed coil and a moving coil. The fixed coil is split into two equal parts, which are placed close together and parallel to each other. The moving coil is pivoted between the two-fixed coils. The fixed and moving coils may be excited separately or they may be connected in series depending upon the use to which the measurement is put. The moving coil is attached to the moving system so that the action of deflecting torque the pointer moves over the scale.

The controlling torque is provided by two springs which also serve the additional purpose of leading the current into and out of the moving coil. Air friction damping is provided in such instruments.

Working: When the instrument is connected in the circuit operating currents flow the coils. Due to this mechanical force exists between the coils. The result is that the moving the coil (which is capable of moving moves the pointer over the scale. The pointer comes to rest at a position where deflecting torque is equal to the controlling torque.

By reversing the current the field due to the fixed coils is reversed as well as the current in the moving coil so that the direction of deflecting torque remain unchanged. Therefore such instrument can be used for both d.c. and a.c. measurement.

Deflecting Torque

Let I_r = Current through fixed coil

I_m = Current through moving coil

$$T_d \propto I_t I_m$$

Since $I_t - I_m$, the fixed and moving coils being in series,

$$T_d \propto I^2 \quad \text{...}(i)$$

Since the control is by springs therefore controlling torque is proportional to the angle of deflection θ *i.e.*,

$$T_e \propto \theta \quad \text{...}(ii)$$

The pointer will come to rest at a position when $T_d = T_e$

From equations (*i*) and (*ii*), we get

$$\theta \propto I^2$$

It is clear that of the pointer is directly proportional to the square of the operating current. Hence, the scale of these instruments is non-uniform being crowded in their lower parts and spread out at the top (see Fig. 16.10).

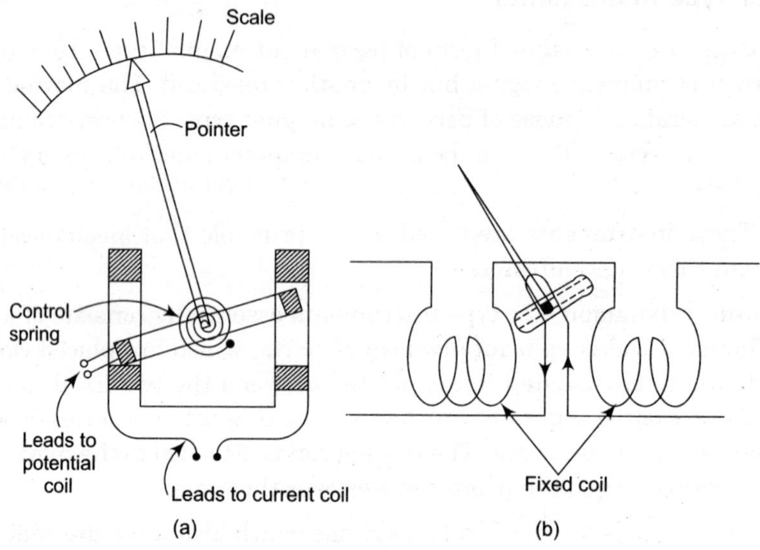

Fig. 16.10 *Constructional feature of dynamometer instrument.*

Advantages:

1. These instruments can be for both a.c. and d.c. measurements.
2. Such instruments are free from hysteresis and eddy current errors.

Disadvantages:

1. Since torque/weight ratio is small, therefore, such instruments have frictional errors which reduce sensitivity
2. Scale is not uniform (in case of ammeters and voltmeters)
3. A good amount of screening of the instruments is required avoid the effect of stray field.
4. These instruments are costlier than other types and therefore, they are rarely used as ammeters or voltmeters.

16.7 SHUNTS

A resistance placed in parallel with an instrument to control the current passing through it when placed in a circuit carrying a fairly large current is called a shunt.

The shunt resistance used with the basic movement may consist of a length of constant temperature resistance wire within the box of the instrument. Alternatively there may be an external shunt having very low resistance.

General requirements of a shunt are as follows:

1. The temperature coefficients of the shunt and instrument should be low and nearly identical.
2. The resistance of the shunt should not vary with time.
3. It should carry the current without excessive temperature rise.
4. It should have a low thermal e.m.f.

Manganin is usually used as a shunt for d.c. instruments since it gives a low value of thermal e.m.f. with copper. Constantan is a useful material for a.c. circuit since its comparatively high thermal e.m.f. being unidirectional is ineffective on these circuits.

Shunts for low current are enclosed in the meter casing but for currents above 200 A, they are mounted separately.

1. **Shunt with ammeter:** Low resistance shunt is connected across its coil as shown in Fig. 16.11.

Let I be the circuit current to be measured

Let R_m = Ammeter resistance

S = Shunt resistance

I_m = Full scale deflection current of ammeters

I_s = Shunt current

$$I = I_s + I_m$$

or $$I_s = I - I_m$$

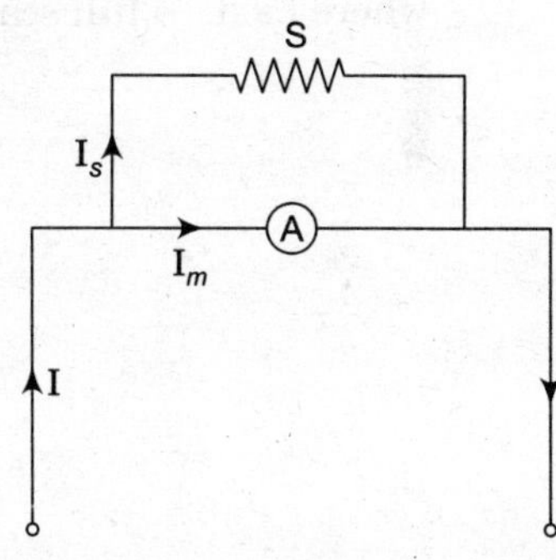

Fig. 16.11 *Shunt with ammeter.*

Since the voltage across the shunt and ammeter is the same,

$$I_m R_m = (I - I_m) R_s$$

$$I_m R_m + I_m R_s = I R_s \quad \text{or} \quad I_m (R_m + R_s) = I R_s$$

$$\frac{I}{I_m} = \frac{R_m + R_s}{R_s}$$

$$I = I_m \cdot \left(\frac{R_m + R_s}{R_s}\right)$$

i.e. Circuit current = f.s.d. Current. $\left(\frac{R_m + R_s}{R_s}\right)$

Instrument constant: The ratio of current to be measured to full-scale deflection current is called instrument constant *i.e.*,

$$\text{Instrument constant} = \frac{I}{I_m} = \frac{R_m + R_s}{R_s}$$

It is clear that with different shunts the same instruments will have different instrument constant.

2. **Shunt with voltmeter:** High resistance is connected in series with its coil as shown in Fig. 16.2.

Let R_m = Voltmeter resistance

R = High resistance in series with voltmeter coil

I_m = Full scale deflection current for voltmeter.

Since voltmeter is connected in parallel with circuit AB,

Voltage across AB = Voltage across voltmeter + Voltage cross high resistance

$$V = I_m (R + R_m)$$

$$R + R_m = \frac{V}{I_m}$$

$$R = \frac{V}{I_m} - R_m$$

i.e., required high resistance,

$$R = \frac{\text{Max. voltage be measured}}{\text{f.s.d. voltmeter current}} - \text{Voltmeter current}$$

where f.s.d. → full scale deflection

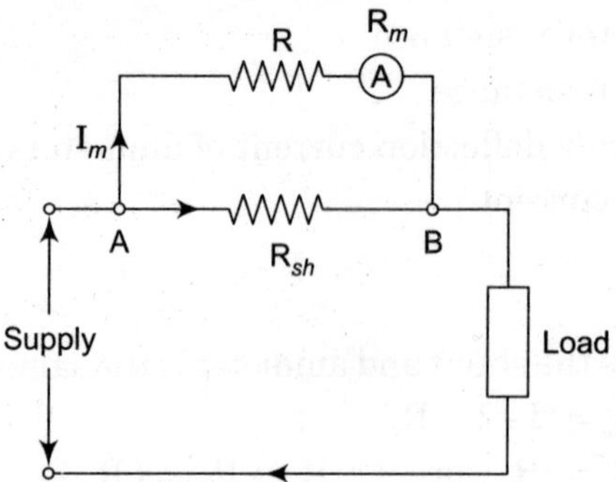

Fig. 16.12 *Shunt with voltmeter*

16.8 SHUNTS FOR A.C. INSTRUMENTS

In order to maintain the current division between the shunt and instrument constant for all frequencies the ratio of impedances of the instrument and leads to that of the shunt must remain constant.

If L_m and L_s are the inductances of the instrument and the shunt respectively then the ratio

$$\frac{\sqrt{R_m^2 + \omega^2 L_m^2}}{\sqrt{R_s^2 + \omega^2 L_s^2}} = \text{Const.}$$

should remain constant for all the frequencies which is possible only when time constants of the shunt and instrument are same.

$$\frac{L_m}{R_m} = \frac{L_s}{R_s} = k$$

Multiplying factor,

$$N = \frac{I}{I_m} = \frac{I_s + I_m}{I_m} = \frac{I_s}{I_m} + 1$$

$$= \frac{\sqrt{R_m^2 + \omega^2 L_m^2}}{\sqrt{R_s^2 + \omega^2 L_m^2}} + 1 = \frac{R_m \sqrt{1 + \omega^2 \left(\frac{L_m}{R_s}\right)^2}}{R_s \sqrt{1 + \omega^2 \left(\frac{L_s}{R_s}\right)^2}} + 1$$

$$\frac{R_m \sqrt{1 + \omega^2 k^2}}{R_s \sqrt{1 + \omega^2 k^2}} + 1 = \frac{R_m}{R_s} + 1 = Mf.$$

16.9 MEASUREMENT OF ELECTRIC VOLTAGE, CURRENT POWER AND ENERGY

Voltage and Current:

Moving iron instrument are used as ammeters and voltmeter only. They work both on a.c. and d.c. system.

(*i*) *Ammeter:* An instruments which is used to measure electric current in a circuit is called an ammeter.

An ammeter is always connected in series with the circuit and carries the current to be measured. This current flowing through the operating coil produces the desired deflecting torque(see Fig. 16.13) since an ammeter to be connected in series therefore it should have a low resistance. Hence, when a moving iron instrument is used as ammeter the operating coil is provided with a few turns of thick wire so that it has low resistance.

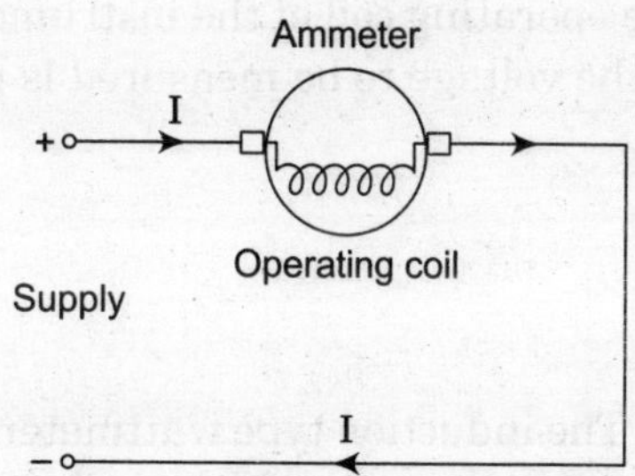

Fig. 16.13 *Current measurement*

(*ii*) *Voltmeter:* An instrument which is used to measure voltage between two points in a circuit is called a voltmeter.

A voltmeter is always connected in parallel in the portion whose voltage is to be measured. This current flowing through the operating coil of the meter produced the deflecting torque (see Fig. 16.14). Since voltmeter is connected in parallel in order that its connection does not disturb the circuit condition therefore the resistance of the voltmeters should be very high. To do so a high resistance (of the order of kilo ohms) is connected in series with the coil of the instrument as shown.

Let R_V be the total resistance (resistance of operating coil + High series resistance) of the instrument. The small current I_V flowing through the instrument is given by

$$I_V = \frac{V}{R_V}$$

Clearly deflection

$$\theta \propto I_V \propto \frac{V}{R_V}$$

$$\theta \propto V$$

Therefore deflection of proportional to the voltage and hence the instrument measures voltage.

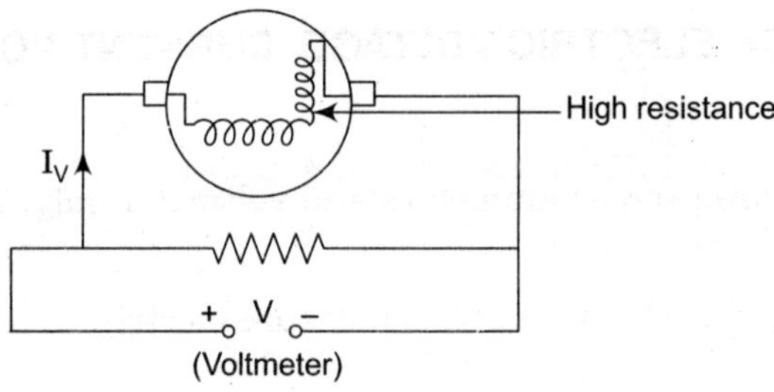

Fig. 16.14 *Voltage measurement*

Induction Voltmeters and Ammeters:

Induction type voltmeters and ammeters are used for a.c. measurements only. They can be either of Ferrari's type or shaded pole type. When used as an ammeter current to be measured or a part of it is passed through the operating coil of the instrument. However, when used as a voltmeter current proportional to the voltage to be measured is passed through the operating oil.

$$T_d \propto I^2$$

$$T_d \propto V^2$$

Power:

(*a*) ***Induction type wattmeter:*** The induction type wattmeter is used to measure a.c. power only in contrast to dynamometer wattmeter which can be used to measure d.c. as well as a.c. power.

Principle: The principle of operation of an induction wattmeter is the same as that of induction ammeters and voltmeters *i.e.* induction principle. However, it differs from induction ammeter or voltmeter in so far that separate two coils are used to produce the rotating flux in place of one coil with phase split arrangement.

Construction: Figure 16.15 shows the principal parts of an induction wattmeter. It consists of two laminated electromagnets. One electromagnet called shunt magnet is connected across supply and carries current proportional to the applied voltage. The coil of the magnet is made highly inductive so that the current (and hence the flux) in it lags behind the supply voltage by 90°. The order electromagnetic, called series magnet is connected in series and carries the load current. The coil of this magnet is made highly non-inductive so that the angle of lag or lead is determined wholly by the load.

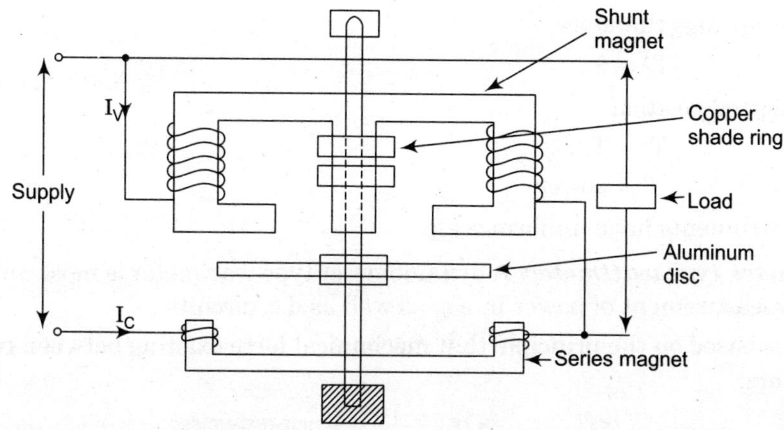

Fig. 16.15 *Induction type wattmeter.*

A thin aluminium disc mounted on the spindle is placed in between the two magnets so that it cuts the fluxes of both the magnets. The controlling torque is provided by spiral springs. The damping is-electromagnetic and is usually provided by a permanent magnet embracing the aluminum disc. Two or more closed copper rings, called shading rings are provided on the central limb of the shunt magnet. By adjusting the position of these rings the shunt magnet flux can be made to lag behind supply voltage by exactly 90º.

Working: When the wattmeter is connected in the circuit to measure a.c. power the shunt magnet carries current proportional to the supply voltage and the series magnet carries the load current. The two fluxes produced by the magnets induce eddy currents in the aluminium disc. The interaction between the fluxes and eddy currents produce the deflecting torque on the disc, causing the pointer connected to the moving system to move over the scale.

Deflecting Torque:

Let

$$V = \text{Applied voltage}$$
$$I_C = \text{Load current carried by the series magnet}$$
$$I_V = \text{Current carried by the shunt magnet}$$
$$\cos\phi = \text{Lagging power factor of the load.}$$

The current I_V in the shunt magnet lags the applied voltage V by 90º and so does the flux ϕ_V produced by it the current I_C in the series magnet is the load current and hence lag (*i.e.*, I_C) is in phase with it. Therefore, the two currents I_C in the current coil and I_V in the voltage coil and also the corresponding fluxes ϕ and ϕ_C are $(90 - \phi)$ apart.

The flux ϕ_V induces eddy currents I_V the aluminum disc which lags behind ϕ_V by 90° similarly ϕ_C induces eddy currents I_C which again lag behind ϕ_C by 90°.

Mean deflecting torque,

$$T_d \propto \phi_V\,\phi_C \sin(90° - \phi)$$
$$\propto VI\cos\phi$$
$$\propto \text{a.c. power}$$

since control is by springs therefore,

$$T_C \propto \theta$$

For steady deflected position,

$$T_d = T_c$$

$$\theta \propto \text{power}$$

Hence such instruments have uniform scale.

(b) *Dynamometer type wattmeter:* A dynamometer type wattmeter is most commonly employed for measurement of power in a.c. as well as d.c. circuits.

Principle: It is based on the principle that mechanical force existing between two current carrying conductors.

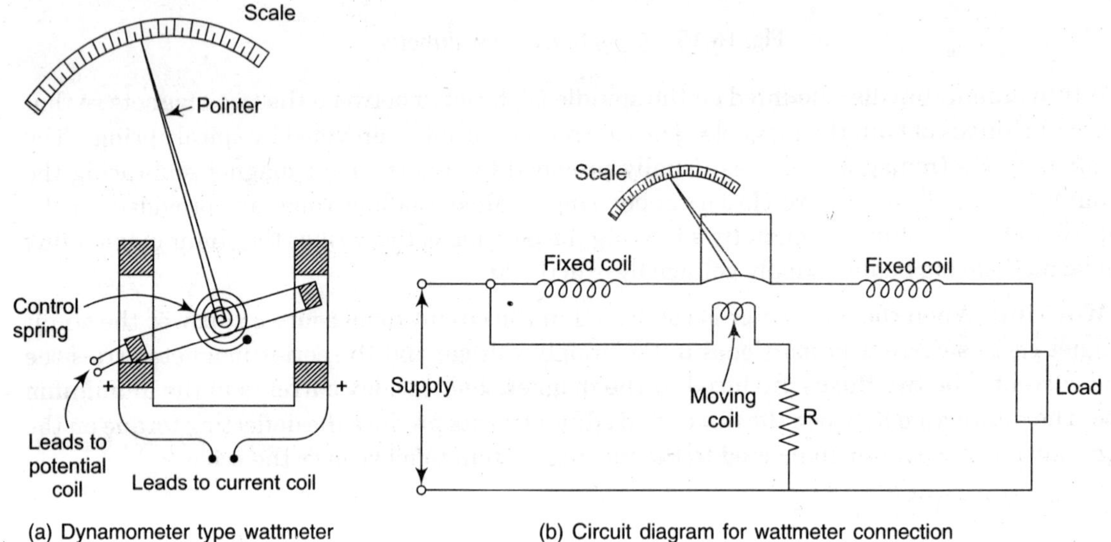

(a) Dynamometer type wattmeter

(b) Circuit diagram for wattmeter connection

Fig. 16.16

Construction: It essentially consists of two coils namely; fixed coil and moving coil. The fixed coil is split into two equal parts which are placed closed together and parallel of each other. The moving coils is pivoted between the two fixed coils and is on the spindle to which is attached the pointer as shown in Fig. 16.16(*a*).

The fixed coils are connected in the series with the load and carry the circuit current. It is therefore called current coil. The moving coil is connected across the load and carries current proportional to the voltage. It is therefore called *potential coil*. Generally a high resistance is connected in series with potential coil to limit the current through it as shown in Fig. 16.16(*b*).

The controlling torque is provided by springs which is also serve the additional purpose of leading current into and out of the moving coil. Air friction damping is employed in such instruments.

Working: When power is to be measured in a circuit the instruments is suitable connected in the circuit. The current coil is connected in series with load so that is carried the circuit current. The potential coil is connected across the load so that is carries current proportional to the voltage.

Due to the current in the coil mechanical force exists between them. The result is that the moving coil which is capable of moving, moves the pointer over the scale. The pointer comes to rest at a position where deflecting torque is equal to the controlling torque.

Reversing the current reverses the field due to fixed coil as well as the current in the moving coil so that the direction of the deflection torque remains unchanged therefore such instruments can be used for the measurement of a.c. as well as d.c. power.

Deflecting Torque:

It can be easily proved that deflecting torque is proportional to the power in the circuit.

d.c. work: Let us suppose that in a d.c. circuit,

$$V = \text{Voltage across load}$$
$$I = \text{Current through load}$$

Current through fixed coil,

$$I_f \propto I$$

Current through moving coil,

$$I_m \propto V$$

Deflecting torque T_d is due to currents I_f and I_m.

$$T_d \propto I_m I_f \propto VI \propto \text{power}$$

a.c. work: Let us suppose that in an a.c. circuit,

$$E = \text{Instantaneous voltage across load}$$
$$I = \text{Instantaneous current through load}$$

If the load has a lagging power factor of cos ϕ then equations become:

$$E = E_m \sin \omega t$$
$$I = I_m \sin (\omega t - \phi)$$

Current through fixed coil

$$I_f \propto i$$

Current through moving coil,

$$i_m \propto e$$

due to large inertia of the moving system the deflection will be proportional to the average torque.

Mean deflecting torque α average of $i_m i_f$

$$\propto \text{Average of } e_i$$
$$\propto \text{Average of } E_m \sin \omega t \; i_m \sin (\omega t - \phi)$$
$$\propto EI \cos \phi$$
$$\propto \text{Power}$$

hence dynamometer type wattmeter can be used for the measurement of a.c. as well as d.c. power.

Scale: We have seen that

$$T_d \propto \text{power}$$
$$T_O \propto \theta$$
$$\theta \propto \text{power}$$

Hence, such instruments have uniform scale.

Advantages:

(*i*) It has uniform scale

(*ii*) By careful design high accuracy can be obtained

(*iii*) It can be used for a.c. as well as d.c. work.

Disadvantages:

(*i*) At low power factors the inductance of the potential coil causes serious errors.

(*ii*) The readings of the instrument may be affected by the stray fields acting on the moving coil. In order to prevent it the instrument is shielded from external fields by enclosing it in a soft iron case.

Energy Meter

Induction type single phase energy meter: Single-phase induction type energy meter is extensively used to measure the electrical energy supplied to a single-phase circuit.

Operating principle: The operation of induction type energy meter depends on the passage of alternating current through two suitable located coils producing rotating magnetic field which interact with a metallic disc suspended near to the coils and causes the disc to rotate.

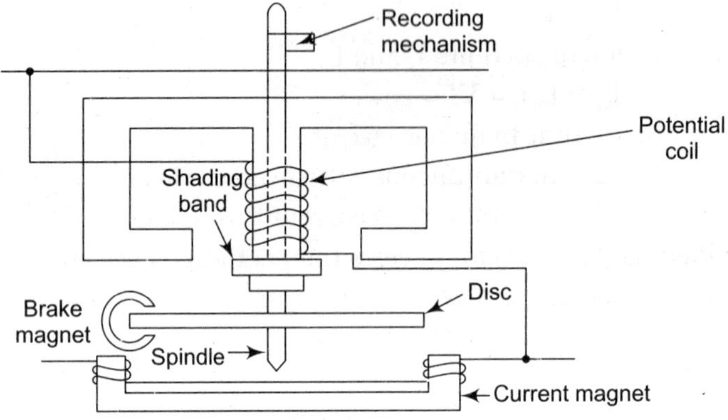

Fig. 16.17 *Energy meter.*

The current coil carries the line current and produces magnetic field in phase with the line current. The pressure coil is made highly inductive so that the current through it lags behind the supply voltage by 90° thus a phase difference of 90° exists between the fluxes produced by the two coils. This sets up a rotating field which interacts with the disc to cause it to rotate.

Construction: A single-phase induction type energy meter generally has the following details:

1. Moving system
2. Operating mechanism
3. Recording mechanism.

1. **Moving system:** The moving system consists of a light aluminium disc on a vertical spindle. The spindle is supported by a cup-shaped jewelled bearing at the bottom end and has a spring journal bearing at the top end.

There is no pointer and control spring so that the disc makes continuous rotation under the action of deflecting torque as shown in Fig. 16.17.

2. **Operating mechanism:** The operating mechanism consists of
   (*i*) series circuit
   (*ii*) shunt magnet
   (*iii*) breaking magnet.
   (*i*) ***Series magnet:*** This series magnet consist a number of U-shaped laminations assemble together to form a core. A thick wire of few turns is wound on both legs of the U-shaped laminated core. The wound coil is known as current coil and is connected in series will the load so that is carried the load current. The series magnet is placed underside the aluminium disc and produces magnetic field proportional to and in phase with the current.
   (*ii*) ***Shunt magnet:*** The shunt magnet consists of a number of M-shaped laminations assembled together to form a core. A fine wire of large turns is wound on the central limb of this magnet. The wound coil is known as pressure coil and is connected across the load so that is carried current proportional to supply voltage. The shunt magnet is placed above the aluminium disc as shown.

   In order to obtain deflecting torque current in the pressure coil must lag behind the supply voltage by 90°. This necessary phase shift is obtained by placing a copper ring (also known as compensating loop) over central limb of shunt magnet. This copper ring as a short circuited transformer secondary. As its inductance is high as compared with its resistance the current circulating in the ring will lag by nearly 90° behind the voltage producing it.
   (*iii*) ***Breaking magnet:*** The speed of aluminium disc is controlled to the required value by the C-shaped permanent breaking magnet. The magnet is mounted is so that the disc resolves in the air gap between the polar extremities. As the disc rotates currents are induced in the disc because it cuts the flux produced by the breaking magnet. The direction of the current in the disc is such that it oppose the rotation of the disc. Since the induced currents in the disc are proportional to the speed of the disc therefore breaking torque is proportional to the disc speed *i.e.* $T_B \propto N$.
3. **Recording mechanism:** When the energy meter is connected in the circuit to measure electrical energy the current coil carries the load current whereas the pressure coil carries current proportional to the supply voltage. The magnetic field due to current coil is in phase with line current whereas the magnetic produced due to pressure coil lags approximately 90° behind the supply voltage.

   The current coil field produces eddy currents in the disc which react with the field due to the pressure coil. Thus a driving force is created which causes the disc to rotate.

   The breaking magnets provide the breaking torque on the disc. By altering the position of this magnet desired speed can be obtained. The spindle is geared to the recording mechanism so that electrical energy consumed in the circuit is directly registered in kWh.

Theory: The current coil carries current proportional to load current whereas the pressure coil carries current proportional to voltage.

SOLVED NUMERICAL PROBLEMS

Example 1. *The torque of an ammeter varies on the square of the current through it. If a current of 6A produces a deflection of 90°, what will be the deflection for a current of 3A when the instrument is*

(*a*) *sping-controlled,*

(*b*) *gravity controlled?*

Solution: We have,

$$T_d \propto I^2$$

(*a*) For spring control,

$$T_C \propto \theta \quad \text{and} \quad \theta \propto I^2$$

$$\text{So, } 90° \propto (6)^2 \text{ and } \theta \propto (3)^2 = 90° \times \frac{(3)^2}{(6)^2} = 90° \times \frac{9}{36}$$

$$= 90° \times 2.5 = 22.5°.$$

(*b*) For gravity control,

$$T_C \propto \sin\theta \quad \therefore \quad \sin\theta \propto I^2$$

$$\sin 90° \propto 6^2 \text{ and } \theta \propto 3^2$$

$$\sin\theta = \frac{9}{36}$$

$$\theta = \sin^{-1} 2.5$$

$$\theta = 14.3°.$$

Example 2. *A 230 V moving-iron voltmeter takes a current of 0.03A when connected to a 230 V d.c. supply. The coil has an inductance of 2 henry. Calculate the reading in the meter when connected to a 230 V, 80 Hz a.c. supply.*

Solution: When used in d.c. supply, the instrument offers ohmic resistance only. Therefore,

$$\text{Resistance of Instrument} = \frac{230}{0.03}$$

$$= \frac{230 \times 100}{3} = \frac{23 \times 1000}{3} = 7666.7\ \Omega$$

When used in a.c. supply,

$$\text{Impedance at 80 Hz} = \sqrt{(7666.7)^2 + (2\pi \times 80 \times 2)}$$

$$= \sqrt{5877 + 2 \times 3.14 \times 80 \times 2} = 7676.8\ \Omega.$$

Instrument reading

$$= 230 \times \frac{7666.7}{7676.8} = 229.6\ \text{V}$$

Example 3. *A moving coil instrument having a resistance of 40 ohms has a full scale deflection of 1 mA. Calculate :*

(*a*) *Shunt resistance to convert the instrument into an ammeter of 2A range.*

(*b*) *Total resistance of the meter.*

Solution: The equivalent circuit is shown in Fig. 16.17:

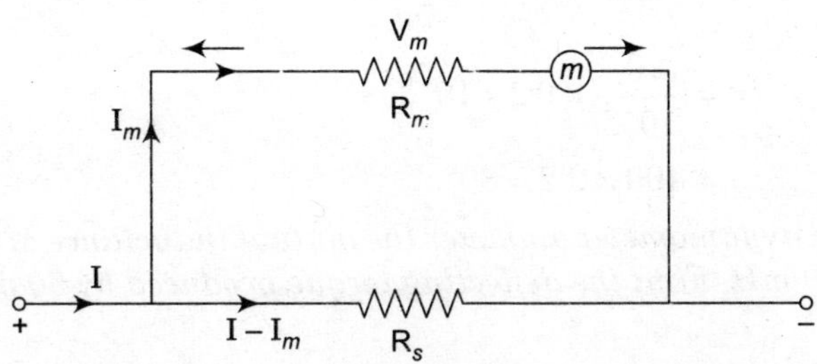

Fig. 16.18

Given,

$$I = 2A$$

$$I_m = 1 \times 10^{-3} \text{ A}$$

$$R_m = 50\ \Omega.$$

$$I_m \times R_m = (I - I_m) R_S$$

$$(a)\ R_S = \frac{1 \times 10^{-3} \times 40}{2 - 1 \times 10^{-3}} = \frac{50 \times 10^{-3}}{2 - 1 \times 10^{-3}} = 0.0250\ \Omega.$$

$$(b)\ \text{Total resistance} = \frac{(0.0250)\ 50}{0.0250 + 50} = 0.025 \text{ ohm.}$$

Example 4. *A moving coil instrument has a full scale deflection of 1 mA and a resistance of 50 Ω. Find the resistance R_1, R_2 and R_3 for converting it into a multi-range voltmeter of ranges 1.5, 5 and 10 volts.*

Solution: For 1.5 V range,

$$I = \frac{V}{R_1 + R_m}$$

or

$$R_1 = \frac{V}{I} - R_m = \frac{1.5}{1 \times 10^{-3}} - 50 = 1450\ \Omega.$$

For 5 V range,

$$R_2 = \frac{5}{1 \times 10^{-3}} - 50 = 4950\ \Omega$$

and for 10 V range,

$$R_3 = \frac{10}{1 \times 10^{-3}} - 50 = 9950\ \Omega.$$

Example 5. *The inductance of an 20 A electrodynamometer ammeter changes uniformly at the rate of 0.0035 μH / degree. The spring constant is 10^{-6} N–m/degree. Determine the angular deflection at full scale.*

Solution: Final steady-state deflection $\theta = \frac{I^2}{K} \cdot \frac{dm}{d\theta}$

$$\frac{dm}{d\theta} = \frac{0.0035}{\frac{\pi}{180}} \times 10^{-6} = 0.2 \times 10^{-6} \text{ Nm/rad.}$$

∴ Deflection
$$\theta = \frac{(20)^2}{10^{-6}} \times 0.2 \times 10^{-6}$$
$$= 400 \times 0.2 = 80°.$$

Example 6. *For a certain dynamometer ammeter the mutual inductance M varies with deflection θ as m = – 6 cos (θ + 30°) mH. Find the deflecting torque produced by 50 mA d.c. corresponding to a deflection of 60°.*

Solution: Rate of change of mutual inductance with deflection.

$$\frac{dM}{d\theta} = \frac{d}{d\theta}\left[-6\cos(\theta + 30°)\right]$$
$$= 6 \sin(\theta + 30°) \text{ mH}$$

Now, $\frac{dM}{d\theta}$ at a deflection of 60° is $\left(\frac{dM}{d\theta}\right)_{\theta = 60°}$

$$= 6 \sin(60° + 30°) \text{ mH}$$
$$= 6 \times 10^{-3} \text{ H/degree.}$$

Deflecting torque,

$$T_d = I^2 \frac{dM}{d\theta}$$
$$= (50 \times 10^{-3})^2 \times 6 \times 10^{-3}$$
$$= 15 \times 10^{-6} \text{ Nm} = 15\ \mu\text{Nm.}$$

Example 7. *The inductive reactance of the pressure coil circuit of a dynamometer wattmeter is 0.4% of its resistance at normal frequency and the capacitance is negligible. Calculate the percentage error and correction factor due to the reactance for loads at*

(*i*) *0.707 P.f. lagging*

(*ii*) *0.5 P.f. leading.*

Solution: Given,

$$\frac{X_p}{R} = 0.4\% = 0.004$$

$$\tan\alpha = \frac{X_p}{R} = 0.004$$

$$\alpha = 0°14'$$

and
$$\sin\alpha = 0.004$$

(*i*) When P.f. = 0.707 (*i.e.*, $\phi = 45°$)

Correction factor $= \dfrac{\cos\phi}{\cos(\phi - \alpha)} = \dfrac{\cos 45°}{\cos 44°46'} = 0.996$

Percentage error $= \dfrac{\sin\alpha}{\cos\phi + \sin\alpha} \times 100 = \dfrac{\sin 0°14'}{\cos 45 + \sin 0°14'} \times 100$

$= \dfrac{0.004 \times 100}{1 + 0.004} = \dfrac{0.4}{1.004} = 0.4.$

(*ii*) When, P.f. = 0.5 (*i.e.*, $\phi = 60°$)

Correction factor $= \dfrac{\cos 60°}{\cos 59°46'} = 0.993$

Percentage error $= \dfrac{\sin 0°14}{\cos 60° + \sin 0°14'} \times 100$

$= \dfrac{0.004 \times 100}{0.577 + 0.004} = \dfrac{0.4}{0.581} = 0.7.$

Example 8. *An energy meter is designed to make 100 revolution of the disc for one unit of energy. Calculate the number of revolutions made by it when connected to a load carrying 40 A at 230 V and 0.4 power factor for an hour. If it actually makes 360° revolutions, find the percentage error.*

Solution: Energy utilised in 1 hour $= 230 \times 40 \times 0.4 \times \dfrac{1}{1000}$

$= 3.68$ kWh.

Number of revolution the disc should made if it is correct $= 3.68 \times 100 = 368$

Number of revolution actually made = 360

$\therefore$ Percentage error $= (368 - 360) \times \dfrac{100}{368} = 2.17\%.$

Example 9. *The disc of an energy meter makes 600 revolution per unit of energy. When a 1000 watt load is connected the disc rotates at 10.2 r.ps. If the load is given for 12 hours, how many units are recorded as error?*

Solution: Energy consumed when load power is 1 kW

$= 1 \times 12 = 12$ kWh.

Total number of revolution made by disc. at this period

$= 10.2 \times 60 \times 12 = 7344$

Energy recorded by meter $= \dfrac{7344}{600} = 12.24$ kWh.

Hence 0.24 unit is recorded extra.

EXERCISES

1. What do you understand by the term 'measurement'? Explain the lord Kelvin's statement for electrical measurement.
2. How many types of measuring instruments are there? Explain in brief.
3. What are the various differences between analog and digital instruments which instrument is most accurate?
4. Explain the following terms in measurement:
   (*i*) True value (*ii*) Error
   (*iii*) Accuracy (*iv*) Precesion
   (*v*) Sampling.
5. Explain the various types of analog instruments with examples.
6. How many types of indicating instruments are there in measurement?
7. Explain various torques required for the indicating instruments.
8. What do you understand by 'dampins'?
9. Explain construction and working of moving iron instruments.
10. At what condition of torque, the pointer of indicating instrument attains equilibrium position?
11. Why the scale of moving iron instrument is non-uniform?
12. Explain construction and working of permanent magnet moving coil instrument.
13. What do you understand by shunt and multipliers in measuring instrument?
14. Why the scale of permanent magnet moving coil instrument is uniform?
15. How the range of ammeter in permanent magnet moving coil instrument is extended?
16. Why multipliers are used in extending the range of voltmeters?
17. Give construction and working of a dynamometer type instrument.
18. Explain how an electrodynamometer type instrument is able to measure the true rms value of a voltage or current irrespective of its waveform?
19. Describe the constructional details of an Electrodynamometer type wattmeter. Derive the expression for torque when the instrument is used in a.c. Explain why it is necessary to make the potential coil circuit purely resistive?
20. Derive the expression for the capacitance to be connected across the resistor in the pressure coil circuit so as to neutralize the effect of inductance of pressure coil circuit.
21. When two wattmeter methods is used for measurement of power in a three phase balanced circuit, comment upon the readings of the two wattmeters under following condition assuming system to be star connected:
   (*i*) When the power factor is unity.
   (*ii*) When the power factor is zero lagging.
   (*iii*) When the power factor is 0.5 lagging.
   (*iv*) When the power factor is 0.3 lagging.

22. Explain construction and working of single-phase energy meter and also explain the adjustment made in energy meter with respect to following points:

(*i*) Lag adjustment

(*ii*) Friction compensation

(*iii*) Creeping.

23. What are the various types of single-phase energy meter? Explain each in brief.

24. A moving coil instrument has the following data:

Number of turns = 100,

Width of coil = 30 mm,

Depth of coil = 40 mm, flux density in the air

gap = 0.2 Wb/m^2.

Calculate the deflecting torque when carrying a current of 15 mA. Also calculate the deflection if the control spring constant is 2×10^{-6} Nm/degree.

25. A moving-coil instrument has at normal temperature a resistance of 15 Ω and a current of 35 millionmphere gives full scale deflection. If its resistance rises to 15.2 Ω due to temperature change, calculate the reading when a current of 2000 A is measured by means of a 0.235×10^{-3} A shunt of constant resistance. What is the percentage error?

26. The inductance of a certain moving iron ammeter is $\left(6 + 2\theta - \frac{1}{2}\theta^2\right)$ μH, where θ is the deflection in radian from the zero position. The control spring torque is 8×10^{-6} N-m/rad. Calculate the scale position in radian for current 1, 2, 3, 4 and 5 A and disucss the scale shape obtain.

27. A 240 V, 8 A electrodynamometer wattmeter has a resistance of current and potential coils of 0.5 Ω and 12000 Ω respectively. Find the percentage error due to each of the two methods of connection. With unit power factor loads at 24 V with current of 6 A.

28. Calculate the monthly bill in the case of a consumer whose maximum demand is 110 kW, average monthly load factor is 20 %, and where tariff in use is ₹30 per kW of maximum demand and 30 paise per kWh consumed.

29. The declared constant of a 5 A, 220 V d.c. watt hour meter is 3285 revolutions per kWh. Calculate the speed of the disc at full load.

17

Introduction to Earthing and Electrical Safety

17.1 INTRODUCTION

Earthing practices adopted at generating stations, sub-stations, distribution structures and lines are of great importance. It is however observed that this item is most often neglected. The codes of practice, technical reference books, and hand books contain a chapter on this subject but they are often skipped treating them as too elementary or even as unimportant. Many reference books on this subject are referred to and such of those points, which are most important, are compiled in the following paragraphs. These are important to every practicing engineer and electricity user.

Grounding is conducting connection by which an electrical circuit or equipment is connected to earth or to some conducting body of relatively large extent that serves in place of earth.

The earthing is provided for

(*a*) Safety of personnel.

(*b*) Prevent or at least minimize damage to equipment as a result of flow of heavy fault currents.

(*c*) Improve reliability of Power Supply.

The earthing is broadly divided as

(*a*) System earthing (connection between parts of plant in an operating system like LV neutral of a power transformer winding) and earth.

(*b*) Equipment earthing (safety grounding) connecting bodies of equipment (like motor body, transformer tank, switch gear box, operating rods of air break switches, LV breaker body, HV breaker body, feeder bearer bodies, etc.) to earth.

The system earthing and safety earthing are interconnected and therefore fault current flowing through system ground rise the potential of the safety ground and also causes step potential gradient in and around the sub-station. But separating the two earthing systems have disadvantages like higher short circuit current, low current flows through relays and long distance to be covered to separate the two earths. After weighing the merits and demerits in each case, the common practice of common and solid (direct) grounding system designed for effective earthing and safe potential gradient is being adopted.

17.2 SHOCK HAZARDS

Grounding of electrical devices and systems are vital to ensuring that people living or working in the environment will be adequately protected. The grounding is very important from personal safety point of views because in case any passage of electrical current through an average human body may cause physiological hazards. In Table 17.1 the effect of current flow through human body and its hazard is given:

Table 17.1 Effect of Current Flow through Human Body

Current level	*Shock hazard*
100 μA	Threshold of perception
1-5 mA	Sensation of pain
5-10 mA	Increased pain
10-20 mA	Intense pain: unable to release grip
30 mA	Breathing affected
40-60 mA	Feeling of asphyxiation
75 mA	Ventricular fibrillation, irregular heart beat

It is obvious that it does not take much current to cause injury and even death. The resistance of an average human in dry condition is about 100 kΩ or higher. When the skin is wet, the resistance drops to 10 kΩ or lower. It is not difficult to see how susceptible human are to shock hazard.

Figures 17.1(*a*) and 17.1(*b*) illustrate what would happen if a person came in contact with the frame of an electric motor where, due to insulation deterioration, a 440 V phase is in contact with the frame. Figure 17.1(*a*) is the condition of the frame being bonded to the service ground terminal, which in turn is connected to the building ground electrode. If the power source feeding the motor is a grounded source, this condition in all likelihood would cause the over current protection to operate and open the circuit to the motor. If power source feeding the motor

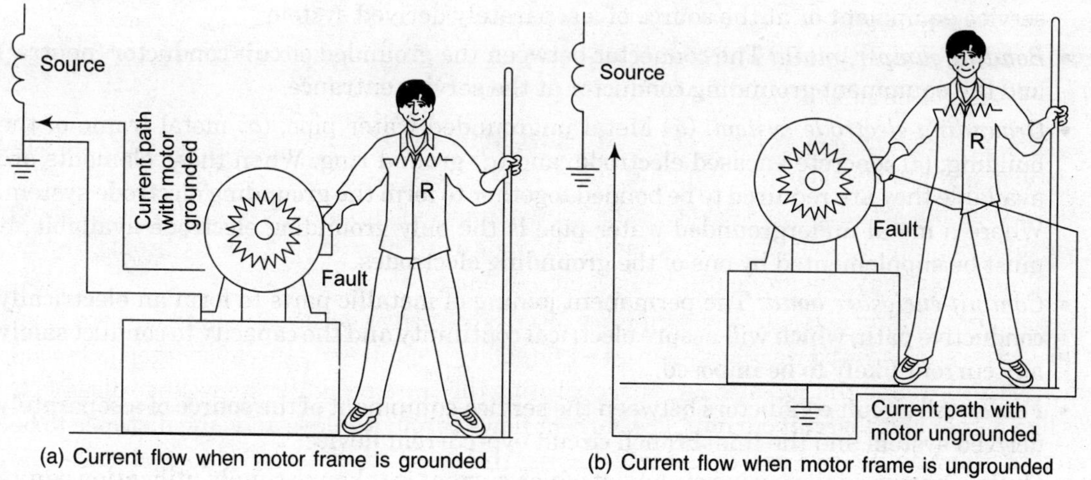

(a) Current flow when motor frame is grounded

(b) Current flow when motor frame is ungrounded

Fig. 17.1

is an ungrounded source, no over current protection is likely to operate; however, the phase that is contacting the frame will be brought to the ground potential and the person-touching the frame is not in danger of receiving an electric shock.

On the other hand, consider Fig. 17.1(*b*), where the motor frame is not bonded to a ground. If source feeding the motor were a grounded source, considerable leakage current would flow through the body of the person. The current levels can reach values high enough to cause the death. If the source is ungrounded, the current flow through the body will be completed by the stray capacitance of cable used to connect the motor to the source. For a 1/0 cable the stray capacitance is of order of 0.17 μF for 1 100-ft cable. The cable reactance is approximately 15700 Ω, currents significant enough to cause a shock would flow through the person in contact with the motor body.

17.3 ESSENTIALS OF GROUNDED SYSTEM

It is best to have a clear understanding of components of ground system to fully grasp the importance of grounding for safety and power quality. The elements are defined as follows:

- *Grounded:* Connected to earth or td some conducting body that serves in place of the earth.
- *Grounded conductor:* A system or circuit conductor that is intentionally grounded.
- *Grounding conductor:* A conductor used to connect equipment or the grounded circuit of wiring system to a grounding electrode or electrodes.
- *Grounding conductor, equipment:* The conductor used to connect the noncurrent-carrying metal parts of equipment, raceways, and other enclosures to the system grounded conductor and/or at the source of a separately derived system.
- *Branch circuit:* The circuit conductors between the final over current devices protecting the circuit and the outlets.
- *Grounding electrode conductor:* The conductor used to connect the grounding electrode to the equipment grounding conductor and /or to the grounded conductor of the circuit at the service equipment or at the source of a separately derived system.
- *Bonding jumper, main:* The connector between the grounded circuit conductor (neutral) and the equipment grounding conductor at the service entrance.
- *Grounding electrode system:* (*a*) Metal ungrounded water pipe; (*b*) metal frame of the building; (*c*) concrete-encased electrode; and (*d*) ground ring. When these elements are available they are required to be bonded together to form the grounding electrode system. Where a metal undergrounded water pipe is the only grounding electrode available, it must be supplemented by one of the grounding electrodes.
- *Conduit enclosure bond:* The permanent joining of metallic parts to form an electrically conductive path, which will assure electrical continuity and the capacity to conduct safely any current likely to be imposed.
- *Feeder:* All circuit conductors between the service equipment of the source of a separately derived system and the final branch circuit overcurrent device.
- *Outlet:* A point on the wiring system at which current is taken to supply utilization equipment.

- *Overcurrent:* Any current in excess of the rated current of equipment or the capacity of a conductor. It may result from overload, short circuit, or grounded fault.
- *Ufer ground:* A method of grounding or connection to the earth in which the reinforcing steel (rebar) of the building, especially at the ground floor, serves as a grounding electrode.
- *Panel board:* A single panel or ground of panel units designed for assembly in the form of a single panel; including buses, automatic overcurrent devices, and with or without switches for the control of light, heat, or power circuits ; designed to be placed in a cabinet or cutout box placed in or against a wall or partition and accessible only from the front.
- *Separately derived systems:* A premise wiring system whose power is derived from a generator, a transformer, or converter windings and has no direct electrical connection, including a solidly connected grounded circuit conductor, to supply conductors originating in another system.
- *Service equipment:* The necessary equipment, usually consisting of a circuit breaker switch and fuses, and their accessories, located near the point of entrance of supply conductors to a building or other structure, or an otherwise defined area, and intended to constitute the main control and means of cutoff of the supply.

17.4 GROUND ELECTRODES

Consist of three basic components:

1. Ground conductor.
2. The connection of the conductor to the electrode.
3. The ground electrode itself.

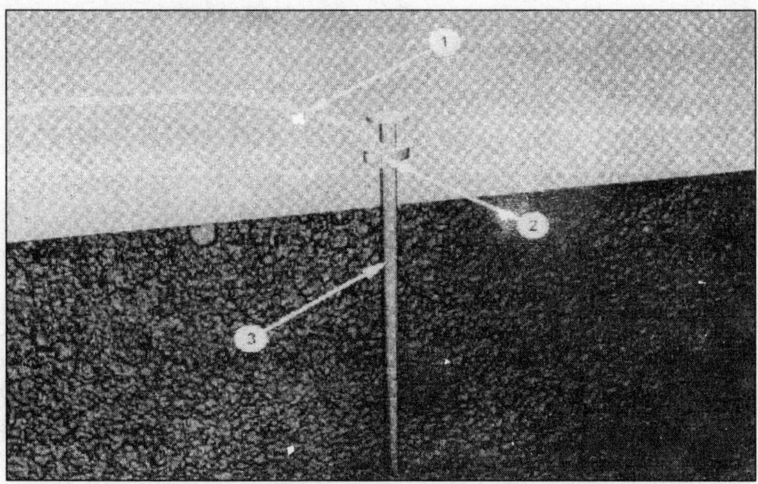

Fig. 17.2

The earth resistance (R) has three basic components:

R_A — The resistance of the ground electrode itself and the connections to the electrode

R_B — The contact resistance of the surrounding earth to the electrode

R_C — The resistance of the surrounding body of earth around the ground electrode

$$R = R_A + R_B + R_C$$

(R_A) — The electrode resistance depends on :

- Length/ depth of the ground electrode
- Electrode material
  - Solid copper, stainless steel (high conductivity, high corrosion resistance but low strength and high cost)
  - Copper clad steel (high strength, high corrosion resistance and low cost)
- The diameter of the rod
  - Has little effect
  - Resistance would only decrease by 10% by double the diameter

(R_B) — The contact resistance of the surrounding earth to the electrode is negligible.

(R_C) — The resistance of the surrounding body of earth around the ground electrode

- Depends on soil conditions such as soil resistivity.

17.5 TYPES OF GROUNDING SYSTEMS

1. Single rod
2. Multiple rods
3. Copper plates
4. Conductor mesh

17.6 EARTHING MATERIALS

(a) Conductors

(*a*) The conductors used must be capable of carrying the anticipated fault current and cope with corrosion over the lifetime of the installation.

(*b*) Bare copper is normally used for a substation earthing grid, being buried at depths between 0.6 and 1.0 m in rectangles of between 3 and 7 m side length.

(*c*) Equipment connections are generally laid at a shallower depth of about 0.2 m. Because of mechanical and thermal criteria, it is unusual for copper of less than 70 mm square to be used.

(*d*) Aluminum is often used for above ground connections and could be used below ground if it is certain that the soil will not cause corrosion problems, but most standards prohibit this.

(*e*) Some protection, such as painting with bitumastic paint, is recommended in the area where the conductor emerges from the ground, as corrosion may occur here and just below. Where not connected directly to the electrode, all metallic substation plant is bonded via above ground conductors.

(b) Connections

- The connection methods used for below ground application are welded, brazed, compression or exothermic. For above ground use, bolted connections are used in addition to these.

- Soldering is not permitted, because the heat generated during fault conditions could cause failure. Bolted joints should normally have their contact faces tinned and particular care is necessary for connections between dissimilar metals, such as copper and aluminum.

(c) Earth Rods

- The earth grid's horizontal electrodes may be supplemented by vertical rods to assist the dissipation of earth fault current, further reduce the overall substation earthing resistance and provide some stability against seasonal changes.
- This is especially useful for small area substation sites (such as GIS substations) or where the rod is of sufficient length to enter the water table.
- Rods may be of solid copper or copper clad steel and are usually of 1.2 m length with screw threads and joints for connecting together in order to obtain the required length for installation in the soil.
- The formula for the effective resistance, R_{ROD} ohms of a single earth rod is given by:

$$R_{ROD} = \frac{\sigma}{2pl} \cdot [\log (8l/d) - 1]$$

where R_{ROD} = Bare vertical earthing rod effective resistance (ohms)

σ = Resistivity of soil (ohm metres)

l = Length of earthing rod (metres)

d = Diameter of earthing rod (metres)

- Long rods are often fitted with test facilities so that their resistance can be measured during regular maintenance checks. Individual rods are also specified for line traps, current and voltage transformers (CVTs) and surge arresters.

17.7 MEASUREMENT OF EARTH RESISTANCE

The measurement of earth resistance is done using three terminal earth meggar or four terminal earth meggar.

Three Terminals: Two temporary electrodes are spikes are driven in straight line one for current and the other voltage at a distance of 150 feet and 75 feet from the earth electrode under test and ohmic values of earth electrode are read in the meggar.

Four Terminals: Four spikes are driven in straight line into the ground at equal intervals. The two outer spikes are connected to current terminals of earth meggar and the two inner spikes to potential terminals of the meggar. Then the earth resistance is measured by rotating the meggar till a steady value is obtained.

17.8 MAINTENANCE OF EARTHING SYSTEM

The following maintenance schedule is mandatory at each of the sub-stations:

Sl. No.	Item	Periodicity
1.	Watering of earth pits	(Not required for Bentonite treated earth pits)
2.	Measurement of earth resistance of individual earth pits	Half yearly @
3.	Measurement of combined earth resistance at all the pits	Half yearly.
4.	Checking of interconnections between earth pits pits and tightness of bolts and nuts.	Quarterly.

Earth resistance of individual earth pits can be measured up by disconnecting the earth connections to the electrode. This is possible if the connections are made to a common clamp which is in turn is fixed round the pipe.

17.9 BENTONITE

- (Bentonite + Water = Moisture Retaining Clay)
- This Moisture Retaining Clay used as an earth-electrode backfill to reduce soil resistivity
- It can absorb moisture from surrounding soil
- It has the ability to hold its moisture content for a considerable period of time
- (50 kg packet = ₹3000/=), (1 Rod = 1 kg = ₹60/=).

17.10 MARCONITE

- (Marconite in place of sand + Cement = Conductive concrete
- This conductive concrete is used as electrode backfill
- It increases effective electrode area thus reducing earth resistance.

17.11 EARTH-GROUND GRID SYSTEMS

Ground grids can take different forms and shapes. The ultimate purpose is to provide a metal grid of sufficient area of contact with the earth so as to derive low resistance between the ground electrode and the earth. Two main requirements of any ground grid are to ensure that it will be stable with the time and that it will not form chemical reactions with other metal objects in the vicinity, such as buried water pipes, building reinforcement bars, etc., and cause corrosion either in the ground grid or the neighbouring metal objects.

A. Ground Rods

The ground rods should be 8-ft in length and consist of following: (*i*) Electrodes of conduits or pipes that are no smaller than ¾-inch trade size; when these conduits are made of steel, the outer surface should be galvanized or otherwise metal-coated for corrosion protection (*ii*) Electrodes of rods of iron or steel that are at least 5/8 inches in diameter; the electrodes should be insulted so that at least an 8-ft length is in contact with soil. The effect of ground rod length on earth resistance is given in the following Table 17.2, and effect of ground rod diameter on ground resistance is given in Table 17.3, considering the soil resistivity equal to 10,000 Ω-cm.

Table 17.2 Effect of Ground Rod Length on Earth Resistance

Ground rod length (ft)	*Earth resistance (Ω)*
5	40
8	25
10	21
12	18
15	17

Table 17.3 Effect of Ground Rod Diameter on Earth Resistance

Ground rod length (ln)	*% Earth resistance*
0.5	100
0.75	90
1.0	85
1.5	78
2.0	76

B. Plates

Rectangular or circulator plates should present an area of at least 2 square-feet to the soil. Electrodes of iron and steel shall be at least ¼ inch in thickness; electrodes of nonferrous metal should have a minimum thickness of 0.6 inch. Plate electrodes are to be installed at a minimum distance of 2.5 ft below the surface of the earth. Table 17.4 gives the earth resistance values for circular plates buried 3 ft below the surface in the soil with resistivity of 10,000 Ω-cm.

Table 17.4 Resistance of Circular Plates Buried 3 feet below surface

Plate area (sq-ft)	*Earth resistance (Ω)*
2	30
4	23
6	18
10	15
20	12

C. Ground Rings

The ground ring encircling a building in direct contact with the earth should be installed at a depth of not than 2.5 ft below the surface of earth. The ground ring should consist of at least 20 ft of bare copper conductor sized not less than #2 AWG. Typically, ground rings are installed in trenches around the building, and wire tails are brought out for connection to the grounded service conductor at the service disconnect panel or switchboard. It is preferred that a continuous piece of wire be used. Table 17.5 gives the resistance of two conductors buried 3 ft below the surface for various conductor lengths at soil resistivity of 10,000 Ω-cm.

Table 17.5 Earth Resistance of buried Conductors

Wire size (AWG)	*Resistance (Ω) for total buried wire length*				
	20 ft	*40 ft*	*60 ft*	*100 ft*	*200 ft*
#6	23	14	7	5	3
#1/0	18	12	6	4	2

17.12 DO'S AND DON'TS FOR ELECTRICAL SAFETY

Table 17.6 Do's and Don'ts for Electrical Safety for Industrial Consumers

Sl. No.	*Do's*	*Don'ts*
1.	Place "Men working" signboards on all switches before commencing work.	Don't close switches unless you are familiar with the circuit, which it controls and also know the reason for its being kept open.
2.	Ensure that all the controlling switches are opened and locked or fuse withdrawn before working on any circuit or apparatus.	Don't touch or tamper with any electrical gear or conductor. Unless you have made sure it is dead and earthed. High Voltage apparatus may give leakage shock or flash over even without touching, without being connected to a visible source.
3.	Treat circuit as live until they are proven to be dead.	Don't test a circuit with bare fingers or hand or other make shift devices to determine whether or not it is alive.
4.	Turn away your face whenever an arc or flash is expected.	Don't close or open a switch or fuse slowly or hesitatingly. Do it quickly, positive and firmly.
5.	Do place rubber mats in front of Electrical switchboards.	Don't use fire extinguishers on electrical equipment unless it is clearly marked as suitable for that purpose.
6.	Use sand or blanket to control fire involving electrical accidents.	Don't bring a naked flame near oil filled equipment and battery. Smoking in battery room is prohibited.
7.	Do examine before use, all the safety appliances such as gloves, safety belt, mats, ladders, goggles, ropes, etc. For their soundness.	Don't allow visitor and unauthorised persons to touch or handle electrical apparatus or come within the danger zone of HV apparatus.
8.	Please concentrate on the work you are doing.	Don't remove danger boards and other warning signs without instruction or interfere with safety barriers or go beyond them without the mandatory precautions.

Table 17.7 Do's and Don'ts for Electrical Safety for Commercial Consumers

Sl. No.	*Do's*	*Don'ts*
1.	All wiring should be undertaken by licensed wiring controllers.	Don't travel on vehicles loaded with goods beyond the permissible height; This may cause electrical fatal accident due to coming into contact with overhead electrical lines.
2.		Don't tie advertisement boards, flags etc. to the electric post.

Table 17.8 Do's and Don'ts for Electrical Safety for Domestic Consumers

Sl. No.	*Do's*	*Don'ts*
1.	Use Standard plug to tap supply from a plug point.	Avoid tapping of supply by inserting bare wires.
2.	Always use standard materials with ISI marks, even if it costs more.	Don't use brackets to tie wire or ropes. Do not dry clothes on wires or cables.
3.	Fused bulbs may be replaced only after the switch off.	Changing the fused bulbs when the switch is on, is dangerous.
4.	Use always properly earthed three pin plugs to connect refrigerators, wet grinders, mixies, washing machines, iron boxes and geysers etc.	Don't touch an electric switch or appliance when your hands are wet or bleeding from a cut. Don't keep the lamp holders without lamp.
5.	Use only lamp holders with 1 amps rating.	Don't purchase substandard electrical fitting to save money which may result in serious accidents.
6.		Don't connect refrigerators, wet grinders, mixies, washing machines, iron boxes and geysers through unearthed plug pin, which may cause serious accidents.

EXERCISES

1. Explain why earthing is provided in the electrical system.
2. What would happens if a person touch an electric up which is not earthed?
3. Define the following:
   (*i*) grounded conductor (*ii*) grounding conductor
   (*iii*) grounding electrode conductor (*iv*) bounding jumper
4. With the help of suitable diagram dozen be various components of grounding electrodes.
5. Explain earth-ground grid system in details.
6. Write the essentials of grounded system.

7. Write the DO's and DONT's for electrical safety for
   (*i*) industrial consumer
   (*ii*) commercial consumer
   (*iii*) domestic consumer.
8. Write the short notes on
   (*i*) measurement of Earth resistance
   (*ii*) benthonik
   (*iii*) marconik
9. Briefly discuss the following:
   (*i*) ground rod
   (*ii*) plate
   (*iii*) ground ring
10. Explain why a bird sitting on a conductor on overhead line does not get any shock.

Experiments
1. Network Theorems

Experiment 1 (*a*): To verify Thevenin's theorem.

1. **Object:** To verify Thevenin's theorem.

2. **Apparatus Required:** Voltmeter (0 – 300 V), D.C. batteries, D.C. ammeter (0-3/6 amp), rheostats, resistors, keys and connecting wires etc.

Apparatus Details

S.No.	*Name of apparatus*	*Type*	*Range/rating*	*Quantity*	*Make*	*Remarks*

3. **Theory/Principle:** According to the Thevenin's theorem, any linear network containing energy sources (generator) and resistances (impedances) can be replaced by an equivalent circuit consisting of a voltage source V_{th} in series with R_{th}. The value of 'V_{th}' is the open circuit voltage between the terminals of the network and 'R_{th}' is the impedance measured between the terminals with all energy sources replaced by their internal impedances (Voltage source treated as short circuit while current source treated as open circuit).

Current across the Load Resistance (R_L) is given by the equation

$$I_L = \frac{V_{th}}{R_{th} + R_L}$$

where V_{th} = Open circuit voltage across the terminal

R_{th} = Equivalent resistance across the terminal

R_L = Load resistance

4. **Circuit Diagram:** The circuit diagram and the Thevenin's equivalent circuit is shown.

5. **Procedure:** Method for finding Thevenin's equivalent circuit.

1. Remove the resistance (called Load resistance R_L), whose current is required.

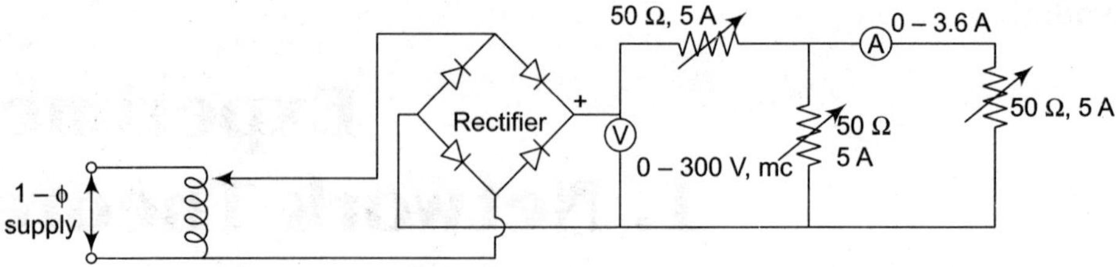

Fig. E.1

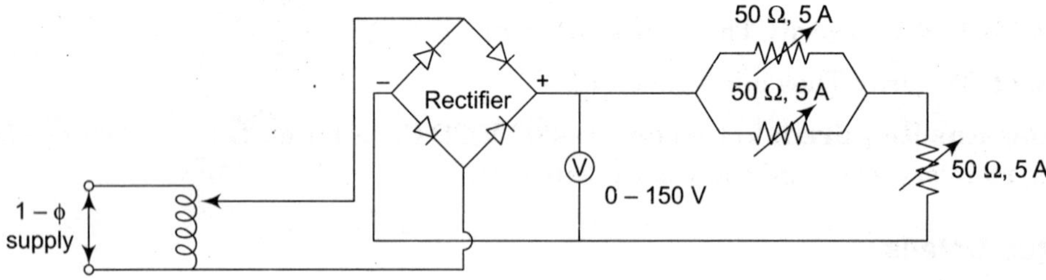

Fig. E.2

2. Find the open circuit voltage V_{OC} which appears across the two terminals from where resistance is removed. It is also called Thevenin's voltage V_{th}.
3. Compute the resistance of the whole network as looked into from these two terminals after all sources of e.m.f. are treated as short circuited while all the current sources are treated as open circuited.
4. Connect R_L back to its terminals from where previously it was removed.
5. Finally, calculate the current flowing through R_L using the equation

$$I_L = \frac{V_{th}}{R_{th} + R_L} \text{ or } I_L = \frac{V_{OC}}{R_{th} + R_L}$$ and compare with meter readings.

6. **Observation Table:**

S.No.	*Load current in amperes* (I_L)	*Voltage across load in volts* (V_L)	*Load resistance in ohms* $R_L = \frac{V_L}{I_L}$	*Open circuit voltage across terminals A and B* (V_{OC} or V_{th})	*Equivalent resistance across the terminals A and B,* R_{th}	*Load current by* $I_L = \frac{V_{th}}{R_{th} + R_L}$
1.						
2.						
3.						
.						
.						
.						
10.						

7. **Calculation:** The load current

$$I_L = \frac{V_{Th}}{R_{Th} + R_L}$$

$$I_L = \ldots.. \text{ amp.}$$

8. **Result**

1. The value of open circuit voltage (V_{OC}) is ...... volts.
2. The value of Thevenin's resistance is ...... ohms.
3. The value of current across load is ...... amps.

It will found that measured value of current flowing through the load I_L are the same as determined by Thevenin's theorem.

9. **Precautions:**

1. All connections should be tight.
2. All steps should be followed carefully.
3. Reading and calculations should be taken carefully.
4. Don't touch the live terminals.

Experiment 1 (*b*): To verify Norton's theorem.

1. **Object:** To verify Norton's theorem.

2. **Apparatus Required:** Voltmeter, D.C. batteries, D.C. ammeter, rheostats, resistors, keys and connecting wires etc.

Apparatus Details

S.No.	*Name of apparatus*	*Type*	*Range/rating*	*Quantity*	*Make*	*Remarks*

3. **Theory Principle:** In reality, Norton's theorem is an alternative of Thevenin's theorem. According to Norton's theorem any two terminals of active network containing voltage sources and resistances when viewed from its output terminals is equivalent to a constant current source with a parallel resistance called Norton's Resistance (R_N).

The current across the Load resistance (R_L) is given by the equation

$$I_L = \frac{R_N \times I_{SC}}{R_N + R_L}$$

where I_L = Load current

I_{SC} = Short circuit current across the terminals

R_N = Norton's resistance (*i.e.*, Thevenin's resistance R_{th})

R = Load resistance

4. **Circuit Diagram:** The circuit diagram and the Norton's equivalent circuit is shown below in the Fig. E.3.

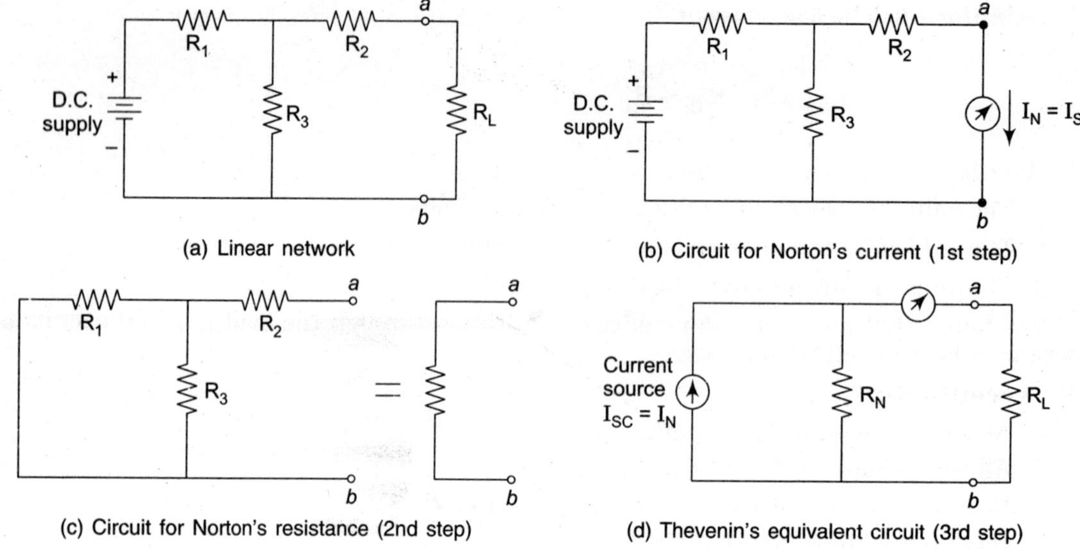

Fig. E.3

5. **Procedure:** Method for finding Norton's equivalent circuit.

1. Remove the resistance (if any) across the two given terminals and put a short circuit across them.
2. Calculate the short circuit current I_{SC}.
3. Remove all voltage source but return their internal resistances, if any.
4. Now find the resistance R_N of the network as looked into from the given terminal, for treated voltage sources as short circuit and current sources as open circuit.
5. The current source (I_{SC}) joined in parallel across R_N between the two terminals of given Norton's equivalent circuit.
6. Compare the result from calculation with meter readings.

6. **Observation Table:**

S.No.	*Load current in amperes* (I_L)	*Voltage across load in volts* (V_L)	*Load resistance in ohms* $R_L = \frac{V_L}{I_L}$	*Open circuit voltage across terminals A and B in volts* (V_{OC})	*Short circuit through terminals A and B* I_{SC}	$RN = \frac{V_{OC}}{I_{SC}}$	*Load current by Norton's theorem* $I_L = \frac{R_N \times I_{SC}}{(R_N + R_L)}$
1.							
2.							
3.							
.							
.							
.							
10.							

7. **Calculation:** The load current

$$I_L = \frac{R_N \times I_{SC}}{R_N + R_L}$$

$$I_L = \ldots \text{Amp.}$$

8. **Result:**

1. The value of open circuit voltage (V_{OC}) is .... volts.
2. The value of short circuit current is .... amps.
3. The value of Norton's resistance is .... ohms.
4. The value of load current is .... amps.

It will found that measured value of current flowing through the load I_L are the same as determined by Norton's theorem.

9. **Precautions:**

1. All connections should be tight.
2. Reading and calculations should be taken carefully.
3. Never touch the live terminals.

Experiment 1 (*c*): To verify the maximum power transfer theorem.

1. **Object:** To verify the maximum power transfer theorem.

2. **Apparatus Required:** Accumulator, D.C. Voltmeter (0–15V), D.C. ammeter (0–2.5 Amp.), Resistance box, Rheostat of suitable rating and connecting wires.

Apparatus Details

S.No.	*Name of apparatus*	*Type*	*Range/rating*	*Quantity*	*Make*	*Remarks*

3. **Theory/Principle:** According to the maximum power transfer theorm as applied to D.C. network, a resistive load will abstract maximum power from a network when the load resistance is equal to the resistance of the network as viewed from the output terminals, with all energy sources replaced by their internal resistance. In the case of A.C. network load impedance should be complex conjugate of source impedance.

4. **Circuit Diagram:** The circuit diagram and graph power and load resistance is shown below in the Fig. E.4.

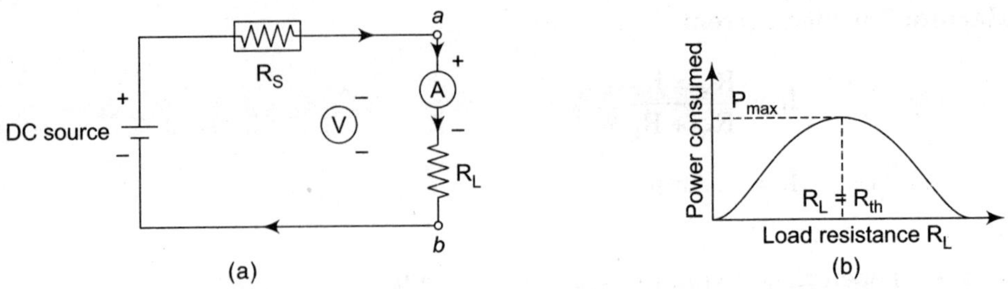

Fig. E.4 *(a) Circuit diagram for maximums Power Transfer theorem, (b) Graph between P_m V_S R_L.*

5. **Procedure:**

1. Connection diagram is shown above in Fig. E.4, where R is fixed at some suitable value, load resistance R_L is kept maximum and ammeter and voltmeter readings are noted. Load resistances R_L is reduced to a very low value in steps and each time ammeter and voltmeter readings are noted.
2. This process is repeated for different values of R.
3. For every value of R, curve is plotted between power consumed in the load resistance and load resistance R_L and from the curve so drawn the value of R_L for maximum power drawn is determined.

6. **Observation Table:**

S.No.	*Resistance R in ohms*	*Current in amperes (I_L)*	*Voltage V in volts*	*Power consumed $P = V_L I_L$ watts*	*Load resistance $RL = V_L I_L$ ohms*

7. **Calculations:**

1. The value of current in load resistance is .... ohms.
2. The value of voltage in load resistance is .... volt.
3. The value of power consumed as .... watts.
4. The value of load resistance is .... ohms.

8. **Result:** It will found that power consumed ($V_L I_L$) will be maximum when R_L becomes equal to R. This verifies the maximum power transfer theorem.

9. **Precautions:**

1. All connections should be tight.
2. Reading should be taken carefully for accurate result.
3. Do not touch the live terminals because touching the live terminals is injurious to health.

2. Power Measurement

Experiment: Study of power measurement.

1. **Object:** To study the power measurement in a three-phase a.c. circuit by two wattmeter method and to determine the power factor of the load.

2. **Apparatus Required:** Dynamometer type wattmeters. Tripple pole iron clad switch (TPIC), three-phase balanced load, moving iron voltmeter, moving iron ammeter, connecting wires, etc.

Apparatus Details

S.No.	*Name of apparatus*	*Type*	*Range/rating*	*Quantity*	*Make*	*Remarks*
1.						
2.						
3.						

3. **Theory/Principle:** The load wattmeters W_1 and W_2, ammeter A and voltmeter V are connected to the three-phase a.c. supply through a TPIC, the readings of wattmeters W_1 and W_2, ammeter A and voltmeter V are noted with different loads. If anyone of the wattmeters gives down scale reading (*i.e.*, when the load factor is below 0.5), the supply is switched off and the connections of current coil (CC) or potential coil (PC) are reserved and the reading is considered as negative.

If two wattmeters W_1 and W_2 are connected as per circuit diagram shown, the power consumption P_1 and P_2 are calculated as :

If current taken by wattmeter W_1 is I_R.

And potential difference across pressure circuit of W_1 is

$$E_{RB} = E_R - E_B \text{ (vector difference)}$$

Power read of wattmeter $W_1 \Rightarrow P_1$ (say) $= I_R (E_R - E_B)$

Similarly power read by wattmeter $W_2 \Rightarrow P_2$ (say) $= I_Y (E_Y - E_B)$

Hence power read by wattmeters W_1 and W_2 is given by

$$P_1 + P_2 = I_R (E_R - E_B) + I_Y (E_Y - E_B)$$

$$P_1 + P_2 = I_R E_R + I_Y E_Y - (I_R + I_Y) E_B \quad \text{...}(a)$$

Also we know that

$$I_R + I_Y + I_B = 0$$

$$I_R + I_Y = -I_B$$

Put the value in equation (*a*), we get

$$P_1 + P_2 = I_R E_R + I_Y E_Y + I_B E_B$$

$$P_1 + P_2 = P$$

Total power $$P = P_1 + P_2 = \sqrt{3}\ V_1 V_L \cos\phi$$

Power factor $$(\cos\phi) = \frac{P}{\sqrt{3}\ V_L \cdot I_L}$$

4. **Circuit Diagram:**

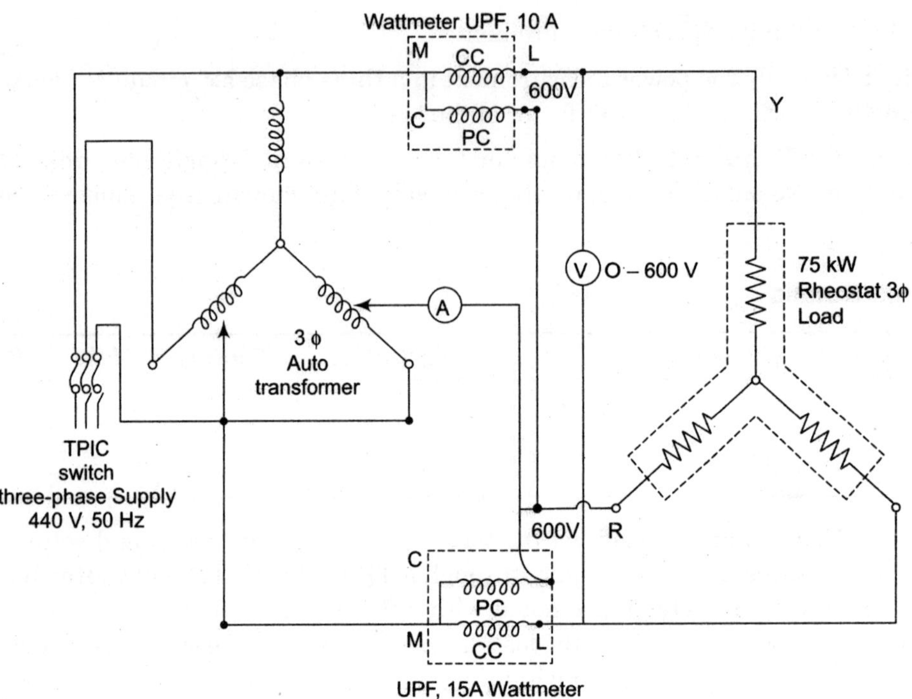

Fig. E.5 *Circuit Diagram for power measurement by two wattmeter method.*

5. **Procedure:**
   1. Connect the circuit as per circuit diagram.
   2. Vary the inductive load.
   3. Note down all the reading carefully.
   4. If one wattmeter reads negative or gives reverse reading, the reading of the wattmeter is taken by reversing the current coil terminal.

6. **Observation Table:**

S.No.	*Voltage V_L (in volts)*	*Current I_L (in amp.)*	*Power P_1 (watt)*	*Power P_2 (watt)*	*Total power P*	*Power factor cos ϕ*
1.						
2.						
3.						
4.						

7. **Calculation:** Total power

$$P = P_1 + P_2$$

or $$P = \text{........ watt}$$

Power factor $$(\cos \phi) = \frac{P}{\sqrt{3}\, V_L \cdot I_L}$$

or $$\cos \phi = \text{........}$$

Result: Net power consumption (P) = ... watt

Power factor (cos ϕ) = ...

9. **Precautions:**
   1. All connections should be tight.
   2. All apparatus should be of suitable range and ratings.
   3. Reading should be taken accurately.
   4. Never touch the live terminals and wires.
   5. Before reversing the connection of CC or PC, switch off the supply.

3. Load Test of Transformer

Experiment: Direct load test on 1-ϕ transformer.

1. **Object:** Measurement of efficiency of a single-phase transformer by load test.

2. **Apparatus Required:** a.c. ammeters, a.c. voltmeters, Wattmeters, Single-phase transformer. Lamp load in series with an inductor DPIC switch, connecting wires.

Apparatus Details

S.No.	Name of apparatus	Type	Range/rating	Quantity	Make	Remarks
1.						
2.						
3.						

3. **Theory/Principle :** The efficiency of a transformer is given by the expression

$$\eta = \frac{\text{Output power}}{\text{Input power}} \times 100$$

or

$$\eta = \frac{P_2}{P_1} \times 100$$

The method of determination of transformer efficiency by direct measurement of output and input does not give accurate result, as the power losses are quite low (or the order of 1-4%). The difference between the reading of output and output instruments is then so small that an instrument error as low as 0.5% would cause an error of the order of 15-20% in the power losses. There is a wastage of large amount of power and no information is available from such a test about the proportion of copper and iron losses.

4. **Circuit Diagram:**

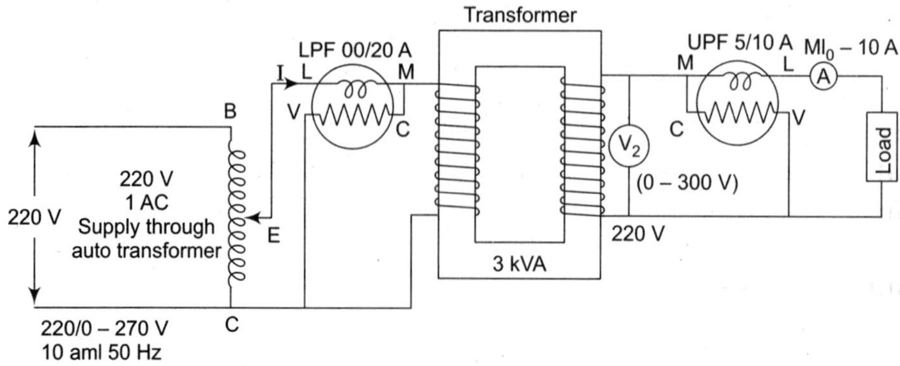

Fig. E.6 *(a) Circuit diagram.*

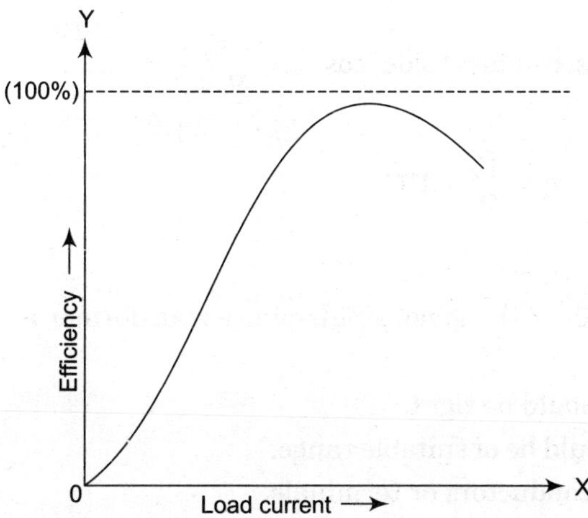

Fig. E.6 *(b) Efficiency curve.*

5. **Theory/Principle:**

1. Make the connection according to the circuit diagram as shown above in figure.
2. Connect the wattmeters, ammeters and voltmeters both in primary as well as in secondary winding and connect a load across the output of the winding under test according to the circuit diagram.
3. The supply is switched on to the circuit through DPIC switch.
4. The readings of voltmeters, ammeters and wattmeters, on input (primary) and output (secondary) are noted for different loads and repeat these readings at least three time and then take their mean.

6. **Observation Table:**

S.No.	*Input side*				*Output side*				$\eta = \frac{P_2}{P_1} \times 100$
	V_1 in volts	*I_1 in amps.*	*P_1 in watts*	*PF P_1/V_1V_1*	*V_2 in volts*	*I_2 in amps.*	*P_2 in watts*	*P_F P_1/V_1V_1*	
1.									
2.									
3.									
4.									
5.									
6.									

7. **Calculations:**

Power input (primary) side, $P_2 = V_1I_1 =$ .... watts.

Power factor in input (primary) side, $\cos \phi_1 = \frac{P_1}{V_1 I_1} =$ ....

Power in output (secondary) side $P_2 = V_2I_2 =$ .... watts.

Power factor in input (secondary) side, $\cos \phi_2 = \dfrac{P_2}{V_2\, I_2} = \ldots\ldots$.

Transformer efficiency $\eta = \dfrac{P_2}{P_1} \times 100$

$$\eta = \ldots\ldots.$$

8. **Result:** The efficiency of the given single-phase transformer is ....
9. **Precautions:**
   1. All connections should be tight.
   2. All apparatus should be of suitable range.
   3. Never touch live conductors or terminals.
   4. Reading should be taken carefully.

4. O.C./S.C. Test of Transformer

Experiment: Study of the single-phase transformer's circuit parameters.

1. **Object:** Determination of circuit parameters and losses in a single-phase transformer by open circuit and short circuit test and obtain the following :

   (*a*) Voltage regulation in percentage.

   (*b*) Equivalent circuit diagram (using circuit parameters).

2. **Apparatus Required:** Single-phase transformer, auto transformer (for variable supply). voltmeters, ammeters, wattmeters, screw driver, combination plier, connecting wires, etc.

Apparatus Details

S.No.	*Name of apparatus*	*Type*	*Range/rating*	*Quantity*	*Make*	*Remarks*

3. **Theory/Principle:** Transformer is a high efficiency, static machine. It transfers the power from one circuit to other circuit at same frequency. The efficiency of small rating transformers can be found by direct loading method, but in the case of large rating transformers, it can not. Because it is impossible to full load the large transformers in laboratory and it is also costly. So we perform the open circuit and short circuit tests on a transformer to measure its losses by which we can calculate efficiency and voltage regulation of the transformer. The main losses occurs in a transformer are as follows:

(A) *No load losses:* These are also known as iron or core losses in a transformer and these losses consists the following :

(*i*) Eddy-current losses

(*ii*) Hysteresis losses.

(B) *Full load losses:* These losses are also known as copper losses and occurs due to the resistance of winding.

Open circuit test: This test perform to find out the no load losses (iron losses) and no load current. Since, at the no-load condition or open circuited secondary condition, load current in primary is very less, so copper losses can be neglected.

In this test, primary winding (generally L.V. side) energised with rated voltage and secondary winding (H.V. side) is open circuited. The connected apparatuses, *i.e.*, voltmeter, ammeter and wattmeters, in primary side measures, voltage across primary terminals, no-load current and no-load losses respectively.

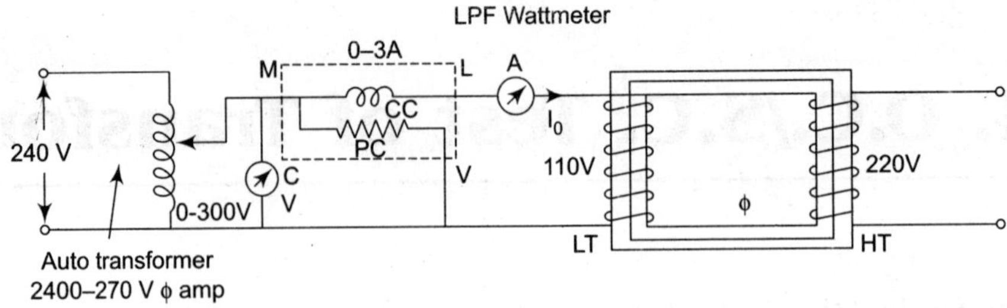

Fig. E.7 *Open circuit test circuit diagram.*

Short circuit test: The main purpose of this test is to find out the copper losses. This test is performed, keeping L.V. winding short circuited with a wire or ammeter and energised H.V. side with its full load current. Since, this full load current is started flowing at the low voltage (generally 5 to 10% of rated voltage), so less magnetic flux produced; due to this reason core losses or iron losses can be neglected. The apparatus, *i.e.*, ammeter, voltmeter and wattmeter, connected in H.V. side, measures the full load current, short circuit voltage and full load copper losses respectively.

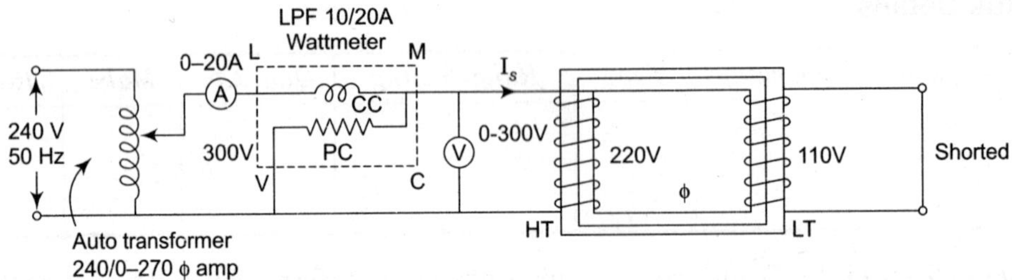

Fig. E.8 *Short circuit test circuit diagram.*

Equivalent Circuit of Transformer

(*a*) **Under no load condition:**

Let rated applied voltage in input = V_1

Current flowing in L.V. side = I_0

No load losses = P_0

Conductance $Y_0 = I_0/V_1$ mho

$$(V_1)^2 g_1 = P_0$$

∴ $g_1 = P_0/(V_1)^2$ mho

and $b_m = \sqrt{(Y_0^2 - g_1^2)}$ mho

where $Y_0 \rightarrow$ Admittance $= \dfrac{1}{Z_0}$

$g_1 \rightarrow$ Conductance $= \dfrac{1}{R_1}$

$b_m \rightarrow$ Conductance $= \dfrac{1}{X_L}$

(*b*) **Under short circuit condition:**

Let short circuit current $= I_{SC}$

Applied voltage in input $= V_{SC}$

Short circuit losses $= P_{SC}$ = Full load Cu-losses

Since, $(I_{SC})^2 R = P_{SC}$

$\therefore$ $R = P_{SC}/(I_{SC})^2$

and $Z = V_{SC}/I_{SC}$

$\therefore$ $X = \sqrt{Z^2 - R^2}$

Voltage regulation: Voltage regulation can be defined as the ratio of voltage drop, between input and output ends to the input or no-load voltage. So,

$$\text{Voltage regulation} = \frac{\text{Voltage drop}}{\text{no-load voltage}}$$

or, percent voltage regulation $(\%R_{LT}) = \dfrac{\text{Voltage drop}}{\text{no-load voltage}} \times 100$

where, in case of transformer,

$$\text{Voltage drop} = I_{SC}\,(R \cos \phi \pm X \sin \phi)$$

Note: Here, +ve sign shows lagging power factor when -ve sign shows leading power factor.

4. **Circuit diagram:**

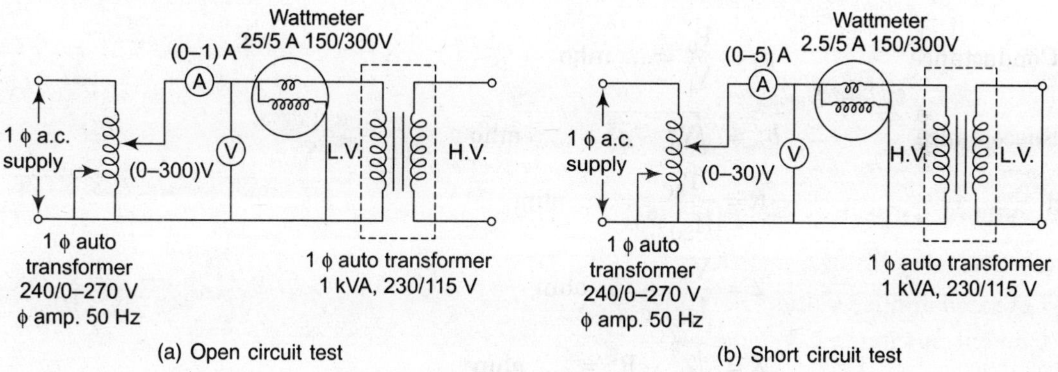

(a) Open circuit test (b) Short circuit test

Fig. E.9

5. **Procedure:**

(*a*) **Open circuit test:**

1. Connect all the apparatus as the circuit diagram shown in Fig. E.9(*a*)
2. ON the switch of power supply and adjust rated voltage across L.V. side
3. Record no-load current, voltage applied and no-load power of the transformer winding in observation table.
4. Switch OFF the supply after recording the readings.

(*b*) **Short circuit test:**

1. Connect all the apparatus and instruments according to circuit diagram shown in Fig. E.9(*a*).

2. Adjust the variac at zero.
3. On the switch of power a.c. supply.
4. Increase the applied voltage slowly, till the full load current in corresponding winding.
5. Note down the ammeter, voltmeter and wattmeter reading in the observation table.
6. Switch OFF the supply after taking the readings.

   cos ϕ of transformer = 0.8 (lag).

6. **Observation Table:**

S.No.	*Open circuit test*			*Short circuit test*		
	Input voltage (V_1) *volts*	*No-load current* (I_0) *amp.*	*Iron losses* (W_0) *watts or* P_0	*Short circuit voltage* (V_{SC}) *volts*	*Full load current* (I_{SC}) *amp.*	*Cu-losses* (W_{SC}) *watts or* P_{SC}

7. **Calculation:**

Admittance $\quad Y_0 = \dfrac{I_0}{V_1} = \ldots\ldots$ mho

Conductance $\quad g_1 = \dfrac{P_0}{V_1} = \ldots\ldots$ mho

Susceptance $\quad b_m = \sqrt{Y_0^2 - g_1^2} = \ldots\ldots$ mho

and $\quad R = \dfrac{P_{SC}}{(I_{SC})^2} = \ldots\ldots$ ohm

$$Z = \frac{V_{SC}}{I_{SC}} = \ldots\ldots \text{ ohm}$$

$$X = \sqrt{Z^2 - R^2} = \ldots\ldots \text{ ohm}$$

Voltage drop $\quad = I_{SC} (R \cos \phi + X \sin \phi)$

$$\% \text{ voltage regulation } (\% R_V) = \frac{\text{Voltage drop}}{\text{Input voltage}} \times 100$$

8. **Result:** Voltage regulation (R_V) of transformer = ... %
9. **Precautions:**
   1. All connections should be tight.
   2. All instruments and apparatus should be suitable range and ratings.
   3. Don't touch any bare conductor or terminal without off the main switch.
   4. Reading should be taken carefully.
   5. Never start the experiment without inspection of lab incharge.
   6. Voltage should be increased gradually under short circuit test.

5. D.C. Generator

Experiment: d.c. generator characteristics.

1. **Object:** Determination of load characteristics of d.c. shunt generator.

2. **Apparatus Required:** d.c. shunt generator, coupled with d.c. shunt motor, ammeters, voltmeters, loading rheostat, connecting wire, tachometer, etc.

Apparatus Detail

S.No.	*Name of apparatus*	*Type*	*Range/rating*	*Quantity*	*Make*	*Remarks*
1.						
2.						
3.						

3. **Theory/Principle:** For the d.c. shunt generator we plot a curve between load current and terminal voltage. In this generator terminal voltage reduces due to :

   1. Armature reaction.
   2. Resistive drop in windings and brushes.
   3. Weak excitation.

   by increasing load current as shown in graph.

In the d.c. series generator, the terminal voltage increases by increasing the load current because flux produced by per pole is proportional to load current as total load current flows through the series winding.

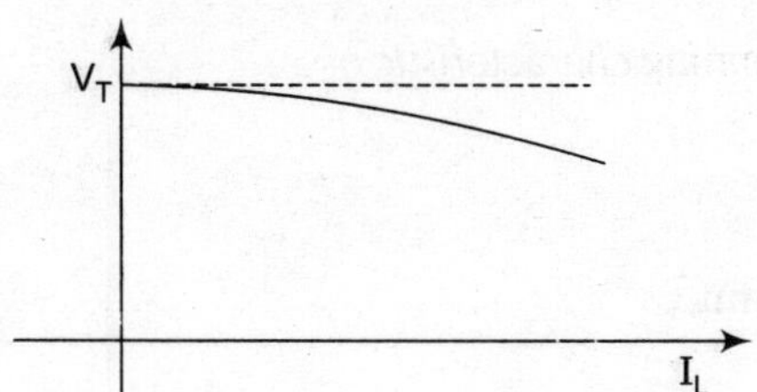

Fig. E.10 *Characteristic curve of d.c. shunt generator between load current and terminal voltage.*

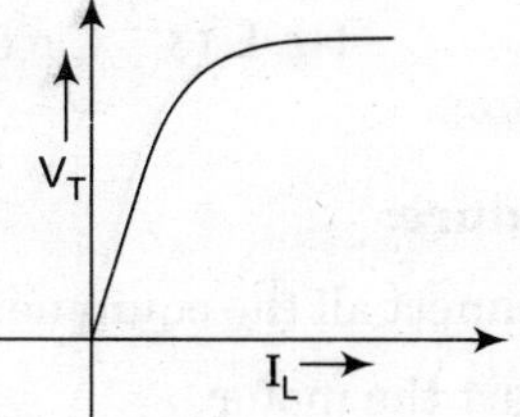

Fig. E.11 *Characteristic curve of d.c. series generator between load current and terminal voltage.*

4. **Circuit diagram:**

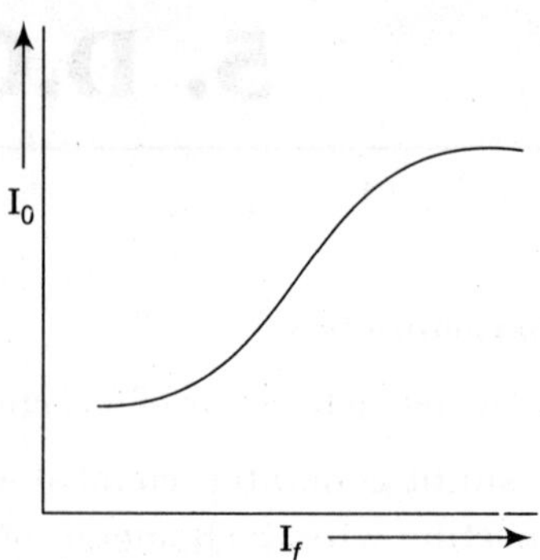

Fig. E.12 *No-Load characteristic of separately excited D.C. shunt generator.*

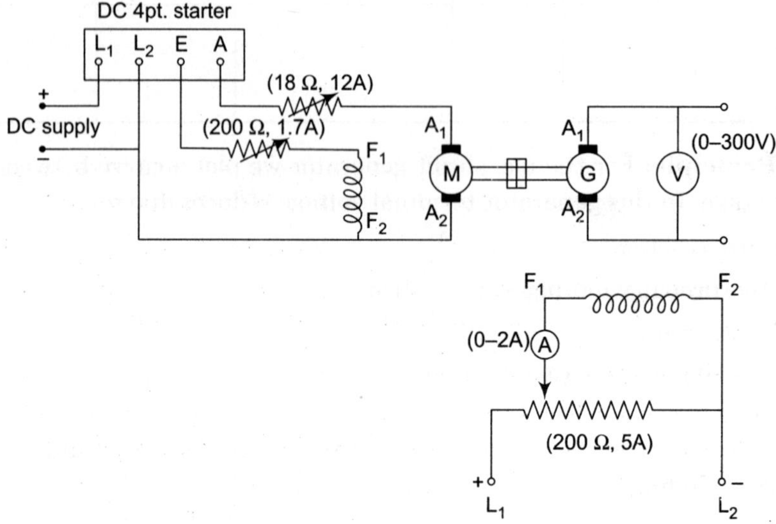

Fig. E.13 *Circuit diagram for determining characteristic of D.C. shunt generator.*

5. **Procedure:**

   1. Connect all the equipments as circuit diagram.
   2. Start the motor.
   3. Now, read the ammeter and voltmeter readings varying the load resistance.
   4. Note the readings in observation table and plot a graph.
   5. Switch OFF the main supply.
   6. Repeat this procedure to obtain the characteristics of D.C. series generator.

6. **Observation Table**

S.No.	*D.C. shunt generator*		*D.C. series generator*	
	Load current I_L *amp.*	*Terminal voltage* V_T *volt*	*Load current* I_L *amp*	*Terminal voltage* V_T *volt*
1.				
2.				
3.				
.				
.				
.				
10.				

Note: Plot the curves upon graph papers for above readings.

7. **Result:** The characteristics of D.C. shunt generator and D.C. series generator are found as observed in observation table and corresponding graph.

8. **Precautions:**

1. All connections should be tight.
2. Readings should be taken carefully.
3. Readings should be taken, increasing the load current.

6. D.C. Shunt Motor

Experiment: Speed control of D.C. shunt motor.

1. **Object:** To study the speed control of a D.C. shunt motor and to draw the speed variation w.r.t. change of: (*a*) field current, (*b*) armature current.

2. **Apparatus Required:** D.C. shunt motor, ammeters, voltmeters, rheostats, DPIC switch and 3-point starter, techometer, connecting leads, etc.

Apparatus Details

S.No.	*Name of apparatus*	*Type*	*Range/rating*	*Quantity*	*Make*	*Remarks*
1.						
2.						
3.						

3. **Theory/Principle :** We know that, in D.C. shunt motors, the back e.m.f. is given by

$$I_b = V - I_a R_a = \frac{Z\phi\, PN}{60\ A}$$

or

$$N = \frac{(V - I_a R_a)}{Z\phi\, PN} \times 60\ A \qquad ...(1)$$

Hence, $$N \propto \frac{1}{\phi} \text{ and } N \propto \frac{1}{R_a} \qquad (\because\ N \text{ is speed in r.p.m.})$$

where ϕ is flux and R_a is armature resistance.

(*a*) *Field control method:* This method is applied, where we have to obtain the speed greater than normal rated speed. If we insert a resistance in field winding, field current becomes less, consequently flux also becomes less and speed increased by the equation (1).

(*b*) *Armature control method:* If we need the speed less than rated speed, we use this method. If we insert a resistance in the series of armature, $I_a R_a$ will increase, therefore the speed of shunt motor reduces by the equation (1).

4. **Procedure:**

(*a*) *Field Control Method:*

1. Make a proper connection according to circuit diagram.
2. Switch on the supply and start the motor with the help of starter.
3. Now, insert the resistance in field winding slowly.

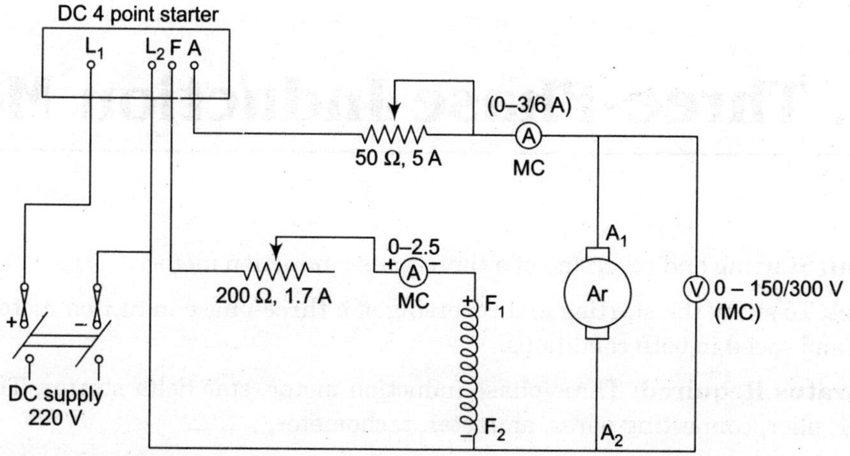

Fig. E.14 *Circuit diagram for speed control of DC shunt Motor.*

4. **Tabulate the readings of field current, voltmeter and speed by techometer in observation table.**
5. **Take some readings varying the field resistance and plot a graph between I_f and N on graph motor.**

7. Three-Phase Induction Motor

Experiment: Starting and reversing of a three-phase induction motor.

1. **Object:** To study the starting and reversing of a three-phase induction motor. Measure the current and speed in both conditions.

2. **Apparatus Required:** Three-phase induction motor, star delta starter, TPIC switch, screw driver, plier, connecting wires, ammeter, tachometer.

Apparatus Details

S.No.	*Name of apparatus*	*Type*	*Range/rating*	*Quantity*	*Make*	*Remarks*
1.						
2.						
3.						

3. **Theory/Principle:** A starter used for keeping the current within reasonable limits for a three-phase induction motor, otherwise the large starting current may cause undesirable voltage dips.

The method of starting a three-phase induction motor is as follows:

1. *D.O.L. starter:* In this method, we can start three-phase inductor motor directly connected to the supply. We use over load relay (OLR) in this starter to remove the overloading and under voltage relay (UVR) to protect the motor against under voltage in running condition by setting the current and voltage limits in relay. Over load relay may be made of bi-metallic type which works on increasing the temperature due to increase the load current of motor. This starter is used for under 5 H.P. motor.
2. *Rotor resistance starter:* A star connected resistance is inserted in the rotor winding at the instant of starting. In this method, the resistance must be connected in star condition so rotor of motor becomes star-circuited by resistane starter. So we can say each three phase induction rotor works as a single-phase induction motor in running condition. As motor gains speed the resistance is cut out in steps. This method is used only in slip ring induction motor.
3. *Star-delta starter:* A three-phase winding of a motor is connected in star at the instant starting. In this condition the induction motor takes reduced voltage and start safely. The applied voltage at starting is $\frac{1}{\sqrt{3}}$ times the line voltage. When motor has picked up the speed the connection changes to delta by means of change over switch. Motor takes rated voltage and rated current in running condition.

4. **Auto transformer starter:** The input voltage to the induction motor's stator is given through an auto transformer. The voltage is applied gradually till its rated voltage. In this method, we increase the voltage gradually so the induction motor takes current slowly-slowly. Normally the 60% taping on the auto-transformer can be used for safe starting current when the induction motor has picked up speed, the full rated voltage is applied by the change over switch.

4. **Circuit diagram:**

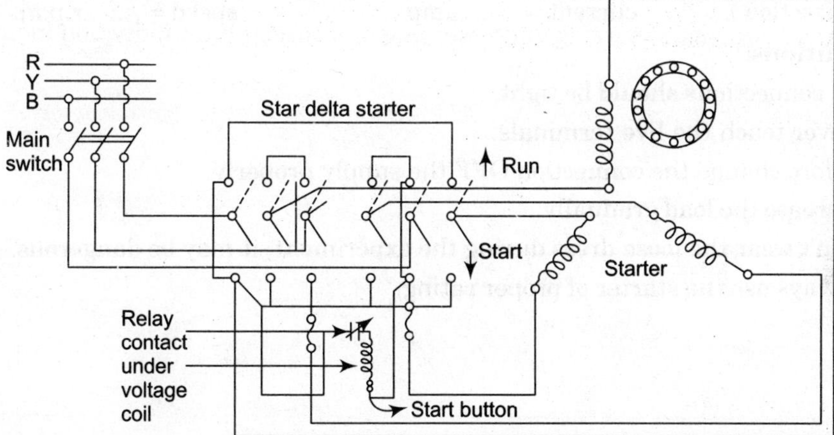

Fig. E.15 *Circuit diagram for star delta starting method of three-phase induction motor.*

5. **Procedure:**
   1. Connect the circuit according to the circuit diagram as shown in Fig. E.15.
   2. First, put the starter as 'start' position, when it gained approximate 80 to 90% of the rated speed, throw it to 'run' position.
   3. Note the starting current and speed of motor.
   4. Load the machine by tightening the belt, till its rated current start flowing. Note down the load current and speed of the motor.
   5. Interchange the connection of any two phases and repeat the above procedure.

6. **Observation Table:**

S.No.	*Forward direction*				*Reverse direction*			
	No-load		*On-load*		*No-load*		*On-load*	
	Current (a.m.p.)	*Speed (r.p.m.)*	*Current (a.m.p.)*	*Speed (r.p.m.)*	*Current (a.m.p.)*	*Speed (r.p.m.)*	*Current (a.m.p.)*	*Speed (r.p.m.)*
1.								
2.								
3.								
4.								
5.								

7. **Result:** When motor run in both reverse and forward direction following comparitive statement will be obtained.

No-Load

Forward direction :	current = ... amp.	speed = ... r.p.m.
Reverse direction :	current = ... amp.	speed = ... r.p.m.

Full Load

Forward direction :	current = ... amp.	speed = ... r.p.m.
Reverse direction :	current = ... amp.	speed = ... r.p.m.

8. **Precautions:**
   1. All connections should be tight.
   2. Never touch the live terminals.
   3. Before change the connection OFF the supply properly.
   4. Increase the load gradually.
   5. Don't wear the loose dress during the experiment, it may be dangerous.
   6. Always use the starter of proper rating.

8. Calibration of Energy Meter

Experiment: Study and calibration of single-phase energy meter.

1. **Object:** To study the construction and principle of single-phase induction type energy meter.

2. **Apparatus Required:** DPIC switch, single-phase energy meter, lamp load and connecting loads wattmeter.

Apparatus Details

S.No.	*Name of apparatus*	*Type*	*Range/rating*	*Quantity*	*Make*	*Remarks*

3. **Theory/Principle:** Induction type energy meters are the most common form of a.c. kWh meters used everyday in domestic and industrial installations. These meters measure electrical energy in kilo-watt hours. The principle of these meter is practically the same as that of the induction wattmeters. Constructionally, the two are similar except that the control spring and pointer of the wattmeter are replaced in the case of watt hour meter, by a break magnet and by a spindle of the meter.

The induction type energy meter consists of laminated electromagnets one of them excited by the current and the other by current proportional to the voltage across the load. The pressure coil is made highly inductive so that the current lag behind the voltage by 90°. A thin Aluminium disc is mounted so as to cut the flux from both the magnets and it is free to rotate.

In all induction instruments we have two fluxes produced by current flowing in the winding of instrument. These fluxes are alternating in nature and so they produce e.m.f. in a metalic disc. Consequently currents, therefore two torque are produced. Total torque is the sum of two torque.

There are four parts of the operting mechanism.

1. Driving system
2. Moving system
3. Braking system
4. Registering system

At constant angular speed the power $P = VI \cos f$ is proportional to the angular sped in r.p.s.

If N_t = Total number of revolution per kWh of meter under test.

W = Wattmeter reading at particular load.

t = Time in second for 'n' revolution of meter under test.

Reading of meter = W × t watt-sec

$$\% \text{ error} = \frac{\text{Meter constant} - \text{Practical reading}}{\text{Meter constant}} \times 100$$

4. **Circuit Diagram:**

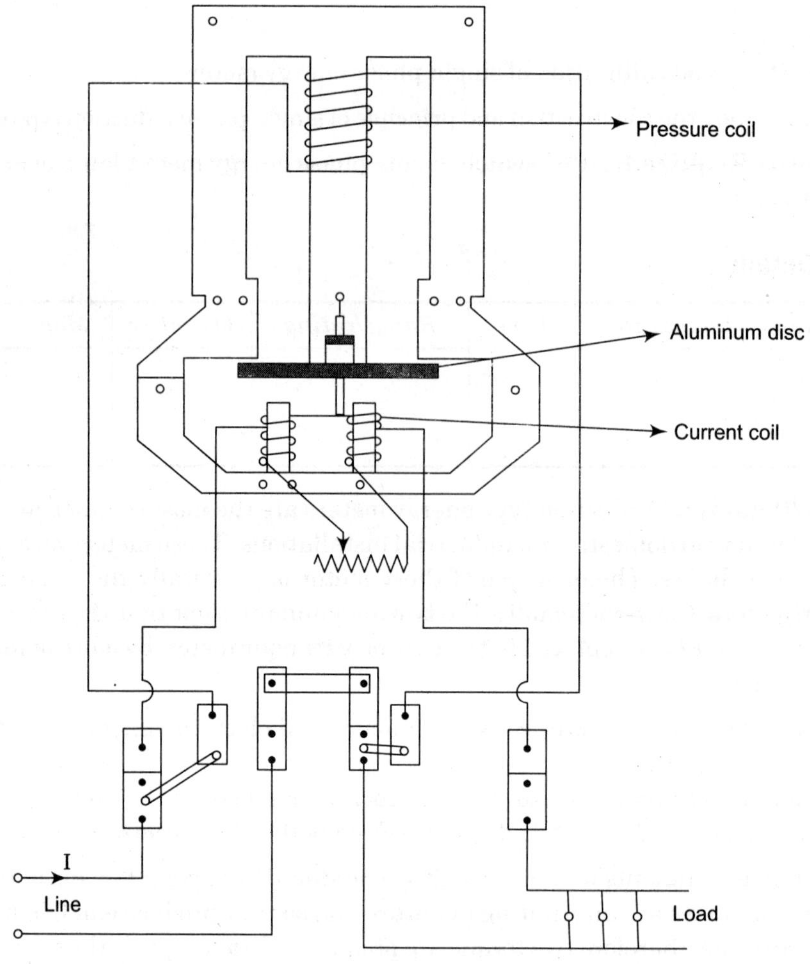

Fig. E.16 *Circuit Diagram for calibration of one-phase energy meter.*

5. **Procedure:**
   1. Connect the circuit according to circuit diagram.
   2. Apply the rated voltage by auto-transformer at no-load.
   3. The current coil of the energy meter is connected in series with load, while the pressure coil across the supply.
   4. Connect the variable resistor or lamp load between phase and neutral. Note the reading of all meters of their corresponding readings.
   5. Note down the time for particular revolution of the disc and reading of the energy meter.

6. Observation Table:

S.No.	*Voltage (in volt)*	*Current (in amp.)*	*Power (P) watts*	*No of rev. (r.p.s.)*	*Energy (Et) kWh*	*Energy (E8) kWh*	*% error*

7. Calculation:

$$\% \text{ error} = \frac{E_s - E_t}{E_s} \times 100$$

8. **Result:** % error = …….. %

9. **Precautions:**

1. All connections should be tight.
2. Never touch the live terminals during the experiment.
3. Before, read the energy meter, it should be connected to the load for minimum 15 minutes, that's why the friction error and temperature error can be removed.
4. Observe the readings carefully.
5. To find out the more accurate error, take some readings and find the mean of error.

Appendix

OBJECTIVE QUESTION

1. Find the odd one from the following elements
   (*a*) Inductor (*b*) Capacitor (*c*) Resistor (*d*) Transistor
2. Kirchhoff's law fails in case of
   (*a*) Linear networks (*b*) Non-linear networks
   (*c*) Dual networks (*d*) Distributed parameter networks
3. For a d.c. voltage, an inductor
   (*a*) Is virtually a short-circuit (*b*) Is an open-circuit
   (*c*) Depends on polarity (*d*) Depends on voltage value
4. Which of the following is an active element?
   (*a*) Voltage source (*b*) Current source
   (*c*) Inductor (*d*) Both (*a*) and (*b*)
5. Which of the following is non-linear element?
   (*a*) Inductor (*b*) Capacitor (*c*) Resistor (*d*) Diode
6. Which of the following is passive element?
   (*a*) Resistor (*b*) Inductor (*c*) Capacitor (*d*) All of these
7. Which of the following is active network?
   (*a*) RLC network without both current source and voltage source
   (*b*) RLC network with both current source and voltage source
   (*c*) RLC network with current source or voltage source
   (*d*) Both (*b*) and (*c*)
8. A passive network has
   (*a*) Current source but no voltage source
   (*b*) Voltage source but no current source
   (*c*) Both current source and voltage source
   (*d*) No current source and no voltage source
9. A wire conductor of resistance 1 ohm is doubled in length. its resistance becomes
   (*a*) The same as that old resistance (*b*) Higher than old resistance
   (*c*) Lower than old resistance (*d*) Non of above
10. A circuit whose properties are not the same in either direction is called
    (*a*) Bilateral circuit (*b*) Unilateral circuit
    (*c*) Reversible circuit (*d*) Inverted circuit

11. The node method of circuit analysis is based on
    (*a*) KVL and Ohm's law (*b*) KCL and KVL
    (*c*) KCL, KVL and Ohm's Law (*d*) KCL and ohm's law
12. The loop method of circuit analysis is based on
    (*a*) KVL and Ohm's law (*b*) KCL and KVL
    (*c*) KCL, KVL and Ohm's Law (*d*) KCL and ohm's law
13. The superposition theorem is applicable to
    (*a*) Current only (*b*) Voltage only
    (*c*) Both current and voltage (*d*) Current, voltage and power
14. The superposition theorem is applicable to
    (*a*) Linear responses only
    (*b*) Linear and non-linear responses
    (*c*) Power calculations
    (*d*) Linear, non-linear and time-variant responses
15. Thevenin's theorem can be applied to calculate the current in
    (*a*) Any load (*b*) A passive load only
    (*c*) A linear load only (*d*) A bilateral load only
16. Suppose Double the voltage in a simple dc circuit and cut the resistance in half, then the current will
    (*a*) Become four times as great (*b*) Become twice as great
    (*c*) Stay the same as it was before (*d*) Become half as great
17. The unit which indicates the rate at which energy is expended?
    (*a*) Watt (*b*) Ampere-hour (*c*) Coulomb (*d*) Volt
18. The current through the 2 kΩ resistance in the circuit shown is
    (*a*) 0 mA (*b*) 1 mA (*c*) 2 mA (*d*) 6 mA

1 kΩ 1 kΩ C A B D 1 kΩ 1 kΩ 6V

19. A star circuit has each element of resistance R/2. The equivalent delta elements will be
    (*a*) R (*b*) 3 R (*c*) 3/2 R (*d*) R/6
20. A delta circuit has each element of value R/2. The equivalent elements of star circuit will be
    (*a*) 2R (*b*) 6R (*c*) 3/2 R (*d*) R/6
21. Application of Norton's theorem to a circuit yields
    (*a*) equivalent current source
    (*b*) equivalent impedance
    (*c*) equivalent current source and impedance in series
    (*d*) equivalent current source and impedance in parallel

22. A 2 Ω resistor carrying 2 ampere current will dissipated power equal to
 (*a*) 4 W (*b*) 8 W (*c*) 16 W (*d*) 32 W
23. The maximum power that can be dissipated in the load in the circuit shown in figure is
 (*a*) 9.1 W (*b*) 8.0 W (*c*) 9.2 W (*d*) 9.6 W

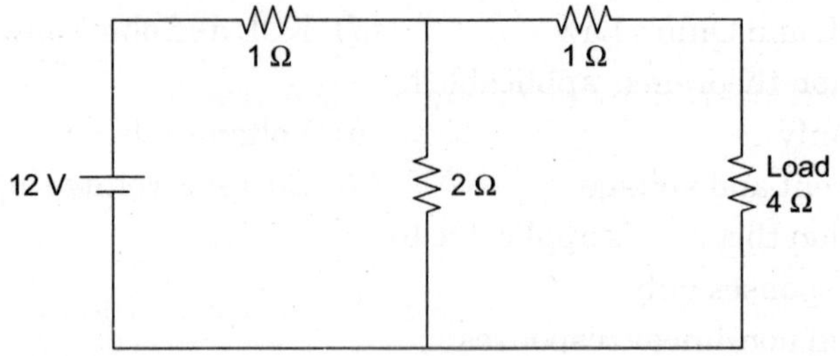

1 Ω 1 Ω
12 V 2 Ω Load 4 Ω

24. The period of a wave is
 (*a*) the same as frequency
 (*b*) time required to complete one cycle
 (*c*) expressed in amperes
 (*d*) none of the above
25. The form factor is the ratio of
 (*a*) peak value to r.m.s. value
 (*b*) r.m.s. value to average value
 (*c*) average value to r.m.s. value
 (*d*) none of the above
26. The period of a sine wave is ______ seconds.
 Its frequency is
 (*a*) 20 Hz (*b*) 30 Hz (*c*) 40 Hz (*d*) 50 Hz
27. A heater is rated as 230 V, 10 kW, A.C. The value 230 V refers to
 (*a*) average voltage
 (*b*) r.m.s. voltage
 (*c*) peak voltage
 (*d*) none of the above
28. If two sinusoids of the same frequency but of different amplitudes and phase angles are subtracted, the resultant is
 (*a*) a sinusoid of the same frequency
 (*b*) a sinusoid of half the original frequency
 (*c*) a sinusoid of double the frequency
 (*d*) not a sinusoid
29. The peak value of a sine wave is 200 V. Its average value is
 (*a*) 127.4 V (*b*) 141.4 V (*c*) 282.8 V (*d*) 200 V
30. If two sine waves of the same frequency have a phase difference of JT radians, then
 (*a*) both will reach their minimum values at the same instant
 (*b*) both will reach their maximum values at the same instant
 (*c*) when one wave reaches its maximum value, the other will reach its minimum value
 (*d*) none of the above
31. The voltage of domestic supply is 220 V. This figure represents
 (*a*) mean value (*b*) r.m.s. value (*c*) peak value (*d*) average value

32. Two waves of the same frequency have opposite phase when the phase angle between them is

    (*a*) 360° (*b*) 180° (*c*) 90° (*d*) 0°

33. The power consumed in a circuit element will be least when the phase difference between the current and voltage is

    (*a*) 180° (*b*) 90° (*c*) 60° (*d*) 0°

34. The r.m.s. value and mean value is the same in the case of

    (*a*) Triangular wave (*b*) Sine wave
    (*c*) Square wave (*d*) Half wave rectified sine wave

35. For the same peak value which of the following wave will 'have the highest r.m.s. value ?

    (*a*) square wave (*b*) half wave rectified sine wave
    (*c*) triangular wave (*d*) sine wave

36. Form Factor is the ratio of

    (*a*) average value/r.m.s. value (*b*) average value/peak value
    (*e*) r.m.s. value/average value (*d*) r.m.s. value/peak value

37. Form factor for a sine wave is

    (*a*) 1.414 (*b*) 0.707 (*c*) 1.11 (*d*) 0.637

38. The phase difference between voltage and current wave through a circuit element is given as 30°. The essential condition is that

    (*a*) both waves must have same frequency
    (*b*) both waves must have identical peak values
    (*c*) both waves must have zero value at the same time
    (*d*) none of the above

39. The r.m.s. value of a sinusoidal A.C. current is equal to its value at an angle of _______ degrees.

    (*a*) 90 (*b*) 60 (*c*) 45 (*d*) 30

40. In a series resonant circuit, the impedance of the circuit is

    (*a*) minimum (*b*) maximum (*c*) zero (*d*) none of the above

41. Power factor of an electrical circuit is equal to

    (*a*) R/Z
    (*b*) cosine of phase angle difference between current and voltage
    (*c*) kW/kVA
    (*d*) ratio of useful current to total current Iw/I
    (*e*) all above

42. The best place to install a capacitor is

    (*a*) very near to inductive load (*b*) across the terminals of the inductive load
    (*c*) far away from the inductive load (*d*) any where

43. Poor power factor

    (*a*) reduces load handling capability of electrical system
    (*b*) results in more power losses in the electrical system

(*c*) overloads alternators, transformers and distribution lines
(*d*) results in more voltage drop in the line
(*e*) results in all above

44. Capacitors for power factor correction are rated in
(*a*) kW (*b*) kVA (*c*) kV (*d*) kVAR

45. In series resonant circuit, increasing inductance to its twice value and reducing capacitance to its half value
(*a*) will change the maximum value of current at resonance
(b) will change the resonance frequency
(*c*) will change the impedance at resonance frequency
(*d*) will increase the selectivity of the circuit

46. Pure inductive circuit
(*a*) consumes some power on average
(*b*) does not take power at all from a line
(*c*) takes power from the line during some part of the cycle and then returns back to it during other part of the cycle
(*d*) none of the above

47. Inductance affects the direct current flow
(*a*) only at the time of turning off (*b*) only at the time of turning on
(*c*) at the time of turning on and off (*d*) at all the time of operation

48. Inductance of a coil Varies
(*a*) directly as the cross-sectional area of magnetic core
(*b*) directly as square of number of turns
(*c*) directly as the permeability of the core
(*d*) inversely as the length of the iron path
(*e*) as (*a*) to (*d*)

49. All the rules and laws of D.C. circuit also apply to A.C. circuit containing
(*a*) capacitance only (*b*) inductance only
(*c*) resistance only (*d*) all above

50. Power factor of an inductive circuit is usually improved by connecting capacitor to it in
(*a*) parallel (*b*) series
(*c*) either (*a*) or (*b*) (*d*) none of the above

51. In a highly capacitive circuit the
(*a*) apparent power is equal to the actual power
(*b*) reactive power is more than the apparent power
(*c*) reactive power is more than the actual powerful
(*d*) actual power is more than its reactive power

52. Power factor of the following circuit will be zero
(*a*) resistance (*b*) inductance (*c*) capacitance (*d*) both (*b*) and (*c*)

53. Power factor of the following circuit will be unity
(*a*) inductance (*b*) capacitance (*c*) resistance (*d*) both (*a*) and (*b*)
54. Power factor of the system is kept high
(*a*) to reduce line losses
(*b*) to maximise the utilization of the capacities of generators, lines and transformers
(*c*) to reduce voltage regulation of the line
(*d*) due to all above reasons
55. In a loss-free R-L-C circuit the transient current is
(*a*) oscillating (*b*) square wave (*c*) sinusoidal (*d*) non-oscillating
56. The r.m.s. value of alternating current is given by steady (D.C.) current which when flowing through a given circuit for a given time produces
(*a*) the more heat than produced by A.C. when flowing through the same circuit
(*b*) the same heat as produced by A.C. when flowing through the same circuit
(*c*) the less heat than produced by A.C. flowing through the same circuit
(*d*) none of the above
57. The double energy transient occur in the
(*a*) purely inductive circuit (*b*) R-L circuit
(*c*) R-C circuit (*d*) R-L-C circuit
58. The power factor at resonance in R-L- C parallel circuit is
(*a*) zero (*b*) 0.08 lagging (*c*) 0.8 leading (*d*) unity
59. In a pure resistive circuit
(*a*) current lags behind the voltage by 90°
(*b*) current leads the voltage by 90°
(*c*) current can lead or lag the voltage by 90°
(*d*) current is in phase with the voltage
60. In a pure inductive circuit
(*a*) the current is in phase with the voltage
(*b*) the current lags behind the voltage by 90°
(*c*) the current leads the voltage by 90°
(*d*) the current can lead or lag by 90°
61. In a circuit containing R, L and C, power loss can take place in
(*a*) C only (*b*) L only (*c*) R only (*d*) all above
62. In any A.C. circuit always
(*a*) apparent power is more than actual power
(*b*) reactive power is more than apparent power
(*c*) actual power is more than reactive power
(*d*) reactive power is more than actual power
63. In a purely inductive circuit
(*a*) actual power is zero (*b*) reactive power is zero
(*c*) apparent power is zero (*d*) none of above is zero

64. Magnitude of current at resonance in R-L-C circuit
   (*a*) depends upon the magnitude of R
   (*b*) depends upon the magnitude of L
   (*c*) depends upon the magnitude of C
   (*d*) depends upon the magnitude of R, L and C

65. In a R-L-C circuit
   (*a*) power is consumed in resistance and is equal to I R
   (*b*) exchange of power takes place between inductor and supply line
   (*c*) exchange of power takes place between capacitor and supply line
   (*d*) exchange of power does not take place between resistance and the supply line
   (*e*) all above are correct

66. In R-L-C series resonant circuit magnitude of resonance frequency can be changed by changing the value of
   (*a*) R only (*b*) L only (*c*) C only (*d*) L or C
   (*e*) R, L or C

67. In a series L-C circuit at the resonant frequency the
   (*a*) current is maximum (*b*) current is minimum
   (*c*) impedance is maximum (*d*) voltage across C is minimum

68. If resistance is 20 Q. and inductance is 27 in a R-L series circuit, then time constant of this circuit will be
   (*a*) 0.001 s (*b*) 0.1 s (*c*) 10 s (*d*) 100 s

69. Which of the following coil will have large resonant frequency?
   (*a*) A coil with large resistance
   (*b*) A coil with low resistance
   (*c*) A coil with large distributed capacitance
   (*d*) A coil with low distributed capacitance

70. If a sinusoidal wave has frequency of 50 Hz with 30 A r.m.s. current which of the following equation represents this wave ?
   (*a*) 42.42 sin 3141 (*b*) 60 sin 25 t (*c*) 30 sin 50 t (*d*) 84.84 sin 25 t

71. The input of an A.C. circuit having power factor of 0.8 lagging is 40 kVA. The power drawn by the circuit is
   (*a*) 12 kW (*b*) 22 kW (*c*) 32 kW (*d*) 64 kW

72. In an AC. circuit, a low value of kVAR compared with kW indicates
   (*a*) low efficiency (*b*) high power factor
   (*c*) unity power factor (*d*) maximum load current

73. All definitions of power factor of a series R-L-C circuit are correct except
   (*a*) ratio of net reactance and impedance
   (*b*) ratio of kW and kVA
   (*c*) ratio of J and Z
   (*d*) ratio of W and VA

74. A delta circuit has each element of value R/2. The equivalent elements of star circuit will be
    (*a*) 2 R (*b*) 6 R (*c*) 3/2 R (*d*) R/6
75. In case of delta connected circuit, when one resistor is open, power will be
    (*a*) zero (*b*) reduced to 1/3 (*c*) reduced by 1/3 (*d*) unaltered
76. For three phase star connected circuit,
    (*a*) Line voltage = phase voltage (*b*) Line current = phase current
    (*c*) Line current = $\sqrt{3}$ phase current (*d*) None of the above is true
77. Which of the following is true for a three phase delta connected circuit
    (*a*) Line voltage = phase voltage (*b*) Line current= phase current
    (*c*) Line current = $1/\sqrt{3}$ phase voltage (*d*) Line current = $\sqrt{3}$ phase current
78. In three phase star connections, line voltage is the same as
    (*a*) phase voltage (*b*) $1/\sqrt{3}$ phase voltage
    (*c*) $\sqrt{3}$ phase voltage (*d*) 3 phase current
79. Power in a Three Phase Circuit = ______.
    (*a*) $P = 3\, V_{Ph}\, I_{Ph} \cos \Phi$ (*b*) $P = \sqrt{3}\, V_L\, I_L \cos \Phi$
    (*c*) Both 1 & 2. (*d*) None of The Above
80. The use of ______ instruments is merely confined within laboratories as standardizing instruments.
    (*a*) absolute (*b*) indicating (*c*) recording (*d*) integrating
    (*e*) none of the above
81. Which of the following instruments indicate the instantaneous value of the electrical quantity being measured at the time at which it is being measured ?
    (*a*) Absolute instruments (*b*) Indicating instruments
    (*c*) Recording instruments (*d*) Integrating instruments
82. ______ instruments are those which measure the total quantity of electricity delivered in a particular time.
    (*a*) Absolute (*b*) Indicating (*c*) Recording (*d*) Integrating
83. Which of the following are integrating instruments ?
    (*a*) Ammeters (*b*) Voltmeters
    (*c*) Wattmeters (*d*) Ampere-hour and watt-hour meters
84. Resistances can be measured with the help of
    (*a*) wattmeters (*b*) voltmeters
    (*c*) ammeters (*d*) ohmmeters and resistance bridges
    (*e*) all of the above
85. According to application, instruments are classified as
    (*a*) switch board (*b*) portable (*c*) both (*a*) and (*b*) (*d*) moving coil
    (*e*) moving iron (f) both (*d*) and (*e*)
86. Which of the following essential features is possessed by an indicating instrument ?
    (*a*) Deflecting device (*b*) Controlling device
    (*c*) Damping device (*d*) All of the above

87. A _____ device prevents the oscillation of the moving system and enables the latter to reach its final position quickly
   (a) deflecting (b) controlling (c) damping (d) any of the above
88. The spring material used in a spring control device should have the following property.
   (a) Should be non-magnetic (b) Most be of low temperature co-efficient
   (c) Should have low specific resistance (d) Should not be subjected to fatigue
   (e) All of the above
89. Which of the following properties a damping oil must possess ?
   (a) Must be a good insulator
   (b) Should be non-evaporating
   (c) Should not have corrosive action upon the metal of the vane
   (d) The viscosity of the oil should not change with the temperature
   (e) All of the above
90. A moving-coil permanent-magnet instrument can be used as _______ by using a low resistance shunt.
   (a) ammeter (b) voltmeter (c) flux-meter (d) ballistic galvanometer
91. A moving-coil permanent-magnet instrument can be used as flux-meter
   (a) by using a low resistance shunt
   (b) by using a high series resistance
   (c) by eliminating the control springs
   (d) by making control springs of large moment of inertia
92. Which of the following devices may be used for extending the range of instruments ?
   (a) Shunts (b) Multipliers
   (c) Current transformers (d) Potential transformers
   (e) All of the above
93. An induction meter can handle current upto
   (a) 10 A (b) 30 A (c) 60 A (d) 100 A
94. For handling greater currents induction wattmeters are used in conjunction with
   (a) potential transformers (b) current transformers
   (c) power transformers (d) either of the above
   (e) none of the above
95. Induction type single phase energy meters measure electric energy in
   (a) kW (b) Wh (c) kWh (d) VAR
   (e) None of the above
96. Most common form of A.C. meters met with in every day domestic and industrial installations are
   (a) mercury motor meters
   (b) commutator motor meters
   (c) induction type single phase energy meters
   (d) all of the above

97. Which of the following meters are not used on D.C. circuits
    (*a*) Mercury motor meters (*b*) Commutator motor meters
    (*c*) Induction meters (*d*) None of the above
98. Which of the following is an essential part of a motor meter ?
    (*a*) An operating torque system (*b*) A braking device
    (*c*) Revolution registering device (*d*) All of the above
99. A potentiometer may be used for
    (*a*) measurement of resistance (*b*) measurement of current
    (*c*) calibration of ammeter (*d*) calibration of voltmeter
    (*e*) all of the above
100. The household energy meter is
    (*a*) an indicating instrument (*b*) a recording instrument
    (*c*) an integrating instrument (*d*) none of the above
101. The pointer of an indicating instrument should be
    (*a*) very light (*b*) very heavy (*c*) either (*a*) or (*b*) (*d*) neither (*a*) nor (*b*)
102. The chemical effect of current is used in
    (*a*) D.C. ammeter hour meter (*b*) D.C. ammeter
    (*c*) D.C. energy meter (*d*) none of the above
103. In majority of instruments damping is provided by
    (*a*) fluid friction (*b*) spring (*c*) eddy currents (*d*) all of the above
104. An ammeter is a
    (*a*) secondary instrument (*b*) absolute instrument
    (*c*) recording instrument (*d*) integrating instrument
105. In a portable instrument, the controlling torque is provided by
    (*a*) spring (*b*) gravity (*c*) eddy currents (*d*) all of the above
106. An air gap is usually inserted in magnetic circuits to
    (*a*) increase m.m.f. (*b*) increase the flux
    (*c*) prevent saturation (*d*) none of the above
107. The relative permeability of a ferromagnetic material is
    (*a*) less than one (*b*) more than one
    (*c*) more than 10 (*d*) more than 100 or 1000
108. The unit of magnetic flux is
    (*a*) henry (*b*) weber
    (*c*) ampereturn/weber (*d*) ampere/metre
109. Permeability in a magnetic circuit corresponds to______ in an electric circuit.
    (*a*) resistance (*b*) resistivity (*c*) conductivity (*d*) conductance
110. Point out the wrong statement. Magnetic leakage is undesirable in electric machines because it
    (*a*) lowers their power efficiency (*b*) increases their cost of manufacture
    (*c*) leads to their increased weight (*d*) produces fringing

111. Relative permeability of vacuum is

(*a*) 1 (*b*) 1 H/m (*c*) 1/4 JI (*d*) $4n \times 10^{-1}$ H/m

112. Permanent magnets are normally made of

(*a*) alnico alloys (*b*) aluminium (*c*) cast iron (*d*) wrought iron

113. Energy stored by a coil is doubled when its current is increased by percent.

(*a*) 25 (*b*) 50 (*c*) 41.4 (*d*) 100

114. Those magnetic materials are best suited for making armature and transformer cores which have_permeability and hystersis loss.

(*a*) high, high (*b*) low, high (*c*) high, low (*d*) low, low

115. The rate of rise of current through an inductive coil is maximum

(*a*) at 63.2% of its maximum steady value

(*b*) at the start of the current flow

(*c*) after one time constant

(*d*) near the final maximum value of current

116. When both the inductance and resistance of a coil are doubled the value of

(*a*) time constant remains unchanged (*b*) initial rate of rise of current is doubled

(*c*) final steady current is doubled (*d*) time constant is halved

117. The initial rate of rise of current through a coil of inductance 10 H when suddenly connected to a D.C. supply of 200 V is ______ Vs

(*a*) 50 (*b*) 20 (*c*) 0.05 (*d*) 500

118. A material for good magnetic memory should have

(*a*) low hysteresis loss (*b*) high permeability

(*c*) low retentivity (*d*) high retentivity

119. Conductivity is analogous to

(*a*) retentivity (*b*) resistivity

(*c*) permeability (*d*) inductance

120. In a magnetic material hysteresis loss takes place primarily due to

(*a*) rapid reversals of its magnetisation

(*b*) flux density lagging behind magnetising force

(*c*) molecular friction

(*d*) it high retentivity

121. Those materials are well suited for making permanent magnets which have ______ retentivity and ______ coercivity.

(*a*) low, high (*b*) high, high (*c*) high, low (*d*) low, low

122. If the area of hysteresis loop of a material is large, the hysteresis loss in this material will be

(*a*) zero (*b*) small (*c*) large (*d*) none of the above

123. Hard steel is suitable for making permanent magnets because

(*a*) it has good residual magnetism (*b*) its hysteresis loop has large area

(*c*) its mechanical strength is high (*d*) its mechanical strength is low

124. Silicon steel is used in electrical machines because it has
(*a*) low co-ercivity (*b*) low retentivity
(*c*) low hysteresis loss (*d*) high co-ercivity

125. Conductance is analogous to
(*a*) permeance (*b*) reluctance (*c*) flux (*d*) inductance

126. The property of a material which opposes the creation of magnetic flux in it is known as
(*a*) reluctivity (*b*) magnetomotive force
(*c*) permeance (*d*) reluctance

127. The unit of retentivity is
(*a*) weber (*b*) weber/sq. m
(*c*) ampere turn/metre (*d*) ampere turn

128. Reciprocal of reluctance is
(*a*) reluctivity (*b*) permeance (*c*) permeability (*d*) susceptibility

129. While comparing magnetic and electric circuits, the flux of magnetic circuit is compared with which parameter of electrical circuit ?
(*a*) E.m.f. (*b*) Current (*c*) Current density (*d*) Conductivity

130. The unit of reluctance is
(*a*) metre/henry (*b*) henry/metre (*c*) henry (*d*) 1/henry

131. A ferrite core has less eddy current loss than an iron core because
(*a*) ferrites have high resistance (*b*) ferrites are magnetic
(*c*) ferrites have low permeability (*d*) ferrites have high hysteresis

132. Hysteresis loss least depends on
(*a*) volume of material (*b*) frequency
(*c*) steinmetz co-efficient of material (*d*) ambient temperature

133. Laminated cores, in clectrical machines, are used to reduce
(*a*) copper loss (*b*) eddy current loss
(*c*) hysteresis loss (*d*) all of the above

134. Which of the following does not change in a transformer ?
(*a*) Current (*b*) Voltage (*c*) Frequency (*d*) All of the above

135. In a transformer the energy is conveyed from primary to secondary
(*a*) through cooling coil (*b*) through air
(*c*) by the flux (*d*) none of the above

136. A transformer core is laminated to
(*a*) reduce hysteresis loss (*b*) reduce eddy current losses
(*c*) reduce copper losses (*d*) reduce all above losses

137. The degree of mechanical vibrations produced by the laminations of a transformer depends on
(*a*) tightness of clamping (*b*) gauge of laminations
(*c*) size of laminations (*d*) all of the above

138. The no-load current drawn by transformer is usually what per cent of the full-load current?
(a) 0.2 to 0.5 per cent (b) 2 to 5 per cent
(c) 12 to 15 per cent (d) 20 to 30 per cent

139. The path of a magnetic flux in a transformer should have
(a) high resistance (b) high reluctance
(c) low resistance (d) low reluctance

140. No-load on a transformer is carried out to determine
(a) copper loss (b) magnetising current
(c) magnetising current and loss (d) efficiency of the transformer

141. The dielectric strength of transformer oil is expected to be
(a) 1 kV (b) 33 kV (c) 100 kV (d) 330 kV

142. Sumpner's test is conducted on trans-formers to determine
(a) temperature (b) stray losses
(c) all-day efficiency (d) none of the above

143. The permissible flux density in case of cold rolled grain oriented steel is around
(a) 1.7 Wb/m^2 (b) 2.7 Wb/m^2 (c) 3.7 Wb/m^2 (d) 4.7 Wb/m^2

144. The efficiency of a transformer will be maximum when
(a) copper losses = hysteresis losses (b) hysteresis losses = eddy current losses
(c) eddy current losses = copper losses (d) copper losses = iron losses

145. No-load current in a transformer
(a) lags behind the voltage by about 75°
(b) leads the voltage by about 75°
(c) lags behind the voltage by about 15°
(d) leads the voltage by about 15°

146. The purpose of providing an iron core in a transformer is to
(a) provide support to windings
(b) reduce hysteresis loss
(c) decrease the reluctance of the magnetic path
(d) reduce eddy current losses

147. Which of the following is not a part of transformer installation ?
(a) Conservator (b) Breather
(c) Buchholz relay (d) Exciter

148. While conducting short-circuit test on a transformer the following side is short circuited
(a) High voltage side (b) Low voltage side
(c) Primary side (d) Secondary side

149. In the transformer following winding has got more cross-sectional area
(a) Low voltage winding (b) High voltage winding
(c) Primary winding (d) Secondary winding

150. A transformer transforms
(a) voltage (b) current (c) power (d) frequency

151. A transformer cannot raise or lower the voltage of a D.C. supply because
    (*a*) there is no need to change the D.C. voltage
    (*b*) a D.C. circuit has more losses
    (*c*) Faraday's laws of electromagnetic induction are not valid since the rate of change of flux is zero
    (*d*) none of the above
152. Primary winding of a transformer
    (*a*) is always a low voltage winding
    (*b*) is always a high voltage winding
    (*c*) could either be a low voltage or high voltage winding
    (*d*) none of the above
153. Which winding in a transformer has more number of turns ?
    (*a*) Low voltage winding (*b*) High voltage winding
    (*c*) Primary winding (*d*) Secondary winding
154. Efficiency of a power transformer is of the order of
    (*a*) 100 per cent (*b*) 98 per cent (*c*) 50 per cent (*d*) 25 per cent
155. In a given transformer for given applied voltage, losses which remain constant irrespective of load changes are
    (*a*) friction and windage losses (*b*) copper losses
    (*c*) hysteresis and eddy current losses (*d*) none of the above
156. A common method of cooling a power transformer is
    (*a*) natural air cooling (*b*) air blast cooling
    (*c*) oil cooling (*d*) any of the above
157. The no load current in a transformer lags behind the applied voltage by an angle of about
    (*a*) 180° (*b*) 120" (*c*) 90° (*d*) 75°
158. In a transformer routine efficiency depends upon
    (*a*) supply frequency (*b*) load current
    (*c*) power factor of load (*d*) both (*b*) and (*c*)
159. In the transformer the function of a conservator is to
    (*a*) provide fresh air for cooling the transformer
    (*b*) supply cooling oil to transformer in time of need
    (*c*) protect the transformer from damage when oil expends due to heating
    (*d*) none of the above
160. Natural oil cooling is used for transformers upto a rating of
    (*a*) 3000 kVA (*b*) 1000 kVA (*c*) 500 kVA (*d*) 250 kVA
161. Power transformers are designed to have maximum efficiency at
    (*a*) nearly full load (*b*) 70% full load
    (*c*) 50% full load (*d*) no load

162. The maximum efficiency of a distribution transformer is

(*a*) at no load (*b*) at 50% full load
(*c*) at 80% full load (*d*) at full load

163. Transformer breaths in when

(*a*) load on it increases (*b*) load on it decreases
(*c*) load remains constant (*d*) none of the above

164. No-load current of a transformer has

(*a*) has high magnitude and low power factor
(*b*) has high magnitude and high power factor
(*c*) has small magnitude and high power factor
(*d*) has small magnitude and low power factor

165. Spacers are provided between adjacent coils

(*a*) to provide free passage to the cooling oil
(*b*) to insulate the coils from each other
(*c*) both (*a*) and (*b*)
(*d*) none of the above

166. Greater the secondary leakage flux

(*a*) less will be the secondary induced e.m.f.
(*b*) less will be the primary induced e.m.f.
(*c*) less will be the primary terminal voltage
(*d*) none of the above

167. The purpose of providing iron core in a step-up transformer is

(*a*) to provide coupling between primary and secondary
(*b*) to increase the magnitude of mutual flux
(*c*) to decrease the magnitude of mag-netizing current
(*d*) to provide all above features

168. The power transformer is a constant

(*a*) voltage device (*b*) current device
(*c*) power device (*d*) main flux device

169. Two transformers operating in parallel will share the load depending upon their

(*a*) leakage reactance (*b*) per unit impedance
(*c*) efficiencies (*d*) ratings

170. If R2 is the resistance of secondary winding of the transformer and K is the transformation ratio then the equivalent secondary resistance referred to primary will be

(*a*) R2/VK (*b*) R2IK2 (*c*) R22!K2 (*d*) R22/K

171. What will happen if the transformers working in parallel are not connected with regard to polarity ?

(*a*) The power factor of the two trans-formers will be different from the power factor of common load
(*b*) Incorrect polarity will result in dead short circuit

(*c*) The transformers will not share load in proportion to their kVA ratings
(*d*) none of the above

172. If the percentage impedances of the two transformers working in parallel are different, then
(*a*) transformers will be overheated
(*b*) power factors of both the trans-formers will be same
(*c*) parallel operation will be not possible
(*d*) parallel operation will still be possible, but the power factors at which the two transformers operate will be different from the power factor of the common load

173. In a transformer the tappings are generally provided on
(*a*) primary side (*b*) secondary side
(*c*) low voltage side (*d*) high voltage side

174. The use of higher flux density in the transformer design
(*a*) reduces weight per kVA (b) reduces iron losses
(*c*) reduces copper losses (*d*) increases part load efficiency

175. The chemical used in breather for transformer should have the quality of
(*a*) ionizing air (*b*) absorbing moisture
(*c*) cleansing the transformer oil (*d*) cooling the transformer oil.

176. The chemical used in breather is
(*a*) asbestos fibre (*b*) silica sand (*c*) sodium chloride (*d*) silica gel

177. An ideal transformer has infinite values of primary and secondary inductances. The statement is
(*a*) true (*b*) false

178. The transformer ratings are usually expressed in terms of
(*a*) volts (*b*) amperes (*c*) kW (*d*) kVA

179. The noise resulting from vibrations of laminations set by magnetic forces, is termed as
(*a*) magnetostrication (*b*) boo
(*c*) hum (*d*) zoom

180. Hysteresis loss in a transformer varies as Bmax = maximum flux density)
(*a*) B max (*b*) B max 1-6 (*c*) B max 1-83 (*d*) B max

181. Material used for construction of transformer core is usually
(*a*) wood (*b*) copper (*c*) aluminium (*d*) silicon steel

182. The thickness of laminations used in a transformer is usually
(*a*) 0.4 mm to 0.5 mm (*b*) 4 mm to 5 mm
(*c*) 14 mm to 15 mm (*d*) 25 mm to 40 mm

183. The function of conservator in a transformer is
(*a*) to project against'internal fault
(*b*) to reduce copper as well as core losses
(*c*) to cool the transformer oil
(*d*) to take care of the expansion and contraction of transformer oil due to variation of temperature of sur-roundings

184. The highest voltage for transmitting electrical power in India is
(*a*) 33 kV (*b*) 66 kV (*c*) 132 kV (*d*) 400 kV

185. In a transformer the resistance between its primary and secondary is
(*a*) zero (*b*) 1 ohm (*c*) 1000 ohms (*d*) infinite

186. A transformer oil must be free from
(*a*) sludge (*b*) odour (*c*) gases (*d*) moisture

187. A Buchholz relay can be installed on
(*a*) auto-transformers (*b*) air-cooled transformers
(*c*) welding transformers (*d*) oil cooled transformers

188. Gas is usually not liberated due to dissociation of transformer oil unless the oil temperature exceeds
(*a*) 50°C (*b*) 80°C (*c*) 100°C (*d*) 150°C

189. The main reason for generation of harmonics in a transformer could be
(*a*) fluctuating load (*b*) poor insulation
(*c*) mechanical vibrations (*d*) saturation of core

190. Distribution transformers are generally designed for maximum efficiency around
(*a*) 90% load (*b*) zero load (*c*) 25% load (*d*) 50% load

191. Which of the following property is not necessarily desirable in the material for transformer core?
(*a*) Mechanical strength (*b*) Low hysteresis loss
(*c*) High thermal conductivity (*d*) High permeability

192. The magnetising current of a transformer is usually small because it has
(*a*) small air gap (*b*) large leakage flux
(*c*) laminated silicon steel core (*d*) fewer rotating parts

193. Which of the following does not change in an ordinary transformer ?
(*a*) Frequency (*b*) Voltage
(*c*) Current (*d*) Any of the above

194. Which of the following properties is not necessarily desirable for the material for transformer core?
(*a*) Low hysteresis loss (*b*) High permeability
(*c*) High thermal conductivity (*d*) Adequate mechanical strength

195. The leakage flux in a transformer depends upon
(*a*) load current
(*b*) load current and voltage
(*c*) load current, voltage and frequency
(*d*) load current, voltage, frequency and power factor

196. The path of the magnetic flux in transformer should have
(*a*) high reluctance (*b*) low reactance
(*c*) high resistance (*d*) low resistance

197. Noise level test in a transformer is a
(*a*) special test (*b*) routine test (*c*) type test (*d*) none of the above

198. Which of the following is not a routine test on transformers ?
(*a*) Core insulation voltage test (*b*) Impedance test
(*c*) Radio interference test (*d*) Polarity test

199. A transformer can have zero voltage regulation at
(*a*) leading power factor (*b*) lagging power factor
(*c*) unity power factor (*d*) zero power factor

200. Helical coils can be used on
(*a*) low voltage side of high kVA trans¬formers
(*b*) high frequency transformers
(*c*) high voltage side of small capacity transformers
(*d*) high voltage side of high kVA rating transformers

201. Harmonics in transformer result in
(*a*) increased core losses
(*b*) increased I2R losses
(*c*) magnetic interference with communication circuits
(*d*) all of the above

202. The core used in high frequency transformer is usually
(*a*) copper core (*b*) cost iron core (*c*) air core (*d*) mild steel core

203. The full-load copper loss of a trans¬former is 1600 W. At half-load, the copper loss will be
(*a*) 6400 W (*b*) 1600 W (*c*) 800 W (*d*) 400 W

204. The value of flux involved m the e.m.f. equation of a transformer is
(*a*) average value (*b*) r.m.s. value
(*c*) maximum value (*d*) instantaneous value

205. Silicon steel used in laminations mainly reduces
(*a*) hysteresis loss (*b*) eddy current losses
(*c*) copper losses (*d*) all of the above

206. Which winding of the transformer has less cross-sectional area ?
(*a*) Primary winding (*b*) Secondary winding
(*c*) Low voltage winding (*d*) High voltage winding

207. Power transformers are generally designed to have maximum efficiency around
(*a*) no-load (*b*) half-load (*c*) near full-load (*d*) 10% overload

208. Which of the following is the main advantage of an auto-transformer over a two winding transformer ?
(*a*) Hysteresis losses are reduced (*b*) Saving in winding material
(*c*) Copper losses are negligible (*d*) Eddy losses are totally eliminated

209. During short circuit test iron losses are negligible because
(*a*) the current on secondary side is negligible
(*b*) the voltage on secondary side does not vary

(c) the voltage applied on primary side is low
(d) full-load current is not supplied to the transformer

210. Two transformers are connected in parallel. These transformers do not have equal percentage impedance. This is likely to result in
(a) short-circuiting of the secondaries
(b) power factor of one of the trans¬formers is leading while that of the other lagging
(c) transformers having higher copper losses will have negligible core losses
(d) loading of the transformers not in proportion to their kVA ratings

211. The changes in volume of transformer cooling oil due to variation of atmospheric temperature during day and night is taken care of by which part of transformer
(a) Conservator (b) Breather
(c) Bushings (d) Buchholz relay

212. An ideal transformer is one which has
(a) no losses and magnetic leakage
(b) interleaved primary and secondary windings
(c) a common core for its primary and secondary windings
(d) core of stainless steel and winding of pure copper metal
(e) none of the above

213. When a given transformer is run at its rated voltage but reduced frequency, its
(a) flux density remains unaffected (b) iron losses are reduced
(c) core flux density is reduced (d) core flux density is increased

214. In an actual transformer the iron loss remains practically constant from noload to fullload because
(a) value of transformation ratio remains constant
(b) permeability of transformer core remains constant
(c) core flux remains practically constant
(d) primary voltage remains constant
(c) secondary voltage remains constant

215. An ideal transformer will have maximum efficiency at a load such that
(a) copper loss = iron loss (b) copper loss < iron loss
(c) copper loss > iron loss (d) none of the above

216. If the supply frequency to the transformer is increased, the iron loss will
(a) not change (b) decrease
(c) increase (d) any of the above

217. Negative voltage regulation is indicative that the load is
(a) capacitive only (b) inductive only
(c) inductive or resistive (d) none of the above

218. Iron loss of a transformer can be measured by
(a) low power factor wattmeter (b) unity power factor wattmeter
(c) frequency meter (d) any type of wattmeter

219. When secondary of a current transformer is open-circuited its iron core will be
   (*a*) hot because of heavy iron losses taking place in it due to high flux density
   (*b*) hot because primary will carry heavy current
   (*c*) cool as there is no secondary current
   (*d*) none of above will happen
220. The transformer laminations are insulated from each other by
   (*a*) mica strip (*b*) thin coat of varnish
   (*c*) paper (*d*) any of the above
221. Which type of winding is used in 3phase shell-type transformer ?
   (*a*) Circular type (*b*) Sandwich type
   (*c*) Cylindrical type (*d*) Rectangular type
222. During open circuit test of a transformer
   (*a*) primary is supplied rated voltage
   (*b*) primary is supplied full-load current
   (*c*) primary is supplied current at reduced voltage
   (*d*) primary is supplied rated kVA
223. Open circuit test on transformers is conducted to determine
   (*a*) hysteresis losses (*b*) copper losses
   (*c*) core losses (*d*) eddy current losses
224. Short circuit test on transformers is conducted to determine
   (*a*) hysteresis losses (*b*) copper losses
   (*c*) core losses (*d*) eddy current losses
225. For the parallel operation of single phase transformers it is necessary that they should have
   (*a*) same efficiency
   (*b*) same polarity
   (*c*) same kVA rating
   (*d*) same number of turns on the secondary side.
226. The transformer oil should have _______ volatility and ______ viscosity.
   (*a*) low, low (*b*) high, high (*c*) low, high (*d*) high, low
227. The function of breather in a transformer is
   (*a*) to provide oxygen inside the tank
   (*b*) to cool the coils during reduced load
   (*c*) to cool the transformer oil
   (*d*) to arrest flow of moisture when outside air enters the transformer
228. The secondary winding of which of the following transformers is always kept closed ?
   (*a*) Step-up transformer (*b*) Step-down transformer
   (*c*) Potential transformer (*d*) Current transformer
229. The size of a transformer core will depend on
   (*a*) frequency (*b*) area of the core
   (*c*) flux density of the core material (*d*) (*a*) and (*b*) both

230. Natural air cooling is generally restricted for transformers up to
(*a*) 1.5 MVA (*b*) 5 MVA (*c*) 15 MVA (*d*) 50 MVA

231. A shell-type transformer has
(*a*) high eddy current losses (*b*) reduced magnetic leakage
(*c*) negligibly hysteresis losses (*d*) none of the above

232. A transformer can have regulation closer to zero
(*a*) on full-load (*b*) on overload
(*c*) on leading power factor (*d*) on zero power factor

233. A transformer transforms
(*a*) voltage (*b*) current
(*c*) current and voltage (*d*) power

234. Which of the following is not the standard voltage for power supply in India ?
(*a*) 11 kV (*b*) 33 kV (*c*) 66 kV (*d*) 122 kV

235. Reduction in core losses and increase in permeability are obtained with transformer employing
(*a*) core built-up of laminations of cold rolled grain oriented steel
(*b*) core built-up of laminations of hot rolled sheet
(*c*) either of the above
(*d*) none of the above

236. In a power or distribution transformer about 10 per cent end turns are heavily insulated
(*a*) to withstand the high voltage drop due to line surge produced by the shunting capacitance of the end turns
(*b*) to absorb the line surge voltage and save the winding of transformer from damage
(*c*) to reflect the line surge and save the winding of a transformer from damage
(*d*) none of the above

237. For given applied voltage, with the increase in frequency of the applied voltage
(*a*) eddy current loss will decrease
(*b*) eddy current loss will increase
(*c*) eddy current loss will remain unchanged
(*d*) none of the above

238. Losses which occur in rotating electric machines and do not occur in trans formers are
(*a*) friction and windage losses (*b*) magnetic losses
(*c*) hysteresis and eddy current losses (*d*) copper losses

239. In a given transformer for a given applied voltage, losses which remain constant irrespective of load changes are
(*a*) hysteresis and eddy current losses (*b*) friction and windage losses
(*c*) copper losses (*d*) none of the above

240. Which of the following statements regarding an idel single-phase transformer having a turn ratio of 1 : 2 and drawing a current of 10 A from 200 V A.C. supply is incorrect?

(*a*) Its secondary current is 5 A (*b*) Its secondary voltage is 400 V
(*c*) Its rating is 2 kVA (*d*) Its secondary current is 20 A
(*e*) It is a step-up transformer

241. The secondary of a current transformer is always short-circuited under operating conditions because it
(*a*) avoids core saturation and high voltage induction
(*b*) is safe to human beings
(*c*) protects the primary circuit
(*d*) none of the above

242. In a transformer the resistance between its primary and secondary should be
(*a*) zero (*b*) 10 Q (*c*) 1000 Q (*d*) infinity

243. A good voltage regulation of a transformer means
(*a*) output voltage fluctuation from no load to full load is least
(*b*) output voltage fluctuation with power factor is least
(*c*) difference between primary and secondary voltage is least
(*d*) difference between primary and secondary voltage is maximum

244. For a transformer, operating at constant load current, maximum efficiency will occur at
(*a*) 0.8 leading power factor (*b*) 0.8 lagging power factor
(*c*) zero power factor (*d*) unity power factor

245. Laminations of core are generally made of
(*a*) case iron (*b*) carbon (*c*) silicon steel (*d*) stainless steel

246. Which of the following could be lamina-proximately the thickness of laminations of a D.C. machine?
(*a*) 0.005 mm (*b*) 0.05 mm (*c*) 0.5 m (*d*) 5 m

247. The armature of D.C. generator is laminated to
(*a*) reduce the bulk (*b*) provide the bulk
(*c*) insulate the core (*d*) reduce eddy current loss

248. The resistance of armature winding depends on
(*a*) length of conductor (*b*) cross-sectional area of the conductor
(*c*) number of conductors (*d*) all of the above

249. The field coils of D.C. generator are usually made of
(*a*) mica (*b*) copper (*c*) cast iron (*d*) carbon

250. The commutator segments are connected to the armature conductors by means of
(*a*) copper lugs (*b*) resistance wires
(*c*) insulation pads (*d*) brazing

251. In a commutator
(*a*) copper is harder than mica (*b*) mica and copper are equally hard
(*c*) mica is harder than copper (*d*) none of the above

252. In D.C. generators the pole shoes are fastened to the pole core by
   (*a*) rivets (*b*) counter sunk screws
   (*c*) brazing (*d*) welding
253. According to Fleming's right-hand rule for finding the direction of induced e.m.f., when middle finger points in the direction of induced e.m.f., forefinger will point in the direction of
   (*a*) motion of conductor (*b*) lines of force
   (*c*) either of the above (*d*) none of the above
254. Fleming's right-hand rule regarding direction of induced e.m.f., correlates
   (*a*) magnetic flux, direction of current flow and resultant force
   (*b*) magnetic flux, direction of motion and the direction of e.m.f. induced
   (*c*) magnetic field strength, induced voltage and current
   (*d*) magnetic flux, direction of force and direction of motion of conductor
255. While applying Fleming's right-hand rule to And the direction of induced e.m.f., the thumb points towards
   (*a*) direction of induced e.m.f.
   (*b*) direction of flux
   (*c*) direction of motion of the conductor if forefinger points in the direction of generated e.m.f.
   (*d*) direction of motion of conductor, if forefinger points along the lines of flux
256. The bearings used to support the rotor shafts are generally
   (*a*) ball bearings (*b*) bush bearings
   (*c*) magnetic bearmgs (*d*) needle bearings
257. In D.C. generators, the cause of rapid brush wear may be
   (*a*) severe sparking (*b*) rough commutator surface
   (*c*) imperfect contact (*d*) any of the above
258. In lap winding, the number of brushes is always
   (*a*) double the number of poles (*b*) same as the number of poles
   (*c*) half the number of poles (*d*) two
259. For a D.C. generator when the number of poles and the number of armature conductors is fixed, then which winding will give the higher e.m.f. ?
   (*a*) Lap winding (*b*) Wave winding
   (*c*) Either of (*a*) and (*b*) above (*d*) Depends on other features of design
260. In a four-pole D.C. machine
   (*a*) all the four poles are north poles (*b*) alternate poles are north and south
   (*c*) all the four poles are south poles (*d*) two north poles follow two south poles
261. Copper brushes in D.C. machine are used
   (*a*) where low voltage and high currents are involved
   (*b*) where high voltage and small cur-rents are involved
   (*c*) in both of the above cases
   (*d*) in none of the above cases

262. A separately excited generator as compared to a self-excited generator
   (*a*) is amenable to better voltage control
   (*b*) is more stable
   (*c*) has exciting current independent of load current
   (*d*) has all above features
263. In case of D.C. machines, mechanical losses are primary function of
   (*a*) current (*b*) voltage (*c*) speed (*d*) none of above
264. Iron losses in a D.C. machine are independent of variations in
   (*a*) speed (*b*) load (*c*) voltage (*d*) speed and voltage
265. In D.C. generators, current to the external circuit from armature is given through
   (*a*) commutator (*b*) solid connection (*c*) slip rings (*d*) none of above
266. Brushes of D.C. machines are made of
   (*a*) carbon (*b*) soft copper (*c*) hard copper (*d*) all of above
267. If B is the flux density, I the length of conductor and v the velocity of conductor, then induced e.m.f. is given by
   (*a*) Blv (*b*) Blv2 (*c*) Bl2v (*d*) Bl2v2
268. In case of a 4-pole D.C. generator provided with a two layer lap winding with sixteen coils, the pole pitch will be
   (*a*) 4 (*b*) 8 (*c*) 16 (*d*) 32
269. The material for commutator brushes is generally
   (*a*) mica (*b*) copper (*c*) cast iron (*d*) carbon
270. The insulating material used between the commutator segments is normally
   (*a*) graphite (*b*) paper (*c*) mica (*d*) insulating varnish
271. In D.C. generators, the brushes on commutator remain in contact with conductors which
   (*a*) lie under south pole (*b*) lie under north pole
   (*c*) lie under interpolar region (*d*) are farthest from the poles
272. If brushes of a D.C. generator are moved in order to bring these brushes in magnetic neutral axis, there will be
   (*a*) demagnetisation only
   (*b*) cross-magnetisation as well as magnetisation
   (*c*) cross-magnetisation as well as demagnetising
   (*d*) cross-magnetisation only
273. Armature reaction of an unsaturated D.C. machine is
   (*a*) cross-magnetising (*b*) demagnetising
   (*c*) magnetising (*d*) none of above
274. D.C. generators are connected to the busbars or disconnected from them only under the floating condition
   (*a*) to avoid sudden loading of the prime mover
   (*b*) to avoid mechanical jerk to the shaft
   (*c*) to avoid burning of switch contacts
   (*d*) all above

275. No-load speed of which of the following motor will be highest ?
   (*a*) Shunt motor (*b*) Series motor
   (*c*) Cumulative compound motor (*d*) Differentiate compound motor
276. The direction of rotation of a D.C. series motor can be changed by
   (*a*) interchanging supply terminals (*b*) interchanging field terminals
   (*c*) either of (*a*) and (*b*) above (*d*) None of the above
277. Which of the following application requires high starting torque ?
   (*a*) Lathe machine (*b*) Centrifugal pump
   (*c*) Locomotive (*d*) Air blower
278. If a D.C. motor is to be selected for conveyors, which rriotor would be preferred ?
   (*a*) Series motor (*b*) Shunt motor
   (*c*) Differentially compound motor (*d*) Cumulative compound motor
279. Which D.C. motor will be preferred for machine tools ?
   (*a*) Series motor (*b*) Shunt motor
   (*c*) Cumulative compound motor (*d*) Differential compound motor
280. Differentially compound D.C. motors can find applications requiring
   (*a*) high starting torque (*b*) low starting torque
   (*c*) variable speed (*d*) frequent on-off cycles
281. Which D.C. motor is preferred for elevators ?
   (*a*) Shunt motor (*b*) Series motor
   (*c*) Differential compound motor (*d*) Cumulative compound motor
282. According to Fleming's left-hand rule, when the forefinger points in the direction of the field or flux, the middle finger will point in the direction of
   (*a*) current in the conductor aovtaat of conductor
   (*b*) resultant force on conductor
   (*c*) none of the above
283. If the field of a D.C. shunt motor gets opened while motor is running
   (*a*) the speed of motor will be reduced %
   (*b*) the armature current will reduce
   (*c*) the motor will attain dangerously high speed 1
   (*d*) the motor will continue to nuvat constant speed
284. Starters are used with D.C. motors because
   (*a*) these motors have high starting torque
   (*b*) these motors are not self-starting
   (*c*) back e.m.f. of these motors is zero initially
   (*d*) to restrict armature current as there is no back e.m.f. while starting
285. In D.C. shunt motors as load is reduced
   (*a*) the speed will increase abruptly
   (*b*) the speed will increase in proportion to reduction in load

(c) the speed will remain almost/constant

(d) the speed will reduce

286. A D.C. series motor is that which

(a) has its field winding consisting of thick wire and less turns

(b) has a poor torque

(c) can be started easily without load

(d) has almost constant speed

287. For starting a D.C. motor a starter is required because

(a) it limits the speed of the motor (b) it limits the starting current to a safe value

(c) it starts the motor (d) none of the above

288. The type of D.C. motor used for shears and punches is

(a) shunt motor (b) series motor

(c) differential compoutid D.C. motor (d) cumulative compound D.C. motor

289. If a D.C. motor is connected across the A.C. supply it will

(a) run at normal speed

(b) not run

(c) run at lower speed

(d) burn due to heat produced in the field winding by .eddy currents

290. To get the speed of D.C, motor below the normal without wastage of electrical energy is used.

(a) Ward Leonard control (b) rheostatic control

(c) any of the above method (d) none of the above method

291. When two D.C. series motors are connected in parallel, the resultant speed is

(a) more than the normal speed (b) loss than the normal speed

(c) normal speed (d) zero

292. The speed of a D.C. shunt motor more than its full-load speed can be obtained by

(a) decreasing the field current (b) increasing the field current

(c) decreasing the armature current (d) increasing the armature current

293. In a D.C. shunt motor, speed is

(a) independent of armature current

(b) directly proportional to the armature current

(c) proportional to the square of the current

(d) inversely proportional to the armature current

294. A direct on line starter is used: for starting motors

(a) iip to 5 H.P. (b) up to 10 H.P.

(c) up to 15 H.P. (d) up to 20 H.P.

295. What will happen if the back e.m.f. of a D.C. motor vanishes suddenly?

(a) The motor will stop (b) The motor will continue to run

(c) The armature may burn (d) The motor will run noisy

296. In case of D.C. shunt motors the speed is dependent on back e.m.f. only because
   (*a*) back e.m.f. is equal to armature drop
   (*b*) armature drop is negligible
   (*c*) flux is proportional to armature current
   (*d*) flux is practically constant in D:C. shunt motors

297. In a D.C. shunt motor, under the conditions of maximum power, the current in the
   (*a*) almost negligible (*b*) rated full-load current
   (*c*) less than full-load current (*d*) more than full-load current

298. These days D.C. motors are widely used in
   (*a*) pumping sets (*b*) air compressors
   (*c*) electric traction (*d*) machine shops

299. By looking at which part of the motor, it can be easily confirmed that a particular motor is D.C. motor?
   (*a*) Frame (*b*) Shaft (*c*) Commutator (*d*) Stator

300. In which of the following applications D.C. series motor is invariably tried?
   (*a*) Starter for a car (*b*) Drive for a water pump
   (*c*) Fan motor (*d*) Motor operation in A.C. or D.C.

301. In D.C. machines fractional pitch winding is used
   (*a*) to improve cooling (*b*) to reduce copper losses
   (*c*) to increase the generated e.m.f. (*d*) to reduce the sparking

302. A three point starter is considered suitable for
   (*a*) shunt motors (*b*) shunt as well as compound motors
   (*c*) shunt, compound and series motors (*d*) all D.C. motors

303. In case-the conditions for maximum power for a D.C. motor are established, the efficiency of the motor will be
   (*a*) 100% (*b*) around 90%
   (*c*) anywhere between 75% and 90% (*d*) less than 50%

304. The ratio of starting torque to full-load torque is least in case of
   (*a*) series motors (*b*) shunt motors
   (*c*) compound motors (*d*) none of the above

305. In D.C. motor which of the following can sustain the maximum temperature rise?
   (*a*) Slip rings (*b*) Commutator
   (*c*) Field winding (*d*) Armature winding

306. Which of the following law/rule can he used to determine the direction of rotation of D.C. motor?
   (*a*) Lenz's law (*b*) Faraday's law
   (*c*) Coloumb's law (*d*) Fleming's left-hand rule

307. Which of the following load normally needs starting torque more than the rated torque?
   (*a*) Blowers (*b*) Conveyors
   (*c*) Air compressors (*d*) Centrifugal pumps

308. The starting resistance of a D.C. motor is generally
(*a*) low (*b*) around 500 Q
(*c*) 1000 Q (*d*) infinitely large
309. The speed of a D.C. series motor is
(*a*) proportional to the armature current
(*b*) proportional to the square of the armature current
(*c*) proportional to field current
(*d*) inversely proportional to the armature current
310. In a D.C. series motor, if the armature current is reduced by 50%, the torque of the motor will be equal to
(*a*) 100% of the previous value (*b*) 50% of the previous value
(*c*) 25% of the previous value (*d*) 10% of the previous value
(*e*) none of the above
311. The current drawn by the armature of D.C. motor is directly proportional to
(*a*) the torque required (*b*) the speed of the motor
(*c*) the voltage across the terminals (*d*) none of the above
312. The power mentioned on the name plate of an electric motor indicates
(*a*) the power drawn in kW (*b*) the power drawn in kVA
(*c*) the gross power (*d*) the output power available at the shaft
313. Which D.C. motor has got maximum self loading property?
(*a*) Series motor (*b*) Shunt motor
(*c*) Cumulatively compounded 'motor (*d*) Differentially compounded motor
314. In a split phase motor, the running winding should have
(*a*) high resistance and low inductance
(*b*) low resistance and high inductance
(*c*) high resistance as well as high inductance
(*d*) low resistance as well as low inductiance
315. If the capacitor of a single-phase motor is short-circuited
(*a*) the motor will not start
(*b*) the motor will run
(*c*) the motor will run in reverse direction
(*d*) the motor will run in the same direction at reduced r.p.m.
316. In capacitor start single-phase motors
(*a*) current in the starting winding leads the voltage
(*b*) current in the starting winding lags the voltage
(*c*) current in the starting winding is in phase with voltage in running winding
(*d*) none of the above
317. In a capacitor start and run motors the function of the running capacitor in series with the auxiliary winding is to
(*a*) improve power factor (*b*) increase overload capacity
(*c*) reduce fluctuations in torque (*d*) to improve torque

318. In a capacitor start motor, the phase displacement between starting and running winding can be nearly
   (*a*) 10° (*b*) 30° (*c*) 60° (*d*) 90°
319. In a split phase motor
   (*a*) the starting winding is connected through a centrifugal switch
   (*b*) the running winding is connected through a centrifugal switch
   (*c*) both starting and running windings are connected through a centrifugal switch
   (*d*) centrifugal switch is used to control supply voltage
320. The rotor developed by a single-phase motor at starting is
   (*a*) more than i.he rated torque (*b*) rated torque
   (*c*) less than the rated torque (*d*) zero
321. Which of the following motor will give relatively high starting torque ?
   (*a*) Capacitor start motor (*b*) Capacitor run motor
   (*c*) Split phase motor (*d*) Shaded pole motor
322. Which of the following motor will have relatively higher power factor ?
   (*a*) Capacitor run motor (*b*) Shaded pole motor
   (*c*) Capacitor start motor (*d*) Split phase motor
323. In a shaded pole motor, the shading coil usually consist of
   (*a*) a single turn of heavy wire which is in parallel with running winding
   (*b*) a single turn of heavy copper wire which is short-circuited and carries only induced current
   (*c*) a multilayer fine gauge copper wire in parallel with running winding
   (*d*) none of the above
324. In a shaded pole single-phase motor, the revolving field is produced by the use of
   (*a*) inductor (*b*) capacitor (*c*) resistor (*d*) shading coils
325. A centrifugal switch is used to dis- connect 'starting winding when motor has
   (*a*) run for about 1 minute
   (*b*) run for about 5 minutes
   (*c*) picked up about 50 to 70 per cent of rated speed
   (*d*) picked up about 10 to 25 per cent of rated speed
326. If a particular application needs high speed and high starting torque, then which of the following motor will be preferred ?
   (*a*) Universal motor (*b*) Shaded pole type motor
   (*c*) Capacitor start motor (*d*) Capacitor start and run motor
327. The value of starting capacitor of a fractional horse power motor will be
   (*a*) 100 uF (*b*) 200 uF (*c*) 300 uF (*d*) 400 Uf
328. In repulsion motor direction of rotation of motor
   (*a*) is opposite to that of brush shift
   (*b*) is the same as that of brush shift
   (*c*) is independent of brush shift

329. In a single phase motor the centrifugal switch
   (*a*) disconnects auxiliary winding of the motor
   (*b*) disconnects main winding of the motor
   (*c*) reconnects the main winding the motor
   (*d*) reconnects the auxiliary winding of the motor
330. The running winding of a single phase motor on testing with meggar is found to be ground. Most probable location of the ground will be
   (*a*) at the end connections
   (*b*) at the end terminals
   (*c*) anywhere on the winding inside a slot
   (*d*) at the slot edge where coil enters or comes out of the slot
331. A capacitor-start single phase induction motor is switched on to supply with its capacitor replaced by an inductor of equivalent reactance value. It will
   (*a*) start and then stop (*b*) start and run slowly
   (*c*) start and run at rated speed (*d*) not start at all
332. Which of the following motors is used in mixies ?
   (*a*) Repulsion motor (*b*) Reluctance motor
   (*c*) Hysteresis motor (*d*) Universal motor
333. Which of the following motors is inherently self starting ?
   (*a*) Split motor (*b*) Shaded-pole motor
   (*c*) Reluctance motor (*d*) None of these
334. The direction of rotation of an hysteresis motor is determined by
   (*a*) interchanging the supply leads
   (*b*) position of shaded pole with respect to main pole
   (*c*) retentivity of the rotor material
   (*d*) none of these
335. Burning out of windings is due to
   (*a*) short circuited capacitor (*b*) capacitor value hiving changed
   (*c*) open circuiting of capacitor (*d*) none of the above
336. The shaft of an induction motor is made of
   (*a*) stiff (*b*) flexible (*c*) hollow (*d*) any of the above
337. The shaft of an induction motor is made of
   (*a*) high speed steel (*b*) stainless steel
   (*c*) carbon steel (*d*) cast iron
338. In an induction motor, no-load the slip is generally
   (*a*) less than 1% (*b*) 1.5% (*c*) 2% (*d*) 4%
339. In medium sized induction motors, the slip is generally around
   (*a*) 0.04% (*b*) 0.4% (*c*) 4% (*d*) 14%
340. In squirrel cage induction motors, the rotor slots are usually given slight skew in order to

(*a*) reduce windage losses
(*b*) reduce eddy currents
(*c*) reduce accumulation of dirt and dust
(*d*) reduce magnetic hum

341. In case the air gap in an induction motor is increased
(*a*) the magnetising current of the rotor will decrease
(*b*) the power factor will decrease
(*c*) speed of motor will increase
(*d*) the windage losses will increase

342. Slip rings are usually made of
(*a*) copper (*b*) carbon (*c*) phospor bronze (*d*) aluminium

343. A 3-phase 440 V, 50 Hz induction motor has 4% slip. The frequency of rotor e.m.f. will be
(*a*) 200 Hz (*b*) 50 Hz (*c*) 2 Hz (*d*) 0.2 Hz

344. In Ns is the synchronous speed and s the slip, then actual running speed of an induction motor will be
(*a*) Ns (*b*) s.N, (*c*) (l-s)Ns (*d*) (Ns-l)s

345. The efficiency of an induction motor can be expected to be nearly
(*a*) 60 to 90% (*b*) 80 to 90% (*c*) 95 to 98% (*d*) 99%

346. The number of slip rings on a squirrel cage induction motor is usually
(*a*) two (*b*) three (*c*) four (*d*) none

347. The starting torque of a squirrel-cage induction motor is
(*a*) low (*b*) negligible
(*c*) same as full-load torque (*d*) slightly more than full-load torque

Magnetism and Electromagnetism

348. A double squirrel-cage induction motor has
(*a*) two rotors moving in oppsite direction
(*b*) two parallel windings in stator
(*c*) two parallel windings in rotor
(*d*) two series windings in stator

349. Star-delta starting of motors is not possible in case of
(*a*) single phase motors (*b*) variable speed motors
(*c*) low horse power motors (*d*) high speed motors

350. The term 'cogging' is associated with
(*a*) three phase transformers (*b*) compound generators
(*c*) D.C. series motors (*d*) induction motors

351. In case of the induction motors the torque is
(*a*) inversely proportional to (V slip) (*b*) directly proportional to (slip)2
(*c*) inversely proportional to slip (*d*) directly proportional to slip

352. An induction motor with 1000 r.p.m. speed will have
(*a*) 8 poles (*b*) 6 poles (*c*) 4 poles (*d*) 2 poles

353. The good power factor of an induction motor can be achieved if the average flux density in the air gap is
(*a*) absent (*b*) small (*c*) large (*d*) infinity

354. An induction motor is identical to
(*a*) D.C. compound motor (*b*) D.C. series motor
(*c*) synchronous motor (*d*) asynchronous motor

355. The injected e.m.f. in the rotor of induction motor must have
(*a*) zero frequency (*b*) the same frequency as the slip fre-quency
(*c*) the same phase as the rotor e.m.f. (*d*) high value for the satisfactory speed control

356. Which of the following methods is easily applicable to control the speed of the squirrelcage induction motor ?
(*a*) By changing the number of stator poles
(*b*) Rotor rheostat control
(*c*) By operating two motors in cascade
(*d*) By injecting e.m.f. in the rotor circuit

357. The crawling in the induction motor is caused by
(*a*) low voltage supply (*b*) high loads
(*c*) harmonics develped in the motor (*d*) improper design of the machine
(*e*) none of the above

358. The auto-starters (using three auto transformers) can be used to start cage induction motor of the following type
(*a*) star connected only (*b*) delta connected only
(*c*) (*a*) and (*b*) both (*d*) none of the above

359. The torque developed in the cage induction motor with autostarter is
(*a*) *k*/torque with direct switching (*b*) K × torque with direct switching
(*c*) K2 × torque with direct switching (*d*) *k*2/torque with direct switching

360. When the equivalent circuit diagram of doouble squirrel-cage induction motor is constructed the two cages can be considered
(*a*) in series (*b*) in parallel
(*c*) in series-parallel (*d*) in parallel with stator

361. It is advisable to avoid line-starting of induction motor and use starter because
(*a*) motor takes five to seven times its full load current
(*b*) it will pick-up very high speed and may go out of step
(*c*) it will run in reverse direction
(*d*) Starting torque is very high

362. Stepless speed control of induction motor is possible by which of the following methods?
(*a*) e.m.f. injection in rotor eueuit (*b*) Changing the number of poles
(*c*) Cascade operation (*d*) None of the above

363. Rotor rheostat control method of speed control is used for
(*a*) squirrel-cage induction motors only (*b*) slip ring induction motors only
(*c*) both (*a*) and (*b*) (*d*) none of the above

364. In the circle diagram for induction motor, the diameter of the circle represents
 (a) slip (b) rotor current (c) running torque (d) line voltage

365. For which motor the speed can be controlled from rotor side ?
 (a) Squirrel-cage induction motor (b) Slip-ring induction motor
 (c) Both (a) and (b) (d) None of the above

366. If any two phases for an induction motor are interchanged
 (a) the motor will run in reverse direction
 (b) the motor will run at reduced speed
 (c) the motor will not run
 (d) the motor will burn

367. An induction motor is
 (a) self-starting with zero torque (b) self-starting with high torque
 (c) self-starting with low torque (d) non-self starting

368. The maximum torque in an induction motor depends on
 (a) frequency (b) rotor inductive reactance
 (c) square of supply voltage (d) all of the above

369. In three-phase squirrel-cage induction motors
 (a) rotor conductor ends are short-circuited through slip rings
 (b) rotor conductors are short-circuited through end rings
 (c) rotor conductors are kept open
 (d) rotor conductors are connected to insulation

370. In a three-phase induction motor, the number of poles in the rotor winding is always
 (a) zero (b) more than the number of poles in stator
 (c) less than number of poles in stator (d) equal to number of poles in stator

371. DOL starting of induction motors is usually restricted to
 (a) low horsepower motors (b) variable speed motors
 (c) high horsepower motors (d) high speed motors

372. The speed of a squirrel-cage induction motor can be controlled by all of the following except
 (a) changing supply frequency (b) changing number of poles
 (c) changing winding resistance (d) reducing supply voltage

373. The 'crawling" in an induction motor is caused by
 (a) high loads (b) low voltage supply
 (c) improper design of machine (d) harmonics developed in the motor

374. The power factor of an induction motor under no-load conditions will be closer to

375. The 'cogging' of an induction motor can be avoided by
 (a) proper ventilation
 (b) using DOL starter
 (c) auto-transformer starter
 (d) having number of rotor slots more or less than the number of stator slots (not equal)

376. If an induction motor with certain ratio of rotor to stator slots, runs at 1/7 of the normal speed, the phenomenon will be termed as
(*a*) humming (*b*) hunting (*c*) crawling (*d*) cogging

377. Slip of an induction motor is negative when
(*a*) magnetic field and rotor rotate in opposite direction
(*b*) rotor speed is less than the syn-chronous speed of the field and are in the same direction
(*c*) rotor speed is more than the syn-chronous speed of the field and are in the same direction
(*d*) none of the above

378. Size of a high speed motor as compared to low speed motorfor the same H.P. will be
(*a*) bigger (*b*) smaller (*c*) same (*d*) any of the above

379. A 3-phase induction motor stator delta connected, is carrying full load and one of its fuses blows out. Then the motor
(*a*) will continue running burning its one phase
(*b*) will continue running burning its two phases
(*c*) will stop and carry heavy current causing permanent damage to its winding
(*d*) will continue running without any harm to the winding

380. A 3-phase induction motor delta connected is carrying too heavy load and one of its fuses blows out. Then the motor
(*a*) will continue running burning its one phase
(*b*) will continue running burning its two phase
(*c*) will stop and carry heavy current causing permanent damage to its winding
(*d*) will continue running without any harm to the winding

381. Synchronous motors are generally not self-starting because
(*a*) the direction of rotation is not fixed
(*b*) the direction of instantaneous torque reverses after half cycle
(*c*) startes cannot be used on these machines
(*d*) starting winding is not provided on the machines

382. In case one phase of a three-phase synchronous motor is short-circuited the motor will
(*a*) not start (*b*) run at 2/3 of synchronous speed
(*c*) run with excessive vibrations (*d*) take less than the rated load

383. A pony motor is basically a
(*a*) small induction motor (*b*) D.C. series motor
(*c*) D.C. shunt motor (*d*) double winding A.C./D.C. motor

384. A synchronous motor can develop synchronous torque
(*a*) when under loaded (*b*) while over-excited
(*c*) only at synchronous speed (*d*) below or above synchronous speed

385. A synchronous motor can be started by
(*a*) pony motor (*b*) D.C. compound motor
(*c*) providing damper winding (*d*) any of the above

386. A three-phase synchronous motor will have
    (*a*) no slip-rings (*b*) one slip-ring (*c*) two slip-rings (*d*) three slip-rings
387. Under which of the following conditions hunting of synchronous motor is likely to occur ?
    (*a*) Periodic variation of load (*b*) Over-excitation
    (*c*) Over-loading for long periods (*d*) Small and constant load
388. When the excitation of an unloaded salient pole synchronous motor suddenly gets disconnected
    (*a*) the motor stops
    (*b*) it runs as a reluctance motor at the same speed
    (*c*) it runs as a reluctance motor at a lower speed
    (*d*) none of the above
389. When V is the applied voltage, then the breakdown torque of a synchronous motor varies as
    (*a*) V (*b*) V312 (*c*) V2 (*d*) 1/V
390. The power developed by a synchronous motor will be maximum when the load angle is
    (*a*) zero (*b*) 45° (*c*) 90° (*d*) 120°
391. A synchronous motor can be used as a synchronous capacitor when it is
    (*a*) under-loaded (*b*) over-loaded (*c*) under-excited (*d*) over-excited
392. A synchronous motor is running on a load with normal excitation. Now if the load on the motor is increased
    (*a*) power factor as well as armature current will decrease
    (*b*) power factor as well as armature current will increase
    (*c*) power factor will increase but armature current will decrease
    (*d*) power factor will decrease and armature current will increase
393. Mostly, synchronous motors are of
    (*a*) alternator type machines (*b*) induction type machines
    (*c*) salient pole type machines (*d*) smooth cylindrical type machines
394. The synchronous motor is not inherently self-starting because
    (*a*) the force required to accelerate the rotor to the synchronous speed in an instant is absent
    (*b*) the starting device to accelerate the rotor to near synchronous speed is absent
    (*c*) a rotating magnetic field does not have enough poles
    (*d*) the rotating magnetic field is produced by only 50 Hz frequency currents
395. As the load is applied to a synchronous motor, the motor takes more armature current because
    (*a*) the increased load has to take more current
    (*b*) the rotor by shifting its phase backward causes motor to take more current
    (*c*) the back e.m.f. decreases causing an increase in motor current
    (*d*) the rotor strengthens the rotating field causing more motor current

396. Synchronous motor always runs at
   (*a*) the synchronous speed
   (*b*) less than synchronous speed
   (*c*) more than synchronous speed
   (*d*) none of the above
397. An over-excited synchronous motor takes
   (*a*) leading current
   (*b*) lagging current
   (*c*) both (*a*) and (*b*)
   (*d*) none of the above
398. The working of a synchronous motor is similar to
   (*a*) gear train arrangement
   (*b*) transmission of mechancial power by shaft
   (*c*) distribution transformer
   (*d*) turbine
   (*e*) none of the above
399. The minimum armature current of the synchronous motor corresponds to operation at
   (*a*) zero power factor leading
   (*b*) unity power factor
   (*c*) 0.707 power factor lagging
   (*d*) 0.707 power factor leading
400. In a synchronous motor, the magnitude of stator back e.m.f. £& depends on
   (*a*) d.c. excitation only
   (*b*) speed of the motor
   (*c*) load on the motor
   (*d*) both the speed and rotor flux
401. If load (or torque) angle of a 4-pole synchronous motor is 6° electrical, its value in mechanical degrees is
   (*a*) 2
   (*b*) 3
   (*c*) 4
   (*d*) 6
402. For V-curves for a synchronous motor the graph is drawn between
   (*a*) field current and armature current
   (*b*) terminal voltage and load factor
   (*c*) power factor and field current
   (*d*) armature current and power factor
403. The back e.m.f. of a synchronous motor depends on
   (*a*) speed
   (*b*) load
   (*c*) load angle
   (*d*) all of the above
404. A synchronous motor can operate at
   (*a*) lagging power factor only
   (*b*) leading power factor only
   (*c*) unity power factor only
   (*d*) lagging, leading and unity power factors
405. In a synchronous motor which loss varies with load ?
   (*a*) Windage loss
   (*b*) Bearing friction loss
   (*c*) Copper loss
   (*d*) Core loss
406. A synchronous motor can be made self starting by providing
   (*a*) damper winding on rotor poles
   (*b*) damper winding on stator
   (*c*) damper winding on stator as well as rotor poles
   (*d*) none of the above
407. The oscillations in a synchronous motor can be damped out by
   (*a*) maintaining constant excitation
   (*b*) running the motor on leading power factors
   (*c*) providing damper bars in the rotor pole faces
   (*d*) oscillations cannot be damped

408. The shaft of synchronous motor is made of
   (*a*) mild steel (*b*) chrome steel (*c*) alnico (*d*) stainless steel
409. When the field of a synchronous motor is under-excited, the power factor will be
   (*a*) leading (*b*) lagging (*c*) unity (*d*) zero
410. The speed regulation of a synchronous motor is always
   (*a*) 1% (*b*) 0.5% (*c*) positive (*d*) zero

ANSWER

1. (*d*)	2. (*d*)	3. (*a*)	4. (*d*)	5. (*d*)	6. (*d*)	7. (*d*)	8. (*d*)	9. (*b*)	10. (*b*)
11. (*d*)	12. (*d*)	13. (*c*)	14. (*a*)	15. (*a*)	16. (*a*)	17. (*a*)	18. (*a*)	19. (*c*)	20. (*d*)
21. (*c*)	22. (*b*)	23. (*d*)	24. (*b*)	25. (*b*)	26. (*d*)	27. (*b*)	28. (*a*)	29. (*a*)	30. (*c*)
31. (*a*)	32. (*b*)	33. (*b*)	34. (*c*)	35. (*a*)	36. (*c*)	37. (*c*)	38. (*a*)	39. (*c*)	40. (*a*)
41. (*e*)	42. (*b*)	43. (*e*)	44. (*d*)	45. (*d*)	46. (*c*)	47. (*c*)	48. (*e*)	49. (*c*)	50. (*a*)
51. (*c*)	52. (*d*)	53. (*c*)	54. (*d*)	55. (*c*)	56. (*b*)	57. (*d*)	58. (*d*)	59. (*d*)	60. (*b*)
61. (*c*)	62. (*a*)	63. (*a*)	64. (*a*)	65. (*e*)	66. (*d*)	67. (*a*)	68. (*b*)	69. (*c*)	70. (*a*)
71. (*c*)	72. (*b*)	73. (*a*)	74. (*d*)	75. (*c*)	76. (*b*)	77. (*d*)	78. (*c*)	79. (*b*)	80. (*a*)
81. (*b*)	82. (*b*)	83. (*d*)	84. (*d*)	85. (*c*)	86. (*d*)	87. (*c*)	88. (*e*)	89. (*e*)	90. (*a*)
91. (*c*)	92. (*e*)	93. (*d*)	94. (*b*)	95. (*c*)	96. (*c*)	97. (*c*)	98. (*d*)	99. (*e*)	100. (*c*)
101. (*a*)	102. (*a*)	103. (*c*)	104. (*a*)	105. (*a*)	106. (*c*)	107. (*d*)	108. (*b*)	109. (*c*)	110. (*a*)
111. (*a*)	112. (*a*)	113. (*c*)	114. (*c*)	115. (*b*)	116. (*a*)	117. (*b*)	118. (*d*)	119. (*c*)	120. (*d*)
121. (*b*)	122. (*c*)	123. (*a*)	124. (*c*)	125. (*a*)	126. (*d*)	127. (*b*)	128. (*b*)	129. (*b*)	130. (*d*)
131. (*d*)	132. (*d*)	133. (*b*)	134. (*c*)	135. (*c*)	136. (*b*)	137. (*d*)	138. (*b*)	139. (*d*)	140. (*c*)
141. (*b*)	142. (*a*)	143. (*a*)	144. (*d*)	145. (*a*)	146. (*c*)	147. (*d*)	148. (*b*)	149. (*a*)	150. (*c*)
151. (*c*)	152. (*c*)	153. (*b*)	154. (*b*)	155. (*c*)	156. (*c*)	157. (*d*)	158. (*d*)	159. (*c*)	160. (*a*)
161. (*a*)	162. (*b*)	163. (*b*)	164. (*d*)	165. (*a*)	166. (*a*)	167. (*c*)	168. (*d*)	169. (*b*)	170. (*b*)
171. (*b*)	172. (*d*)	173. (*c*)	174. (*a*)	175. (*b*)	176. (*d*)	177. (*b*)	178. (*d*)	179. (*c*)	180. (*b*)
181. (*d*)	182. (*a*)	183. (*d*)	184. (*d*)	185. (*d*)	186. (*d*)	187. (*d*)	188. (*d*)	189. (*d*)	190. (*d*)
191. (*c*)	192. (*a*)	193. (*a*)	194. (*c*)	195. (*a*)	196. (*b*)	197. (*c*)	198. (*c*)	199. (*a*)	200. (*a*)
201. (*d*)	202. (*c*)	203. (*d*)	204. (*c*)	205. (*a*)	206. (*d*)	207. (*c*)	208. (*b*)	209. (*c*)	210. (*d*)
211. (*a*)	212. (*a*)	213. (*d*)	214. (*c*)	215. (*a*)	216. (*c*)	217. (*a*)	218. (*a*)	219. (*a*)	220. (*b*)
221. (*b*)	222. (*a*)	223. (*c*)	224. (*b*)	225. (*b*)	226. (*a*)	227. (*d*)	228. (*d*)	229. (*d*)	230. (*a*)
231. (*b*)	232. (*c*)	233. (*c*)	234. (*d*)	235. (*a*)	236. (*a*)	237. (*c*)	238. (*a*)	239. (*a*)	240. (*d*)
241. (*a*)	242. (*d*)	243. (*a*)	244. (*d*)	245. (*c*)	246. (*c*)	247. (*d*)	248. (*d*)	249. (*b*)	250. (*a*)
251. (*c*)	252. (*b*)	253. (*b*)	254. (*b*)	255. (*d*)	256. (*a*)	257. (*d*)	258. (*b*)	259. (*b*)	260. (*b*)
261. (*a*)	262. (*d*)	263. (*c*)	264. (*b*)	265. (*a*)	266. (*a*)	267. (*a*)	268. (*b*)	269. (*d*)	270. (*c*)
271. (*c*)	272. (*c*)	273. (*a*)	274. (*d*)	275. (*b*)	276. (*b*)	277. (*c*)	278. (*a*)	279. (*b*)	280. (*b*)
281. (*d*)	282. (*a*)	283. (*c*)	284. (*d*)	285. (*c*)	286. (*a*)	287. (*b*)	288. (*d*)	289. (*d*)	290. (*a*)
291. (*c*)	292. (*a*)	293. (*a*)	294. (*a*)	295. (*c*)	296. (*d*)	297. (*d*)	298. (*c*)	299. (*c*)	300. (*a*)

301. (*d*) 302. (*b*) 303. (*d*) 304. (*b*) 305. (*c*) 306. (*d*) 307. (*b*) 308. (*a*) 309. (*d*) 310. (*c*)
311. (*a*) 312. (*d*) 313. (*d*) 314. (*b*) 315. (*a*) 316. (*a*) 317. (*a*) 318. (*d*) 319. (*a*) 320. (*d*)
321. (*a*) 322. (*a*) 323. (*b*) 324. (*d*) 325. (*c*) 326. (*a*) 327. (*c*) 328. (*b*) 329. (*a*) 330. (*d*)
331. (*d*) 332. (*d*) 333. (*b*) 334. (*b*) 335. () 336. (*a*) 337. (*c*) 338. (*a*) 339. (*c*) 340. (*d*)
341. (*b*) 342. (*c*) 343. (*c*) 344. (*c*) 345. (*b*) 346. (*d*) 347. (*a*) 348. (*c*) 349. (*a*) 350. (*d*)
351. (*d*) 352. (*b*) 353. (*b*) 354. (*d*) 355. (*b*) 356. (*a*) 357. (*c*) 358. (*c*) 359. (*c*) 360. (*b*)
361. (*a*) 362. (*b*) 363. (*b*) 364. (*b*) 365. (*b*) 366. (*a*) 367. (*c*) 368. (*d*) 369. (*b*) 370. (*d*)
371. (*a*) 372. (*c*) 373. (*d*) 374. () 375. (*d*) 376. (*c*) 377. (*c*) 378. (*b*) 379. (*a*) 380. (*c*)
381. (*b*) 382. (*a*) 383. (*a*) 384. (*c*) 385. (*d*) 386. (*c*) 387. (*a*) 388. (*a*) 389. (*a*) 390. (*c*)
391. (*d*) 392. (*d*) 393. (*c*) 394. (*a*) 395. (*b*) 396. (*a*) 397. (*a*) 398. (*b*) 399. (*b*) 400. (*a*)
401. (*b*) 402. (*a*) 403. (*c*) 404. (*d*) 405. (*c*) 406. (*d*) 407. (*c*) 408. (*a*) 409. (*b*) 410. (*d*)